The IMA Volumes
in Mathematics
and its Applications

Volume 71

Series Editors
Avner Friedman   Willard Miller, Jr.

Institute for Mathematics and
its Applications
IMA

The **Institute for Mathematics and its Applications** was established by a grant from the National Science Foundation to the University of Minnesota in 1982. The IMA seeks to encourage the development and study of fresh mathematical concepts and questions of concern to the other sciences by bringing together mathematicians and scientists from diverse fields in an atmosphere that will stimulate discussion and collaboration.

The IMA Volumes are intended to involve the broader scientific community in this process.

Avner Friedman, Director
Willard Miller, Jr., Associate Director

* * * * * * * * * *

IMA ANNUAL PROGRAMS

1982–1983	Statistical and Continuum Approaches to Phase Transition
1983–1984	Mathematical Models for the Economics of Decentralized Resource Allocation
1984–1985	Continuum Physics and Partial Differential Equations
1985–1986	Stochastic Differential Equations and Their Applications
1986–1987	Scientific Computation
1987–1988	Applied Combinatorics
1988–1989	Nonlinear Waves
1989–1990	Dynamical Systems and Their Applications
1990–1991	Phase Transitions and Free Boundaries
1991–1992	Applied Linear Algebra
1992–1993	Control Theory and its Applications
1993–1994	Emerging Applications of Probability
1994–1995	Waves and Scattering
1995–1996	Mathematical Methods in Material Science

IMA SUMMER PROGRAMS

1987	Robotics
1988	Signal Processing
1989	Robustness, Diagnostics, Computing and Graphics in Statistics
1990	Radar and Sonar (June 18 - June 29)
	New Directions in Time Series Analysis (July 2 - July 27)
1991	Semiconductors
1992	Environmental Studies: Mathematical, Computational, and Statistical Analysis
1993	Modeling, Mesh Generation, and Adaptive Numerical Methods for Partial Differential Equations
1994	Molecular Biology

* * * * * * * * * *

SPRINGER LECTURE NOTES FROM THE IMA:

The Mathematics and Physics of Disordered Media

Editors: Barry Hughes and Barry Ninham
(Lecture Notes in Math., Volume 1035, 1983)

Orienting Polymers

Editor: J.L. Ericksen
(Lecture Notes in Math., Volume 1063, 1984)

New Perspectives in Thermodynamics

Editor: James Serrin
(Springer-Verlag, 1986)

Models of Economic Dynamics

Editor: Hugo Sonnenschein
(Lecture Notes in Econ., Volume 264, 1986)

Frank P. Kelly   Ruth J. Williams
Editors

Stochastic Networks

With 68 Illustrations

Springer-Verlag
New York Berlin Heidelberg London Paris
Tokyo Hong Kong Barcelona Budapest

Frank P. Kelly
Statistical Laboratory
University of Cambridge
16 Mill Lane
Cambridge CB2 1SB, England

Ruth J. Williams
Department of Mathematics
University of California — San Diego
9500 Gilman Drive
La Jolla, CA 92093-0112 USA

Series Editors:
Avner Friedman
Willard Miller, Jr.
Institute for Mathematics and Its
  Applications
University of Minnesota
Minneapolis, MN 55455 USA

Mathematics Subject Classifications (1991): 60-06, 60J65, 60K25, 60K30, 68M20, 90B12, 90B15

Library of Congress Cataloging-in-Publication Data
Stochastic networks / Frank P. Kelly, Ruth J. Williams, editors.
       p.   cm. — (The IMA volumes in mathematics and its
    applications ; v. 71)
      Includes bibliographical references (p.   ).
      ISBN 0-387-94531-8 (acid-free paper)
      1. Queuing theory — Congresses.   2. Stochastic analysis —
    Congresses.   I. Kelly, F. P. (Frank P.)   II. Williams, R. J. (Ruth
    J.), 1955- .   III. Series.
    QA274.8.S76   1995
    519.8′2 — dc20   95-17145

Printed on acid-free paper.

Production managed by Hal Henglein; manufacturing supervised by Jacqui Ashri.
Camera-ready copy prepared by the IMA.
Printed and bound by Braun-Brumfield, Ann Arbor, MI.
Printed in the United States of America.

9 8 7 6 5 4 3 2 1

ISBN 0-387-94531-8 Springer-Verlag New York Berlin Heidelberg

The IMA Volumes
in Mathematics and its Applications

Current Volumes:

Volume 67: Mathematics in Industrial Problems, Part 7
    by Avner Friedman

Volume 68: Flow Control
    Editor: Max D. Gunzburger

Volume 69: Linear Algebra for Signal Processing
    Editors: Adam Bojanczyk and George Cybenko

Volume 70: Control and Optimal Design of Distributed Parameter
    Systems
    Editors: John E. Lagnese, David L. Russell, and Luther W. White

Volume 71: Stochastic Networks
    Editors: Frank P. Kelly and Ruth J. Williams

Forthcoming Volumes:

1992 Summer Program: *Environmental Studies: Mathematical,
    Computational, and Statistical Analysis*

1992–1993: *Control Theory*

    Robotics

    Nonsmooth Analysis & Geometric Methods in Deterministic Optimal
    Control

    Adaptive Control, Filtering and Signal Processing

    Discrete Event Systems, Manufacturing, Systems, and Communication
    Networks

1993 Summer Program: *Modeling, Mesh Generation, and
    Adaptive Numerical Methods for Partial Differential Equations*

1993-1994: *Emerging Applications of Probability*

    Discrete Probability and Algorithms   Random Discrete Structures

    Mathematical Population Genetics

    Stochastic Problems for Nonlinear Partial Differential Equations

    Image Models (and their Speech Model Cousins)

    Stochastic Models in Geosystems

    Classical and Modern Branching Processes

1994 Summer Program: *Molecular Biology*

1994-1995: *Waves and Scattering*

Computational Wave Propagation

Wavelets, Multigrid and Other Fast Algorithms (Multipole, FFT) and Their Use In Wave Propagation

Waves in Random and Other Complex Media

FOREWORD

This IMA Volume in Mathematics and its Applications

STOCHASTIC NETWORKS

is based on the proceedings of a workshop that was an integral part of the 1993–94 IMA program on "Emerging Applications of Probability." We thank Frank P. Kelly and Ruth J. Williams for organizing the workshop and for editing the proceedings. We also take this opportunity to thank the National Science Foundation, the Air Force Office of Scientific Research, the Army Research Office, and the National Security Agency, whose financial support made the workshop possible.

Avner Friedman

Willard Miller, Jr.

PREFACE

Research on stochastic networks has powerful driving applications in the modelling of manufacturing, telecommunications, and computer systems. These various applications have raised common mathematical issues of some subtlety, and a notable feature of the workshop was the way in which experts in different areas such as operations research, systems science and engineering, and applied mathematics have been attacking important problems from different viewpoints.

The workshop was timely for both applications and theory. In the past decade the proliferation of local and global communication networks for computer and human communication, the development of parallel computers with large numbers of processors, and the design of flexible and robust manufacturing systems have spurred major advances in our understanding of queueing networks, and the workshop provided an opportunity to review recent progress. While research on queueing networks uses many of the traditional queueing theory insights, it is more concerned with how network components interact than with detailed models of how an individual queue behaves. In the last few years there have been some surprises, in particular with regard to the conditions for stability of multiclass queueing networks, and this area formed a major theme of the workshop. Other important themes concerned the challenges reflected Brownian motion has set both as a mathematical object and as a modelling paradigm; the usefulness of ideas from the interacting particle system world; the application of large deviation theory; and the developing connections with optimization and dynamical systems theory.

Fluid models, and their connections with the stability and performance of multiclass queueing networks, were vigorously discussed in the opening days of the workshop. Intense interest in this subject has been fuelled by recent examples due to Lu and Kumar, Rybko and Stolyar, Bramson and others which show that even if the nominal traffic intensity at each station is strictly less than one, a multiclass network need not be stable, in the sense that the total number of customers in the system may grow towards infinity. Reports on recent efforts at determining conditions for stability of such networks included the work of Kumar, Meyn et al. on finding Lyapunov functions using a linear programming approach to obtain bounds on performance and conditions for stability of mostly re-entrant lines. An interesting feature is that the linear programs for performance and stability are in duality; this is reminiscent of the duality between the partial differential equations for stability and for steady state distributions of reflected diffusions. In recent work Jim Dai and Gideon Weiss have used piecewise linear Lyapunov functions to show stability of some fluid models which in turn guarantees positive recurrence for the queueing model. An interesting problem has been to determine the conditions under which the fluid model is stable for any work conserving service discipline. Maury Bramson described his striking two station example where there is instability even though the arrivals are Poisson, service is exponential and the service discipline is the egalitarian FIFO. Dimitris Bertsimas showed how bounds on the achievable performance of stochastic networks could be obtained by solving linear and nonlinear optimization problems whose formulation is motivated by conservation law ideas. His wide-ranging talk included some interesting comparisons between the fields of stochastic modelling and mathematical programming.

Fluid limits for closed multiclass queueing networks were introduced by Viên Nguyen (joint work with Michael Harrison). By means of a seemingly simple two station example (a closed analogue of the Lu and Kumar example), she illustrated that fluid limits for closed networks can exhibit very interesting and unexpected behaviour. The behaviour seems to have links with the issue of stability for open multiclass networks.

Assuming the fluid model for a network is stable, one can then propose a Brownian model as an approximation to the network when it is heavily loaded. (The justification of this approximation and what one means by heavily loaded are issues that still need to be resolved.) Another way to view the Brownian model is as a description of deviations from the fluid model or a long time view of a balanced fluid model. Michael Harrison gave a splendid overview of the current state of knowledge in this area and introduced some bold conjectures which fueled discussion in the early part of the workshop. Harrison's paper in volume 10 of these IMA proceedings has been immensely useful as a guide to (then) known and conjectured results on Brownian networks, and his paper here will, we expect, be similarly helpful to workers in the field.

Ruth Williams followed Harrison's talk with an overview of the current state of knowledge regarding reflecting Brownian motions as approximate models of multiclass open queueing networks. She largely focussed on the geometric conditions for existence and uniqueness of these diffusion processes, conditions for positive recurrence and characterization of the stationary distribution. The result of Dupuis and Williams that stability of all solutions of the Skorokhod problem for the drift path is sufficient for positive recurrence of the associated reflecting Brownian motion is similar to (and indeed motivated) the result of Dai that stability of the fluid model is sufficient for positive recurrence of the associated queueing model.

Frank Kelly initiated the discussion of problems of optimal scheduling and routing in stochastic networks, focussing on loss networks and queueing networks. He described various bounds on performance, and limiting regimes involving either heavy traffic or diverse routing under which the bounds may be approached. This was followed by talks of Neil Laws on trunk reservation as a good control mechanism in loss networks, and of Marty Reiman on polling systems and the use of a heavy traffic approximation and averaging principle to approximate the total unfinished work, and the waiting time of an arriving customer, by a one-dimensional diffusion.

Larry Wein expanded on the analysis of polling models by discussing the optimal scheduling of a multiclass single server station, which can be used to model a polling system or a stochastic economic lot scheduling problem. In both the talks of Reiman and Wein, heavy traffic diffusion approximations provided the intuitive motivation for the proposed policies. In his talk Mete Soner provided a rigorous connection between the optimal scheduling problem for a two station, two customer class (criss-cross) queueing network and the optimal control problem for a multidimensional diffusion obtained via a heavy traffic limit from the original queueing problem. The theme of polling models recurred in the talk of Guy Fayolle, where certain state-dependent polling models were analysed for ergodicity or transience and, in the case of symmetric ergodic systems, various polling strategies could be compared.

Another approach to optimization problems was described by Harold Kushner with the objective being to facilitate efficient numerical computations. His method is to approximate a system in heavy traffic by a diffusion and then to approximate the diffusion by a Markov chain with small step size. This discretizes the problem and allows rapid numerical computations to be performed on the Markov chains corresponding to networks with up to four nodes. Kushner's experience suggests that for such systems, the optimal controls for the Markov chains are good controls for the original queueing network.

A further perspective on the stability of queueing networks was presented by François Baccelli with his saturation rule for stability of open queueing networks under stationary ergodic assumptions. This has application to Jackson-type networks and stochastic Petri networks. Matrix product form solutions allowed Bhaskar Sengupta to analyse a multiclass queue with LCFS service discipline and non-exponential service or interarrival times. Sergei Berezner reported on his work with Suhov and Rose on synchronization rules for star-like networks.

Balaji Prabhakar sketched his recent proof with Tom Mountford of the long standing Reiman-Simon conjecture that a stationary ergodic arrival process of rate $\alpha < 1$ fed through a sequence of independent rate one memoryless queues yields an output that tends to a Poisson process of rate α as the number of queues tends to infinity. The usefulness of ideas from the field of interacting particle systems was well illustrated in this talk and in the talk of Bramson.

Avi Mandelbaum described some of his experiences modelling real service networks involving people as servers and customers and applying existing tools in novel ways to enhance the performance of these networks. The term 'Queueing Science,' which came up in the talk, seems to be an especially appropriate one for this activity. Many of the early operational research applications of queueing theory involved service networks, but in more recent times the focus of applications has shifted to technological applications where customers are jobs in a manufacturing system or messages in a telecommunication system. In some respects the technological applications are easier in that there is no need to worry about predicting the various ways people may react to disparate circumstances. Avi emphasised that for service networks the application of queueing theory results must involve various other disciplines such as psychology and organizational behaviour.

Walter Willinger's talk on data he has collected from telecommunications networks stimulated thought on what might be the best model for the primitive processes (service times, etc.) associated with these networks. The self-similar properties of his data raised the question of whether fractional Brownian motion might not be a useful model in some circumstances. The talk provoked valuable discussion of the assumptions underlying many of the existing models in the field.

Jean Walrand talked about design and control of ATM (asynchronous transfer mode) networks, the high speed telecommunication networks currently under development. He described two ideas for this, with the objective of having simple design rules and robust control strategies for networks. ATM networks are rapidly becoming a very important application area for stochastic networks. There remains much further work to be done on the formulation of basic models, but the central role of large deviations is already clear.

Large deviation was a common theme in several of the talks of the last two days of the workshop. These included Alan Weiss' analysis of Aloha-type protocols for multiple access channels, the analysis of infinite server queues of Peter Glynn and Paul Dupuis' general large deviation principle for queueing systems. The use of dynamical systems ideas in the evaluation of the rate function indicates another connection between deterministic dynamical systems with non-smooth constraints and stochastic networks. A. Ganesh reported on the stationary tail probabilities of a tandem of exponential servers fed by a renewal process (joint work with Venkat Anantharam).

Much of the workshop was concerned with coarse approximate models (primarily diffusion and fluid approximations) and coarse performance estimates (for example large deviation estimates, or limits as a network becomes large), rather than with high-fidelity models and exact performance analysis methods. This search for more robust and general insights is a salient feature of much contemporary research in the field.

The workshop provided an immensely valuable opportunity for participants to learn from and argue with each other, and to boldly conjecture. It was an exciting, intense week, and we have included in this volume the workshop's schedule and list of participants as a reminder and to help contacts to be pursued. Many of the speakers have contributed papers to the volume related to their own talks, to later developments, or to other topics in the area of the workshop.

The workshop was funded by the IMA. We also acknowledge financial support from the National Science Foundation, the Air Force Office of Scientific Research, the Army Research Office, and the National Security Agency. We thank Avner Friedman, Willard Miller, Jr. and Michael Steele for their invitation to organize the workshop and for their support. The excellent local arrangements were cheerfully and efficiently administered by Pam Rech and Kathi Polley. Finally we thank Patricia V. Brick for her unfailing assistance in the production of this volume.

Frank P. Kelly
Ruth J. Williams

CONTENTS

WORKSHOP SCHEDULE

OVERVIEW
(Chairs: Frank Kelly, Tom Kurtz)

Michael Harrison. *Multiclass queueing networks: towards an asymptotic theory of stability and approximation*

P.R. Kumar. *Scheduling queueing networks: stability, performance analysis and design*

Frank Kelly. *Dynamic routing in stochastic networks*

Ruth Williams. *Brownian models of multiclass queueing networks*

MODELLING AND STABILITY
(Chairs: Elena Krichagina, Dimitris Bertsimas)

Avi Mandelbaum. *Service networks: modeling, analysis and inference*

Viên Nguyen. *A closed queueing network with discontinuous fluid limits, wandering bottlenecks and other peculiar behavior*

Jim Dai. *Stability and instability of fluid approximations for re-entrant line queueing networks*

Maury Bramson. *Two badly behaved queueing networks*

François Baccelli. *On the saturation rule for the stability of queues*

Sergei Berezner. *Star-like networks with synchronization rules*

OPTIMIZATION AND MODELING
(Chairs: Richard Weber, Gideon Weiss)

Neil Laws. *Loss networks under optimal trunk reservation*

Martin Reiman. *Polling systems in heavy traffic*

Lawrence Wein. *Dynamic scheduling of a two-class queue with setups*

Guy Fayolle. *A state-dependent polling model with Markovian routing*

Mete Soner. *Heavy traffic convergence of a controlled, multi-class queueing system*

Dimitris Bertsimas. *A mathematical programming approach to stochastic optimization*

LIMITS AND LARGE DEVIATIONS
(Chairs: Bruce Hajek, Richard Serfozo)

Harold Kushner. *Controlled and optimally controlled multiplexing systems: heavy traffic limits and a numerical exploration*

Paul Dupuis. *The large deviation principle for a general class of queueing systems*

A. Ganesh. *On the stationary tail probability of a tandem of exponential servers fed by a renewal process*

Bhaskar Sengupta. *Matrix product-form solutions and a LCFS queue*

Balaji Prabhakar. *Convergence of an infinite sequence of queues*

CONTRIBUTED TALKS
(Chairs: Ruth Williams, Frank Kelly)

Walter Willinger. *Traffic modeling: theory versus practice*

Robert Maier. *High-precision asymptotics for Brownian models*

Stephen Turner. *Dynamic routing in multiparented networks*

Yakov Kogan. *State-dependent systems with large service rates*

Peter Taylor. *A discussion of the correspondence between batch movement networks of queues and single movement networks of queues*

Volker Schmidt. *On the structure of insensitive GSMP with reallocation and with point-process input*

Ravi Mazumdar. *Forward equations for reflected diffusions with jumps*

LARGE DEVIATIONS AND CONTROL
(Chair: Michael Harrison)

Peter Glynn. *Approximations and limit theorems for infinite-server queues*

Alan Weiss. *Aloha, aloha*

Jean Walrand. *Design and control of ATM networks*

LIST OF PARTICIPANTS

ALANYALI, MURAT
C&SBL
U of Illinois-Urbana
1308 West Main St.
Urbana, IL 61801
alanyali@uicslth.csl.uiuc.edu

ANANTHARAM, VENKAT
School of Electrical Engineering
Cornell University
Phillips Hall
Ithaca, NY 14853
ananth@camus.ee.cornell.edu

BACCELLI, FRANCOIS
INRIA
Sophia-Antipolis
2004 Route des Lucioles, Valbonne
Cedex FRANCE 06565
baccelli@sophia.inria.fr

BAN, JUNHWA
Department of Mathematics
Seoul National University
Shinlim-Dong Kwanak-Gu
Seoul, 151-742
KOREA

BENJAAFAR, SAIFALLAH
Mechanical Engineering
Univerisity of Minnesota
125 Mech Engineering
Minneapolis, MN 55455
saif@maroon.tc.umn.edu

BENJAMIN, JEFF
Dept. of Mech. Eng.
University of Minnesota
Minneapolis, MN 55455

BEREZNER, SERGEI
Mathematical Statistics
University of Natal-South Africa
King George V Avenue
Durban 4001
SOUTH AFRICA
berezner@ph.und.ac.za

BERTSIMAS, DIMITRIS
Sloan School of Management
MIT
Room E53-359
Cambridge MA 02139-4307
dbertsim@math.mit.edu

BRAMSON, MAURY
Dept. of Mathematics
University of Wisconsin
Van Vleck Hall
Madison WI 53706
bramson@vanvleck.math.wisc.edu

CHEN, ZHEN-QING
Dept. of Mathematics
U of California-San Diego
La Jolla, CA 92093-0112
zchen@euclid.ucsd.edu

CHUNG, DONG MYUNG
Sogang University
Korea
dmchung@ccs.sogang.ac.kr

CVITANIC, JAKSA
Department of Statistics
Columbia University
New York, NY 10027

DAI, JIM
School of Mathematics
Georgia Tech
Atlanta GA 30332-0205
dai@isye.gatech.edu

DAI, WANYANG
School of Mathematics
Georgia Tech
Atlanta, GA 30332
wdai@math.gatech.edu

DASGUPTA, SONIA
Dept. of Statistics
University of Minnesota
270 Vincent Hall
Minneapolis, MN 55455
sonia@stat.umn.edu

DOUGHERTY, ANNE M.
Dept. of Mathematics
University of Wisconsin
Madison, WI 53706
doughert@math.wisc.edu

DOWN, DOUG
Cordinated Science Lab.
U of Illinois-Urbana
Urbana, IL 61801
down@decision.csl.uiuc.edu

DUPUIS, PAUL
Division of Applied Mathematics
Brown University
Providence RI 02912
dupuis@cfm.brown.edu

FAYOLLE, GUY
INRIA
Domaine de Voluceau -B.P. 105
78153 Le Chesnay Cedex
FRANCE
guy.fayolle@inria.fr

GANESH, A.
School of Electrical Engineering
Cornell University
Phillips Hall
Ithaca, NY 14853
ganesh@phaeton.ee.cornell.edu

GIBBENS, RICHARD
Dept. of Math & Mathematical Statistics
University of Cambridge
16 Mill Lane
Cambridge
England CB2 1SB
R.J.Gibbens@statslab.cam.ac.uk

GLYNN, PETER W.
Dept. of Operations Research
Stanford University
Stanford CA 94305-4022
glynn@leland.stanford.edu

GU, JUN-MIN
Math Dept.
University of Wisconsin-Madison
Van Vleck Hall
Madison, Wi 53706
gu@math.wisc.edu

HAJEK, BRUCE
Electrical & Computer Engineering
U of Illinois-Urbana
Coordinated Science Lab
1308 W. Main
Urbana IL 61801
b-hajek@uiuc.edu

HALE, WILLIAM
Dept. of Mechanical Engineering
University of Minnesota
whale@maroon.tc.umn.edu

HARRISON, MICHAEL
Graduate School of Business
Stanford University
Stanford CA 94305-5015
fharrison@gsb-lira.stanford.edu

HEIM, GREG
OMS
University of Minnesota
gheim@vx.cis.umn.edu

HILDEBRAND, MARTIN
Department of Mathematics
University of Texas at Austin
Austin, TX 78712

HU, YIMING
Center for Stochastic and Chaotic Processes
Case Western Reserve University
Cleveland, OH 44106

JIN, HAIMING
Computer & Systems Research Lab
U of Illinois-Urbana
Urbana, IL 61801
jin@bellman.csl.uiuc.edu

JOHNSON, TIM
Dept. of Mech. Eng.
University of Minnesota
Minneapolis, MN 55455

KELLY, FRANK
Dept. of Mathematics & Math Stat
Cambridge University
16 Mill Lane
Cambridge CB2 1SB
ENGLAND
F.P.Kelly@statslab.cam.ac.uk

KI, HO SAM
Deparment of Mathematics
Seoul National University
Shinlim-Dong Kwanak-Gu
Seoul 151-742
KOREA

KOGAN, YAKOV
AT&T Bell Laboratories
101 Crawford Corner Road
Holmdel, NJ 07733-3030
yakov@buckaroo.att.com

KONSTANTOPOULOS, TAKIS
Department of ECE
University of Texas-Austin
Austin TX 78712
takis@alea.ece.utexas.edu

KRICHAGINA, ELENA
c/o Smirnov, Functional Analysis Sector
SISSA/ISAS
via Beirut 2-4
34014 Trieste
ITALY
krich@tsmi19.sissa.it

KRISHNAMURTHY, SATISH
University of Minnesota
127 Vincent Hall
206 Church St. S.E.
Minneapolis, MN 55455
krish005@maroon.tc.umn.edu

KUMAR, P.R.
Coordinated Science Laboratory
U of Illinois-Champaign
1308 West Main St.
Urbana IL 61801
prkumar@markov.csl.uiuc.edu

KUMAR, SUNIL
Coordinated Science Lab.
U of Illinois-Urbana
1308 W. Main St.
Urbana, IL 61801
sunil@monk.csl.uius.edu

KURTZ, THOMAS G.
Center for Mathematical Sciences
U of Wisconsin-Madison
610 Walnut Street
Madison WI 53706
Kurtz@math.wisc.edu

KUSHNER, HAROLD
Division of Applied Mathematics
Brown University
Providence RI 02912-0001
hjk@dam.brown.edu

LAWS, NEIL   Dept. of Statistics
University of Oxford
South Parks Road
Oxford OX1 3TG
ENGLAND
laws@stats.ox.ac.uk

LEHOCZKY, J.
Dept. of Statistics
Carnegie Mellon University
Pittsburgh PA 15213
jpl@stat.cmu.edu

LU, SHENGLIN
University of Michigan
317 West Engineering Bldg.
Ann Arbor, MI 48109
shenglu@msri.org

MAIER, ROBERT
Dept. of Mathematics
University of Arizona
Tucson, AZ 85721
rsm@math.arizona.edu

MALYSHEV, V.
INRIA
Domaine de Voluceau
Rocquencourt, BP 105
78153 Le Chesnay, Cedex FRANCE
vadim.malyshev@inria.fr

MANDELBAUM, AVI
Dept. of Industrial Engineering
Technion
Israel Institute of Technology
32000 Haifa
ISRAEL
avim@techunix.technion.ac.il

MARGOLIUS, BARBARA
Applied Mathematics
Case Western Reserve University
220 Yost Hall
Cleveland OH 44106
bhm@po.cwru.edu

MAZUMDAR, RAVI
INRS-Telecommunications
University of Quebec
Ile des Soeurs
Quebec
CANADA H3E 1H6
mazum@inrs-telecom.uquebec.ca

MCCLURE, PETER
Dept. of Mathematics & Astronomy
University of Manitoba
Winnipeg, Manitoba
CANADA R3T 2N2
joptmc@ccu.umanitoba.ca

MCDONALD, DAVID
Dept of Mathematics
University of Ottawa
585 King Edward, PO Box 450 STN A
Ottawa Ontario
CANADA K1N 6N5
dmdsg@omid.mathstat.uottawa.ca

MENCHIKOV, MIKHAIL
Deptartment of Probability
Moscow State University
Faculty of Mechanics and Mathematics
Moscow 119899
RUSSIA
menshikov@11rs.math.msu.su

NAGAEV, SERGEI
Institute of Mathematics
Siberian Branch of the Russian Academy
Universitetskii pr. 4
630090 Novosibirsk
RUSSIA 630090
nagaev@math.nsk.su

NARAYANAN, BABU O.
DIMACS/CORE Bldg.
PO BOX 1179
Piscataway NJ08854-1179
(908) 445 4579

NEUHAUSER, CLAUDIA
University of Southern California
DRB 306
1042 West 36th Place
Los Angeles CA 90089-1113
neuhause@math.wisc.edu

NGUYEN, VIEN
Sloan School of Management
MIT
50 Memorial Drive E53-353
Cambridge MA 02142-1347
vien@mit.edu

ORMECI, EGEMEN LERZAN
Dept. of Operations Research
Case Western Reserve University
11408 Bellflower Rd.
Cleveland, OH 44106
elo@po.cwru.edu

PAPADAKIS, SPYROS
Dept. of EECS
U of California-Berkeley
307 Cory Hall, #146
Berkeley, CA 94720
spyros@eecs.berkeley.edu

PETERSON, WILLIAM
Dept. of Mathematics & Computer Science
Middlebury College
Middlebury VT 05753
billp@mit.edu

PRABHAKAR, BALAJI
Department of Electrical Engineering
U of California-Los Angeles
Los Angeles CA 90024
balaji@shockley.ee.ucla.edu

RAMAKRISHNAN, RAJESH
Dept. of Mech. Eng.
University of Minnesota
Minneapolis, MN 55455

REIMAN, MARTIN
AT&T Bell Labs
Room 2C-117
600 Mountain Avenue
Murray Hill NJ 07974
marty@research.att.com

SAICHEV, ALEKSANDER
Dept. of Radio Physics
University of Nizhni-Russia
Novgorod RUSSIA

SCHIEF, ANDREAS
ZFE BT 13
Siemens
Corporate Research and Development
D-81730 Munich
GERMANY
andreas.schief@zfe.siemens.de

SCHMIDT, VOLKER
Abteilung Stochastik
Universitat Ulm
Abteilung Stochastik
Helmholtzstrasse 18
D-89069 Ulm, GERMANY
schmidt@mathematik.uni-ulm.de

SCHWERER, ELIZABETH
Stanford
P.O. Box 9634
Stanford, CA 94309
pschwerer@gsb-lira.stanford.edu

SENGUPTA, BHASKAR
NEC Research Institute
4 Independence Way
Princeton NJ 08540
bhaskar@ccrl.nj.nec.com

SEPPALAINEN, TIMO
Institute for Mathematics
University of Minnesota
514 Vincent Hall
206 Church St. S.E.
Mpls, MN 55455
seppalai@ima.umn.edu

SERFOZO, RICHARD
School of Industrial & Systems Eng
Georgia Tech
Atlanta GA 30332-0001
rserfozo@isye.gatech.edu

SONER, METE
  Department of Mathematics
  Carnegie Mellon University
  Pittsburgh PA 15213-3890
  mete@fermat.math.cmu.edu

SRINIVASAN, RAJ
  Dept. of Mathematics
  University of Saskatchewan
  Saskatoon Saskatchewan
  CANADA S7N 0W0
  raj@snoopy.usask.ca

TAAFFE, MICHAEL
  Dept. of O.M.S.
  University of Minnesota
  332 Mgmt/Econ Bldg
  Minneapolis, MN 55455
  taaffe@maroon.tc.umn.edu

TAKAHARA, GLEN
  Dept. of Mathematics
  Queens University
  Kingston, Ontario
  CANADA K7L3N6
  takahara@glen.mast.queensu.ca

TAKSAR, MICHAEL
  Dept. of Applied Mathematics
  SUNY
  Stony Brook NY 11794-3600
  taksar@ams.sunysb.edu

TAYLOR, PETER
  Dept. of Applied Mathematics
  University of Adelaide
  GPO Box 498
  Adelaide SA 5001
  AUSTRALIA

TIER, CHARLES
  Dept. of Mathematics (M/C 249)
  U of Illinois-Chicago
  322 Science and Engineering Offices
  851 S. Morgan
  Chicago, IL 60607-7045
  tier@math.uic.edu

TURNER, STEPHEN
  Statistical Laboratory
  University of Cambridge
  16 Mill Lane
  Cambridge
  ENGLAND CB2 1SB
  s.r.e.turner@statslab.cam.ac.uk

WADHWANI, DINESH
  OMS
  University of Minnesota
  332-M&E
  wadhwani@msi.umn.esu

WALRAND, J.
  U of California-Berkeley
  Cory Hall
  Berkeley CA 94720
  wlr@eecs.berkeley.edu

WEBER, RICHARD
  Dept. of Engineering
  University of Cambridge
  Trumpington Street
  Cambridge CB2 1PZ
  UNITED KINGDOM
  rrw1@cus.cam.ac.uk

WEIN, LAWRENCE M.
  Sloan School of Management
  MIT
  Room E53-343
  Cambridge MA 02139-4307
  lwein@eagle.mit.edu

WEISS, ALAN
  AT&T Bell Labs
  Room 2C-118
  600 Mountain Avenue
  Murray Hill NJ 07974-2010
  apdoo@research.att.com

WEISS, GIDEON
  Dept. of Industrial & Systems Eng
  Georgia Tech
  Atlanta GA 30332
  gweiss@isye.gatech.edu

WILLIAMS, RUTH
Department of Mathematics
University of California-San Diego
9500 Gilman Drive
La Jolla CA 92093-0112
williams@russel.ucsd.edu

WILLINGER, WALTER
Bellcore
Room 2P-372
445 South Street
Morristown NJ 07960-1901
walter@bellcore.com

WOYCZYNSKI, WOJBOR
Center for Stochastic & Chaotic Processes
in Science and Technology
Case Western Reserve University
Cleveland OH 44106
waw@po.cwru.edu

WU, CHIA-LIN
Math Dept.
U of California-Irvine
4823 Verano Place
Irvine, CA 927-3175
cwu@math.uci.edu

ZIEDINS, ILZE
Department of Mathematics & Statistics
University of Auckland
Private Bag 92019
Auckland
NEW ZEALAND
ziedins@mat.auckland.ac.nz

ZIRBEL, CRAIG L.
Department of Mathematics and Statistics
Lederle GRC
University of Massachusetts at Amherst
Amherst, MA 01003-4515
zirbel@math.umass.edu

BALANCED FLUID MODELS OF MULTICLASS QUEUEING NETWORKS: A HEAVY TRAFFIC CONJECTURE

J. MICHAEL HARRISON*

Abstract. An open multiclass network is said to be *well behaved in heavy traffic*, or simply *well behaved*, if it satisfies a conventional heavy traffic limit theorem. In such a theorem the heavy traffic limit is derived from (or simply is) a regulated or reflected Brownian motion whose dimension equals the number of stations in the network. Recent examples show that not all multiclass networks are well behaved in this sense, and a top priority for heavy traffic researchers is to understand the exceptional set of ill behaved network models. In this paper we consider a large family of open multiclass queueing networks, associate with each network a deterministic fluid analog, and develop the following conjecture: A multiclass network is well behaved in heavy traffic if and only if its balanced fluid analog exhibits a certain stability property. This general conjecture resonates with other contemporary developments in queueing network theory, where fluid models play an increasingly important role. However, our view of fluid models differs from the conventional one in ways that seem to be important for heavy traffic theory.

Key words. queueing networks, multiclass networks, heavy traffic, fluid models

AMS(MOS) subject classifications. 90B15, 60K25

1. Introduction. In the last few years research in queueing network theory has been redirected by a series of brilliant and sobering examples, one of which is depicted in Figure 1. In this example, developed by Dai and Wang [9], there are three single-server stations and arriving customers follow a deterministic route that requires seven distinct services or operations. Each operation is performed at a specific station, and the mean service times for the various operations are shown in Figure 1. The external arrival rate is denoted by λ, and for the sake of concreteness let us consider the case where arrivals are Poisson and all service time distributions are exponential. Finally, the order of service at each station is first-in-first-out (FIFO). As readers may verify, the total expected service time required at each station by each customer is 0.95, so for each station j the associated traffic intensity parameter is $\rho_j = 0.95\lambda$. One naturally expects the network to be stable (that is, positive recurrent) if $\rho_j < 1$ for each station j, or equivalently, if $\lambda < \lambda^* \equiv 1/0.95$.

The network pictured in Figure 1 is an example of what Kumar [13] has called a *re-entrant line*, which is a special type of multiclass open queueing network distinguished by a single external arrival process and deterministic routing. We define a different *customer class* for each stop on the route, and we allow each class to have its own service time distribution. In a general multiclass open network one also associates customer classes with service stations in a many-to-one fashion, allowing each class to have its own

---

* Graduate School of Business, Stanford University, Stanford, CA 94305-5015.

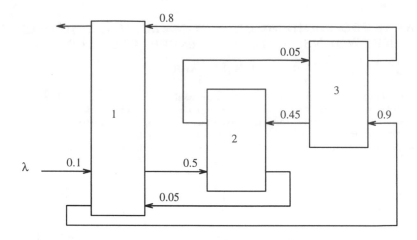

FIG. 1.1. *The Three-Station FIFO Network of Dai and Wang*

service time distribution, but customers may change class in Markovian fashion upon completing service, and there may be external arrivals into each class.

In 1990 Harrison and Nguyen [11] proposed a general scheme for approximating an open multiclass network by an associated "Brownian system model." For an open network with d service stations, the proposed Brownian approximation is derived from a d-dimensional regulated or reflected Brownian motion (RBM). Harrison and Nguyen [11] gave concrete procedures for calculating parameters of the approximating RBM from parameters of the original queueing network, and they conjectured that the proposed approximation could be justified by a conventional heavy traffic limit theorem. For the network pictured in Figure 1 this would mean the following, for example. Consider a sequence of systems indexed by $n = 1, 2, \ldots$ which are identical in all regards except that their external arrival rates λ^n approach the presumed critical value λ^* as $n \to \infty$. Let $W_i^n(t)$ be the workload backlog at time t for server i (that is, the sum of the impending service times for customers queued at station i, plus the remaining service time for the customer in service there, if any) in the nth system, and form a three-dimensional workload process $W^n(t)$ in the obvious way. Without loss of generality (see section 4) assume that

$$(1.1) \qquad \sqrt{n}(\lambda^n - \lambda^*) \to c \quad \text{(a finite constant)} \qquad \text{as } n \to \infty,$$

and consider the sequence of normalized workload processes

$$(1.2) \qquad \tilde{W}^n(t) = \frac{1}{\sqrt{n}} W^n(nt) \quad \text{for } n = 1, 2, \ldots$$

The heavy traffic limit theorem referred to earlier would state that

$$(1.3) \qquad \tilde{W}^n \Rightarrow \tilde{W} \quad \text{as} \quad n \to \infty,$$

where the symbol $\Rightarrow$ denotes weak convergence in the space of Skorohod sample paths, and $\tilde{W}$ is an RBM in the three-dimensional non-negative orthant, with parameters calculated according to the recipe of Harrison and Nguyen.

To repeat, (1.3) is an example of what we call a *conventional* heavy traffic limit theorem for the workload process W. An unconventional limit theorem would be one that used a scaling different from (1.2), given the same heavy traffic hypothesis (1.1), or that obtained a weak limit different from the RBM specified by Harrison and Nguyen [11], or both.

Dai and Wang [9] showed that the conventional heavy traffic limit theorem (1.3) is not valid for the network pictured in Figure 1: the putative Brownian limit $\tilde{W}$ simply does not exist. That is, there exists no stochastic process satisfying the conditions that Harrison and Nguyen [11] used to "define" $\tilde{W}$. Dai and Nguyen [8] subsequently showed that if the scaled processes $\tilde{W}^n$ converged to any limit process, that limit would have to satisfy the Harrison-Nguyen description, so we are left to ponder (at least) the following questions: Is (1.1) the wrong notion of "heavy traffic"? Is (1.2) the wrong scaling for W? To get a heavy traffic limit theorem, must one focus on some other process which is either coarser or finer than W?

Events subsequent to the Dai-Wang paper have tended to suggest that the first of these questions is the crucial one. In particular, Bramson [1,2] developed examples of FIFO re-entrant lines where the network is unstable although $\rho_j < 1$ for each station j. By analogy, it seems likely that the Dai-Wang network pictured in Figure 1 is stable only when $\lambda < \lambda_* < \lambda^*$. That is, one is led to conjecture that the "critical arrival rate" λ_* is actually less than the naive value λ^*, and hence our "heavy traffic" condition (1.1) represents a case of supersaturation inconsistent with the scaling (1.2). No theoretical techniques have been developed as yet for determining the stability region of this network, but an unpublished simulation study [7] has shown that it is unstable when $\lambda = 1$ (and hence $\rho_j = 0.95$ for each station j). Of course, determining the *exact* stability region of a network via simulation is a practical impossibility.

Perplexing examples like the one described above have led to renewed interest in rigorous heavy traffic limit theory, as opposed to heuristic derivation of Brownian approximations. A central problem is to identify the set of multiclass network models for which the "conventional" heavy traffic limit theorem fails to hold, and explain why. For "bad networks" in the indicated set, one can ask further questions and seek alternative approximations, but it seems likely that conventional heavy traffic analysis fails only for networks that are in some sense "pathological" or "uninteresting." If such a case can be made, then refined analysis of the exceptional set may not be worthwhile. For example, Bramson's [1,2] examples can be reasonably described as queueing systems in which FIFO scheduling is disastrously ineffective. Taking this point of view, it is more interesting to understand when FIFO "fails" and what scheduling rules outperform it than it is to analyze system performance under the inferior discipline.

A general conjecture to be developed in this paper is that the key to deeper understanding of multiclass networks lies in the behavior of their fluid analogs (also called fluid limits or fluid models). It is conjectured that the conventional heavy traffic limit theorem is valid for a multiclass network if and only if its balanced fluid analog exhibits a property that we call asymptotic stability. This conjecture is made more plausible by recent work of Dai [6] showing an intimate relationship between fluid model behavior and network stability.

Two features of our treatment are notable. First, because fluid models are considered only in relation to heavy traffic analysis, our attention is directed toward *balanced* fluid models, in which the total load imposed on each service station is precisely equal to its capacity. These balanced fluid models exhibit more delicate limiting behavior than the stable fluid models emphasized by Dai [6]. Second, the scaling by which we obtain balanced fluid models as heavy traffic limits of multiclass networks is more delicate than the fluid scaling that has been customary in open network theory. As a consequence we view fluid models as intermediate between queueing network models and their Brownian approximations, whereas the conventional view is that Brownian models are intermediate between queueing networks and fluid models [5].

The remainder of the paper is organized as follows. Section 2 contains a general description of an open multiclass network with FIFO scheduling at each station, and of the formal fluid analog for such a network. Section 3 introduces the notion of asymptotic stability for balanced fluid models. Section 4 reviews the assumptions and scaling that lead to Brownian models as heavy traffic limits, and Section 5 discusses the alternative scaling that leads to balanced fluid models as heavy traffic limits. Section 6 contains the general conjecture referred to above, with some rather sketchy supporting arguments. Attention is restricted throughout the paper to FIFO scheduling, although priority scheduling is in many ways simpler to analyze, because so much of the antecedent literature focuses on FIFO networks.

Sections 2 and 3 do contain some new results. For example, we show how the "reflection matrix" R that plays such a central role in the Brownian approximation for a multiclass network also arises naturally in conjunction with the network's balanced fluid analog. Most of the paper, however, consists of definitions, conjectures and rough heuristic arguments. Even the definitions and conjectures lack complete mathematical precision, but to complement the current research focus on special network structures and concrete examples, we hope to suggest that a useful *general* theory for multiclass networks may indeed be attainable. In particular, the balanced fluid models emphasized here are much easier to think about than Brownian limits of queueing networks, because all rates of flow are bounded in a fluid model, and hence all associated processes are absolutely continuous. Thus a reduction of heavy traffic problems to questions about fluid model behavior would have to be reckoned a victory, although the theory of fluid models is itself very new and as yet undeveloped.

In this introduction there have been frequent references to the 1990 paper by Harrison and Nguyen [11] in which Brownian approximations for general multiclass networks were first proposed. In 1993, chastened by the Dai-Wang counterexamples, those same authors published a more careful, more comprehensive, and more accessible survey [12] of heavy traffic theory for open multiclass networks. In the remainder of this paper the later survey will be used as the standard reference. In particular, the notation used here agrees with that used in [12] except as explicitly noted, and readers should consult [12] for a more thorough discussion of standard topics (such as representation of network processes, or conventional heavy traffic limit theory) that are treated in brief outline here.

2. An open multiclass network and its fluid analog. The network is composed of d single-server stations (the letter d is mnemonic for *dimension*), each of which serves arriving customers on a FIFO basis. Each server is assumed here to be perfectly reliable. Thus, in the notation of [12], we have the special case where $S_j(t) \equiv t$ for each $j = 1, \ldots, d$. (The more general situation where servers are subject to random breakdowns or other interruptions does not add anything to the issues under consideration here, and the additional notation required to treat that case would represent a needless distraction.) There are c customer classes ($c \geq d$), and each class $k = 1, \ldots, c$ is served at a unique station $s(k)$. Let $\mathcal{C}(j)$ be the set of all classes k such that $s(k) = j$, calling this the *constituency* of server j, and define the $d \times c$ *constituency matrix* C via

$$(2.1) \qquad C_{jk} = \begin{cases} 1 & \text{if } k \in \mathcal{C}(j) \\ 0 & \text{otherwise.} \end{cases}$$

To avoid trivialities assume that $\mathcal{C}(j)$ is non-empty for each server j, so the matrix C has dimension d. Taken as primitive is a c-dimensional external arrival process $E = \{E(t), t \geq 0\}$ whose kth component process $E_k(t)$ records cumulative arrivals into class k. It is assumed that E satisfies a functional central limit theorem (FCLT) with asymptotic mean vector α and asymptotic covariance matrix K. One interprets α_k as the long-run *external* arrival rate into class k ($k = 1, \ldots, c$), while elements of K give the variance-covariance characteristics of the external arrival processes. It is *not* required that external arrivals to different classes be independent, and this makes for a much more powerful theory.

Service times for each class k form an independent and identically distributed (iid) sequence with mean $m_k > 0$ and finite variance. Thus service time distributions are associated with customer classes rather than service stations, which again makes for a more powerful theory. Upon completion of service at station $s(k)$, a class k customer turns next into a customer of class l with probability P_{kl}, and exits the network with probability $1 - \sum_{l=1}^{c} P_{kl}$. The transition matrix $P = (P_{kl})$ is taken to be transient (that is, all customers eventually leave the network), so we can define

$$(2.2) \qquad Q = (I - P')^{-1} = (I + P + P^2 + \ldots)'.$$

Let us denote by λ_k the *total* arrival rate into class k, including both external arrivals and internal transitions, and by λ the corresponding c-vector. (All vectors should be envisioned as column vectors unless something is said to the contrary.) These total arrival rates satisfy the traffic equations

$$(2.3) \qquad \lambda_k = \alpha_k + \sum_{l=1}^{c} \lambda_l P_{lk} \qquad (k = 1, \ldots, c).$$

One may express (2.3) in vector form as $\lambda = \alpha + P'\lambda$, so the desired solution is

$$(2.4) \qquad \lambda = Q\alpha.$$

The *traffic intensity parameter* for server j is

$$(2.5) \qquad \rho_j = \sum_{k \in \mathcal{C}(j)} m_k \lambda_k \qquad (j = 1, \ldots, d).$$

Defining $M = \mathrm{diag}(m_1, \ldots, m_c)$, we can rewrite (2.5) more compactly as

$$(2.6) \qquad \rho = CM\lambda.$$

The data for our open network model are: the distribution of the external arrival process E; the service time distributions for the c different customer classes; the switching matrix P; and the network's initial state. Deviating from the notation used in [12], we shall denote by $Z_k(t)$ the total number of class k customers in existence at time t, either waiting or being served, calling Z_k the *queue length process* for class k. (In [12] the c-dimensional queue length process was denoted by N rather than Z.) To specify the network's initial state one must specify not only the initial queue length vector $Z(0)$ but also the *ordering* of customers in the initial queue at each station (remember that FIFO scheduling is assumed) and the expended service time of the customer in service at each station, if any.

Later we shall focus on the heavy traffic situation where ρ_j is near one for each server j, but no such restriction need be imposed at this point. In the deterministic fluid analog of the multiclass open network described above, the servers process continuous fluids rather than discrete customers, and we distinguish among fluids of classes $1, \ldots, c$. Class k fluid arrives at deterministic rate α_k from outside the network, and when server $s(k)$ devotes its full attention to processing class k fluid, the inventory Z_k is decreased at rate $\mu_k \equiv 1/m_k$ $(k = 1, \ldots, c)$ while other fluid inventories Z_l are increased at rate $\mu_k P_{kl}$ $(l = 1, \ldots, c)$. (Here and later we refer to "inventories" rather than "queue lengths" when describing the fluid model, in order to emphasize the continuous nature of material flowing through the system.) In general, a server may split its attention among the fluid classes in its constituency: if we denote by r_k the fraction of time devoted by server $s(k)$ to class k, then the instantaneous processing rate for class k is $r_k \mu_k$ and we have the obvious constraint

$$(2.7) \qquad \sum_{k \in \mathcal{C}(j)} r_k \leq 1 \qquad (j = 1, \ldots, d).$$

The initial state of the fluid model is described by a non-negative initial inventory vector $Z(0)$ plus additional information to be specified shortly. The fluid model then consists of continuous vector functions $A(t), W(t),$ $Y(t), D(t), Z(t),$ and $F(t), t \geq 0$ (the dimensions of these various functions will soon be apparent) that jointly satisfy equations and auxiliary conditions displayed below. Readers should interpret $F_k(t)$ as the total amount of class k fluid created by internal processing up to time t (the letter F is mnemonic for *feedback* flow), $A_k(t)$ as the total class k input to server $s(k)$ over that same time interval, including both feedback flows and external input (the letter A is mnemonic for *arrivals*), $D_k(t)$ as the total amount of class k fluid departing station $s(k)$, $Y_j(t)$ as the cumulative idleness or unused capacity at station j, and $W_j(t)$ as the amount of time required for server j to complete processing of all fluids in inventory at station j at time t (the letter W is mnemonic for immediate *workload*). As initial conditions, set

$$(2.8) \qquad F(0) = 0 \quad \text{and} \quad W(0) = CMZ(0).$$

Also, let e denote the d-vector of ones. Then the relationships defining the fluid model are:

(2.9) $A(t) = \alpha t + F(t),$

(2.10) $W(t) = W(0) + CMA(t) - et + Y(t),$

(2.11) Y is continuous and nondecreasing with $Y(0) = 0,$

(2.12) Y_j can only increase when $W_j = 0,$ $(j = 1, \ldots, d),$

(2.13) $D_k(t + W_j(t)) = Z_k(0) + A_k(t)$ $(j = 1, \ldots, d$ and $k \in \mathcal{C}(j)),$

(2.14) $Z(t) = Z(0) + A(t) - D(t),$ and

(2.15) $F(t) = P'D(t).$

Actually, this specification of the fluid model is not quite complete, but before going into that matter let us be clear as to the interpretation of (2.9)–(2.15). Equations (2.9), (2.14), and (2.15) are more or less self-explanatory. Of course, (2.10) says that

$$(2.16) \qquad W_j(t) \; = \; \left[W_j(0) + \sum_{k \in \mathcal{C}(j)} m_k A_k(t) - t \right] + Y_j(t)$$

for each station j. The term in square brackets on the right side of (2.16) represents cumulative workload input to station j minus cumulative *potential* workload output, and thus (2.10), (2.11), and (2.12) together say that cumulative idleness Y_j at each station j increases in the minimum amounts necessary to keep the immediate workload process W_j non-negative. It is (2.13) that specifies the service discipline to be FIFO at each station j, as follows. There are $W_j(t)$ units of work for server j at time t, and this work will be completed at future time $\tau_j(t) \equiv t + W_j(t)$. Under FIFO, then, the cumulative output up to that future time $\tau_j(t)$ is precisely equal, for each class k served at the station, to the initial inventory of class k fluid plus the cumulative input up to t.

What is missing from (2.9)–(2.15) is a specification, for each station j, of the output or departure flows $D_k(\cdot)$, $k \in \mathcal{C}(j)$, over the initial time interval $[0, W_j(0)]$. This specification tells us how fluids in the initial inventory vector $Z(0)$ are ordered in the queues at the various stations when the system begins operation. To complete the fluid model description (2.9)–(2.15), we simply assume given for each $j = 1, \ldots, d$ a set of continuous, non-decreasing, non-negative functions $D_k(\cdot)$, $k \in \mathcal{C}(j)$, such that

$$(2.17) \qquad \sum_{k \in \mathcal{C}(j)} m_k D_k(t) = t \quad \text{for } 0 \leq t \leq W_j(0).$$

(That is, the departure flows from station j over the initial time interval are consistent with full server utilization.)

The initial state of the fluid model is specified by the inventory vector $Z(0)$ plus the initial departure flows described in the previous paragraph. More generally, the "state of the system" at any time $t \geq 0$ can be described by a vector function $x(\cdot) = (x_1(\cdot), \ldots, x_c(\cdot))$ that satisfies the following two restrictions. First, each component $x_k(\cdot)$ is non-negative, continuous, non-decreasing and bounded on $[0, \infty)$. Thus $x(\cdot)$ has a finite limit $z = x(\infty) \geq 0$, and it will be convenient to write $w = CMz$. Second, for each $j = 1, \ldots, d$,

$$(2.18) \qquad \sum_{k \in \mathcal{C}(j)} m_k x_k(s) \; = \; \begin{cases} s & \text{if } 0 \leq s \leq w_j \\ w_j & \text{if } s \geq w_j. \end{cases}$$

One interprets $x(\cdot)$ as the vector of cumulative output flows that will be generated from the network's current fluid inventories, ignoring future arrivals to the stations, extended to the time domain $[0, \infty)$ in a natural way.

(That extension is not necessary, of course, but it makes for a less cluttered mathematical development.) We shall denote by S the set of all *state descriptions* x satisfying the restrictions above, endowed with the topology of uniform convergence on compact intervals. As the preceding development suggests, the c-dimensional inventory vector corresponding to state x is $z = x(\infty)$, and the d-dimensional immediate workload vector corresponding to x is $w = CMz$. That is, if a multiclass fluid model is in state x at some time t, then $W(t) = w$ and $Z(t) = z$.

Given a solution of (2.9)–(2.15), which necessarily includes a specification of the initial departure flows described above, we can construct a state-space description $X = (X(t), t \geq 0)$ of the fluid model's trajectory by defining, for each $j = 1, \ldots, d$ and $k \in \mathcal{C}(j)$,

$$(2.19) \qquad X_k(t, s) = \begin{cases} D_k(t+s) - D_k(t) & \text{if } 0 \leq s \leq W_j(t) \\ Z_k(t) & \text{if } s \geq W_j(t). \end{cases}$$

In this notational system, $X(t) = X(t, \cdot)$ is a function in S analogous to the generic state x described earlier.

The question naturally arises as to whether the initial data of a fluid model plus the equations and auxiliary conditions (2.9)–(2.15) uniquely determine the functions A, W, Y, D, Z, and F. At the time of this writing, the answer is simply not known, although it *is* known that for some multiclass networks with *priority* scheduling disciplines, the analogous fluid equations have multiple solutions [10]. As a hedge against that possibility, when we speak of the *fluid analog* or *fluid model* of a multiclass queueing network, this is understood to mean the set of *all* solutions for (2.9)–(2.15), given the specified initial conditions.

PROPOSITION 2.1. *For any initial state and any corresponding solution of (2.9)–(2.15), we have*

$$(2.20) \qquad\qquad W(t) = CM\,Z(t), \qquad t \geq 0.$$

REMARK.   This relationship is "obvious" from the interpretations of W and Z. Its proof, which is not trivial, represents a check on the fluid model formulation (2.9)–(2.15).

PROOF.   Multiplying (2.14) by CM, and using the fact that $CMZ(0) = W(0)$ by (2.8), we have

$$(2.21) \qquad CM\,Z(t) = W(0) + CM\,A(t) - CM\,D(t).$$

Thus, by (2.10) it suffices to show that

$$(2.22) \qquad\qquad CM\,D(t) = et - Y(t) \quad \text{for } t \geq 0,$$

or equivalently,

$$(2.23) \qquad \sum_{k \in \mathcal{C}(j)} m_k D_k(t) = t - Y_j(t) \quad \text{for } t \geq 0 \quad \text{and } j = 1, \ldots, d.$$

To prove (2.23) let us fix a service station j. From (2.10) we know that $W_j(t) > 0$ if $0 \leq t < W_j(0)$, and hence from (2.11) and (2.12) that $Y_j(\cdot) = 0$ over the initital interval $[0, W_j(0)]$. Over that interval, then, (2.23) follows from the condition (2.17) that initial outflow data are required to satisfy. From (2.16) we have that $t \to t + W_j(t)$ is a continuous non-decreasing function. Furthermore, that function goes to infinity as $t \to \infty$, so to complete the proof of (2.23) it will suffice to show that

$$(2.24) \qquad \sum_{k \in \mathcal{C}(j)} m_k D_k(t + W_j(t)) = (t + W_j(t)) - Y_j(t + W_j(t))$$

for all $t \geq 0$. From (2.10)–(2.12) it follows that $Y_j(\cdot)$ is constant over the interval $[t, t + W_j(t)]$ and hence (2.24) is equivalent to

$$(2.25) \qquad \sum_{k \in \mathcal{C}(j)} m_k D_k(t + W_j(t)) = t + W_j(t) - Y_j(t).$$

To prove (2.25), first multiply (2.13) by m_k, then sum over $k \in \mathcal{C}(j)$, then substitute in the obvious way from (2.10), again using the initial condition $W(0) = CMZ(0)$ in (2.8). □

PROPOSITION 2.2. *Given any initial state and any corresponding solution of (2.9)–(2.15), define* $\xi(t) \equiv CMQZ(t)$, $t \geq 0$. *Then*

$$(2.26) \qquad \xi(t) = \xi(0) - (e - \rho)t + Y(t), \quad t \geq 0.$$

REMARK. In terms of the original queueing network, one may interpret Q_{kl} as the expected number of visits to class k made by a customer that starts in class l. In the fluid analog, then, it follows that $\xi_j(t)$ represents the total amount of time required from server j to complete all processing of fluid existing anywhere in the network at time t, assuming that external inputs are simply shut off.

PROOF. By substituting (2.9) and (2.15) into (2.14) we obtain $Z(t) = Z(0) + \alpha t - (I - P')D(t)$. Premultiplying this equation by CMQ, and recalling that $\rho = CMQ\alpha$ by definition, we have $\xi(t) = \xi(0) + \rho t - CMD(t)$. The proof is completed by substituting from (2.22). □

3. Asymptotic stability of balanced fluid models. For reasons explained later in Section 4, we now restrict attention to the *balanced* fluid model with $\rho = e$. Specializing (2.26) to this case, one has

$$(3.1) \qquad \xi(t) = \xi(0) + Y(t), \quad t \geq 0.$$

Because each component of the cumulative idleness process Y is non-decreasing, the same is true for the total workload process ξ. Thus $\xi(t)$ converges to some limit $\xi(\infty)$ as $t \to \infty$, and that limit is finite if and only if $Y(\infty) < \infty$. That is, the vector of total workloads (one component for each server or station) approaches a finite limit if and only if total idleness over the infinite horizon is finite for each server. Recalling that $\xi(t) = CMQZ(t)$ by definition, we see that the asymptotic inventory vector $Z(\infty)$, if it exists, can only be zero if $Z(0) = 0$ (that is, the system starts empty). Thus a balanced fluid model cannot be stable in the sense of Dai [6], but it may still be stable in the following sense.

DEFINITION 3.1.   *A balanced fluid model is said to be* asymptotically stable *if, for each initial state $X(0) \in S$ and each solution of (2.9)–(2.15), we have $X(t) \rightarrow X(\infty) \in S$ as $t \rightarrow \infty$.*

Of course, the limit state $X(\infty)$ referred to in this definition will depend in general on the initial state $X(0)$. The remainder of this section is devoted to describing limit states and the way in which they depend on initial data. Based on general experience with dynamic system models, one expects that only "equilibrium points" of the fluid model will be achievable as limit states. To rigorously prove that in the current context is an imposing technical task, so we shall simply conjecture the equivalence and explore its implications.

DEFINITION 3.2.   *A state $x \in S$ is said to be an* equilibrium point *for the balanced fluid model if, starting from state x, there exists a solution of (2.9)–(2.15) whose state-space description is $X(t) = x$, $t \geq 0$.*

PROPOSITION 3.1.   *Consider a state $x \in S$, denoting by $z = x(\infty)$ and $w = CMz$ the associated inventory vector and immediate workload vector, respectively. Then x is an equilibrium point if and only if, for each station $j = 1, \ldots, d$ with $w_j > 0$,*

$$(3.2) \qquad x_k(s) = \lambda_k s \quad \text{for all } k \in \mathcal{C}(j) \text{ and } 0 \leq s \leq w_j.$$

REMARK.   As a consequence of (3.2) we have that $z_k = \lambda_k w_j$ for all $j = 1, \ldots, d$ and $k \in \mathcal{C}(j)$. Defining a $c \times d$ matrix Λ via

$$(3.3) \qquad \Lambda_{kj} = \begin{cases} \lambda_k & \text{if } k \in \mathcal{C}(j) \\ 0 & \text{otherwise,} \end{cases}$$

we can write this more compactly as

$$(3.4) \qquad z = \Lambda w.$$

One may paraphrase (3.4) by saying that, in an equilibrium state, the fluid inventories for the classes served at a station are necessarily proportional to the classes' total arrival rates. In addition, (3.2) tells us that the fluid classes are uniformly mixed in the queue or storage buffer at the station, and hence will flow out in the same fixed proportions.

PROOF.   Remembering that $\lambda = \alpha + P'\lambda$ and $\rho = CM\lambda = e$, it is easy to verify that any x satisfying (3.2) is an equilibrium point. For the converse, let x be an equilibrium point, and set $z = x(\infty)$ and $w = CMz$ as usual. Let $D = (D(t), t \geq 0)$ be the departure process in the solution of

(2.9)–(2.15) whose state space description satisfies $X(t) = x$, $t \geq 0$. From (2.19) we then have

$$(3.5) \qquad D_k(t + s) - D_k(t) = X_k(t, s) = x_k(s)$$

for each $t \geq 0$, each station j with $w_j > 0$, each $k \in \mathcal{C}(j)$, and each $s \in [0, w_j]$. Setting $t = 0$ and noting that $D_k(0) = 0$, we have $D_k(s) = x_k(s)$ for $0 \leq s \leq w_j$. Therefore, $D_k(t + s) = D_k(t) + D_k(s)$ for $t \geq 0$ and $0 \leq s \leq W_j$. Because departure flows $D_k(\cdot)$ are absolutely continuous in the fluid model this implies $D_k(t + s) - D_k(t) = \delta_k s$ for some $\delta_k > 0$, and hence $x_k(s) = \delta_k s$, $0 \leq s \leq w_j$. It is then an easy matter to show that the only vector δ of constant outflow rates that is consistent with constant inventory levels $(Z(t) = z$ for all $t \geq 0)$ is $\delta = \lambda$. $\square$

CONJECTURE 3.1. *For any initial state and any solution of (2.9)– (2.15), if a limit state $X(\infty)$ exists, it is an equilibrium point.*

PROPOSITION 3.2. *Suppose that a balanced fluid model is asymptotically stable and satisfies Conjecture 3.1 (that is, every limit state is an equilibrium point). If $R \equiv (CMQ\Lambda)^{-1}$ exists, then for any initial state and any solution of (2.9)–(2.15), the limiting immediate workload vector $W(\infty)$ and limiting cumulative idleness vector $Y(\infty)$ are related by*

$$(3.6) \qquad W(\infty) = RCMQ\, Z(0) + RY(\infty).$$

REMARK. It will be seen in the next section that R is the so-called reflection matrix for the RBM that appears in the conventional heavy traffic limit theorem for the open multiclass network under discussion. If the inverse defining R does not exist, it is known that the network does not satisfy a conventional heavy traffic limit theorem. Similarly, one may plausibly conjecture that existence of R is necessary for asymptotic stability of the balanced fluid model, but no proof has been found thus far.

PROOF. Given that the limit state $X(\infty)$ must be an equilibrium point, it is immediate from Proposition 3.1 that $Z_k(\infty) = \lambda_k W_j(\infty)$ for all $k \in \mathcal{C}(j)$. One can write this in vector form as

$$(3.7) \qquad Z(\infty) = \Lambda W(\infty).$$

Now (3.1) says that the total workload process $\xi(t) \equiv CMQZ(t)$ satisfies $\xi(t) = \xi(0) + Y(t)$ in the balanced case, hence $\xi(\infty) = \xi(0) + Y(\infty)$, or equivalently,

$$(3.8) \qquad CMQZ(\infty) = CMQZ(0) + Y(\infty).$$

Substituting (3.7) into (3.8) gives $CMQ\Lambda Z(\infty) = CMQZ(0) + Y(\infty)$, and then premultiplying both sides of that equation by R gives the desired conclusion. $\square$

4. A conventional heavy traffic limit theorem. Informally, we say that an open multiclass network is *in heavy traffic*, or that heavy traffic conditions prevail, if the traffic intensity parameter ρ_j is near one (either less than, greater than, or actually equal to one) for each station $j = 1, \ldots, d$. Long experience suggests that under such conditions network behavior can be approximated by means of the central limit theorem (CLT). To formally justify such an approximation, it is customary to consider a sequence of models indexed by $n = 1, 2, \ldots$ such that (here and later, a superscript n signifies a parameter or process associated with the nth network model)

$$(4.1) \qquad \lambda^n \;\to\; \lambda \;>\; 0 \qquad \text{as} \quad n \;\to\; \infty$$

$$(4.2) \qquad m_k^n \;\to\; m_k \;>\; 0 \qquad \text{as} \quad n \;\to\; \infty \quad (k = 1, \ldots, c)$$

$$(4.3) \qquad \sqrt{n}(1 - \rho_j^n) \;\to\; c_j \qquad \text{as} \quad n \;\to\; \infty \quad (j = 1, \ldots, d)$$

where c_j is a finite constant (either positive, negative or zero). In addition, if the external arrival process E and the service time distributions are allowed to change as one passes through the sequence of systems, then additional assumptions are needed (variance parameters must converge, Lindeberg conditions must be imposed to get CLT convergence, and so forth). For all practical purposes, however, it is sufficient to suppose that the service time distributions are rescaled by their mean values m_k^n as one passes through the sequence, with all else fixed, and then no additional assumptions are required. Of course, (4.1)–(4.3) imply that the limiting parameters satisfy

$$(4.4) \qquad \rho_j \;\equiv\; \sum_{k \in \mathcal{C}(j)} \lambda_k m_k \;=\; 1 \quad \text{for all} \quad j = 1, \ldots, d.$$

Condition (4.3) further demands that $\rho_1^n, \ldots, \rho_d^n$ approach unity at the same rate, but the quantification of that rate by means of the factor $\sqrt{n}$ is simply a convenient convention. We could replace $\sqrt{n}$ in (4.3) by a positive factor b_n such that $b_n \to \infty$, and then b_n would simply replace $\sqrt{n}$ in all that follows. To obtain central limit theorem convergence, consider the scaled processes

$$(4.5) \qquad \tilde{Z}^n(t) \;=\; \frac{1}{\sqrt{n}} Z^n(nt) \quad \text{and} \quad \tilde{Y}^n(t) \;=\; \frac{1}{\sqrt{n}} Y^n(nt).$$

That is, compressing the time scale by a factor of n under hypothesis (4.3), one expects to see queue lengths and cumulative idleness quantities of order $\sqrt{n}$, at least over moderate periods of scaled time.

To repeat, (4.5) is the sort of scaling associated with CLT convergence, and a "conventional" heavy traffic limit theorem has the form

$$(4.6) \qquad (\tilde{Y}^n, \tilde{Z}^n) \;\Rightarrow\; (\tilde{Y}, \tilde{Z}) \quad \text{as} \quad n \;\to\; \infty,$$

where $\tilde{Y}$ and $\tilde{Z}$ are continuous processes associated with an "approximating Brownian system model." To be more precise, let us define the obvious sequence of normalized immediate workload processes (indexed by $n = 1, 2, \ldots$)

$$(4.7) \qquad \tilde{W}^n(t) = \frac{1}{\sqrt{n}} W^n(nt).$$

The complete form of the conventional heavy traffic limit theorem is

$$(4.8) \qquad (\tilde{W}^n, \tilde{Y}^n, \tilde{Z}^n) \Rightarrow (\tilde{W}, \tilde{Y}, \Lambda \tilde{W}) \quad \text{as } n \to \infty$$

where $\tilde{W}$ is an RBM in $\mathbb{R}_+^d$ and $\tilde{Y}$ is the associated "pushing process on the boundary." The data of the RBM are (i) a $d \times d$ covariance matrix derived from the switching matrix and variance parameters of the original queueing network, (ii) a d-dimensional drift vector derived from the vector of constants c appearing in (4.3), and (iii) a $d \times d$ reflection matrix R that Harrison and Nguyen [12] define as

$$(4.9) \qquad R = (I + G)^{-1} \quad \text{where} \quad G = CMQP'\Lambda.$$

(Actually, the Harrison-Nguyen definition of G uses another matrix Δ in place of Λ, but the two are equivalent when one specializes their network formulation to the case of perfectly reliable servers, as we do throughout.) To see that (4.9) is equivalent to the definition of R given in Proposition 3.2, note the following. First, using the fact that $\rho \equiv CM\lambda = e$ by (4.4), it is easy to verify that $CM\Lambda = I$. Thus, remembering that $Q = (I + P + P^2 + \ldots)'$,

$$(4.10) \qquad I + G = CM(I + QP')\Lambda = CMQ\Lambda,$$

which establishes the desired equivalence. As explained in [12], Taylor and Williams [15] have shown that the "approximating Brownian system model" on the right side of (4.8) is well defined if and only if R exists and is what is called a *completely-S* matrix. That condition is obviously necessary, then, if the conventional heavy traffic limit theorem (4.8) is to hold, but is it sufficient? The answer is not known, but it may very well be negative.

Readers should note that the weak limit on the right side of (4.8) is characterized by what Reiman [14] called a "state space collapse." The c-dimensional scaled queue length processes $\tilde{Z}^n$ converge weakly to a limit process $\tilde{Z}$ that is effectively d-dimensional: For each station j one has $\tilde{Z}_k(t) = \lambda_k \tilde{W}_j(t)$ for all $t \geq 0$ and all $k \in \mathcal{C}(j)$, so the customer classes served at a station live in fixed proportions in the heavy traffic limit. In the remark following Proposition 3.1 it was noted that the equilibrium states (and hence, if one accepts Conjecture 3.1, the limit states) of a network's balanced fluid analog have this property, but for the Brownian limit it holds at every instant in time.

Thus far nothing has been said about the initial conditions assumed in the "conventional heavy traffic limit theorem" (4.8). It has been customary in past heavy traffic research to assume that each system in the sequence satisfying (4.1)–(4.3) starts empty, which obviously leads to a Brownian limit $\tilde{W}$ with $\tilde{W}(0) = 0$. If one wishes to obtain a "general" initial state for the Brownian limit, what assumptions are appropriate for the sequence of queueing networks? In light of (4.7), the obvious guess is

$$(4.11) \qquad \frac{1}{\sqrt{n}} W^n(0) \;\to\; w \qquad \text{as } n \to \infty.$$

One naturally supposes that if (4.1)–(4.3) are augmented by (4.11), plus possibly some minor technical conditions, then (4.8) will hold with the limit processes satisfying $\tilde{W}(0) = w$ and $\tilde{Y}(0) = 0$. But that conclusion is almost certainly wrong in general, as we discuss in the following section. Consideration of Brownian limits with "general initial state" leads one back to the subject of fluid models, and to alternative scalings of time and state space that yield fluid models as heavy traffic limits. This in turn gives at least a strong hint as to what can "go wrong" in heavy traffic analysis of some multiclass networks.

5. Balanced fluid models as heavy traffic limits. Continuing our study of an open network sequence satisfying (4.1)–(4.3), consider now the scaled processes

$$(5.1) \qquad \bar{Z}^n(t) \;=\; \frac{1}{\sqrt{n}} Z^n(\sqrt{n}t) \quad \text{and} \quad \bar{Y}^n(t) \;=\; \frac{1}{\sqrt{n}} Y^n(\sqrt{n}t).$$

As discussed below, one expects that this scaling will give convergence to a balanced fluid model (thus it is called *fluid scaling* hereafter), meaning that

$$(5.2) \qquad (\bar{Y}^n, \bar{Z}^n) \;\Rightarrow\; (\bar{Y}, \bar{Z}) \quad \text{as } n \to \infty,$$

where $(\bar{Y}, \bar{Z})$ are part of a solution for the "fluid equations" (2.9)–(2.15). It is the limiting parameters $\alpha, m_1, \ldots, m_c$ in (4.1)–(4.2) that appear in those equations, so (4.4) tells us that the fluid limit $(\bar{Y}, \bar{Z})$ is necessarily balanced. To get a non-trivial initial state in the fluid limit, one obviously wants to consider initial queue length vectors in the sequence of systems such that

$$(5.3) \qquad \frac{1}{\sqrt{n}} Z^n(0) \;\Rightarrow\; z \quad \text{as } n \to \infty.$$

But the discussion of fluid model initial conditions in Section 2 shows that (5.3) is not sufficiently detailed. That is, information must be added to describe the ordering of customer classes in the initial queues, but before that subject is broached, let us consider the relationship between fluid scaling and the scaling used in Section 4 to obtain a Brownian limit.

The scaling embodied in (5.1) is that associated with the law of large numbers (LLN), where space and time are compressed by the same factor. We choose the factor $\sqrt{n}$ so that our two scalings (4.5) and (5.1) agree in their units of measurement for both queue length and cumulative idleness. As we have defined things, fluid scaling is distinguished from CLT scaling by a milder compression of time. Equivalently, CLT scaling amounts to fluid scaling plus a second compression of the time scale by $\sqrt{n}$. In this sense, a Brownian model obtained as the weak limit after CLT scaling (if there is one) represents a more highly aggregated approximation to the original queueing model than does a fluid model obtained as a weak limit after the scaling (5.1). Another way of saying this is that fluid scaling, when compared with CLT scaling, shows the evolution of the system over relatively short periods of time. Such a view of fluid models is unconventional, as explained at the end of this section.

Returning to the matter of initial conditions, one may take as given, for each system n and each station j, an initial workload level $W_j^n(0)$ and a collection of initial departure processes $\{D_k^n(t),\ 0 \leq t \leq W_j^n(0)\}$ for classes $k \in \mathcal{C}(j)$. To avoid needless complications, we shall treat all of this initial data as deterministic, and then define a vector x^n of right-continuous, non-decreasing functions $x_k^n(\cdot)$ on $[0, \infty)$ via

$$(5.4) \qquad x_k^n(s) \;=\; \begin{cases} D_k^n(s) & \text{if } 0 \leq s \leq W_j^n(0) \\ Z_k^n(0) & \text{if } s \geq W_j^n(0). \end{cases}$$

Note that $x_k^n(\cdot)$ is right continuous at time $W_j^n(0)$ because $D_k^n(W_j^n(0)) = Z_k^n(0)$. In light of (5.1) it is natural to define scaled initial states $\bar{x}^n$ via

$$(5.5) \qquad \bar{x}^n(s) \;=\; \frac{1}{\sqrt{n}} x^n(\sqrt{n}s), \quad s \geq 0,$$

and we then extend (5.3) by assuming (here convergence is in the standard Skorohod topology, and S is the fluid model state space defined in Section 2)

$$(5.6) \qquad \bar{x}^n \;\to\; \bar{x} \in S \quad \text{as } n \to \infty.$$

Hereafter, when we speak of the "limiting fluid model" this is understood to mean the solution set of equations (2.9)–(2.15) with initial state $\bar{x}$ and parameters $\alpha, m_1, \ldots, m_c$ equal to the limiting values in (4.1)–(4.2).

Let us consider now the putative fluid limit theorem (5.2). Arguing as in Section 4 of Dai and Weiss [10], one can show that the sequence of processes associated with our sequence of systems (a vector arrival process, vector workload process, etc. for each system n) is tight under fluid scaling. Moreover, the processes obtained as weak limits for any convergent subsequence necessarily satisfy all of the "fluid equations" (2.9)–(2.15) except condition (2.13), which is specific to FIFO scheduling. (Dai and Weiss treat all non-idling service disciplines simultaneously in their theory.) Finally, it is known that fluid limits of FIFO networks necessarily satisfy (2.13) in certain special cases, and we shall assume that is true in our general setting as well, although the proof is certainly not trivial. If the processes $\bar{Y}$ and $\bar{Z}$ associated with the limiting fluid model were unique, the fluid limit theorem (5.2) would then be established. If the limiting fluid model has multiple solutions, it could still be that a particular one of those solutions emerges as a deterministic limit in (5.2), or that (5.2) holds with $(\bar{Y}, \bar{Z})$ a *stochastic* process whose paths all lie in the solution set of the limiting fluid model.

To repeat an earlier point, our fluid scaling (5.1) is chosen so that the fluid and Brownian limits for a sequence of networks, if both exist, will measure queue lengths and cumulative idleness in the same units, the difference in the two limit models being their time scales. This view of fluid models is unconventional. Given our setup, for example, Chen and Mandelbaum [3,4,5] would define the "fluid scaled processes"

$$(5.7) \qquad \hat{Z}^n(t) = \frac{1}{n} Z^n(nt) \quad \text{and} \quad \hat{Y}^n(t) = \frac{1}{n} Y^n(nt).$$

This too is LLN scaling but of a cruder sort than (5.1). If our fluid limit theorem (5.2) is to hold, regardless of how well or badly the fluid limit $(\bar{Y}, \bar{Z})$ may behave, it is obviously necessary that

$$(5.8) \qquad (\hat{Y}^n, \hat{Z}^n) \Rightarrow (0,0) \quad \text{as } n \to \infty.$$

On the other hand, (5.8) does not imply (5.2), nor does it tell us anything about the fluid limit $(\bar{Y}, \bar{Z})$ if (5.2) *does* hold.

6. Relating the Brownian and fluid limits. Given a sequence of multiclass networks satisfying the heavy traffic conditions (4.1)–(4.3), and with initial data satisfying (5.6), let us suppose that both the fluid limit theorem (5.2) and the Brownian limit theorem (4.7) hold. Let us also assume that the limiting fluid model has a unique solution which is asymptotically stable. In the fluid model, then, there exist both a finite total idleness vector $\bar{Y}(\infty)$ and a finite limiting queue length vector $\bar{Z}(\infty)$. Now comparing the Brownian scaling (4.5) with the fluid scaling (5.1) we see that

$$(6.1) \qquad \tilde{Y}^n(t) = \bar{Y}^n(\sqrt{n}t) \quad \text{and} \quad \tilde{Z}^n(t) = \bar{Z}^n(\sqrt{n}t).)$$

Because the Brownian limit $(\tilde{Y}, \tilde{Z})$ is continuous by definition, (6.1) strongly suggests that (there is an interchange of limits involved, so we cannot claim this as a rigorous conclusion)

$$(6.2) \qquad \tilde{Y}(0+) = \bar{Y}(\infty) \quad \text{and} \quad \tilde{Z}(0+) = \bar{Z}(\infty).$$

For those who have only contemplated limit theorems for systems starting empty, (6.2) is surprising on several levels. First, according to hypothesis

(5.3) the initial inventory vector in the Brownian limit model is $\tilde{Z}(0) = \lim \tilde{Z}^n(0) = z$, and similarly one has $\tilde{Y}(0) = \lim \tilde{Y}^n(0) = 0$. Thus (6.2) says that in general $\tilde{Y}(\cdot)$ and $\tilde{Z}(\cdot)$ may have jumps at $t = 0$, although the sample paths of these Brownian limit processes are presumably continuous thereafter. Hereafter, we shall say that (6.2) describes the "effective initial state" of the Brownian limit model. The effective initial inventory vector $\tilde{Z}(0+)$ is equal to the fluid model's asymptotic inventory vector $\bar{Z}(\infty)$, because the entire approach to equilibrium in the fluid model takes place in an instant on the Brownian time scale, and in general that approach to equilibrium entails positive amounts of idleness. A related surprise is that the effective initial values $\tilde{Y}(0+)$ and $\tilde{Z}(0+)$ in the Brownian model depend not just on the initial inventory vector z in (5.3) but on the entire initial state $\bar{x}$ for the limiting fluid model that was assumed in (5.6).

Continuing in this same vein, let us now suppose that the fluid limit theorem (5.2) holds but that the resulting balanced fluid model fails to satisfy $\bar{Y}(t) \to \bar{Y}(\infty) < \infty$ and $\bar{Z}(t) \to \bar{Z}(\infty)$. In this case it follows from (6.1) that the Brownian limit theorem (4.7) cannot hold because the sequence $\{(\tilde{Y}^n, \tilde{Z}^n)\}$ is not tight. That is, (6.2) implies that after Brownian scaling one or both of these processes will either diverge or oscillate rapidly as n gets large. We now arrive at the conjecture announced in the abstract and introduction of this paper.

CONJECTURE 6.1.   *The following two statements are equivalent.   (a) The Brownian limit theorem (4.7) holds for every choice of initial state $\bar{x}$ in (5.6).   (b)   The network's balanced fluid analog is asymptotically stable.*

The only argument that has actually been advanced in this section is one suggesting that (b) is necessary for (a). The conjecture that it is also sufficient may simply be called an optimistic leap of faith. Actually, it has not really even been argued that (b) is necessary for (a) because asymptotic stability of the balanced fluid model demands that *all* solutions of the "fluid equations" (2.9)–(2.15) approach limits as $t \to \infty$, but if there are multiple solutions, some may be unachievable as fluid limits of network processes. The point of view reflected in Conjecture 6.1 is that if "bad solutions" exist, they will indeed appear as limits under fluid scaling, and hence weak convergence under Brownian scaling cannot occur. Of course, it may actually be that the fluid model is unique for the FIFO networks under discussion here (that is known to be false for priority networks), in which case there is no distinction between the network's fluid limit and its fluid model.

In conclusion, let us return to the notion of fluid model "state descriptions" introduced in Section 2. This concept suggests a potential extension of conventional heavy traffic limit theory under Brownian scaling, as follows. Given a sequence of networks satisfying the heavy traffic assumptions (4.1)–(4.3), and with initial data satisfying (5.6), let us define for each system $n = 1, 2, \ldots$ a function-valued process $X^n = \{X^n(t),\ t \geq 0\}$ via

$$(6.3) \qquad X_k^n(t, s) = \begin{cases} D_k^n(t + s) - D_k^n(t) & \text{if } 0 \leq s \leq W_j^n(t) \\ Z_k^n(t) & \text{if } s \geq W_j^n(t). \end{cases}$$

This is precisely analogous to definition (2.18) for fluid models, but now $X_k^n(t) = X_k^n(t, \cdot)$ takes values in the space of right-continuous functions having left limits (that is, the space of Skorohod paths) on $[0, \infty)$, and hence $X^n(t) = X^n(t, \cdot)$ takes values in the corresponding $c - fold$ product space. The natural scaling for this sequence of "state space descriptions" is given by

$$(6.4) \qquad \tilde{X}_k^n(t, s) = \frac{1}{\sqrt{n}} X^n(nt, \sqrt{n}s),$$

and then one may seek to show that

$$(6.5) \qquad \tilde{X}^n \Rightarrow \tilde{X} \quad \text{as } n \to \infty,$$

where $\tilde{X}$ is a process taking values in the fluid model state space S defined in Section 2. (Of course, the setting for this limit theorem is a very complicated function space, and a rigorous proof would presumably entail great technical complexity.) A heuristic argument analogous to that supporting Conjecture 6.1 suggests that the limit process $\tilde{X}$ must, in fact, take values in the subset S^* of fluid model *equilibrium points* characterized in Proposition 3.1. In that case $\tilde{X}(t)$, the state description of the limiting Brownian system model at time t, would be completely captured by the standard d-dimensional workload vector $\tilde{W}(t)$. What the extended limit theorem (6.5) would add to existing proofs of "state space collapse" is precise mathematical justification for the claim that, in the Brownian limit, the customer classes served at any given station are uniformly mixed in that station's queue at every instant of time. That is, the fixed proportions which are known to characterize the queue length vector $Z(t)$ further hold throughout the various queues at every instant of time.

7. Acknowledgements. Elizabeth Schwerer has been my research assistant throughout the gestation and preparation of this paper, and I have learned a great deal from her in our many hours of conversation, especially concerning the crucial role of state space collapse in heavy traffic theory. A preliminary version of the paper (that is, an even more preliminary version of the ideas discussed here) was presented at the Institute for Mathematics and its Applications (IMA), where I spent four weeks during February and March of 1994. Thanks are due to Frank Kelly and Ruth Williams for organizing the superb IMA workshop on stochastic networks, and to Avner Friedman and Mike Steele for higher-level management contributing to the Institute's stimulating and collegial atmosphere. Conversations with Jim Dai about his work on fluid models, and particularly about the examples analyzed in his joint work with Gideon Weiss [10], have been the most important single influence on my thinking about the subject. Discussions with Frank Kelly, Elena Krichagina, Avi Mandelbaum, and Ruth Williams during our overlapping periods in residence at the IMA were also most helpful, and subsequent e-mail exchanges with Maury Bramson about his current work have reduced the fuzziness in my thinking about fluid models and their relation to Brownian limits, though not as much as readers of this paper might have wished.

REFERENCES

[1] BRAMSON, M., *Instability of FIFO queueing networks*, Ann. Appl. Prob., 4 (1994), 414–431.

[2] BRAMSON, M., *Instability of FIFO queueing networks with low arrival rates*, Ann. Appl. Prob., to appear.

[3] CHEN, H. AND A. MANDELBAUM, *Discrete flow networks: Bottleneck analysis and fluid approximations*, Mathematics of Operations Research, 16 (1991), 408–446.

[4] CHEN, H. AND A. MANDELBAUM, *Stochastic discrete flow networks: diffusion approximations and bottlenecks*, Annals of Probability, 4 (1991), 1463–1519.

[5] CHEN, H. AND A. MANDELBAUM, *Hierarchical modeling of stochastic networks, Part I: Fluid models*, Applied probability in manufacturing systems, D. D. YAO, (ed.), Forthcoming.

[6] DAI, J. G., *On positive Harris recurrence of multiclass queueing networks: a unified approach via fluid limit models*, Ann. Appl. Prob., to appear.

[7] DAI, J. G., personal communication.

[8] DAI, J. G. AND V. NGUYEN, *On the convergence of multiclass queueing networks in heavy traffic*, Ann. Appl. Prob., 4 (1994), 26–42.

[9] DAI, J. G. AND Y. WANG, *Nonexistence of Brownian models for certain multiclass queueing networks*, Queueing Systems: Theory and Applications, 13 (1993), 41–46.

[10] DAI, J. G. AND G. WEISS, *Stability and instability of fluid models for certain multiclass networks*, Preprint.

[11] HARRISON, J. M. AND V. NGUYEN, *The QNET method for two-moment analysis of open queueing networks*, Queueing Systems: Theory and Applications, 6 (1990), 1–32.

[12] HARRISON, J. M. AND V. NGUYEN, *Brownian models of multiclass queueing networks: current status and open problems*, Queueing Systems: Theory and Applications, 13 (1993), 5–40.

[13] KUMAR, P. R., *Re-entrant lines*, Queueing Systems: Theory and Applications: Special issue on queueing networks, 13 (1993), 87–110.

[14] REIMAN, M. I., *Some diffusion approximations with state space collapse*, in Proc. Internat. Seminar on Modeling and Performance Evaluation Methodology, (1983) Springer-Verlag, Berlin and New York.

[15] TAYLOR, L. M. AND R. J. WILLIAMS, *Existence and uniqueness of semimartingale reflecting Brownian motions in an orthant*, Probab. Theory Rel. Fields, 96 (1993), 283–317.

SCHEDULING QUEUEING NETWORKS: STABILITY, PERFORMANCE ANALYSIS AND DESIGN*

P.R. KUMAR[†]

Abstract. Queueing networks are a useful class of models in many application domains, e.g., manufacturing systems, communication networks, and computer systems. Control is typically exercised over such systems by the use of scheduling policies.

However, if one ventures outside a certain special class of systems for which the steady state distribution has a product form, very little is known concerning their performance or even stability. In the first half of this survey paper, we present new theoretical developments on stability analysis and performance evaluation for queueing networks and scheduling policies. We show how one may solve problems in stability and performance analysis by solving linear programs.

In the second half, we address the problem of scheduling a class of queueing networks called re-entrant lines, which model semiconductor manufacturing plants. We propose a new class of scheduling policies based on smoothing all the flows in the system. We also report briefly from an extensive simulation study comparing the performance of the suggested scheduling policies with a range of other scheduling policies, for a variety of semiconductor manufacturing plant models.

1. Introduction. In this paper, we provide a brief survey of some recent results concerning queueing networks. We shall address three sets of issues: stability, performance analysis, and the design of scheduling policies. The results are mainly drawn from [30,27,22,23,32], to which we refer the reader for further details.

After a brief description of a motivating application, we begin by exhibiting a counterexample that shows that even simple queueing networks can be unstable for some scheduling policies. This motivates the problem of how to establish the stability of queueing networks and scheduling policies. We exhibit a new approach based on using linear programming. We show how to construct a linear program (LP) such that every feasible solution is a weighting matrix whose associated quadratic form serves as a stochastic Lyapunov function. Thus, we obtain a simple linear programming procedure for establishing stability. This approach establishes not just the positive recurrence of the underlying Markov chain, but, in fact, a strong form of geometric ergodicity. We mention some extensions of this approach and illustrate its use on several examples.

Next, we turn to the problem of performance evaluation of queueing networks and scheduling policies. Consider a queueing network that has a steady state in which the mean number of customers is finite. (This is the sort of stability issue that is studied in the earlier first part of the

---

* The research reported here has been supported in part by the U.S. Army Research Office under Contract No. DAAL-03-91-G-0182, by the National Science Foundation under Grant No. ECS-92-16487, and by the Joint Services Electronics Program under Contract No. N00014-90-J1270.

† University of Illinois, Department of Electrical and Computer Engineering and the Coordinated Science Lab, 1308 West Main Street, Urbana, IL 61801, USA.

paper.) We show that whenever such stability holds, the mean values of certain random variables determining the performance, satisfy a set of linear equality constraints. In addition to satisfying non-negativity constraints, these mean values also satisfy a set of "non-idling" inequality constraints if the scheduling policy under consideration is non-idling. Further, many systems satisfy additional linear equalities or inequalities, e.g., closed systems, networks under buffer priority policies, etc. Given these linear constraints, the performance of a system can be bounded by solving a linear program. Thus we obtain upper and lower bounds on mean queue lengths, by linear programming. We also exhibit the application of this methodology on a variety of examples, showing the power of the approach.

We also show that one can obtain linear programs to bound the transient (as opposed to steady-state) performance of queueing networks and scheduling policies.

Given two separate linear programs, one for establishing stability, and another for studying performance, one may well wonder what the connection is between these two linear programs. We show that these two linear programs are simply the duals of each other. This close connection allows us to develop a variety of the results relating the boundedness of the performance linear program to the existence of a quadratic stochastic Lyapunov function.

The third issue we address is that of designing good scheduling policies for queueing networks. We focus on the important and topical problem of scheduling semiconductor manufacturing plants. Here, wafers at several stages of their life may be in competition for the same set of machines. Thus, potentially, by properly scheduling the system, i.e., deciding which lot of wafers is served next, one can decrease the mean manufacturing lead time (called mean "cycle-time" in the industry). These systems are very large scale, involving about ten to twenty service stations, and a hundred or more stages of processing (i.e., buffers) for each wafer. One therefore wants to design scheduling policies that are easily implemented on large scale systems, whose decisions can be determined in real-time, and which can cope with random events such as machine failures.

We develop a new set of scheduling policies called "fluctuation smoothing" policies. First, we develop a policy to reduce the variance of lateness of parts. Then, we show how to modify this policy so that it reduces the variance of the cycle-time. Then we report briefly from the results of an extensive comparative simulation study, which shows that our Fluctuation Smoothing Policy for Reducing the Variance of the Cycle-Time (FSVCT), indeed reduces the standard deviation of the cycle-time, by a substantial amount. Surprisingly, as a by product we discover that FSVCT also reduces substantially the *mean* cycle-time, when the arrivals to the plant are periodic (called "deterministic"). In developing a hypothesis for why this is so, we develop yet another Fluctuation Smoothing Policy for Reducing the Mean Cycle-Time (FSMCT), for systems whose arrivals are not periodic.

This policy has been tested extensively, and we briefly report from a study showing that it achieves its purpose. The net result is a new scheduling policy that reduces the mean waiting time by about 20%, and the standard deviation of cycle-time by about 50%, in comparison to the well known First Come First Serve Policy (FCFS or FIFO). These results have been subjected to statistical tests.

The main purpose of the FSMCT policy is to smooth the burstiness of all flows into every buffer in the system, simultaneously. That this can be done by an intuitively rational design is surprising.

There is yet another interpretation of FSMCT as a policy that at each time keeps track of the total downstream shortfall of parts from every buffer (short fall := actual number of parts downstream − mean number of parts downstream), and gives priority to the buffer whose shortfall is greatest. This is an appealing interpretation, which also shows that the scheduling policy is "stationary," in the terminology of Markov Decision Processes.

The rest of this paper is organized as follows. In Section 2, we start with a motivating problem: how to schedule semiconductor manufacturing plants. We describe the features of semiconductor manufacturing which lead to a re-entrant line. We provide an example of a model plant. In Section 3, we place re-entrant lines in perspective by showing their relationship vis-a-vis the traditional dichotomy in manufacturing systems: flow shops and job shops. Next, in Section 4, using Little's Theorem as a starting point, we design a myopic scheduling policy – the Last Buffer First Serve (LBFS) policy. Then, in Section 5, we motivate the problem of stability of queueing networks and scheduling policies by providing an example of a re-entrant line which is unstable under the Longest Processing Time (LPT) buffer priority policy. We also briefly mention some previous results establishing the stability of certain buffer priority policies. In Section 6, we begin by providing a model of a re-entrant line, on which we focus (for brevity only) in the rest of the paper. Then in Sections 6 and 7, we develop the linear programming approach to establishing stability. In Section 8, we provide some instability/non–robustness tests. In Section 9, we turn to the problem of performance evaluation, and show how one can obtain linear equality and inequality constraints on the performance of a system. In Section 10, we show how one can obtain more linear equalities and inequalities for specific modules which are prevalent in many applications. We also provide several examples showing the power of the approach. In Section 11, we show how one can obtain bounds on transient performance. In Section 12, we establish a duality relationship between the performance and stability problems, and examine some of its consequences in Section 13. In Section 14, we turn to the problem of scheduling semiconductor manufacturing plants. We develop the new class of fluctuation smoothing policies for reducing the variance of lateness and variance of cycle-time, and briefly present the results of a simulation study. Then, we develop a hypothesis for the particular behavior observed in the simulations, and, exploiting this, we design the new fluctuation smoothing policy for reducing the mean cycle-time. We report on its simulation performance, and also present another interpretation of it as alleviating total downstream starvation. Finally, in Section 15, we end on a broad note.

2. Semiconductor manufacturing. A semiconductor manufacturing plant (also called fabrication line or fab) takes raw wafers (about 8" or diameter) and imprints several layers of chemical patterns on them. As shown in Figure 1, the processing of a wafer is layer by layer, and the final wafer can be regarded as a multilayered sandwich. Each layer, in turn, requires many steps of processing (deposition, etching, etc.), and many layers repeat many of the same processing steps.

A machine (or station containing a set of identical machines) can perform one particular processing operation. Thus, wafers return several times to many of the machines for the performance of a particular processing step at several layers. The resulting manufacturing system is shown in Figure 2. It shows the route of a single product (called process flow or recipe) in an aggregated model of a full scale production line. There are a large number (60) of stages of processing. Many of the machines are revisited several times; for example, Station 1 is visited 14 times by each wafer. This type of a manufacturing system is accordingly called a "re-entrant line" [21]. Wafers usually travel in groups of about 20 wafers, called a *lot*, which we shall simply refer to hereafter as a single "part."

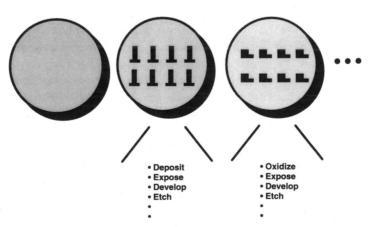

Figure 1: The semiconductor manufacturing process.

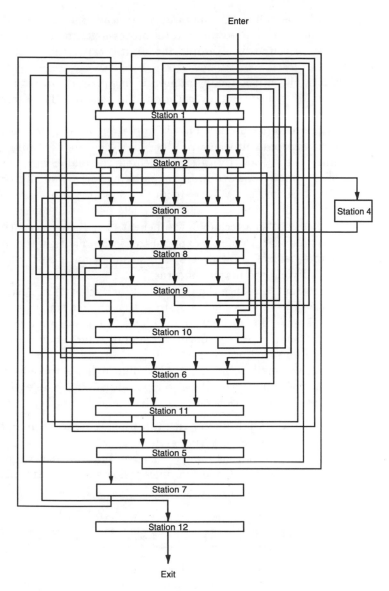

Figure 2: An aggregated model of a semiconductor manufacturing plant.

The key feature of a re-entrant line is that several parts at different stages of production may be in competition for the same machine (or set of machines at a station). These parts (like cars at a traffic light) may have to wait for machines. This can lead to large waiting times, an unfortunate characteristic of semiconductor manufacturing plants. For example, while each part may require a total service time from all machines of only about 10 days, a part may nevertheless take 60 days to complete production. The time to complete production, usually referred to as *sojourn time* in queueing networks, is called the *manufacturing lead time* or *cycle-time*. The ratio of mean cycle-time to mean total processing time can be large, between 2 and 10 (it is 6 in the example above); it is called the "cycle-time multiplier."

A key concern in semiconductor manufacturing plants is to reduce the mean cycle-time. It clearly has several economic benefits. For example, when developing new products it allows faster prototyping, and quicker responsiveness to customer needs. This is important in a rapidly changing technological environment where product life cycles are becoming very short, sometimes just a year. By Little's Theorem, a smaller mean cycle-time also implies a smaller mean number of parts in process, called work-in process or WIP. Reducing the WIP reduces the capital tied up, as well as the clutter in the plant. Moreover, when products become obsolescent, so can the WIP. There is also a technological benefit to reducing cycle-times. The smaller the time that a wafer is exposed to contaminants in the plant, the greater is the yield of good wafers, important in an industry where contamination and yield are major concerns.

It is also important to reduce the variance of the cycle-times. A smaller variation in cycle-times allows managers to better predict when products will be completed, and thus meet their due-dates. Thus, a small variance of cycle-times allows for improved coordination of the plant's output to the downstream operations such as assembly, etc.

In Section 14 we shall design new scheduling policies to reduce the mean and variance of cycle-times. We also report briefly from an extensive comparative study of their performance on plant models.

3. The traditional dichotomy in manufacturing systems: flow shops and job shops. Manufacturing systems usually fall into two categories, flow shops and job shops (though, of course, there are other types of plants, e.g., for ship building). Good examples of flow shops are automobile assembly lines; see Figure 3.

Figure 3: A flow shop.

The key features are that the flow of parts is acyclic (i.e., a part never revisits a machine or worker[1]), and the line is dedicated to a single type of part, which is manufactured in high volume. The second category into which many manufacturing systems fall is job shops. Good examples are metal cutting shops which consist of a number of "resources", e.g., lathes, milling machines, etc. A job shop can accept custom orders, and each order may require a specific sequence of operations, i.e., a route. A caricature is shown in Figure 4, where several "routes" are shown. There may not be

---

[1] It may do so for quality control purposes.

much of a pattern in the different routes; they may look "random." Each route is also typically of low volume.

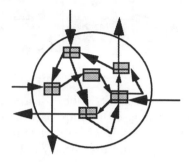

Figure 4: A job shop.

With semiconductor manufacturing plants, we have a new type of system, – a re-entrant line. It combines features of both flow shops and job shops. For example, unlike flow shops, the route is re-entrant; however, unlike job shops, it may be of high volume. Thus, it features the additional complication of competition for machines by parts, not present in flow shops. However, there is more structure to re-entrant lines, than is usually present in job shops. Thus, one wishes to tailor the design of scheduling polices to accommodate the special features of re-entrant lines, and take advantage of whatever structure they possess. (In particular, ideas developed in other environments, e.g., kan-bans, should not be blindly used).

4. Beginning the design of scheduling policies: Little's Theorem and the Myopic LBFS Policy. Consider a stable system. Let

λ := average arrival rate of parts to the system,
L := average number of parts in the system,
W := average time spent by parts in the system.

Then, under mild stability assumptions, Little's Theorem says that

$$L = \lambda W.$$

Thus, for a fixed plant throughput λ, one can reduce the mean cycle time W, by just reducing the mean number of parts L in the system.

This provides a convenient starting point for an initial design of a scheduling policy. Consider the re-entrant line shown in Figure 5. It consists of 3 machines and 6 buffers. Parts visit the buffers in the order b_1, $b_2, \ldots, b_6$.

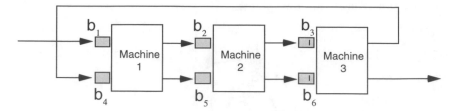

Figure 5: The myopic LBFS policy for a re–entrant line.

Consider Machine 3. Suppose there are parts in each of the buffers b_3 and b_6, and Machine 3 has to decide which part to process next. If it works on a part in buffer b_6, then it can immediately reduce the number of parts in the system – since parts in b_6 immediately leave the system. However, working on a part in b_3 simply sends it to b_4, without changing the total number of parts in the system.

Thus, from a myopic point of view it is better to work on parts in buffer b_6 rather than parts in b_3. Extending this reasoning to the other machines too, it is myopically better to prefer b_5 to b_2, and b_4 to b_1. This suggests ordering the buffers as $\{b_6, b_5, \ldots, b_1\}$, (i.e., in the reverse of the order they are visited), and giving preference according to this ordering. This is called the *Last Buffer First Serve (LBFS) Policy*. It is an example of a *buffer priority policy*.

However, such a myopic policy is not optimal in the long run. In the example above, it may be that Machine 1 is idle since buffers b_1 and b_4 are both empty. In that case, it may be better for Machine 3 to work on the part in b_3, and thereby provide work for Machine 1 to do. Such "starvation avoidance" of machines, especially bottleneck machines, may yield dividends in the long run. While one may qualitatively discern such properties, the determination of an optimal policy, by a methodology such as dynamic programming, which balances both short and long term objectives, is intractable. In the face of such difficulties, several heuristics have been proposed for scheduling; see for example Panwalker and Iskander [36].

5. Possible instability and stability under Bursty arrivals. It is clearly of interest to be able to precisely evaluate the performance of a proposed scheduling policy, such as, for example, the LBFS policy. However, available results in queueing networks are not applicable.

Through the work of Jackson [15,16], Baskett, Chandy, Muntz and Palacios [2], and Kelly [19], explicit solutions have been determined for the steady state distributions of certain queueing networks. Such networks are broadly refereed to as "product form" queueing networks, due to the multiplicative form of their steady state probability distribution. Unfortunately, if the service time distributions for the parts in the buffers at the same machine are different (there are some minor exceptions, unimportant here), or if the scheduling policy gives priority according to the buffers, then the resulting networks generally fall outside the above class.

The types of networks of interest to us possess both these features, and very little is known concerning their behavior. For a deterministic system, we shall say that it is stable, if all its buffer levels are bounded. In this paper if a stochastic system is such that the underlying Markov chain has a steady-state, i.e., it is positive recurrent, we shall say that is is *stable*. It is not even known whether such networks have a steady state. Clearly, it is necessary to first resolve whether a network is stable before one can address other more detailed questions related to steady state performance.

We will now show that even simple networks can be unstable, even when the usual capacity condition, that the nominal load brought to each server is within the server's capacity, is met. This example is drawn from [30]. Other similar examples for different models of systems and other scheduling policies can be found in [24].

Counterexample 1: A system which is unstable under the longest processing time buffer priority policy. Consider the system shown in Figure 6.

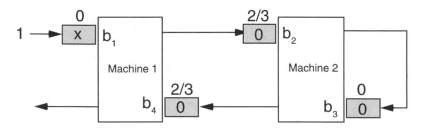

Figure 6: An unstable buffer priority policy.

We shall consider the particular buffer priority policy which uses the ordering $\{b_4, b_2, b_3, b_1\}$ to determine service priority. Thus, at Machine 1, parts in buffer b_1 are processed only when b_4 is empty, and at Machine 2, parts in buffer b_3 are processed only when b_2 is empty. (This policy is different from the earlier LBFS policy, which uses the different ordering $\{b_4, b_3, b_2, b_1\}$).

Let us suppose that parts arrive periodically to the system, with interarrival time of one unit. Parts in b_1 require 0 time units for processing. (This does not however mean that parts can skip b_1 and proceed directly to b_2. They may have to wait for b_4 to become empty, before Machine 1 can process them. There is a difference between 0 and nothing!). Parts in buffers b_2, b_3 and b_4 require 2/3, 0 and 2/3 time units each of processing time, respectively.

Note that the buffer priority policy $\{b_4, b_2, b_3, b_1\}$ corresponds to giving priority to those parts which have the *longest* processing time, and hence is called the *Longest Processing Time (LPT)* buffer priority policy.

We will now show that the system is unstable for some initial conditions. Consider the initial condition $x(0^-) = (\alpha, 0, 0, 0)$ where the i-th component of $x(t)$, denoted $x_i(t)$, represents the number of parts in buffer b_i at time t.

Since b_4 is empty at $t = 0^-$, Machine 1 immediately processes all the α parts in b_1 (since they each take only 0 seconds of processing time), and sends them to b_2. Thus, Machine 2 starts working on b_2. Meanwhile, new parts arrive at b_1, one every time unit, and are immediately sent on to b_2 (since b_4 is still empty). At time $t = 2\alpha^-$, b_2 is empty for the first time. The reason is that a total of $(\alpha + 2\alpha) = 3\alpha$ parts have been sent to b_2 from b_1, and these parts take a total of 2α time units for processing (since each part at b_2 takes 2/3 of a time unit). Thus, at time $t = 2\alpha^-$, Machine 2 is finally able to begin working on the 3α parts in b_3 (recall that b_3 has lower priority than b_2). Since each part in b_3 takes only 0 time units, these 3α parts are immediately sent to b_4. This is the first time that b_4 has any parts, and since b_4 has higher priority over b_1, Machine 1 stops working on b_1, and starts working only on b_4. From $t = 2\alpha^-$ to $t = 4\alpha^-$, Machine 1 works on the 3α parts in b_4. At time $4\alpha^-$, b_4 is empty again, and Machine 1 can turn to b_1. During this time interval, however, 2α new parts have entered b_1. Thus, at time $4\alpha^-$, the new state of the system is $x(4\alpha^-) = (2\alpha, 0, 0, 0)$.

This is a repeat of the situation at time 0^-, except that the number of parts in b_1 has doubled. The cycle of events repeats itself; at time $t = 12\alpha^-$, the state is $x(12\alpha^-) = (4\alpha, 0, 0, 0)$, etc. Thus the number of parts in the system is unbounded, and the system is unstable. (The overall growth rate is linear since parts enter the system linearly in time).   □

If τ_i denotes the processing time for parts in b_i, then the *nominal load* on Machine σ is

(5.1) $$\rho_\sigma := \sum_{\{i : b_i \text{ is served by Machine } \sigma\}} \lambda \tau_i,$$

where λ is the arrival rate. In the above example, $\rho_1 = \rho_2 = 2/3 < 1$, and so the usual capacity condition is satisfied. Yet, as we have seen, the system is unstable as a consequence of the scheduling policy.

It is worth noting that other unstable systems are identified in Kumar and Seidman [24]. In some systems, all clearing policies (also called exhaustive service policies) are unstable. They also show that some systems are unstable even if no part has a re-entrant route. Also, sufficient conditions are provided for stability of some policies.

Motivated by this example, the stability of some buffer priority policies, and the *Least Slack* (LS) policy, is addressed in [30]. The following positive results are established there.

Theorem 1: Stability of FBFS. *Consider the buffer priority policy using the ordering* $\{b_1, b_2, \ldots, b_L\}$, *where the buffers are ordered in the way they are visited. This is called the* First Buffer First Serve (FBFS) *policy. If* $\rho_\sigma < 1$ *for each Machine* σ, *then the system is stable for all arrivals which satisfy the model,*

(5.2)   # *of arrivals in* $[s, t] \leq \lambda(t - s) + \delta$ *for all* $0 \leq s \leq t$,

for some δ.[2]

Theorem 2: Stability of LBFS. *Under the same conditions as in Theorem 1, the LBFS policy is also stable.*

It is worth noting that while the proof of stability of FBFS is simple, the proof of stability of LBFS is quite complex. It relies on a contractive property of the delay experienced by a part, which arrives to find n parts already in the system. In particular, it is shown that the delay of such a part satisfies

(5.3)   Delay $\leq$ $(\bar{w} + \epsilon)n + c(\epsilon)$.

Above $\bar{w} := \max_\sigma \sum_{\{i: b_i \text{ is served by machine } \sigma\}} \tau_i$ is the maximum work brought by an incoming part to a machine, i.e., a bottleneck machine. Also $c(\epsilon)$ is a constant for every $\epsilon > 0$. The result (5.3) shows that under LBFS, even re-entrant lines behave like an acyclic "pipeline."

Let us suppose that incoming parts have a due-date stamped on them. If the parts arrive in the order of their due-dates, then the LBFS policy coincides with the well known *Earliest Due Date (EDD)* policy, and so we see that EDD is also stable.

In fact, more is true. For each part, let us define a slack s, by

Slack s of a part in buffer b_i :=   Due date of part $-\zeta_i$,

where ζ_i is a number associated with buffer i. The *Least Slack* (LS) policy gives priority to a part with the smallest slack. By extending the proof of stability of LBFS, one can establish the stability of all Least Slack policies.

Theorem 3: Stability of all least slack policies. *Assume that the arrivals follow the bursty model (5.2), and the capacity condition* $\rho_\sigma < 1$ *is satisfied for each station* σ. *Suppose that due-dates are assigned in such a way that difference between arrival times and due dates is bounded. Then, for every choice of the parameters* $\{\zeta_i\}$, *the corresponding Least Slack policy is stable.*

For all these results, we refer the reader to [30].

---

[2] Such a model of arrivals is introduced in [6]. The parameter δ allows for some burstiness in arrivals.

6. Stability of queueing networks and scheduling policies. We will now develop methods for establishing the stability of stochastic queueing networks by simply computing the value of a linear program and checking whether it is 1 or 0. The results in this section and the next are drawn from [22].

For simplicity, only, we focus on re-entrant lines. Consider the system shown in Figure 7. There are L buffers, $b_1, b_2, \ldots, b_L$, and several machines. We shall denote by $\sigma(i)$, the machine serving buffer b_i. We allow $\sigma(i) = \sigma(j)$ even if $i \neq j$. Parts arrive as a Poisson process of rate λ to buffer b_1. They move from b_i to b_{i+1} after obtaining service at b_i. After completing service at b_L, they leave the system. Parts at buffer b_i require a processing time which is exponentially distributed with mean $\frac{1}{\mu_i}$. As is usual, we assume that the interarrival times and service times are independent.

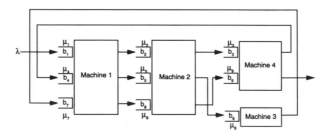

Figure 7: A re–entrant line.

Let $x_i(t)$ denote the number of parts in buffer b_i at time t. Since t is a continuous time variable, this is a continuous time system, which we take to be right continuous with left limits. However we prefer to study a discrete time system, obtained by sampling as follows. If a part in buffer b_i is not being processed by the corresponding machine, we shall suppose that there is a fictitious or virtual part which is being so processed. We sample the system at all arrival times, and all real or virtual service completion times. The rates at which these events occur are λ and μ_i, respectively. We shall now assume (by rescaling time, if necessary) that,

$$\lambda + \sum_{i=1}^{L} \mu_i = 1.$$

Let $\{\tau_n\}$ be the sequence of sampling times, obtained above, with $\tau_0 := 0$. When applied to a continuous time Markov chain, this procedure yields a discrete-time Markov chain whose steady state distribution is the same; it is called "uniformization;" see [28]. (However we have not so far assumed that the scheduling policy is stationary, and so the system may not be a Markov chain).

We shall suppose that the scheduling policy does not change the buffer that it is working on during the time interval $[\tau_n, \tau_{n+1})$. We shall call such scheduling polices as "non-interruptive," and restrict attention to them.

Define the random variable $w_i(\tau_n) := 1$ if the scheduling policy is actually working on a real part in buffer b_i in the time interval $[\tau_n, \tau_{n+1})$, and $:= 0$ otherwise.

If e_i is the i-th coordinate vector, then the discrete time system $x(\tau_n) = (x_1(\tau_n), \ldots, x_L(\tau_n))$ is described by

$$
x(\tau_{n+1}) = \begin{cases}
x(\tau_n) + e_1 & \text{w.p. } \lambda, \\
x(\tau_n) - e_i + e_{i+1} & \text{w.p. } w_i(\tau_n)\mu_i \text{ for } 1 \leq i \leq L-1, \\
x(\tau_n) - e_L & \text{w.p. } w_L(\tau_n)\mu_L, \\
x(\tau_n) & \text{otherwise.}
\end{cases}
$$

(6.1)

Let us now restrict our attention to scheduling policies which are *non-idling*. These are policies where a machine is not allowed to stay idle if any one of its buffers is non-empty. Thus,

(6.2) $$\sum_{\{i:\sigma(i)=\sigma\}} w_i(\tau_n) = 1 \text{ whenever } \sum_{\{i:\sigma(i)=\sigma\}} x_i(\tau_n) \geq 1.$$

We will now examine whether all non-idling policies yield a stable system, i.e., if the system is stable for all choices of $w_i(\tau_n)$, subject only to the requirement that they depend only on the past, and satisfy (6.2).

Consider a quadratic functional, $x^T(\tau_n)Qx(\tau_n)$, where

$$Q = Q^T.$$

Let $\mathcal{F}_{\tau_n}$ be the past σ-algebra. Suppose that we can find a Q such that

(6.3) $$E[x^T(\tau_{n+1})Qx(\tau_{n+1}) \mid \mathcal{F}_{\tau_n}] \leq x(\tau_n)Qx(\tau_n) - \gamma|x(\tau_n)| + c,$$

where $\gamma > 0$, $|x| := \sum x_i$ = number of parts in the system, and c is some constant. (Note that the conditional expectation above exists, since $x(\tau_n)$ can grow only linearly with n). Then, we can take unconditional expectations to obtain,

$$E(x^T(\tau_{n+1})Qx(\tau_{n+1})) \leq E(x^T(\tau_n)Qx(\tau_n)) - \gamma E|x(\tau_n)| + c.$$

By summing, and dividing by N, we obtain

(6.4) $$\frac{\gamma}{N}\sum_{n=0}^{N-1} E|x(\tau_n)| \leq \frac{x^T(0)Qx(0)}{N} - \frac{E(x^T(\tau_N)Qx(\tau_N))}{N} + c.$$

Suppose now that Q satisfies the condition,

(6.5) $$x^T(\tau_N)Qx(\tau_N) \geq 0 \text{ for all } N.$$

Then, (6.4) yields

(6.6) $$\frac{\gamma}{N}\sum_{n=0}^{N-1} E|x(\tau_n)| \leq \frac{x^T(0)Qx(0)}{N} + c \text{ for all } N.$$

Since the average of the mean number of parts in the system is then bounded, we can say that the system is "stable-in-the-mean."

Above, we have not assumed that the system is Markovian. If the scheduling policy chooses $w_i(\tau_n)$ purely as a function of $x(\tau_n)$, then, in the terminology of Markov Decision Processes, we have a *stationary* scheduling policy, and $\{x(\tau_n)\}$ is a time-inhomogeneous Markov chain. Note also that the resulting Markov chain has a single closed communicating class, since the origin can be reached from every state. Moreover, it is aperiodic since the system can remain at the origin for two or more consecutive time periods.

For Markov chains, stability-in-the-mean implies positive recurrence. The reason for this is that if it were not positive recurrent, then

$$\text{Prob}\left(|x(\tau_n)| \leq c/\gamma\right) \to 0 \text{ as } n \to \infty,$$

since the set of such states is compact (see [18]), which however contradicts (6.6).

In fact, more is true. Let us take $\sqrt{x^T(\tau_n)Qx(\tau_n)}$ as a Lyapunov function. It can then be shown that for some $\epsilon > 0$, and compact set K,

$$E\left[\sqrt{x^T(\tau_{n+1})Qx(\tau_{n+1})} \mid \mathcal{F}_{\tau_n}\right] \leq \sqrt{x^T(\tau_n)Qx(\tau_n)} - \epsilon,$$

$$\text{whenever } x(\tau_n) \in K^c,$$

and takes bounded jumps. Then, it can be concluded (see [22]) that the Markov chain has a *geometrically converging exponential moment*, i.e., there exist $\epsilon > 0$, $r > 1$, and $c < +\infty$, so that for any function $f(y) \leq e^{\epsilon|y|}$, and any initial condition $x(\tau_0) = x_0$,

$$\sum_{n=0}^{+\infty} r^n \left| E(f(x(\tau_n))) - \sum_y \pi(y)f(y) \right| \leq ce^{\epsilon|x_0|},$$

where π is the steady-state distribution of the Markov chain. Thus, all polynomial moments $|x|^p$, and even an exponential moment $e^{\epsilon|x|}$ exist, and converge geometrically fast to their steady state values.

The central issue that remains is this. How can we find, if it exists, a symmetric matrix Q which satisfies (6.5), and for which the Lyapunov function x^TQx has the "negative drift" property (6.3)?

First, let us consider the condition (6.5). It is clearly true if $x^TQx \geq 0$ for every x in the positive orthant, since queue lengths are nonnegative. Such a symmetric matrix is said to be *copositive*. Copositive matrices have been extensively studied in linear complementarity theory [5]. They are characterized by each principle submatrix, for which the cofactors of the last row are nonnegative, having a nonnegative determinant [5]. However, testing whether a matrix is copositive is NP-complete [34,1]. Note, though, that every symmetric *nonnegative* matrix is copositive. Hence, to avoid a copositivity check, one could confine attention to nonnegative matrices, though that would lead to a less powerful stability test.

Now we turn to the crucial issue of finding a symmetric matrix Q for which (6.3) holds. From the probabilities of the various transitions given in (6.1), it is easy to compute $E[x^T(\tau_{n+1})Qx(\tau_{n+1}) \mid \mathcal{F}_{\tau_n}]$. Simple calculations show that (6.3) is equivalent to finding a $Q = Q^T$ such that,

$$2\lambda e_1^T Qx(\tau_n) + \lambda e_1^T Qe_1 + 2\sum_{i=1}^{L-1} \mu_i w_i(\tau_n)(e_{i+1} - e_i)^T Qx(\tau_n)$$

$$+\sum_{i=1}^{L-1} \mu_i w_i(\tau_n)(e_{i+1} - e_i)^T Q(e_{i+1} - e_i)$$
$$-2\mu_L w_L(\tau_n)e_L^T Qx(\tau_n) + \mu_L w_L(\tau_n)e_L^T Qe_L$$
$$\leq -\gamma|x(\tau_n)| + c \text{ for some } \gamma > 0 \text{ and } c.$$

Bounding all the terms not involving $x(\tau_n)$ on the left hand side above by a constant, we see that (6.3) is equivalent to

$$2\lambda e_1^T Qx(\tau_n) + 2\sum_{i=1}^{L-1} \mu_i w_i(\tau_n)(e_{i+1} - e_i)^T Qx(\tau_n)$$

(6.7) $\qquad -2\mu_L w_L(\tau_n)e_L^T Qx(\tau_n) \leq -\gamma|x(\tau_n)|$ for some $\gamma > 0$.

Now we observe that both the left and right hand sides above are linear in $x(\tau_n)$. Hence (6.7) (and therefore (6.3)) holds if the coefficient of every component of $x_j(\tau_n)$ is less than $-\gamma$, whenever $x_j(\tau_n) \geq 1$ (note that $x_j(\tau_n)$ is integral), i.e.,

$$\lambda q_{1j} + \sum_{i=1}^{L} \mu_i(q_{i+1,j} - q_{ij})w_i(\tau_n) \leq -\gamma \text{ for } j = 1, \ldots, L, \text{ if } x_j(\tau_n) \geq 1.$$

(6.8)

(Above we have taken $q_{L+1,j} := 0$). Thus we seek $q_{ij} = q_{ji}$ such that (6.8) holds.

However, the above inequality involves the *random variables* $w_i(\tau_n)$. We can exploit the non–idling nature of the policies under consideration, quantified in (6.2), to bound them.

Let us decompose the sum $\sum_{i=1}^{L} \alpha_{ij}$, above, machine by machine, i.e.,

$$\sum_{i=1}^{L} \alpha_{ij} = \sum_{\{i:\sigma(i)=\sigma(j)\}} \alpha_{ij} + \sum_{\{\sigma:\sigma\neq\sigma(j)\}} \sum_{\{i:\sigma(i)=\sigma\}} \alpha_{ij}.$$

Then, since we are only interested in establishing the inequality when $x_j(\tau_n) \geq 1$, from (6.2) we have,

$$\sum_{\{i:\sigma(i)=\sigma(j)\}} \mu_i(q_{i+1,j} - q_{ij})w_i(\tau_n) \leq \max_{\{i:\sigma(i)=\sigma(j)\}} \mu_i(q_{i+1,j} - q_{ij}),$$

since the $w_i(\tau_n)$'s at Machine $\sigma(j)$ are nonnegative, and add to 1. At the other machines $\sigma \neq \sigma(j)$, the coefficients may add to 1 or 0, since the other machines may or may not be working. Hence,

$$\sum_{\{\sigma:\sigma\neq\sigma(j)\}} \sum_{\{i:\sigma(i)=\sigma\}} \mu_i(q_{i+1,j} - q_{ij})w_i(\tau_n) \leq \sum_{\{\sigma:\sigma\neq\sigma(j)\}}$$

$$\max_{\{i:\sigma(i)=\sigma\}} \mu_i(q_{i+1,j} - q_{ij})^+.$$

Thus we see that (6.3) holds whenever

$$\lambda q_{1j} + \max_{\{i:\sigma(i)=\sigma(j)\}} \mu_i(q_{i+1,j} - q_{ij}) + \sum_{\{\sigma:\sigma\neq\sigma(j)\}} \max_{\{i:\sigma(i)=\sigma\}} \mu_i(q_{i+1,j} - q_{ij})^+$$

$$(6.9) \qquad\qquad\qquad\qquad\qquad\qquad \leq -\gamma \text{ for } j = 1,\dots, L.$$

Note now that if such a $Q = Q^T$ exists, then we can take $\gamma = 1$, since we can always scale Q. Thus, one can search for the largest γ in $0 \leq \gamma \leq 1$, for which (6.9) holds for some Q. This largest γ will either be 0 or 1. If it is 1, then we have established the stability of *all* non-idling scheduling policies. If it is 0, then we have obtained no conclusion.

Theorem 4: A linear programming stability test. *Consider the linear program:*

$$\text{Max } \gamma$$

subject to

$$0 \leq \gamma < 1$$
$$q_{ij} = q_{ji} \text{ for all } i, j$$
$$q_{L+1,j} = 0 \text{ for all } j$$
$$\lambda q_{1j} + r_j + \sum_{\{\sigma:\sigma\neq\sigma(j)\}} s_{\sigma j} \leq -\gamma \text{ for all } j$$
$$r_j \geq \mu_i(q_{i+1,j} - q_{ij}) \text{ for all } i \text{ with } \sigma(i) = \sigma(j), \text{ and all } j$$
$$s_{\sigma j} \geq \mu_i(q_{i+1,j} - q_{ij}) \text{ for all } i \text{ with } \sigma(i) = \sigma \neq \sigma(j), \text{ and all } j, \sigma$$
$$s_{\sigma j} \geq 0 \text{ for all } \sigma, j.$$

(i) *Suppose the value of the LP is 1, and suppose Q is an optimal solution. If an optimal Q is copositive, then all non-idling scheduling policies are stable-in-the-mean. Also, for all non-idling stationary scheduling policies, the system has a geometrically converging exponential moment.*

(ii) *Consider the same linear program with the additional constraints*

$$q_{ij} \geq 0 \text{ for all } i, j.$$

If the value of this LP is 1, then since every optimal Q is nonnegative, the conclusions of (i) above hold without the need for a separate copositivity check of Q.

(iii) *If the value of the first LP is 0, then no conclusion can be drawn.*

Example 1. Consider the system shown in Figure 8.

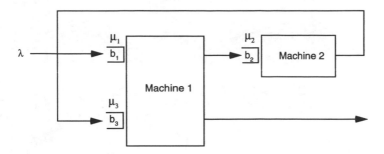

Figure 8: System of Examples 1, 2 and 4.

Suppose that $\mu_1 = \mu_2 = \mu_3 := \mu$. The capacity condition is $\rho := \frac{2\lambda}{\mu} < 1$. It is easy to check, *analytically*, that the LP in Theorem 4.ii has a solution with $\gamma = 1$. Hence, all non-idling policies are stable. □

Example 2. Consider the same system as in Figure 8, except now that while we still consider $\mu_1 = \mu_3$, we let μ_2 be different. Then, the capacity condition is

$$\rho_1 := \frac{2\lambda}{\mu_1} < 1, \text{ and } \rho_2 := \frac{\lambda}{\mu_2} < 1.$$

In Figure 9 we plot the value of the LP in Theorem 4.ii for all (ρ_1, ρ_2) in the capacity region. We see that there is a small region where the value of the LP is zero, and in this region, our test for the stability of all non-idling polices is inconclusive. □

In [12] and [7] it is shown that the system is stable even in the small indeterminate region above.

Example 3. Consider the system shown in Figure 10. Dai and Wang [10] and Dai and Nguyen [9] have shown that this system does not admit a Brownian network approximation in the heavy traffic limit $\rho_1 = \rho_2 =: \rho \nearrow 1$. Using our LP of Theorem 4.ii, we see that for

$$\rho < 0.528,$$

the LP has value 1, while it is zero for all larger ρ.

In [12,7] a slightly larger value of ρ is determined for which stability holds. In fact, simulations in [29] show that the system appears unstable for even larger values.

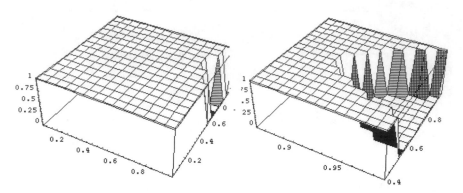

Figure 9: Value of LP for all non–idling, non–interruptive policies in
Example 2. The figure on the right is a more detailed view of one corner
of the figure on the left.

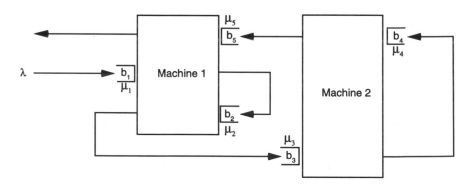

Figure 10: Dai–Wang system of Example 3.

7. Stability of buffer priority policies. In many cases, while all
non-idling policies may not be stable, specific buffer priority or other policies may be stable. One would therefore like to conduct a more specific
test. We illustrate the idea by showing how one can analyze a given buffer
priority policy.

Consider a buffer priority policy using the ordering $\{b_{\theta(1)}, b_{\theta(2)}, \ldots,$
$b_{\theta(L)}\}$. Clearly, if $\sigma(i) = \sigma(j)$ and $\theta(j) < \theta(i)$, then buffer b_j is preferred
to b_i. Thus, if $x_j(\tau_n) \geq 1$, one has $w_i(\tau_n) = 0$. Hence, in (6.8), one can
drop such $w_i(\tau_n)$'s. This allows us to refine (6.9), and obtain the following
theorem.

Theorem 5: Stability of buffer priority policies. *Suppose*
$\{b_{\theta(1)}, b_{\theta(2)}, \ldots, b_{\theta(L)}\}$ *is a buffer priority policy. Suppose there exists a copositive matrix Q such that,*

$$
\lambda q_{1j} + \max_{\{i:\sigma(i)=\sigma(j) \ and \ \theta(i)\leq\theta(j)\}} \mu_i(q_{i+1,j} - q_{ij})
$$

(7.1)
$$
\sum_{\{\sigma:\sigma\neq\sigma(j)\}} \max_{\{i:\sigma(i)=\sigma\}} \mu_i(q_{i+1,j} - q_{ij})^+ \leq -1,
$$

then the system has geometrically converging exponential moment. ☐

The test of (7.1) is conducted, as before, by linear programming.

Example 4. Consider the system shown in Figure 8. We assume that $\mu_1 = \mu_3$, while μ_2 may be different. By computation, one can check that for every $\rho_1 := \frac{2\lambda}{\mu_1} < 1$, and $\rho_2 < 1$, both the FBFS and LBFS buffer priority policies are stable. ☐

In fact, in [8], it has recently been established that FBFS is stable for all re-entrant lines (within their capacity region, of course), while in [7] and [26], it has been established that the LBFS policy is also stable for all re-entrant lines. In [25] it is shown that the FSMCT policy of Section 14.1 is also similarly stable.

8. Instability and non-robustness of systems. We have seen that one can (when possible) use linear programming to construct a matrix $Q = Q^T$ that has the negative drift property (6.3). If the resulting matrix Q is copositive, then we have been able to deduce geometric convergence of an exponential moment, for the stationary policy under consideration. What can we conclude if Q is *not* copositive?

We will show below that if (6.3) holds for a non-copositive $Q = Q^T$, then either the stationary policy under consideration does *not* have a finite first moment, or there exist arbitrarily small perturbations of the policy which do not have a finite first moment. Thus, at best, the stationary policy is highly non-robust.

Let $\mathcal{C}$ denote the closed communicating class of the Markov chain resulting from the stationary policy. We recall that under stationary non–idling scheduling policies, the Markov chain has only a single closed communicating class which is also aperiodic. The following result is from [23].

Theorem 6: Instability or non-robustness when Q is not copositive. *Suppose for a stationary policy there exists a non-copositive $Q = Q^T$ such that the negative drift condition (6.3) holds.*

(i) *If $\inf_{x \in \mathcal{C}} x^T Q x = -\infty$, then the stationary policy does not have a finite first moment.*

(ii) If $\inf_{x \in C} x^T Q x > -\infty$, then either the stationary policy itself, or an arbitrarily small perturbation of it, does not have a finite first moment.

Proof. Consider the function $V(x) := \max\{-x^T Q x, 0\}$. Since Q is not copositive, $V(x) \not\equiv 0$ in the positive orthant. In fact since one can scale x, $\inf_{x \in R_L^+} x^T Q x = -\infty$. Simple calculations (see [23]) show that $V(x)$ has a nonnegative drift outside a compact set; more specifically,

$$E[V(x(\tau_{n+1})) \mid \mathcal{F}_{\tau_n}] \geq V(x(\tau_n)) \text{ for } |x(\tau_n)| \geq \frac{c}{\gamma}.$$

Since $V(x) = O(|x|^2)$, we see that $E[V(x(\tau_{n+1})) \mid \mathcal{F}_{\tau_n}] \geq V(x(\tau_n))$ whenever $V(x(\tau_n)) > m$, for some m. Moreover,

$$V(x(\tau_{n+1})) - V(x(\tau_n)) = O(|x(\tau_n)|),$$

since $x(\tau_n)$ takes bounded jumps by (6.1). From the Optional Sampling Theorem of Doob [11], we can deduce (i) (see [23] for details).

To see (ii) we simply note that one can perturb the stationary policy so that all non-empty buffers get a very small service attention. This makes the whole state space reachable, while preserving the negative drift. Thus the perturbed policy does not have a finite first moment. $\qquad\square$

9. Performance bounds for queueing networks and scheduling policies. We now examine the issue of quantifying the performance of a system. The results in this section and the next are mainly drawn from [27]. Some sharpenings of the results are from [23].

Suppose now that, under a particular stationary non-idling scheduling policy, the system is stable, i.e., positive recurrent. Suppose moreover that the system has a *finite first moment in steady-state*, i.e.,

$$(9.1) \qquad E|x(\tau_n)| < +\infty,$$

in steady state. Then, it can be shown that,

$$(9.2) \qquad E[x^T(\tau_{n+1}) Q x(\tau_{n+1}) - x^T(\tau_n) Q x(\tau_n)] = 0,$$

for *every* matrix Q. This is particularly easy to see if

$$E|x(\tau_n)|^2 < +\infty$$

in steady state, but the result (9.2) also holds when just (9.1) holds, as shown in [23].

Relation (9.2) is equivalent to,

$$(9.3) \qquad E[x_i(\tau_{n+1}) x_j(\tau_{n+1}) - x_i(\tau_n) x_j(\tau_n)] = 0 \text{ for all } i, j.$$

Note that this is a set of $\frac{L(L+1)}{2}$ equalities, one for each distinct q_{ij}.

To compute the left hand side of (9.3), we begin by computing the conditional expectation

$$(9.4) \qquad E[x_i(\tau_{n+1})x_j(\tau_{n+1}) - x_i(\tau_n)x_j(\tau_n) \mid \mathcal{F}_{\tau_n}],$$

using the transition probabilities (6.1). There are several cases. Consider, for example, $i = j = 1$:

$$(9.5) \qquad x_1^2(\tau_{n+1}) = \begin{cases} (x_1(\tau_n) + 1)^2 & \text{w.p. } \lambda, \\ (x_1(\tau_n) - 1)^2 & \text{w.p. } w_1(\tau_n)\mu_1, \\ x_1^2(\tau_n) & \text{otherwise.} \end{cases}$$

For $i = 1$ and $j = 2$,

$$x_1(\tau_{n+1})x_2(\tau_{n+1}) = \begin{cases} (x_1(\tau_n) + 1)x_2(\tau_n) & \text{w.p. } \lambda, \\ (x_1(\tau_n) - 1)(x_2(\tau_n) + 1) & \text{w.p. } w_1(\tau_n)\mu_1, \\ x_1(\tau_n)(x_2(\tau_n) - 1) & \text{w.p. } w_2(\tau_n)\mu_2. \end{cases}$$

It is a simple matter to write the other cases down.

We illustrate the consequence of (9.3), for $i = j = 1$. From (9.5), we see that (9.4) evaluates to

$$(9.6) \quad E[x_1^2(\tau_{n+1}) - x_1^2(\tau_n) \mid \mathcal{F}_{\tau_n}] = \lambda(2x_1(\tau_n)+1) - \mu_1 w_1(\tau_n)(2x_1(\tau_n)-1).$$

Let us define

$$(9.7) \qquad z_{ij} := E[w_i(\tau_n)x_j(\tau_n)].$$

Taking unconditional expectations in (9.6), we see that

$$(9.8) \quad E[x_1^2(\tau_{n+1}) - x_1^2(\tau_n)] = 2\lambda E[x_1(\tau_n)] + \lambda - 2\mu_1 z_{11} + \mu_1 E[w_1(\tau_n)].$$

Now recall that since our scheduling policy is non-idling, we have the condition (6.2). Hence, $x_j(\tau_n) = \sum_{\{i:\sigma(i)=\sigma(j)\}} w_i(\tau_n)x_j(\tau_n)$, and so

$$(9.9) \qquad E[x_j(\tau_n)] = \sum_{\{i:\sigma(i)=\sigma(j)\}} z_{ij}.$$

Also, since the system is stable, every buffer b_i is worked on for a proportion of time that is exactly enough to meet the work arriving for b_i. Hence,

$$(9.10) \qquad E(w_i(\tau_n)) = \frac{\lambda}{\mu_i}.$$

Substituting (9.9) and (9.10) in (9.8), we see that

$$E[x_1^2(\tau_{n+1}) - x_1^2(\tau_n)] = 2\lambda \sum_{\{i:\sigma(i)=\sigma(1)\}} z_{i1} + 2\lambda - 2\mu_1 z_{11}.$$

Hence, from (9.3), for $i = j = 1$, we obtain the equality,

$$2\lambda \sum_{\{i:\sigma(i)=\sigma(1)\}} z_{i1} + 2\lambda - 2\mu_1 z_{11} = 0.$$

Similarly, by considering all the other cases for (i,j), one can obtain $L(L+1)/2$ equality constraints in the L^2 variables $\{z_{ij}\}$. These are shown in equations (9.13-9.18), in Theorem 7 below. We note that these constraints are obtained independently in [3] and [27].

In addition, the variables $\{z_{ij}\}$ satisfy several other inequality constraints. Consider a station $\sigma \neq \sigma(j)$. It may or may not be working when $x_j(\tau_n) \geq 1$. Hence for $\sigma \neq \sigma(j)$, one only has,

$$x_j(\tau_n) \geq 1 \Rightarrow \sum_{\{i:\sigma(i)=\sigma\}} w_i(\tau_n) \leq 1.$$

Hence

(9.11)
$$\sum_{\{i:\sigma(i)=\sigma\}} w_i(\tau_n) x_j(\tau_n) \leq x_j(\tau_n).$$

Thus, upon taking expectations in (9.11), and using (9.9), we obtain the *non-idling inequality constraints* (9.19), shown in Theorem 7 below.

Also, clearly all the z_{ij}'s are non-negative, yielding (9.20).

Note finally that due to (9.9), the mean number in the system can be written in terms of the z_{ij}'s as,

$$E|x(\tau_n)| = \sum_{j=1}^{L} \sum_{\{i:\sigma(i)=\sigma(j)\}} z_{ij}.$$

Thus we see that the mean number is given by a linear expression in the z_{ij}'s, which in turn are constrained by several linear equalities and inequalities. The situation is ripe for linear programming.

Theorem 7: Performance bounds for all non-idling policies.
Consider any non-idling stationary scheduling policy.

(i) *In steady–state, the mean number of parts is bounded above by the value of the linear program:*

(9.12)
$$\text{Max} \sum_{j=1}^{L} \sum_{\{i:\sigma(i)=\sigma(j)\}} z_{ij}$$

subject to

(9.13)
$$2\lambda \sum_{\{i:\sigma(i)=\sigma(1)\}} z_{i1} + 2\lambda - 2\mu_1 z_{11} = 0$$

(9.14) $2\mu_{j-1}z_{j-1,j} + 2\lambda - 2\mu_j z_{jj} = 0$ for $j = 2, \ldots, L$

(9.15) $\lambda \sum_{\{j:\sigma(j)=\sigma(2)\}} z_{j2} - \lambda - \mu_1(z_{12} - z_{11}) - \mu_2 z_{21} = 0,$

$\lambda \sum_{\{i:\sigma(i)=\sigma(j)\}} z_{ij} - \mu_1 z_{1j} - \mu_j z_{j1} + \mu_{j-1}z_{j-1,1} = 0$

(9.16) $\text{for } j = 3, \ldots, L,$

$\mu_{i-1}z_{i-1,i+1} - \mu_i z_{i,i+1} - \lambda + \mu_i z_{ii} - \mu_{i+1}z_{i+1,i} = 0$

(9.17) $\text{for } i = 2, \ldots, L-1,$

$\mu_{i-1}z_{i-1,j} - \mu_i z_{i,j} + \mu_{j-1}z_{j-1,i} - \mu_j z_{ji} = 0$

(9.18) $\text{for } i = 2, \ldots, L-2, \text{ and } j = i+2, \ldots, L$

(9.19) $\sum_{\{j:\sigma(j)=\sigma\}} z_{ji} \leq \sum_{\{j:\sigma(j)=\sigma(i)\}} z_{ji} \text{ for } i = 1, \ldots, L, \text{ and } \sigma \neq \sigma(i)$

(9.20) $z_{ij} \geq 0$ for $i = 1, \ldots, L$ and $j = 1, \ldots L.$

(ii) *The mean number of parts in steady-state is bounded below by the same linear program, with a "min" replacing the "max" in (9.12).*

Proof. The above theorem has already been established if the non–idling policy under consideration has a finite first moment, since that property was used to derive the constraints (9.13-9.18). The only fact that has not yet been established is that the upper bound also applies to policies which may not have a finite first moment. This is done in [23] as follows.

The key idea is that every non-idling policy can be written as the limit of stable policies.

We will consider a policy which uses time sharing. Thus the $w_i(\tau_n)$'s may not be just 0 or 1, but are allowed to take values in between, subject only to the non–idling restriction (6.2). However, our theory so far has not really needed this 0 or 1 property.

Consider the stationary scheduling policy,

$$w_i(\tau_n) = \frac{\frac{1}{\mu_i}}{\sum_{\{j:\sigma(j)=\sigma(i)\}} \frac{1}{\mu_j}}.$$

This policy allocates a fixed proposition of a machine's capacity to each buffer, and the capacity provided is enough to meet the buffers needs.

Thus, under it, the system behaves like a tandem of $M/M/1$ buffers, and is clearly stable, with finite moments of all orders. However, the policy is not non-idling. This is easily remedied. Consider the modified policy

$$w_i(\tau_n) = \frac{\frac{1}{\mu_i}1(x_i(\tau_n) \geq 1)}{\sum_{\{j:\sigma(j)=\sigma(i)\}} \frac{1}{\mu_j}1(x_j(\tau_n) \geq 1)}$$

(taking $0/0 = 0$). This restricts service attention to non-empty buffers only, and is therefore non-idling. Moreover, a simple sample path argument shows that every part has a smaller delay under this policy, than under the earlier sometimes idling policy. Hence the new non-idling stationary policy is also stable, with finite moments of all orders. Call this policy p.

Now suppose that the value (9.12) of the above LP is finite, say M. We will now show that every non-idling policy is then stable, and has a mean number of parts bounded by M.

To see this, consider any non-idling policy $\bar{p}$. This policy may not be stable. However consider the composite policy $\bar{p}_N$ which follows policy $\bar{p}$ whenever $|x| \leq N$, but follows the policy p (which has a finite first moment) whenever $|x| > N$.

The composite policy $\bar{p}_N$ is stable, with a finite first moment, since whenever the state leaves a compact set, it applies a good policy. Hence its performance is bounded by M, uniformly for all N, since the LP bound applies to all stable policies with a finite first moment. However, as $N \to +\infty$, $\bar{p}_N$ converges to $\bar{p}$ pointwise. Applying weak convergence arguments, we see that the mean number under $\bar{p}$ is also bounded by M.

Thus, all non-idling policies are stable, and in fact have a mean number of parts bounded by M.   □

The following examples are drawn from [27].

Example 5: Lower bound on cycle-time multiplier. Consider the re-entrant line shown in Figure 11.

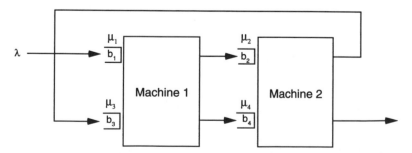

Figure 11: The open re–entrant line of Examples 5, 7, and 8.

We take $\mu_1 = \mu_2 = \mu_3 = \mu_4 =: \mu$. The load factor on either machine is $\rho_1 = \rho_2 = \rho := \frac{2\lambda}{\mu}$. Note that by Little's Theorem, the mean cycle-time of a part is given by $\dfrac{\text{Mean number in system}}{\lambda}$. Note that $\frac{4}{\mu}$ is the mean total processing time of part. Figure 12 plots the lower bound on the attainable cycle–time multiplier ratio $\dfrac{\text{Mean Cycle–Time}}{\text{Mean Total Processing Time}}$ as a function of the system load ρ.

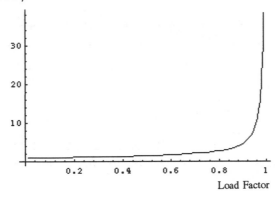

Figure 12: Lower bound on cycle–time multiplier for Example 5.

Example 6: Comparison with lower bounds of Ou and Wein.
Ou and Wein [35] have proposed a different method for obtaining bounds. They determine a system whose cost is lower than all the policies in the original system. Then they simulate the new system, thus obtaining a confidence interval for a lower bound. In [27], the lower bound is compared with the 95% confidence intervals of [35], for the systems shown in Figures 13, 14, and 15.

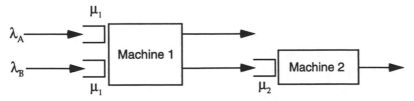

Figure 13: Ou and Wein's system (a) of Example 6.
Balanced: $\mu_1 = 2, \mu_2 = 1$. Imbalanced: $\mu_1 = 2, \mu_2 = 1.5$.
Light: $\lambda_A = \lambda_B = 0.3$. Medium: $\lambda_A = \lambda_B = 0.6$.
Heavy: $\lambda_A = \lambda_B = 0.9$. Very Heavy: $\lambda_A = \lambda_B = 0.99$.

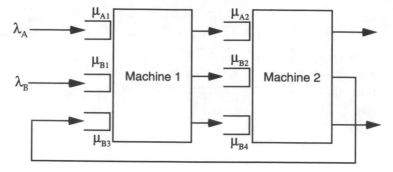

Figure 14: Ou and Wein's system (b) of Example 6.
Balanced: $\mu_{A1} = 1/4$, $\mu_{A2} = 1$, $\mu_{B1} = 1/8$, $\mu_{B2} = 1/6$, $\mu_{B3} = 1/2$,
$\mu_{B4} = 1/7$.
Imbalanced: $\mu_{A1} = 1/4$, $\mu_{A2} = 2/3$, $\mu_{B1} = 1/8$, $\mu_{B2} = 1/4$, $\mu_{B3} = 1/2$,
$\mu_{B4} = 3/14$.
Light: $\lambda_A = \lambda_B = 3/140$. Medium: $\lambda_A = \lambda_B = 6/140$.
Heavy: $\lambda_A = \lambda_B = 9/140$. Very Heavy: $\lambda_A = \lambda_B = 99/1400$.

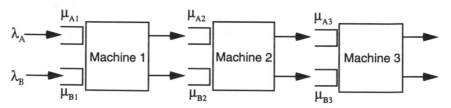

Figure 15: Ou and Wein's system (c) of Example 6.
Balanced: $\mu_{A1} = 1/2$, $\mu_{A2} = 1/4$, $\mu_{A3} = 1/6$, $\mu_{B1} = 1/7$, $\mu_{B2} = 1/5$,
$\mu_{B3} = 1/3$,
Imbalanced: $\mu_{A1} = 1/2$, $\mu_{A2} = 1/2$, $\mu_{A3} = 1$, $\mu_{B1} = 1/7$, $\mu_{B2} = 1/4$,
$\mu_{B3} = 1/2$
Light: $\lambda_A = \lambda_B = 1/30$. Medium: $\lambda_A = \lambda_B = 2/30$.
Heavy: $\lambda_A = \lambda_B = 0.1$. Very Heavy: $\lambda_A = \lambda_B = 0.11$.

The results are given in Figure 16.

System	Scenario	Ou & Wein's Bound	LP Bound
(a)	Balanced Light	0.775 (± 0.010)	0.7286
	Balanced Medium	2.47 (± 0.046)	2.10
	Balanced Heavy	13.2 (± 0.824)	9.90
	Balanced Very Heavy	85.9 (± 23.0)	99.99
	Imbalanced Light	0.621 (± 0.007)	0.6286
	Imbalanced Medium	1.85 (± 0.031)	1.90
	Imbalanced Heavy	9.46 (± 0.544)	9.60
	Imbalanced Very Heavy	72.8 (± 24.6)	99.66
(b)	Balanced Light	0.734 (± 0.011)	0.7087
	Balanced Medium	2.18 (± 0.045)	1.9385
	Balanced Heavy	8.48 (± 0.611)	8.209
	Balanced Very Heavy	43.8 (± 12.9)	72.859
	Imbalanced Light	0.56 (± 0.006)	0.624
	Imbalanced Medium	1.47 (± 0.031)	1.7562
	Imbalanced Heavy	6.94 (± 0.451)	7.6289
	Imbalanced Very Heavy	43.5 (± 11.9)	72.0349
(c)	Balanced Light	1.17 (± 0.013)	1.0116
	Balanced Medium	3.32 (± 0.055)	2.6475
	Balanced Heavy	12.2 (± 0.702)	11.4
	Balanced Very Heavy	71.4 (± 22.0)	118.14
	Imbalanced Light	0.717 (± 0.008)	0.7116
	Imbalanced Medium	1.78 (± 0.023)	1.9718
	Imbalanced Heavy	8.33 (± 0.890)	8.65
	Imbalanced Very Heavy	50.9 (± 14.8)	84.4913

Figure 16: Comparison of LP lower bounds with Ou and Wein's bounds, for systems (a), (b), and (c) of Example 6.

□

10. Additional linear constraints for specific system models.
Queueing networks and scheduling policies often possess much special structure. For example, the system may be a closed network with a fixed population of jobs, machines may fail, the service distributions may not be
exponential, a specific buffer priority policy may be in place, etc. In this
section we wish to show how such specific structures can be exploited to
provide better bounds. We will show that they provide additional linear
constraints which can be appended to the linear programs.

The following examples are drawn from [27], to which we refer the
further details.

10.1. Buffer priority policies. Consider a buffer priority policy us-
ing the ordering $\{b_{\theta(1)}, b_{\theta(2)}, \ldots, b_{\theta(L)}\}$. Clearly, if b_i and b_j share the same
machine, and b_i has higher priority than b_j, then

$$x_i(\tau_n) \geq 1 \quad \Rightarrow \quad w_j(\tau_n) = 0.$$

Hence

$$x_i(\tau_n)w_j(\tau_n) = 0.$$

Thus, we obtain the additional *buffer priority constraints*,

$$z_{ij} = 0 \text{ whenever } \sigma(i) = \sigma(j) \text{ and } \theta(i) < \theta(j).$$

We should note that for buffer priority policies, an upper bound as
given in Theorem 7, with the buffer priority constraints appended to the
LP, does *not* apply unless the policy has a finite first moment. The reason is
that the proof that one can dispense with the finite first moment condition
for the class of all non–idling policies, used the facts that the class of all
non–idling policies contains a stable policy, and that it is closed under
composition. Those facts need not hold for restrictive classes of policies.

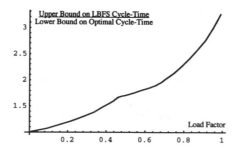

Figure 17: The ratio $\dfrac{\text{Upper bound on LBFS cycle–time}}{\text{Lower bound on optimal cycle–time}}$ in Example 7.

Example 7: Performance of LBFS. Consider the system shown in Figure 11. Let $\mu_i \equiv \mu$. For the LBFS policy, one can determine an upper bound on its performance, and compare it to the lower bound for all non-idling policies. This ratio, which bounds the deviation from optimality, is plotted in Figure 17, as a function of system load. One notes that LBFS is nearly optimal in light traffic, and never worse than about $3\frac{1}{2}$ times optimal, even in heavy traffic. □

Example 8: Comparison of LBFS and FBFS. Based on a myopic application of Little's Theorem, one believes that, for many systems, LBFS may be better than FBFS. Consider the system shown in Figure 11, with $\mu_1 = \mu_2 = \mu_3/7 = \mu_4/7$, and $\rho_1 = \rho_2 = 0.4$.

Under (the pre-emptive) FBFS, from Burke's Theorem, we know that the departure stream leaving b_1 is Poisson, and moreover its past is independent of the present value of x_1. Thus one has independence of w_1 and x_2. Similarly, one has independence of w_2 and x_1. These give the additional constraints,

$$z_{12} = \frac{\lambda}{\mu_1} z_{22}, \text{ and } z_{21} = \frac{\lambda}{\mu_2} z_{11}.$$

Figure 18 provides the lower and upper bounds on FBFS performance, both with and without these "tandem" constraints, and on LBFS performance. We see that LBFS is better than FBFS, since its upper bound is lower than the lower bound for FBFS. □

Policy	Lower Bound on Number in System	Upper Bound on Number in System
LBFS	1.06667	1.60075
FBFS Bounds Without Tandem Constraints	1.6141	2.48964
FBFS Bounds With Tandem Constraints	1.73718	2.48964

Figure 18: Comparison of LBFS and FBFS policies for Example 8.

10.2. Closed queueing networks. If a queueing network is closed, then the constraint,

$$\sum_{j=1}^{L} \sum_{\{i:\sigma(i)=\sigma(j)\}} z_{ij} = N$$

is satisfied, where N is the population size. Moreover, if one samples the system whenever a particular buffer is being worked on, then one always sees N. Hence

$$\sum_{j=1}^{L} z_{ij} = \frac{\lambda}{\mu_i} N,$$

where

$$\lambda \quad := \quad \text{system throughput, i.e., the rate that parts circulate.}$$

The earlier equations (9.13-9.18) need to be modified slightly, since there are no exogenous arrivals. Thus one obtains linear programs for upper and lower bounds on the throughput λ.

Example 9: Comparison of buffer priority policies for a balanced two station system. Consider the system shown in Figure 19, where $\mu_1 = 1/3$, $\mu_2 = 2/7$, $\mu_3 = 1$ and $\mu_4 = 2$. There are four buffer priority policies: LBFS $= \{b_4, b_3, b_2, b_1\}$, FBFS $= \{b_1, b_2, b_3, b_4\}$, Harrison and Wein's Balanced Policy[3] $= \{b_4, b_1, b_3, b_2\}$, and the "Unbalanced" Policy $= \{b_3, b_2, b_1, b_4\}$.

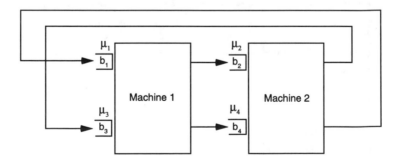

Figure 19: Closed re–entrant line of Examples 9 and 10.

---

[3] Based on the analysis of a Brownian Network model, it is conjectured in [14] that a certain method of ordering the buffers has optimal throughput as $N \nearrow \infty$. In [17] it has recently been shown that the throughput does indeed converge to the maximal throughput.

Figure 20 presents upper and lower bounds on throughput for these four buffer priority policies, as well as the class of all non-idling policies, for two population sizes, $N = 20$ and 100.

Policy	N = 20		N = 100	
	Lower Bound on Throughput	Upper Bound on Throughput	Lower Bound on Throughput	Upper Bound on Throughput
All policies	0.195122	0.240343	0.236686	0.248007
LBFS	0.237624	0.239282	0.247423	0.247780
FBFS	0.195122	0.240343	0.236686	0.248007
Balanced	5/21 ≈ 0.238095	0.240343	25/101 ≈ 0.247525	0.248007
Unbalanced	0.195122	5/21 ≈ 0.238095	0.236686	25/101 ≈ 0.247525

Figure 20: Comparison of policies for closed re–entrant line of Example 9. The Balanced policy has higher throughput than the Unbalanced policy. □

An issue which arises in the study of closed queueing networks is whether, under a particular scheduling policy, the throughput attained converges as $N \nearrow \infty$, to the maximal throughput attainable for that system. For a system with a fixed route, the maximal throughput attainable is the reciprocal of the mean work per part, brought to the bottleneck machine, in one cycle. It is

$$\lambda^* = \min_{\sigma} \frac{1}{\sum_{\{i:\sigma(i)=\sigma\}} \frac{1}{\mu_i}}.$$

(The definition is easily extended when the route is random).

Instead of studying the limit of the bounds of the LP's as $N \nearrow \infty$, one can directly study the bounds on the *limiting LP*. There is an issue of continuity here: Is the limit of the values of a sequence of LP's equal to the value of the limit of the sequence of LP's? This is not always so, and "regularity conditions" under which such equality holds are known in the literature. (See [33] where it is shown that the result holds for perturbations of the RHS's of the constraints, when the set of feasible solutions to the dual is compact).

Let us sidestep the issue and determine a sufficient condition for the convergence of the throughput to λ^* as $N \nearrow \infty$.

We illustrate just the basic idea; the details can be found in [27]. For a population of size N, consider one of the more or less typical constraints,

$$2\mu_{j-1}z_{j-1,j} + 2\lambda - 2\mu_j z_{jj} = 0.$$

Normalizing by N, the population size, and denoting $\bar{z}_{ij} := z_{ij}/N$, we obtain,

$$2\mu_{j-1}\bar{z}_{j-1,j} + \frac{2\lambda}{N} - 2\mu_j \bar{z}_{jj} = 0.$$

As $N \nearrow \infty$, the limiting LP has the constraint,

$$2\mu_{j-1}\bar{z}_{j-1,j} - 2\mu_j \bar{z}_{jj} = 0.$$

In this way one obtains the following Theorem.

Theorem 8: Sufficient condition for a buffer priority policy to have maximal throughput when population size increases. *Consider a closed queueing network with a population of size N, operating under a buffer priority policy using the ordering $\{b_{\theta(1)}, b_{\theta(2)}, \ldots, b_{\theta(L)}\}$. The throughput λ_N converges to λ^* as $N \nearrow +\infty$, if the following linear program has value λ^*:*

$$\text{Min} \quad \lambda$$

subject to

$$\sum_{j=1}^{L} \sum_{\{i:\sigma(i)=\sigma(j)\}} \bar{z}_{ij} = 1,$$

$$\sum_{j=1}^{L} \bar{z}_{ij} = \frac{\lambda}{\mu_i} \text{ for } 1 \leq i \leq L,$$

$$2\mu_{j-1}\bar{z}_{j-1,j} - 2\mu_j \bar{z}_{jj} = 0 \text{ for } 1 \leq j \leq L,$$

$$\mu_{i-1}\bar{z}_{i-1,i+1} - \mu_i \bar{z}_{i,i+1} + \mu_i \bar{z}_{ii} - \mu_{i+1}\bar{z}_{i+1,i} = 0 \text{ for } 1 \leq i \leq L,$$

$$\mu_{i-1}\bar{z}_{i-1,j} - \mu_i \bar{z}_{ij} + \mu_{j-1}\bar{z}_{j-1,i} - \mu_j \bar{z}_{ji} = 0$$
$$\text{for } 1 \leq i \leq L-2 \text{ and } i+2 \leq j \leq L$$
$$\text{with } j \not\equiv i+1 (mod\ L),$$

$$\bar{z}_{ij} = 0 \text{ for } \sigma(i) = \sigma(j) \text{ and } \theta(j) < \theta(i),$$

$$\sum_{\{j:\sigma(j)=\sigma\}} \bar{z}_{ji} \leq \sum_{\{j:\sigma(j)=\sigma(i)\}} \bar{z}_{ji} \text{ for all } \sigma \neq \sigma(i),$$

$$\bar{z}_{ij} \geq 0.$$

□

Example 10. Consider the system of Figure 19. Let $\mu_1 = 2$, $\mu_2 = 1$, $\mu_3 = 4$ and $\mu_4 = 3$. Under the FBFS buffer priority policy using the ordering $\{b_1, b_2, b_3, b_4\}$, the value of the above LP is $\frac{9}{13}$, while $\lambda^* = \frac{3}{4}$. Hence we are unable to conclude that FBFS asymptotically attains λ^*. (A similar result can also hold for a system where all machines are equally loaded, i.e., a balanced system; see [27] for an example).   □

In [13] it has recently been shown that in the closed version of Counterexample 1, the throughput for some policies does not converge to λ^* as $N \nearrow +\infty$. Recently, in [17] several results are established concerning the throughput of closed queueing networks.

10.3. Systems with machine failures. If a machine is subject to failure, it can be modeled as being interrupted by a VIP customer.

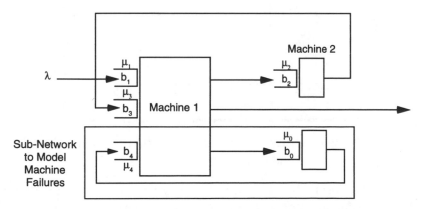

Figure 21: Model of system with machine failures in Example 11.

Example 11: Modeling machine failures. Consider the system in Figure 21. The lower closed loop contains just one VIP part. In addition to all the usual constraints, one has

$$z_{14} = z_{34} = 0,$$

to model the priority of the VIP part. Also,

$$z_{40} = 0, z_{04} = 0 \text{ and } z_{00} + z_{44} = 1,$$

since there is only one part trapped in the lower loop. Finally, since either b_0 or b_4 must always be busy, we have $z_{02}+z_{42} = z_{22}$, $z_{01}+z_{41} = z_{11}+z_{31}+z_{41}$, and $z_{03} + z_{43} = z_{13} + z_{33} + z_{43}$.

For the values $\mu_i \equiv \mu$, and $\lambda = \mu/4$, with the normalization $\lambda + \sum_{i=0}^{4} \mu_i = 1$, consider the LBFS policy. Figure 22 presents the bounds on the mean number of (real) parts in the system, as the ratio $\frac{\mu_0}{\mu_4} = \frac{MTTR}{MTTF}$ increases. It is of interest to note that the upper bound has the desirable property that it stays finite throughout the capacity region. □

MTTR/MTTF	Lower Bound on Number in System	Upper Bound on Number in System
0.2	1.5935	2.3482
0.3	1.9233	2.9356
0.4	2.3575	3.5364
0.5	2.9583	4.1825
0.6	3.8506	5.1354
0.99	174.5401	224.3769

Figure 22: Mean number in system vs. MTTR/MTTF, under the LBFS policy, for Example 11.

10.4. General distributions. Many systems have non-exponential distributions for interarrival and service times. They can be approximated by phase-type distributions.

Example 12: $E_2/M/1$ Queue. An $E_2/M/1$ queue can be modeled as in Figure 23, where the subnetwork on the left models the interarrival time, and a customer splits into two, one entering Machine 1, and the other going back to b_1.

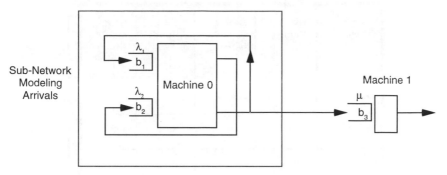

Figure 23: $E_2/M/1$ queue of Example 12.

This subnetwork has one trapped customer.

Since there is only one part in the subnetwork,

$$z_{12} = 0 \text{ and } z_{21} = 0.$$

Since one of the buffers b_1 or b_2 is always busy,

$$z_{13} + z_{23} = z_{33}.$$

In addition, one has all the other usual constraints.

For this example, one can explicitly solve for the bounds $z_{33,min}$ and $z_{33,max}$ on the mean number in the system. They are

$$z_{33,\min} = \text{Max}\left\{ \rho, \frac{\mu\rho^2}{\lambda_2(1-\rho)}, \frac{\rho}{1-\rho} - \frac{\mu^2\rho^2}{\lambda_1\lambda_2(1-\rho)} \right\},$$

$$z_{33,\max} = \text{Min}\left\{ \frac{\rho}{1-\rho}, \frac{\rho}{1-\rho} - \frac{\mu^2\rho^2}{\lambda_1\lambda_2(1-\rho)} + \frac{\lambda_2}{\lambda_1+\lambda_2} \right\}.$$

It can be checked that the upper and lower bounds differ by no more than $\frac{\lambda_2}{\lambda_1+\lambda_2} \geq \frac{1}{2}$ (without loss of generality one can take $\lambda_2 \leq \lambda_1$). Also, the upper bound is always tighter than Kingman's upper bound (see [20]), by at least $\rho/2$. Figure 24 presents some numerical values, for the usual "E_2" case, where $\lambda_1 = \lambda_2$. □

Load Factor ρ	Exact Value	Lower Bound by LP	Upper Bound by LP	King-man's Upper Bound
0.6	1.1901	0.875	1.375	1.675
0.75	2.3229	2.0000	2.5	2.875
0.9	6.8295	6.5	7.0	7.45
0.99	74.333	74.0	74.5	74.995

Figure 24: Mean number in $E_2/M/1$ queue of Example 12.

11. Bounds on transient performance. We can also obtain bounds on transient performance, as shown in [23]. Consider the initial condition $x(0)$. Let us define

$$(11.1) \qquad \bar{z}_{ij}(N) := \frac{1}{N} \sum_{n=0}^{N-1} E(w_i(\tau_n)x_j(\tau_n)).$$

Clearly,

$$\frac{1}{N} E\left[\sum_{n=0}^{N-1} E[x_i(\tau_{n+1})x_j(\tau_{n+1}) \mid \mathcal{F}_{\tau_n}] - x_i(\tau_n)x_j(\tau_n) \right]$$
$$= \frac{E(x_i(\tau_N)x_j(\tau_N)) - x_i(0)x_j(0)}{N}$$
$$\geq -\frac{1}{N}x_i(0)x_j(0).$$

The left hand side above can be computed exactly as in earlier sections, using the transition probabilities given in (6.1), except that we now write the final result using the averages values (11.1). This gives us constraints, and thus bounds, on the transient performance.

Theorem 9: Bounds on transient performance. *The transient performance $\frac{1}{N} \sum_{n=0}^{N-1} E[c^T x(\tau_n)]$ is bounded above by the linear program[4]:*

$$\text{Max} \sum_{j=1}^{L} c_j \bar{x}_j$$

---

[4] We have used $\bar{x}_j := \sum_{\{i:\sigma(i)=\sigma(j)\}} \bar{z}_{ij}(N)$ to simplify the LP.

subject to:

$$\bar{x}_j - \sum_{\{i:\sigma(i)=\sigma(j)\}} \bar{z}_{ij}(N) = 0 \qquad 1 \le j \le L,$$

$$2\lambda\bar{x}_1 + 2\lambda - 2\mu_1\bar{z}_{11}(N) \ge -\frac{1}{N}x_1^2(0),$$

$$2\mu_{j-1}z_{j-1,j} + 2\lambda - 2\mu_j\bar{z}_{jj}(N) \ge -\frac{1}{N}x_j^2(0) \text{ for } 2 \le j \le L,$$

$$\lambda\bar{x}_2 - \lambda - \mu_1\bar{z}_{12}(N) + \mu_1\bar{z}_{11}(N) - \mu_2\bar{z}_{21}(N) \ge -\frac{1}{N}x_1(0)x_2(0),$$

$$\lambda\bar{x}_j - \mu_1\bar{z}_{1j}(N) - \mu_j\bar{z}_{j1}(N) + \mu_{j-1}\bar{z}_{j-1,1}(N) \ge -\frac{1}{N}x_1(0)x_j(0)$$
$$\text{for } 3 \le j \le L,$$

$$\mu_{i-1}\bar{z}_{i-1,i+1}(N) - \mu_i\bar{z}_{i,i+1}(N) - \lambda + \mu_i\bar{z}_{ii}(N) - \mu_{i+1}\bar{z}_{i+1,i}(N) \ge -\frac{1}{N}x_i^2(0)$$
$$\text{for } 2 \le i \le L-1,$$

$$\mu_{i-1}\bar{z}_{i-1,j}(N) - \mu_i\bar{z}_{ij} + \mu_{j-1}\bar{z}_{j-1,i} - \mu_j\bar{z}_{ji}(N) \ge -\frac{1}{N}x_i(0)x_j(0)$$
$$\text{for } 2 \le i \le L-2 \text{ and } i+2 \le j \le L,$$

$$\sum_{\{i:\sigma(i)=\sigma\}} \bar{z}_{ij}(N) - \bar{x}_j \le 0 \text{ for } 1 \le j \le L \text{ and } \sigma \ne \sigma(j),$$

$$\bar{z}_{ij}(N) \ge 0 \text{ for } 1 \le i,j \le L.$$

The transient performance is bounded below by the same LP, with a "Min" replacing the "Max." ☐

We note that additional constraints (e.g., buffer priority constraints) can also be appended to the linear program for specific systems, as earlier.

12. The duality of stability and performance. We have seen, in Sections 6 and 9, two linear programs, one for establishing an upper bound on performance, and another for establishing stability. What is the connection between these two linear programs? In this section we will show that they are simply the duals of each other, in the sense of linear programming. The results in this and the next section are drawn from [23].

Let us begin by considering the class of all non–idling scheduling policies.

Theorem 10: Duality of performance LP and drift LP.

(i) *Consider the* Performance LP[5]:

$$\text{Max} \sum_{i=1}^{L} \bar{x}_i$$

subject to

$$\bar{x}_j - \sum_{\{i:\sigma(i)=\sigma(j)\}} z_{ij} = 0 \text{ for } 1 \leq j \leq L, \qquad (p_j)$$

$$2\lambda\bar{x}_1 + 2\lambda - 2\mu_1 z_{11} = 0 \qquad (r_{11})$$

$$\lambda\bar{x}_2 - \lambda - \mu_1 z_{12} + \mu_1 z_{11} - \mu_2 z_{21} = 0 \qquad (r_{12})$$

$$\lambda\bar{x}_j - \mu_1 z_{1j} - \mu_j z_{j1} + \mu_{j-1} z_{j-1,1} = 0 \text{ for } 3 \leq j \leq L \qquad (r_{1j})$$

$$2\mu_{j-1} z_{j-1,j} + 2\lambda - 2\mu_j z_{jj} = 0 \text{ for } 2 \leq j \leq L \qquad (r_{jj})$$

$$\mu_{i-1} z_{i-1,i+1} - \mu_i z_{i,i+1} - \lambda + \mu_i z_{ii} - \mu_{i+1} z_{i+1,i} = 0 \text{ for } 2 \leq i \leq L-1 \ (r_{i,i+1})$$

$$\mu_{i-1} z_{i-1,j} - \mu_i z_{ij} + \mu_{j-1} z_{j-1,i} - \mu_j z_{ji} = 0 \text{ for } 2 \leq i \leq L-2 \atop \text{and } i+2 \leq j \leq L \qquad (r_{ij})$$

$$\sum_{\{i:\sigma(i)=\sigma\}} z_{ij} - \bar{x}_j \leq 0 \text{ for } 1 \leq j \leq L, \text{ and } \sigma \neq \sigma(j) \qquad (w_{\sigma j})$$

$$z_{ij} \geq 0.$$

---

[5] The Performance LP (9.12-9.20) given earlier, is rewritten here using $\bar{x}_j := \sum_{\{i:\sigma(i)=\sigma(j)\}} z_{ij}$.

Its dual is the LP,

$$\text{Min} \sum_{i=1}^{L-1} \lambda r_{i,i+1} - \sum_{i=1}^{L} \lambda r_{ii}$$

subject to

$$\lambda r_{1i} + p_i - \sum_{\{\sigma:\sigma\neq\sigma(i)\}} w_{\sigma i} \geq 1 \text{ for } 1 \leq i \leq L \qquad (\bar{x}_i)$$

$$-\mu_j r_{jj} + \mu_j r_{j,j+1} - p_j \geq 0 \text{ for } 1 \leq j \leq L \qquad (z_{jj})$$

$$-\mu_j r_{ij} + \mu_i r_{i+1,j} - p_j 1_{\{\sigma(i)=\sigma(j)\}} + w_{\sigma(i),j} 1_{\{\sigma(i)\neq\sigma(j)\}} \geq 0 \atop \text{for } j \geq i+1. \qquad (z_{ij})$$

Above, by r_{ij} for $i > j$, we mean r_{ji}. The dual variables associated with the constraints are shown in parentheses, alongside the constraints.

(ii) *Defining*

$$q_{ij} := -r_{ij},$$

and eliminating the p_i's and $w_{\sigma,i}$'s, we see that the dual of the Performance LP above is the Drift LP:

$$\text{Min} \sum_{i=1}^{L} \lambda q_{ii} - \sum_{i=1}^{L-1} \lambda q_{i,i+1}$$

subject to

$$\lambda q_{1j} + \max_{\{i:\sigma(i)=\sigma(j)\}} \mu_i(q_{i+1,j} - q_{ij})$$

$$+ \sum_{\{\sigma:\sigma\neq\sigma(j)\}} \max_{\{i:\sigma(i)=\sigma\}} \mu_i(q_{i+1,j} - q_{ij})^+ \leq -1, \text{ for } 1 \leq j \leq L,$$

$$q_{ij} = q_{ji} \text{ and } q_{L+1,j} = 0 \text{ for } 1 \leq i,j \leq L.$$

Thus we see that whenever the Performance LP (for the class of all non–idling policies) has a finite upper bound, the Drift LP (for the class of all non–idling policies) has a feasible solution; in particular, there exists a $Q = Q^T$ whose associated quadratic form has the negative drift (6.3). Note however, that we have not yet established that the matrix Q is copositive; that is done in the next section.

Let us now consider briefly the LP for transient performance, given in Theorem 9. There, the earlier equality constraints in (9.12-9.20) are replaced by inequality constraints. Thus, the dual of the transient LP features nonnegativity constraints $q_{ij} \geq 0$ for all i, j. Hence, when the LP for transient performance is bounded, there exists a *nonnegative* matrix $Q = Q^T$ whose associated quadratic form has negative drift. Thus we discover a very interesting connection between nonnegative Q's in the drift condition and transient performance.

It is also true that a similar duality as in Theorem 10, also holds for the class of buffer priority policies; the associated Performance LP of Section 10.1 and the Drift LP of Section 7 are similarly duals; see [23] for details. However, in this latter case, we are unable to conclude that when the Performance LP is bounded, the resulting feasible Q in the dual Drift LP is automatically copositive.

13. Consequences of duality. We can take advantage of duality to considerably sharpen our results.

13.1. Boundedness of performance LP implies that all non-idling policies have a geometrically converging first moment. We will now show that when the Performance LP (9.12-9.20) is bounded, every stationary non-idling policy has a geometrically converging exponential moment.

To see this, we use duality and our instability results, as follows.

Since the Performance LP is bounded, its dual, the Drift LP. is feasible. Hence there exists a symmetric matrix Q for which $x^T(\tau_n)Qx(\tau_n)$ has the negative drift (6.3). We now show that this matrix Q is automatically copositive. Suppose not, then either a non-idling policy, or a small perturbation of it, which is still non-idling, does not have a finite first moment. However that is impossible, since we have already seen that all non-idling policies have performance bounded by M. Thus, Q is copositive. Then however we know that all stationary non-idling policies have a geometrically converging exponential moment.

Theorem 11. *Suppose the Performance LP (9.12-9.20) for the class of all non-idling policies is bounded.*

(i) *Then there exists a $Q = Q^T$ satisfying the negative drift condition. Moreover, every such Q is copositive.*

(ii) *All stationary non-idling polices have a geometrically converging exponential moment.*

It is an important open issue whether Theorem 11 can be extended to Performance LP's incorporating additional constraints to model specific policies, such as buffer priority policies, and others considered in Section 10.

13.2. Uniform bounds, stability and the monotone LP. We note that stability is not a monotone property. For example, an originally stable system can become unstable if some service rates are increased, see [4].

Consider the Drift LP with nonnegativity constraints $q_{ij} \geq 0$ appended to it. From an examination of (6.9), it is easy to see that if negative drift holds for some λ, it also holds for all $\lambda' \leq \lambda$. Hence one has established stability not just for a particular value of the arrival rate λ, but also for *all smaller arrival rates*. However, this requires the existence of a nonnegative matrix Q satisfying the Drift LP.

Now we show how the requirement of nonnegativity for all coefficients q_{ij} can be weakened. Thereby, not only do we establish stability for a range of λ's, but we also obtain bounds for all smaller arrival rates. This is done by obtaining an LP with certain monotonicity properties.

Consider the following "flow control" device. Whenever the number of parts in the system exceeds M, we discard the arrivals. Clearly the system is stable, and so the Performance LP gives an upper bound. Note now that the effective arrival rate is less than λ. Let us define,

$$w_0(\tau_n) = \begin{cases} 1 \text{ if arrivals are "on"} \\ 0 \text{ if arrivals are "off."} \end{cases}$$

Now, if we run through the arguments of Section 9, we simply find that all the earlier terms of the form $E(\lambda x_j(\tau_n))$ are replaced by $E[\lambda w_o(\tau_n)x_j(\tau_n)]$, which is less than $E(\lambda x_j(\tau_n))$. Thus some of the earlier equality constraints become inequality constraints. Now one can take the limit as $M \nearrow \infty$, obtaining an LP which bounds the original system without any flow control.

Theorem 12: The monotone LP. *Consider the LP (9.12-9.20), replacing however the earlier constraints (9.13,9.14,9.15,9.17) by the following constraints:*

$$\lambda \sum_{\{i:\sigma(i)=\sigma(1)\}} z_{i1} - \mu_1 z_{11} \geq -\lambda,$$

$$\mu_{j-1} z_{j-1,j} - \mu_j z_{jj} \geq -\lambda \text{ for } 2 \leq j \leq L,$$

$$\lambda \sum_{\{i:\sigma(i)=\sigma(2)\}} z_{i2} - \mu_1 z_{12} + \mu_1 z_{11} - \mu_2 z_{21} \geq 0,$$

$$\lambda \sum_{\{i:\sigma(i)=\sigma(j)\}} z_{ij} - \mu_1 z_{1j} - \mu_j z_{j1} + \mu_{j-1} z_{j-1,1} \geq 0 \text{ for } 3 \leq j \leq L,$$

$$\mu_{i-1} z_{i-1,i+1} - \mu_i z_{i,i+1} + \mu_i z_{ii} - \mu_{i+1} z_{i+1,i} \geq 0 \text{ for } 2 \leq i \leq L-1.$$

Its value for any λ bounds the mean number of parts in the system for all arrival rates $\lambda' \leq \lambda$.   □

From the way λ enters the constraints above, it is clear that the feasible set is increasing in λ, and hence so is the value of the LP.

One may note that in the dual this corresponds to setting only certain q_{ij}'s nonnegative.

As earlier, this can be extended to systems incorporating other constraints.

14. Design: scheduling of re-entrant lines. Let us return to the problem of scheduling re-entrant lines, introduced in Section 2. The following results are drawn from [32].

We will address the issue of designing good policies for three objectives: (i) reducing the variance of lateness, (ii) reducing the standard deviation of cycle-times, and (iii) reducing the mean cycle-time. In this and later sections, our development will not be mathematical. It will be more akin to physics, with emphasis on developing hypothesis, testing, refining hypotheses etc., all for large system models.

We will generically denote a part by π, and adopt the following notation:

$$
\begin{aligned}
\alpha(\pi) &= \quad \text{release time or arrival time of a part } \pi, \\
e(\pi) &= \quad \text{exit time from system of a part } \pi, \\
\delta(\pi) &= \quad \text{due-date of a part } \pi.
\end{aligned}
$$

We first address the problem of reducing the *variance of lateness*, $var(e(\pi) - \delta(\pi))$. Let,

$\zeta_i \quad := \quad$ estimate of mean remaining time to exit for a part in buffer b_i.

The quantity $\delta(\pi) - t$, where t is the current time, represents the time until due date for a part π. If the part π is located in buffer b_i, then the estimate of the remaining time in the system is ζ_i. Thus, the "slack"

$$
s_1(\pi) := \delta(\pi) - t - \zeta_i
$$

is a measure of urgency of the part.

Let us therefore consider the *Least Slack* (LS) Policy which gives priority to the part with the smallest slack. Such a policy seems intuitively fair. Quantitatively, what such fairness amounts to is an effort to make all parts equally late (or early). If all parts are equally late or equally early, then the variance of lateness is small. Thus, one has reason to believe that the above policy reduces the variance of lateness.

We note that since the current time t is common to all parts, it can be dropped from the comparison, and one can thus redefine the slack more simply as,

$$(14.1) \qquad\qquad s_1(\pi) := \delta(\pi) - \zeta_i.$$

We call this Least Slack policy with slack defined by (14.1), the *Fluctuation Smoothing Policy for the Variance of Lateness* (FSVL).

Now let us turn to the issue of reducing the *variance of the cycle-time*, $var(e(\pi) - \alpha(\pi))$. Note that if we were to *set* the due-date $\delta(\pi)$ as equal to the arrival time $\alpha(\pi)$, i.e.,

$$\delta(\pi) := \alpha(\pi),$$

then,

$$e(\pi) - \delta(\pi) = \text{ lateness } = \text{ cycle-time } = e(\pi) - \alpha(\pi).$$

Thus, the earlier policy of reducing the variance of lateness can be used to reduce the variance of cycle-time. Thus we redefine the slack as,

$$s_2(\pi) := \alpha(\pi) - \zeta_i,$$

and call the corresponding least slack policy the *Fluctuation Smoothing Policy for Variance of Cycle-time* (FSVCT).

Let us see how our FSVCT policy performs. In [31], extensive simulation tests have been conducted on models of a Hewlett Packard Research and Development Fabrication Line Model detailed in [37], and a full scale production line. The scheduling policies compared include all the policies tested in the earlier comprehensive study in [37], as well as others. We present some results for the cases of Poisson releases (arrivals) to the plants, and *deterministic* (i.e., periodic) releases. Other release polcies are also examined in [32]. Define the mean waiting time as the mean cycle time minus the sum of the mean processing times. Let Waiting Time (Policy) denote the average over the simulation runs of the mean waiting time for a particular scheduling policy. The FCFS (aka FIFO) policy is well known. Define the percentage improvement in waiting time of a policy with respect to FCFS as [Waiting Time (FCFS) - Waiting Time (Policy)]/Waiting Time (FCFS). Similarly, the quantity, percentage improvement in standard deviation of cycle time of a policy with respect to FCFS is defined. Below, in Figures 25 and 26, we present the percentage improvements of various policies on a Hewlett Packard Research and Development Fabrication Line Model of [37].

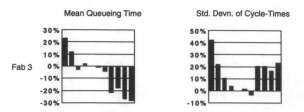

Figure 25: Performance under deterministic releases. Shown are the percentage improvements in the Mean Waiting Time and the Standard Deviation of Cycle-Time, over the baseline FIFO Policy. The FSVCT Policy is shown leftmost.

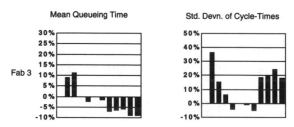

Figure 26: Performance under Poisson releases. Shown are the percentage improvements in the Mean Waiting Time and the Standard Deviation of Cycle-Time, over the baseline FIFO Policy.The FSVCT Policy is shown leftmost.

As observed from the figures, FSVCT provides an improvement in standard deviation of cycle time, over FCFS, of about 40%, for both deterministic and Poisson releases.

Interestingly, while the FSVCT policy was not designed to reduce the mean cycle-time, we note in Figure 25 that it does provide a substantial improvement, about 20%, over FCFS for the *mean waiting time*, when the releases are deterministic. However, there is less of an improvement when the arrivals are Poisson.

To understand why this is so (and to later exploit the understanding and devise a new policy for reducing the mean cycle-time), we recall the *First Law of Queueing Systems – delay is caused by fluctuations.* This is manifested in, for example, the Pollaczek-Khintchine Formula, where one sees that fluctuations in the service time, i.e., the variance of the service time, increase the mean delay in an $M/G/1$ queue. Similarly, Kingman's bounds also show this to be so for the variance in interarrival times.

With this fact – fluctuations cause delays – we can understand why FSVCT reduces the delay under deterministic releases. Clearly the arrivals have no burstiness – they are periodic. Hence, when we reduce the variance of the cycle-times, we have therefore also reduced the *variance of interdeparture times* from the system.

Note now that our definition of ζ_i has the property,

$$(14.2) \qquad\qquad \zeta_i = \zeta_{ij} + \zeta_j,$$

where

$$\zeta_{ij} := \text{ mean time to go from } b_i \text{ to } b_j.$$

Thus, when comparing parts in two buffers b_i and b_k at a machine, both of which precede b_j, we see that we could have equivalently defined the slack as

$$s_2(\pi) = \alpha(\pi) - \zeta_{ij},$$

since the constant term ζ_j in (14.2) can be ignored in comparisons. This, however, simply corresponds to viewing b_{j-1} as the exit from the system, and trying to reduce the variance of *interarrival times* to b_j.

Thus, under deterministic releases, FSVCT tries to reduce the variance of interarrival times to all buffers b_j – *simultaneously*. Keeping in mind the first law of queueing systems, we see that it reduces the mean cycle-time under deterministic releases.

Now the question arises, how can we exploit this to design a scheduling policy that reduces the mean cycle-time not just under deterministic releases, but also under other releases? The answer is simple! We simply *set* deterministic due-dates, i.e.,

$$\delta(\pi) := \frac{n}{\lambda} \quad \text{ if } \quad \pi \text{ is the } n\text{-th part released}$$

into the system, and

$$\frac{1}{\lambda} = \text{ mean interarrival time,}$$

and continue to reduce the variance of lateness. Since the due-dates are periodic, when we reduce the variance of lateness, we will end up reducing the variance of interdeparture times. Due to the definition of the ζ_i's, we will thus be *simultaneously reducing the fluctuations in arrivals to all buffers*, and then we should be able to reduce the mean cycle-time.

To summarize, we redefine the slack as

$$s_3(\pi) := \frac{n}{\lambda} - \zeta_i \quad \text{ where } \pi \text{ is the } n\text{-th part released}$$

into the system, and it is

currently located in b_i.

The corresponding Least Slack policy is called the *Fluctuation Smoothing Policy for Mean Cycle-Time* (FSMCT).

How well does FSMCT perform? In Figure 27 we show the performance for Poisson releases. Shown are Waiting Time (Policy) and Standard Deviation (Policy), for various policies.

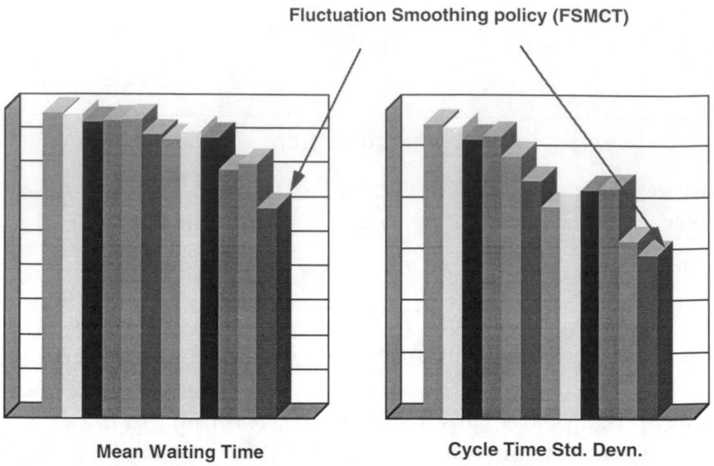

Fluctuation Smoothing policy (FSMCT)

Mean Waiting Time   Cycle Time Std. Devn.

Figure 27: Improvement obtained in mean waiting time by using FSMCT.
The FSMCT policy is rightmost. The FSVCT policy is second from the
right.

FSMCT yields about a further 18% improvement over FSVCT.

Regarding release policies, the *Workload Regulation Release* (WR) Policy is advocated in [37]. The simulations in [32] affirm the choice of this
release policy. While its mean cycle–time under FSMCT is almost the same
as deterministic release, its variance is smaller.

Hence we propose the combination of using WR for releases, and
FSMCT for scheduling. Figures 28 and 29 show the performance improve-
ments obtained by using either or both of these policies in comparison to
Deterministic release and FIFO scheduling.

Fab 3: Mean Queueing Time	FIFO	FSMCT	FSVCT
Deterministic Release	1317.7	1014.7	Same policy as FSMCT
Workload Regulation Release	1292.24	1003.01	1018.74

Figure 28: Improvements in mean waiting time obtained by using WR
Release and FSMCT, over Deterministic Release and FIFO.

Fab 3: Standard Deviation of Cycle-Time	FIFO	FSMCT	FSVCT
Deterministic Release	240.93	139.64	Same policy as FSMCT
Workload Regulation Release	173.31	103.72	83.21

Figure 29: Improvements in variance of cycle time obtained by using WR Release and FSMCT, over Deterministic Release and FIFO.

The improvements are substantial, and, as statistically tested in [32], significant.

14.1. FSMCT is a stationary policy. We now show that FSMCT is a *stationary* scheduling policy, in the sense that it depends only on the vector $x(t)$ of queue lengths. Thus, we can potentially use Markovian methods to study its performance.

Clearly, instead of $s_3(\pi)$, one can use any monotone increasing function $f(s_3(\pi))$ to make decisions. Consider the choice

$$f(s_3(\pi)) := \lambda s_3(\pi) - (\text{number of departures from system up to time } t).$$

The first term on the right hand side above is simply $\lambda s_3(\pi) = n - \lambda \zeta_i$. By Little's Theorem, $\lambda \zeta_i$ = mean number of parts downstream from buffer b_i. Thus,

$$f(s_3(\pi)) = (n - \text{number of departures from system up to time } t)$$
$$- \text{mean number of parts downstream from buffer } b_i.$$

Note now that parts are always served from the head of a buffer, and so parts never overtake each other. Hence parts depart from the system in the order that they arrive. Since part π is the n-th release into the system, and it is located in b_i,

$$(n - \text{number of departures from system up to time } t)$$
$$= \text{number of parts downstream from } b_i.$$

Thus,

$$f(s_3(\pi)) = \text{downstream shortfall from } b_i,$$

where the downstream shortfall from buffer b_i is the difference between the number of parts downstream from b_i and the mean number downstream. In addition to showing that FSMCT is a stationary policy, this also provides a very appealing interpretation of the policy – as attempting to alleviate downstream shortfalls.

15. Concluding remarks. We simply end on a broad note. There is a clear need for a theory of scheduling which is (i) tractable for very large systems, (ii) able to handle random events, e.g., machine failures, and (iii) implementable dynamically in real time. There is strong need for a deeper theory on two fronts, (i) a descriptive theory of performance analysis, and (ii) a prescriptive theory of control or scheduling. There are of course many important open problems.

What we believe is that there is a need for researchers to keep an ultimate application in focus, where the phrase "in focus" is interpreted in a broad but nevertheless purposeful sense. We believe that theory and applications are not antagonistic. A knowledge of the application allows for creativity on modeling issues, as well as problem formulations, and allows an easier development of a richer and deeper theory.

Finally, we believe that industry should make a serious effort in publicizing and exposing researchers and potential researchers to problems, models, data, etc.

REFERENCES

[1] L.-E. Andersson, G. Chang, and T. Elfving. Criteria for copositive matrices and nonnegative Bezier patches. Technical Report LiTH-MAT-R-93-27, Linkoping University and University of Science and Technology, China, August 1993.

[2] F. Baskett, K. M. Chandy, R. R. Muntz, and F. G. Palacios. Open, closed and mixed networks of queues with different classes of customers. *J. Assoc. Comput. Mach.*, 22(2):248–260, April 1975.

[3] D. Bertsimas, I. Ch. Paschalidis, and J. N. Tsitsiklis. Optimization of multiclass queueing networks: Polyhedral and nonlinear characterizations of achievable performance. *Annals of Applied Probability*, 4:43–75, 1994.

[4] M. Bramson. Instability of FIFO queueing networks with quick service times. Technical report, Mathematics Department, University of Wisconsin, Madison, WI, 1993.

[5] R. W. Cottle, G. J. Habetler, and C. E. Lemke. On classes of copositive matrices. *Linear Algebra and Its Applications*, 3:295–310, 1970.

[6] R. L. Cruz. A calculus for network delay, part I: Network elements in isolation. *IEEE Transactions on Information Theory*, 37(1):114–131, January 1991.

[7] J. Dai and G. Weiss. Stability and instability of fluid models for certain re-entrant lines. Preprint, February 1994.

[8] J. G. Dai. On positive Harris recurrence of multiclass queueing networks: A unified approach via fluid limit models. Technical report, Georgia Institute of Technology, 1993. To appear in *Annals of Applied Probability*.

[9] J. G. Dai and Vien Nguyen. On the convergence of multiclass queueing networks in heavy traffic. *The Annals of Applied Probability*, 4(1):26–42, 1994.

[10] J. G. Dai and Y. Wang. Nonexistence of Brownian models for certain multiclass queueing networks. *Queueing Systems: Theory and Applications: Special Issue on Queueing Networks*, 13(1–3):41–46, May 1993.

[11] J. L. Doob. *Stochastic Processes*. John Wiley and Sons, New York, NY, 1953.

[12] D. G. Down and S. P. Meyn. Piecewise linear test functions for stability of queue-

ing networks. In *Proceedings of the IEEE 33th Conference on Decision and Control*, Buena Vista, FL, December 1994. to appear.

[13] J. M. Harrison and V. Nguyen. Some badly behaved closed queueing networks. Technical report, 1994.

[14] J. M. Harrison and L. M. Wein. Scheduling networks of queues: Heavy traffic analysis of a two-station closed network. *Operations Research*, 38(6):1052–1064, 1990.

[15] J. R. Jackson. Networks of waiting lines. *Mathematics of Operations Research*, 5:518–521, 1957.

[16] J. R. Jackson. Jobshop-like queueing systems. *Management Science*, 10:131–142, 1963.

[17] H. Jin, J. Ou, and P. R. Kumar. The throughput of closed queueing networks–uniform bounds, efficient scheduling, heavy traffic behavior, and buffer priority policies. Technical report, University of Illinois, Coordinated Science Laboratory, 1994.

[18] S. Karlin and H. M. Taylor. *A First Course in Stochastic Processes*. Academic Press, New York, NY, 1975.

[19] F. P. Kelly. *Reversibility and Stochastic Networks*. John Wiley and Sons, New York, NY, 1979.

[20] J. F. C. Kingman. Inequalities in the theory of queues. *Journal of the Royal Statistical Society, Series B*, 32:102–110, 1970.

[21] P. R. Kumar. Re-entrant lines. *Queueing Systems: Theory and Applications: Special Issue on Queueing Networks*, 13(1–3):87–110, May 1993.

[22] P. R. Kumar and S. P. Meyn. Stability of queueing networks and scheduling policies. To appear in *IEEE Tranactions on Automatic Control*, February 1995, 1993.

[23] P. R. Kumar and Sean Meyn. Duality and linear programs for stability and performance analysis of queueing networks and scheduling policies. Technical report, C. S. L., University of Illinois, 1993.

[24] P. R. Kumar and T. I. Seidman. Dynamic instabilities and stabilization methods in distributed real-time scheduling of manufacturing systems. *IEEE Transactions on Automatic Control*, AC- 35(3):289–298, March 1990.

[25] S. Kumar and P. R. Kumar. Fluctuation smoothing policies are stable for stochastic re- entrant lines. To appear in *Proceedings of the 33rd IEEE Conference on Decision and Control*, December 1994.

[26] S. Kumar and P. R. Kumar. The last buffer first policy is stable for stochastic re-entrant lines. Technical report, Coordinated Science Laboratory, University of Illinois, Urbana, IL, 1994.

[27] S. Kumar and P. R. Kumar. Performance bounds for queueing networks and scheduling policies. *IEEE Transactions on Automatic Control*, AC- 39:1600–1611, August 1994.

[28] S. Lippman. Applying a new device in the optimization of exponential queueing systems. *Operations Research*, 23:687–710, 1975.

[29] S. C.-H. Lu. *Control policies for scheduling of semiconductor manufacturing plants*. PhD thesis, University of Illinois, Urbana-Champaign, IL, 1994.

[30] S. H. Lu and P. R. Kumar. Distributed scheduling based on due dates and buffer priorities. *IEEE Transactions on Automatic Control*, AC- 36(12):1406–1416, December 1991.

[31] S. C. H. Lu, Deepa Ramaswamy, and P. R. Kumar. Efficient scheduling policies to reduce mean and variance of cycle–time in semiconductor manufacturing plants. Technical report, University of Illinois, Urbana, IL, 1992. To appear in *IEEE Transactions on Semiconductor Manufacturing, 1994*.

[32] S. C. H. Lu, Deepa Ramaswamy, and P. R. Kumar. Efficient scheduling policies to reduce mean and variance of cycle–time in semiconductor manufacturing plants. *IEEE Transactions on Semiconductor Manufacturing*, 7(3):374–385, 1994.

[33] K. G. Murty. *Linear Programming*. John Wiley and Sons, New York, NY, 1983.

[34] K. G. Murty and S. N. Kabadi. Some NP-complete problems in quadratic and nonlinear programming. *Mathematical Programming*, 39:117–129, 1987.

[35] J. Ou and L. M. Wein. Performance bounds for scheduling queueing networks. *Annals of Applied Probability*, 2:460–480, 1992.

[36] S. S. Panwalker and W. Iskander. A survey of scheduling rules. *Operations Research*, 25(1):45–61, January-February 1977.

[37] L. M. Wein. Scheduling semiconductor wafer fabrication. *IEEE Transactions on Semiconductor Manufacturing*, 1(3):115–130, August 1988.

STABILITY OF OPEN MULTICLASS QUEUEING
NETWORKS VIA FLUID MODELS

J.G. DAI*

Abstract. This paper surveys recent work on the stability of open multiclass queueing networks via fluid models. We recapitulate the stability result of Dai [8]. To facilitate study of the converse of the stability result, we distinguish between the notion of *fluid limit* and that of *fluid solution*. We define the stability region of a service discipline and the global stability region of a network. Examples show that piecewise linear Lyapunov functions are powerful tools in determining stability regions.

Key words. Stability, queueing networks, fluid models, scheduling, performance analysis, Harris recurrence, heavy traffic, Brownian models.

1. Introduction. There has been a recent surge in studying stability/instability of multiclass queueing networks. See, for example, Lu and Kumar [21], Rybko and Stolyar [24], Whitt [27], Bramson [2,3] and Seidman [25]. To show that the instability can occur even in a Kelly-type network, a network in which all customers visit a station have a common service time distribution, we consider the three station network pictured in Figure 1.1. Jobs (or customers) arrive at station 1 according to a general renewal process with arrival rate 1. Each job follows a deterministic route, and the station sequence that a job visits is 1, 2, 3, 2, 3, 2, 1, 3 and 1. Following Kelly [19], a job class is defined for each processing stage. Therefore, in this example, each station processes three job classes. Each class may have its own general service time distribution (thus a job may have different processing requirements on different visits to a station), and the service discipline at each station can be general.

Assume that at station 1, priority is given to customer classes in order $(9, 7, 1)$, where class 9 has the highest priority. At station 2, priority is given to customer classes in order $(4, 2, 6)$, and at station 3 the service discipline is first-in–first-out (FIFO). Assume further that the mean arrival rate to class 1 is 1, and the mean service time for each visit to stations 1 and 2 is 0.3 and for each visit to station 3 is 0.1. Therefore the nominal workloads per unit of time for servers 1, 2 and 3 are 90%, 90% and 30%, respectively. Dai and Meyn [9] simulated this network under two distributional assumptions. In the first case (case (M)), all distributions are assumed to be exponential. In the second case (case (D)), all interarrival and service times are constants. Therefore there is no randomness at all in the network. For (M) and (D2), the network is initially empty. For (D1), there are two jobs initially in

---

* School of Industrial and Systems Engineering, and School of Mathematics, Georgia Institute of Technology, Atlanta, GA 30332-0205. Research supported by NSF grants DMS-9203524 and DDM-9215233, and two grants from the Texas Instruments Corporation. Part of this work was done while the author was visiting the *Institute For Mathematics And Its Applications*, whose financial support is also acknowledged.

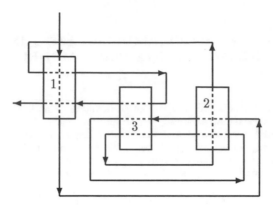

FIG. 1.1. *A three station network that may be unstable under certain priority service disciplines*

front of buffer 1. It appears from Table 1.1 that the average queue lengths in the simulations (M) and (D1) are growing without bound, whereas in simulation (D2) the total customer population seems bounded. Figure 1.2 plots the queue length processes at stations 1 and 2 for system (M) in the first 10,000 units of simulation time. The plot again suggests that the *total* queue length cycles to infinity. Readers are referred to Section 7.1 or Dai and Weiss [12, Remark 3 in Section 6] for the insight into the instability revealed in these simulations (see also Chen [5, Theorem 4.5] and Gu [16].)

It is now known that the stability of a queueing network is closely related to that of the corresponding fluid model as in Rybko and Stolyar [24], Dai [8], Chen [5] and Stolyar [26]. (See Chen and Mandelbaum [6] for a survey on fluid models.) In this paper, we recapitulate the main result in Dai [8] which says that a service discipline in an open queueing network is stable if the corresponding fluid model eventually drains to zero starting from any initial condition. We carefully distinguish the notion of a *fluid limit* from that of a *fluid solution*. We believe that this distinction is helpful in studying the converse of the stability result. We also introduce definitions of the stability region of a service discipline and of the global stability region of a network. We show that the piecewise Lyapunov functions used in Botvitch and Zamyatin [1], Dai and Weiss [12] and Down and Meyn [14] provide a powerful tool in determining a stability region.

2. A multiclass network

2.1. Network model. We consider a network composed of d single server stations, which we index by $i = 1, \ldots, d$. The network is populated by K classes of customers, where customers of class k $(k = 1, \ldots, K)$ arrive to the network via an exogenous arrival process with i.i.d. interarrival times $\{\xi_k(n), n \geq 1\}$. We allow $\xi_k(n) \equiv \infty$ for all n for some k, in which case we say that the external arrival process for customers of class k is

TABLE 1.1

For (M) and (D1), average queue lengths at stations 1 and 2 grow without bound, while the queue length at station 3 nearly zero. The simulation (D2) is well behaved, even though the network differs from (D1) initially by only two jobs.

case	running time	queue length at each station			utilization rate at each station			cycle time
		1	2	3	1	2	3	
(M)	1,000	41.07	91.74	0.10	0.73	0.82	0.25	137.66
	10,000	493.61	772.79	0.10	0.76	0.77	0.26	1289.58
	100,000	4993.16	7106.94	0.11	0.77	0.76	0.25	12446.21
(D1)	1,000	37.62	79.96	0.00	0.73	0.79	0.26	108.29
	10,000	483.49	718.17	0.00	0.77	0.77	0.26	1228.28
	100,000	4534.39	8301.29	0.00	0.74	0.79	0.26	13439.38
(D2)	1,000	0.40	0.30	0.04	0.90	0.90	0.30	2.85
	10,000	0.42	0.29	0.04	0.90	0.90	0.30	2.85
	100,000	0.42	0.29	0.04	0.90	0.90	0.30	2.85

null. We let $\mathcal{E}$ denote the set of classes with non-null exogenous arrivals. Hereafter, whenever external arrival processes are under discussion, only classes with non-null exogenous arrivals are considered. Class k customers require service at station $s(k)$. Their service times are also i.i.d., and are denoted $\{\eta_k(n), n \geq 1\}$. We assume that the buffers at each station have infinite capacity.

Routing is assumed to be *Bernoulli* among classes, so that upon completion of service at station $s(k)$, a class k customer becomes a customer of class ℓ with probability $P_{k\ell}$, and exits the network with probability $1 - \sum_{\ell} P_{k\ell}$, independent of all previous history. To be more precise, let $\phi^k(n)$ be the routing vector for the nth class k customer who finishes service at station $s(k)$. The ℓth component of $\phi^k(n)$ is one if this customer becomes a class ℓ customer and zero otherwise. Therefore, $\phi^k(n)$ is a K-dimensional "Bernoulli random variable" with parameter P'_k, where P_k denotes the kth row of $P = (P_{k\ell})$ (all vectors are envisioned as column vectors, and primes denote transpose). We assume that for each k the sequence $\phi^k = \{\phi^k(n), n \geq 1\}$ is i.i.d., and that $\phi^1, \ldots, \phi^K$ are mutually independent, as well as independent of the arrival and service processes. The transition matrix $P = (P_{k\ell})$ is taken to be transient. That is,

$$(2.1) \qquad I + P + P^2 + \ldots \quad \text{is convergent.}$$

Condition (2.1) implies that all customers eventually leave the network. Hence the systems we consider are open queueing networks, although some more general networks may also be included (cf., Dai and Meyn [9]). This network description is quite standard, and may be found in numerous related papers (see, for example, Harrison and Nguyen [18]).

Throughout this paper, we assume that

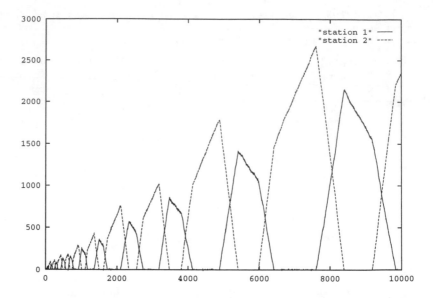

FIG. 1.2. *The queue lengths at each station oscillate with increasing magnitude: Mutual blocking between machines 1 and 2 results in instability.*

(**A1**) $\xi_1, \ldots, \xi_K, \eta_1, \ldots, \eta_K$ are mutually independent, and i.i.d. sequences.

(**A2**) $\mathsf{E}[\xi_\ell(1)] < \infty$ for $\ell \in \mathcal{E}$ and $\mathsf{E}[\eta_k(1)] < \infty$ for $k = 1, \ldots, K$.

(**A3**) For each $k \in \mathcal{E}$, there exists some nonnegative function $q_k(x)$ on $\mathbb{R}_+$ with $\int_0^\infty q_k(x)\,dx > 0$, and some integer j_k, such that

$$\mathsf{P}(\xi_k(1) \geq x) > 0 \quad \text{for all} \quad x > 0,$$
$$\mathsf{P}(\xi_k(1) + \cdots + \xi_k(j_k) \in dx) \geq q_k(x)\,dx.$$

Conditions (A1) and (A2) are quite standard, although the independence assumption (A1) can be relaxed: see the remark after Proposition 2.1 of Dai [8]. Condition (A3) is required to establish ergodicity of the network. Under this condition, the argument used in Lemma 3.4 of Meyn and Down [22] may be applied to deduce that all compact subsets of a state space are *petite*. (For the definition of a petite set, see Section 4.1 of Meyn and Tweedie [23].) Frequently, milder conditions can be invoked to obtain this property for the network (see, for example, Assumption (A3') of Dai and Meyn [9].) Condition (A3) is *not* needed for bounding the moments of queue lengths (see Theorem 8.1.)

For future reference, let $\alpha_k = 1/\mathsf{E}[\xi_k(1)]$ and $\mu_k = 1/\mathsf{E}[\eta_k(1)]$ be the arrival rate and service rate for class k customers, respectively. The set $\mathcal{C}_i = \{k : s(k) = i\}$ is called the *constituency* for station i. We let C denote the $d \times K$ *incidence matrix,*

$$C_{ik} = \begin{cases} 1 & \text{if } s(k) = i \\ 0 & \text{otherwise.} \end{cases}$$

In light of assumption (2.1), $(I - P')^{-1}$ exists and is equal to

$$(I - P')^{-1} = (I + P + P^2 + \ldots)'.$$

Put $\lambda = (I - P')^{-1}\alpha$. One interprets λ_k as the *effective* arrival rate to class k. For each $i = 1, \ldots, d$ we define the *nominal workload* for server i per unit of time as

$$(2.2) \qquad \rho_i = \sum_{k \in C_i} \lambda_k / \mu_k.$$

In vector form, we have $\rho = CM\lambda$, where $M = \text{diag}(m_1, \ldots, m_K)$ and $m_k = 1/\mu_k$.

2.2. Service disciplines. To fully describe a multiclass network, we must also specify how the server chooses among the various classes at a station. A *service discipline* at station i dictates which job will be served next when server i completes a service. We assume that service disciplines are non-idling (work-conserving), which means that a server works continuously whenever there is work to be done at the station. For concreteness, at each station one of the following service disciplines is employed: first-in–first-out (FIFO), static buffer priority among classes (both preemptive and nonpreemptive), and head-of-line processor sharing among classes. Notice that under these service disciplines, the server may split its capacity among classes at a station, and at most one customer in each class can receive partial service time. Readers will see that our approach actually can be applied to far more general service disciplines. In particular, processor sharing at station i among the first $r(k)$ customers in class k with $s(k) = i$ can be treated similarly. However, the true processor sharing discipline among all customers at a station is ruled out.

2.3. A Markovian state. Now we define a state process for the network, which depends upon the particular service discipline employed. Let $Q_k(t)$ be the queue length for class k customers, including the one being serviced. Let $Q(t) = (Q_1(t), \ldots, Q_K(t))'$. Thus, $Q(t)$ is the K-dimensional vector of *class-level* queue lengths at time t. For each station i, define

$$N_i(t) = \sum_{k \in C_i} Q_k(t).$$

Then $N(t) = (N_1(t), \ldots, N_d(t))'$ is the d-dimensional vector of *station-level* queue lengths at time t.

Even when all distributions are exponential, under the FIFO service discipline, $\{Q_k(t) : k \in C_i\}$ does not contain enough information to tell which customer will be served next at station i. In this case, we need to know how customers are lined up at station i. Thus, we define for station i,

$$(2.3) \qquad \mathbb{Q}_i(t) = (k_{i,1}, k_{i,2}, \ldots, k_{i,N_i(t)}),$$

where $k_{i,j}$ is the class number for jth customer at station i. (If $N_i(t) = 0$, $\mathbb{Q}_i(t)$ is defined to be an empty list.) Put

$$\mathbb{Q}(t) = (\mathbb{Q}_1(t), \ldots, \mathbb{Q}_d(t)).$$

Then $\mathbb{Q}(t)$ tells exactly how customers are lined up at each station. It embodies more information than $Q(t)$ does. For static buffer priority and head-of-line processor sharing service disciplines, we simply let

$$(2.4) \qquad \mathbb{Q}(t) = Q(t).$$

Therefore, in general, the state $X(t)$ at time t is

$$X(t) = (\mathbb{Q}(t), U(t), V(t)),$$

where $U(t) = (U_k(t) : k \in \mathcal{E})' \in \mathbb{R}_+^{|\mathcal{E}|}$ and $V(t) = (V_1(t), \ldots, V_K(t))' \in \mathbb{R}_+^K$. For $k \in \mathcal{E}$, $U_k(t)$ is the remaining time before the next class k customer will arrive from outside. For $k = 1, \ldots, K$, $V_k(t)$ is the remaining service time for the class k customer that is in service, which is set to be a fresh class k service time if $Q_k(t) = 0$. Both $U(t)$ and $V(t)$ are taken to be right continuous.

We let X denote the *state space* for the state process, which is by definition equal to the set of possible values for the state $X(t)$, and we let $x = (\mathbb{Q}, U, V)$ denote a generic state in X. Notice that the first component $\mathbb{Q}$ captures the positions of customers in the network. It can be finite dimensional as in (2.4), or infinite dimensional as is the case for the FIFO service discipline. We use $|\mathbb{Q}|$ to denote the total number of jobs in the network, and for a $u \in \mathbb{R}^K$, $|u| = \sum_{k=1}^K |u_k|$. For a state $x = (\mathbb{Q}, U, V) \in \mathsf{X}$, we define the norm of x to be

$$|x| = |\mathbb{Q}| + |U| + |V|,$$

Let X be endowed with the natural induced topology. It is easy to check that the sublevel set

$$C(n) = \{x \in \mathsf{X} : |x| \le n\}$$

is a compact subset of X for any n.

It was shown in Dai [8, Section 2.2] that $X = \{X(t), t \ge 0\}$ is a strong Markov process. This allows us to assume at our disposal the usual elements that constitute a Markovian environment for X. Formally, it is assumed hereafter that $((\Omega, \mathfrak{F}), \mathfrak{F}_t, X(t), \theta_t, \mathsf{P}_x)$ is a Borel right process on the measurable state space $(\mathsf{X}, \mathfrak{B}_\mathsf{X})$. In particular, $X = \{X(t), t \geq 0\}$ has right-continuous sample paths; it is defined on $(\Omega, \mathfrak{F})$ and is adapted to $\{\mathfrak{F}_t, t \geq 0\}$; $\{\mathsf{P}_x, x \in \mathsf{X}\}$ are probability measures on $(\Omega, \mathfrak{F})$ such that for all $x \in \mathsf{X}$,

$$\mathsf{P}_x\{X(0) = x\} = 1,$$

and

$$(2.5) \quad \mathsf{E}_x\{f(X \circ \theta_\tau) \mid \mathfrak{F}_\tau\} = \mathsf{E}_{X(\tau)}f(X) \quad \text{on} \quad \{\tau < \infty\}, \quad \mathsf{P}_x\text{-a.s.},$$

where τ is any $\mathfrak{F}_t$-stopping-time,

$$(X \circ \theta_\tau)(\omega) = \{X(\tau(\omega) + t, \omega), t \geq 0\},$$

and f is any real-valued bounded measurable function (the domain of f is the space of X-valued right-continuous functions on $[0, \infty)$, equipped with the Kolmogorov σ-field generated by cylinders).

3. Discrete system dynamics. Let $x = (\mathbb{Q}(0), U(0), V(0))$ be the initial state of the network under a specified service discipline. In this section, we attach a superscript x to a symbol to explicitly denote the dependence on initial state x. In particular, $Q_k^x(t)$ is the queue length for class k customers at time t. For $\ell \in \mathcal{E}$ and $k = 1, \ldots, K$,

$$
\begin{aligned}
E_\ell^x(t) &= \max\{n \geq 1 : U_\ell(0) + \xi_\ell(1) + \ldots + \xi_\ell(n-1) \leq t\}, \quad t \geq 0, \\
S_k^x(t) &= \max\{n \geq 1 : V_k(0) + \eta_k(1) + \ldots + \eta_k(n-1) \leq t\}, \quad t \geq 0.
\end{aligned}
$$

It is easy to check that $E_\ell^x(t)$ is the number of exogenous arrivals to class ℓ by time t, and $S_k^x(r)$ is the number of service completions of class k customers if server $s(k)$ devotes r units of time to class k customers. Let $T_k^x(t)$ be the cumulative time that server $s(k)$ has devoted to class k customers by time t. Then, $S_k^x(T_k^x(t))$ is the number of service completions for class k by time t. Recall the routing vectors $\phi^k(j)$ defined in Section 2. Let

$$(3.1) \qquad \Phi^k(n) = \sum_{j=1}^{n} \phi^k(j).$$

Then $\Phi_\ell^k(n)$ is the number of class k customers routed to class ℓ among the first n class k service completions. It follows that for $k = 1, \ldots, K$,

$$(3.2) \qquad Q_k^x(t) = Q_k^x(0) + E_k^x(t) + \sum_{\ell=1}^{K} \Phi_k^\ell(S_\ell^x(T_\ell^x(t))) - S_k^x(T_k^x(t)).$$

Let

$$I_i^x(t) = t - \sum_{k \in \mathcal{C}_i} T_k^x(t), \quad i = 1, \ldots, d.$$

Then $I_i^x(t)$ is the cumulative time that server i is idle in $[0, t]$. Besides (3.2), we have

(3.3) $Q^x(t) = (Q_1^x(t), \ldots, Q_K^x(t))' \geq 0, \quad t \geq 0,$

(3.4) $T^x(t) = (T_1^x(t), \ldots, T_K^x(t))'$ is a non-decreasing and starts from 0,

(3.5) $I^x(t) = (I_1^x(t), \ldots, I_d^x(t))'$ is non-decreasing,

Because all of the allowable service disciplines are non-idling, the cumulative idle time $I_i^x(t)$ does not increase when $N_i^x(t) > 0$, where as before

$$N_i^x(t) = \sum_{k \in \mathcal{C}_i} Q_k^x(t).$$

That is,

(3.6) $$\int_0^\infty N_i^x(t) \, dI_i^x(t) = 0.$$

Recall the constituency matrix C defined in Section 2. In vector form, (3.2)–(3.6) can be written as

(3.7) $$Q^x(t) = Q^x(0) + E^x(t) + \sum_{\ell=1}^K \Phi^\ell(S_\ell^x(T_\ell^x(t))) - S^x(T^x(t)),$$

(3.8) $Q^x(t) \geq 0, \quad t \geq 0,$

(3.9) $T^x(0) = 0,$ and $T^x(\cdot)$ is a non-decreasing,

(3.10) $I^x(t) = et - CT^x(t)$ is non-decreasing,

(3.11) $$\int_0^\infty CQ^x(t) \, dI^x(t) = 0,$$

where, as usual, $S^x(T^x(t)) = (S_1^x(T_1^x(t)), \ldots, S_K^x(T_K^x(t)))'$. Notice that (3.7)–(3.11) hold for FIFO, buffer priority disciplines, and the head-of-line processor sharing discipline that are considered in this paper. For each service discipline, there are additional equations that, together with (3.7)–(3.11), describe the network dynamics for the discipline. In the remainder of this section, we derive these extra equations for static buffer priority disciplines.

Under a buffer priority service discipline, one can envision that customers in class k wait in their own buffer. Customers in distinct buffers have different service priorities, and we assume that there are no ties among classes. Within each buffer, customers are served in FIFO discipline. For concreteness, we assume the discipline is preemptive resume. Let H_k denote the set of indices for all classes served at station $s(k)$ which have priority greater than or equal to that of class k, and let

$$T_k^{x,+}(t) = \sum_{\ell \in H_k} T_\ell^x(t)$$

$$I_k^{x,+}(t) = t - T_k^{x,+}(t),$$

$$Q_k^{x,+}(t) = \sum_{\ell \in H_k} Q_\ell^x(t).$$

Then $T_k^{x,+}(t)$ is the cumulative amount of service in $[0,t]$ dedicated to customers whose classes are included in H_k, and $I_k^{x,+}(t)$ is the total unused capacity that is available to serve customers whose class does not belong to H_k. Note that $I_i^x(t)$ is a station level quantity representing the total unused capacity in $[0,t]$ by server i; whereas $I_k^{x,+}(t)$ is a class level quantity. The priority service discipline requires that for every k, all the service capacity of station $s(k)$ is dedicated to classes in H_k, as long as the workload present in these buffers is positive. Thus we may express the additional condition by the integral equation

$$(3.12) \qquad \int_0^\infty Q_k^{x,+}(t)\, dI_k^{x,+}(t) = 0, \qquad 1 \leq k \leq K.$$

4. Fluid limit dynamics.

DEFINITION 4.1. A sequence of functions $f_n(\cdot) : \mathbb{R}_+ \to \mathbb{R}$ is said to be convergent to $f(\cdot)$ *uniformly on compact sets* (u.o.c.) if for every $t > 0$,

$$\sup_{0 \leq s \leq t} |f_n(s) - f(s)| \to 0$$

as $n \to \infty$.

In the following lemma, we present functional strong laws of large numbers for some processes defined. The proof can be found in Dai [8, Lemma 4.2]. Recall that a state $x = (\mathbb{Q}(0), U(0), V(0))$ has three components.

LEMMA 4.1. *Assume that*

$$\lim_{|x| \to \infty} \frac{1}{|x|} U(0) = \bar{U} \quad and \quad \lim_{|x| \to \infty} \frac{1}{|x|} V(0) = \bar{V}.$$

Then as $|x| \to \infty$, almost surely

$$(4.1) \qquad \frac{1}{|x|} \Phi^k([|x|t]) \to P_k' t, \quad u.o.c.$$

$$(4.2) \qquad \frac{1}{|x|} E_k^x(|x|t) \to \alpha_k (t - \bar{U}_k)^+, \quad u.o.c.$$

$$(4.3) \qquad \frac{1}{|x|} S_k^x(|x|t) \to \mu_k (t - \bar{V}_k)^+, \quad u.o.c.,$$

where [t] is the integer part of t.

LEMMA 4.2. *For a fixed sample path ω, for each sequence x_n of initial states with $|x_n| \to \infty$, there is a subsequence $\{x_{n_j}\}$ such that*

$$(4.4) \qquad \frac{1}{|x_{n_j}|}(T_1^{x_{n_j}}(|x_{n_j}|\,t,\omega),\ldots T_K^{x_{n_j}}(|x_{n_j}|\,t,\omega))$$
$$\to (\bar{T}_1(t,\omega),\ldots,\bar{T}_K(t,\omega))$$

uniformly (with respect to t) on compact sets as $j \to \infty$.

Proof. Let ω be a fixed sample path. Let $0 \le s < t$. It is easy to check that

$$\frac{1}{|x|}T_k^x(|x|\,t,\omega) - \frac{1}{|x|}T_k^x(|x|\,s,\omega) \le t - s.$$

The lemma then follows easily.   □

REMARK. In general the limit $\bar{T}(t,\omega) = (\bar{T}_1(t,\omega),\ldots,\bar{T}_K(t,\omega))'$ is random. That is $\bar{T}(t,\omega)$ may indeed depend on ω. However, from now on the dependence of ω is suppressed from the expression.

THEOREM 4.1. *Consider a non-idling service discipline. For almost all sample paths ω and any sequence of initial states $\{x_n\}$ with $|x_n| \to \infty$, there is a subsequence $\{x_{n_j}\}$ such that*

$$(4.5) \qquad \frac{1}{|x_{n_j}|}(Q^{x_{n_j}}(0), U^{x_{n_j}}(0), V^{x_{n_j}}(0)) \to (\bar{Q}(0), \bar{U}, \bar{V}),$$

$$(4.6) \qquad \frac{1}{|x_{n_j}|}(Q^{x_{n_j}}(|x_{n_j}|\,t), T^{x_{n_j}}(|x_{n_j}|\,t)) \to (\bar{Q}(t), \bar{T}(t)) \quad u.o.c.$$

Furthermore, $(\bar{Q}, \bar{T})$ satisfies the following set of equations.

$$(4.7) \qquad \bar{Q}(t) = \bar{Q}(0) + (\alpha t - \bar{U})^+ - (I - P)'M^{-1}(\bar{T}(t) - \bar{V})^+,$$

$$(4.8) \qquad \bar{Q}(t) \ge 0,$$

$$(4.9) \qquad \bar{T}(t) \text{ is non-decreasing and starts from zero,}$$

$$(4.10) \qquad \bar{I}(t) = et - C\bar{T}(t) \text{ is non-decreasing,}$$

$$(4.11) \qquad \int_0^\infty C\bar{Q}(t)\,d\bar{I}(t) = 0.$$

Proof. Recall first that $|x_n| = |Q^{x_n}(0)| + |U^{x_n}(0)| + |V^{x_n}(0)|$. Thus,

$$\frac{1}{|x_n|}|Q^{x_n}(0)| \le 1, \quad \frac{1}{|x_n|}|U^{x_n}(0)| \le 1, \quad \frac{1}{|x_n|}|V^{x_n}(0)| \le 1$$

for all n. Therefore by Lemma 4.2, there is a subsequence $\{n_j\}$ such that (4.5) and (4.4) hold. By Lemma 4.1 and (3.7), we have

$$\frac{1}{|x_{n_j}|}Q^{x_{n_j}}(|x_{n_j}|\,t) \to \bar{Q}(t),$$

where $\bar{Q}(t)$ satisfies (4.7). Conditions (4.8)–(4.10) follow from (3.8)–(3.10), whereas Condition (4.11) follows from (3.11) and Lemma 2.4 of Dai and Williams [13].   ☐

Theorem 4.1 holds for FIFO, buffer priority disciplines and the head-of-line processor sharing discipline, as well as many other non-idling disciplines. For a particular service discipline, a limit $(\bar{Q}(\cdot), \bar{T}(\cdot))$ usually satisfies more equations, in addition to (4.7)–(4.11). For a buffer priority discipline, let

$$\bar{Q}_k^+(t) = \sum_{\ell \in H_k} \bar{Q}_\ell(t),$$

$$\bar{T}_k^+(t) = \sum_{\ell \in H_k} \bar{T}_\ell(t).$$

THEOREM 4.2. *Under the static preemptive resume buffer priority discipline, the limit* $(\bar{Q}(t), \bar{T}(t))$ *in Theorem 4.1 satisfies*

(4.12) $$\int_0^\infty \bar{Q}_k^+(t) \, d(t - \bar{T}_k^+(t)) = 0, \quad k = 1, \ldots, K,$$

in addition to (4.7)–(4.11).

Proof. The theorem follows from (3.12) and Lemma 2.4 of [13].   ☐

DEFINITION 4.2. A limit $(\bar{Q}(\cdot), \bar{T}(\cdot))$ in Theorem 4.1 is called a *fluid limit* under a service discipline with initial fluid level $\bar{Q}(0)$ and *delays* $\bar{U}$ and $\bar{V}$. We use $\mathfrak{L}$ to denote the set of such limits.

DEFINITION 4.3. The *delayed fluid model* of a buffer priority service discipline in a network with delay $(\bar{U}, \bar{V}) \in \mathbb{R}_+^{|\mathcal{E}|+K}$ starting from $\bar{Q}(0)$ is defined to be the set of equations (4.7)–(4.12). Any solution $(\bar{Q}(\cdot), \bar{T}(\cdot))$ to equations (4.7)–(4.12) is called a *fluid solution* of the fluid model for the buffer priority discipline. We use $\mathfrak{M}$ to denote the collection of all solutions $(\bar{Q}(\cdot), \bar{T}(\cdot))$ of the fluid model.

REMARK. For a given discipline, one can define the corresponding delayed fluid model similarly. The only change needed is to replace (4.12) in Definition 4.3 with a condition analogous to (4.12) that is specific to the discipline.

It is obvious that any fluid limit is a fluid solution to the fluid model. Therefore, we have

(4.13) $$\mathfrak{L} \subset \mathfrak{M}.$$

When there is a single customer class served at each station, the fluid model has a unique solution. In this case, there is no need to differentiate between fluid limits and fluid solutions. However, in multiclass networks, it is quite typical that the fluid model has multiple solutions. The reason to distinguish a fluid limit from a fluid solution will be explained in the next section.

5. Stability of the fluid model and the queueing network.

DEFINITION 5.1.   The delayed fluid model of a service discipline is *stable* if there exists a $\delta > 0$ such that for any fluid solution $(\bar{Q}(\cdot), \bar{T}(\cdot)) \in \mathfrak{M}$ with $|\bar{Q}(0)| + |\bar{U}| + |\bar{V}| = 1$, $\bar{Q}(t) \equiv 0$ for $t \geq \delta$.

When the delays $\bar{U}$ and $\bar{V}$ are zeros, the corresponding delayed fluid model is called the *undelayed* fluid model, or simply the fluid model. That is, the undelayed fluid model is defined by adding the following equation to (4.8)–(4.11)

$$(5.1) \qquad \bar{Q}(t) = \bar{Q}(0) + \alpha t - (I - P)' M^{-1} \bar{T}(t).$$

If we let $|x| \to \infty$ while keeping $U(0)$ and $V(0)$ bounded, then the corresponding fluid limit is a solution to the undelayed fluid model. Assume that all distributions in the network are exponential. Because of the memoryless property of the distribution, for many service disciplines including FIFO and buffer priority disciplines, the residual interarrival times and service times are not needed in the state descriptions. Thus the corresponding fluid limit is always undelayed. However, under general distributional assumptions on the network, the corresponding fluid limit is delayed. Chen [5, Theorem 5.3] proved the following proposition.

PROPOSITION 5.1.   *If the undelayed fluid model is stable, then the delayed fluid model is stable.*

The emptying time δ in Definition 5.1 is independent of a particular fluid solution and initial state. Stolyar [26, Proposition 6] proved, however, that an apparent weaker condition implies the stability of a fluid model. The proof also follows from the remark following Proposition 3.3 of Dupuis and Williams [15].

PROPOSITION 5.2.   *The undelayed fluid model is stable if and only if there exists some norm $\| \cdot \|$ on $\mathbb{R}_+^K$ such that for any $\bar{Q} \in \mathfrak{M}$ with $\|\bar{Q}(0)\| = 1$, there exists $t > 0$ such that $\|\bar{Q}(t)\| < 1$.*

The following theorem was proved in Dai [8, Theorem 4.3].

THEOREM 5.1.   *A service discipline is positive Harris recurrent if the corresponding fluid limit model is stable.*

We conjecture that under certain service disciplines there are fluid solutions which cannot be achieved as fluid limits. It is therefore conceivable that there is a queueing network with certain service discipline, whose fluid limit model is stable but whose fluid model is not stable. This distinction is important in formulating a correct converse result of Theorem 5.1. The converse problem is still open although in many specific cases it has been shown that the instability of a fluid limit model implies the instability of the corresponding queueing network.

DEFINITION 5.2.   For a service discipline π in an open multiclass queueing network, its *stability region* is defined to be

$$(5.2) \quad \mathfrak{D}_\pi = \left\{ (\alpha, m) \in \mathbb{R}_+^{|\mathcal{E}|+K} : \begin{matrix} \text{such that the discipline } \pi \text{ is pos-} \\ \text{itive Harris recurrent} \end{matrix} \right\}.$$

REMARK. We believe the stability region for a non-idling discipline depends on α and m only, not on the second or higher moments.

Recall that, for concreteness, at each station one of the following service disciplines is employed: first-in–first-out (FIFO), static buffer priority among classes (both preemptive and nonpreemptive), and head-of-line processor sharing among classes. However, it is easy to see that more disciplines can be considered in the following definition.

DEFINITION 5.3. The *global stability region* for an open multiclass queueing network is defined to be

$$(5.3) \qquad \mathfrak{G} = \cap_\pi \mathfrak{D}_\pi,$$

where π ranges among all allowable service disciplines.

Similarly, using the stability notion defined in Definition 5.1, we can define the stability region $\bar{\mathfrak{D}}_\pi$ of a service discipline π for the fluid model and the global stability region $\bar{\mathfrak{G}}$ of the fluid model. Theorem 5.1 proves that

$$\bar{\mathfrak{D}}_\pi \subset \mathfrak{D}_\pi \quad \text{and} \quad \bar{\mathfrak{G}} \subset \mathfrak{G}.$$

Let

$$(5.4) \qquad \mathfrak{D}_0 = \left\{ (\alpha, m) \in \mathbb{R}_+^{|\mathcal{E}|+K} : \rho_i < 1, \quad i = 1, \ldots, d. \right\}$$

where $\rho = (\rho_1, \ldots, \rho_d)'$ is defined in (2.2). Obviously, we have

$$\mathfrak{D}_\pi \subset \mathfrak{D}_0 \quad \text{and} \quad \bar{\mathfrak{D}}_\pi \subset \mathfrak{D}_0.$$

Bramson [2,3] proved the surprising result that

$$\mathfrak{D}_{\text{FIFO}} \neq \mathfrak{D}_0.$$

Similar results were proved by Lu and Kumar [21], Rybko and Stolyar [24] and Seidman [25].

6. Calculus of fluid models. A nice feature of the fluid model is that one can apply calculus to it. It follows from (4.9) and (4.10) that $\bar{T}(\cdot)$ is Lipschitz continuous, and hence by (5.1) $\bar{Q}(\cdot)$ is Lipschitz continuous. Therefore we have the following proposition.

PROPOSITION 6.1. *The paths $\bar{Q}_k(\cdot)$ and $\bar{T}_k(\cdot)$ are absolutely continuous. Therefore they have derivatives almost everywhere with respect to the Lebesgue measure on $[0, \infty)$.*

A path $f(\cdot)$ is said to be regular at t if it is differentiable at t. We use $\dot{f}(t)$ to denote the derivative of $f(\cdot)$ at a regular point t.

The following simple lemma, which appeared in Dai and Weiss [12], turns out to be very useful for the analysis of stability of the fluid model.

LEMMA 6.1. *Let $f(\cdot)$ be an absolutely continuous nonnegative function on $[0, \infty)$.*

(i) If $f(t) = 0$ and $\dot{f}(t)$ exists, then $\dot{f}(t) = 0$.

(ii) Assume that for some $\epsilon > 0$ and almost every regular point $t > 0$, whenever $f(t) > 0$ then $\dot{f}(t) \leq -\epsilon$. Then $f(t) = 0$ for all $t \geq \delta$, where $\delta = f(0)/\epsilon$. Furthermore, $f(\cdot)$ is nonincreasing, and hence once it reaches zero it stays there forever.

Throughout this paper, whenever the derivative of the fluid model is considered at time t, we assume t is a regular point of $\bar{Q}(\cdot)$ and $\bar{T}(\cdot)$. Let $a \in \mathbb{R}_+^K$ be fixed. Define

$$G(t) = C\mathrm{diag}(a)(I - P')^{-1}\bar{Q}(t),$$

and

$$f(t) = \max\{G_1(t), \ldots, G_d(t)\}.$$

Note that $f(t)$ is a piecewise linear function of $\bar{Q}(t)$. It is easy to check that $f(t)$ is nonnegative and Lipschitz continuous and hence absolutely continuous.

THEOREM 6.1. If there exists an $a = (a_1, \ldots, a_K)' > 0$ and $\epsilon > 0$ such that whenever $|\bar{Q}(t)| > 0$, $\dot{f}(t) \leq -\epsilon$, then the service discipline is stable.

Proof. The proof follows from Lemma 6.1.   ☐

COROLLARY 6.1. Assume that in a two station network there exist $a = (a_1, \ldots, a_K)' > 0$ and $\epsilon_i > 0$ for $i = 1, 2$ such that $\dot{G}_i(t) \leq -\epsilon_i$ whenever $\bar{N}_i(t) > 0$. Furthermore, assume that $G_1(t) \leq G_2(t)$ whenever $\bar{N}_1(t) = 0$ and $G_2(t) \leq G_1(t)$ whenever $\bar{N}_2(t) = 0$, where $\bar{N}_i(t) = \sum_{k \in C_i} \bar{Q}_k(t)$. Then, the service discipline is stable.

7. Examples.

7.1. The Lu-Kumar-Bramson-type network.

A re-entrant line is a multiclass open queueing network, whose routing matrix is of the form $P_{k,k+1} = 1$ for $k = 1, \ldots, K-1$ and $P_{k,\ell} = 0$ otherwise and $\mathcal{E} = \{1\}$. Consider a two station re-entrant line, where all customers visit stations 1, 2, ..., 2, 1, ..., 1. Therefore, $s(1) = 1$, $s(k) = 2$ for $k = 2, \ldots, r$, and $s(k) = 1$ for $k = r+1, \ldots, K$. When $r = 3$ and $K = 4$, the resulting network is the Lu-Kumar network [21]. When $r = K-1$, the resulting network is the Bramson network [2]. We call this network a Lu-Kumar-Bramson-type network. For the Lu-Kumar network, assuming that $\alpha_1 = 1$, Dai and Weiss [12] showed that

$$\bar{\mathfrak{G}} = \{m \in \mathbb{R}_+^4 : m_1 + m_4 < 1, \quad m_2 + m_4 < 1, \quad m_2 + m_3 < 1\}.$$

Furthermore, the global stability region is realized by the Lu-Kumar buffer priority discipline that gives classes 2 and 4 higher priorities. That is,

$$\bar{\mathfrak{G}} = \bar{\mathfrak{D}}_{\mathrm{Lu-Kumar-buffer-priority}}.$$

Therefore, the Lu-Kumar priority discipline is the "worst" one in the sense that it gives the smallest stability region. The approach used in [12] was modified from Botvitch and Zamyatin [1]. The method has been further generalized by Dai and VandeVate [10]. For the Lu-Kumar-Bramson type networks, using Corollary 6.1, they proved that

$$(7.1) \quad \left\{ m \in \mathbb{R}_+^K : \rho_1 < 1, \quad \rho_2 < 1, \quad \sum_{k=2}^{r-1} m_k + \sum_{k=r+1}^{K} m_k < 1 \right\} \subset \bar{\mathfrak{G}}.$$

Now consider the following buffer priority discipline, which generalizes the Lu-Kumar priority discipline. At station 1 priorities, in decreasing order, are given as K, $K-1$, ..., $r+1$ and 1. At station 2 priorities are given as $r-1, r-2, \ldots, 2$ and r. Under this buffer priority discipline, the network is effectively reduced to the original Lu-Kumar network with a new set of parameters $\tilde{m} = (\tilde{m}_1, \tilde{m}_2, \tilde{m}_3, \tilde{m}_4)' \in \mathbb{R}_+^4$, where

$$\tilde{m}_1 = m_1, \quad \tilde{m}_2 = \sum_{k=2}^{r-1} m_k, \quad \tilde{m}_3 = m_r, \quad \tilde{m}_4 = \sum_{k=r+1}^{K} m_k.$$

Therefore, we can apply results on Lu-Kumar network in Dai and Weiss [12, Section 5] to ensure

$$\bar{\mathfrak{G}} = \left\{ m \in \mathbb{R}_+^K : \rho_1 < 1, \rho_2 < 1, \sum_{k=2}^{r-1} m_k + \sum_{k=r+1}^{K} m_k < 1 \right\}.$$

The worst discipline is this generalized Lu-Kumar buffer priority discipline. For the queueing network under discussion, consider the generalized Lu-Kumar buffer priority preemptive resume discipline. Following Theorem 5.1 and (7.1), we have

$$\left\{ m \in \mathbb{R}_+^K : \rho_1 < 1, \quad \rho_2 < 1, \quad \sum_{k=2}^{r-1} m_k + \sum_{k=r+1}^{K} m_k < 1 \right\} \subset \mathfrak{G}.$$

Harrison [17] made the following key observation. Suppose the network initially has customers only in buffers 1 and r. Then no two classes different from 1 and r can ever be worked on simultaneously. Consequently,

$$\sum_{k=2}^{r-1} T_k^x(t) + \sum_{k=r+1}^{K} T_k^x(t) \le t, \quad t \ge 0.$$

Using this observation, it is not difficult to argue that

$$\left\{ m \in \mathbb{R}_+^K : \rho_1 < 1, \quad \rho_2 < 1, \quad \sum_{k=2}^{r-1} m_k + \sum_{k=r+1}^{K} m_k < 1 \right\} = \mathfrak{G}.$$

When the service times at station 3 in the network pictured in Figure 1.1 are zero, the corresponding network is reduced to a Lu-Kumar-Bramson-type network with $r = 4$ and $K = 6$. Therefore $m_k = 0.25$, $k = 1, \ldots, 6$, is

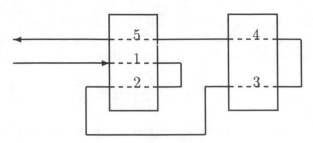

FIG. 7.1. *The Dai-Wang network*

the critical value for stability when all service times have the same mean. This explains why the network in Figure 1.1 is unstable when service time at station 3 are small.

7.2. The Dai-Wang network. For the network pictured in Figure 7.1, again by using Corollary 6.1 and assuming $\alpha_1 = 1$, Dai and Vande-Vate [10] showed that

$$\{m \in \mathbb{R}_+^5 : \rho_1 < 1, \quad \rho_2 < 1, \quad m_5 < (1 - m_1)(1 - m_3)\} \subset \bar{\mathfrak{G}}.$$

Now consider the priority discipline giving priorities, in decreasing order, at station one: 5, 1, 2 and at station two: 3, 4. We show, using the construction similar to the one in [12], that if $\rho_1 < 1$, $\rho_2 < 1$ and

$$(1 - m_1)(1 - m_3) \leq m_5,$$

the discipline is not stable for the fluid model. Indeed, let the fluid model start from $\bar{Q}(0) = (1, 0, 0, 0, 0)'$. Then at $t_1 = 1/(\mu_1 - 1)$, $\bar{Q}(t_1) = (0, 1/(1 - m_1), 0, 0, 0)'$. Let t_2 be the first time that the fluid level at buffer 2 reaches zero. Let ν be the departure rate from buffer 2 between t_1 and t_2. It follows from Dai and Weiss [12, Proposition 3.1] that

$$\nu = (1 - m_1)\mu_2.$$

We first check that $\nu > \mu_3$. In fact, if $\nu \leq \mu_3$, we have

$$m_1 + m_2 + m_5 \geq m_1 + m_5 + \frac{1}{1 - m_1}\frac{1}{m_3}$$

$$\geq m_1 + \frac{1}{1 - m_1} \geq 1,$$

contradicting $\rho_1 < 1$. It is easy to check that $t_2 - t_1 = (1/(\nu - 1))(1/(1 - m_1))$ or equivalently,

$$t_2 = \frac{\nu}{\nu - 1}\frac{1}{\mu_1 - 1} + \frac{1}{\nu - 1}.$$

and

$$\begin{aligned}\bar{Q}(t_2) &= (0,0,1+t_2-\mu_3(t_2-t_1),\mu_3(t_2-t_1),0)' \\ &= (0,0,(\nu-\mu_3)(1/(\nu-1))(1/(1-m_1)), \\ &\qquad \mu_3(1/(\nu-1))(1/(1-m_1)),0)'.\end{aligned}$$

Let t_3 be the first time that buffer 3 reaches zero. At t_3,

$$\bar{Q}(t_3) = (0,0,0,1+t_3,0)',$$

where

$$t_3 - t_2 = \frac{(\nu-\mu_3)(1/(\nu-1))(\mu_1/(\mu-1))}{\mu_3-1}.$$

We can check that

$$\begin{aligned}1+t_3 &= t_3-t_2+t_2+1 \\ &= \frac{1}{\mu_3-1}\frac{\mu_1}{\mu_1-1}\frac{\nu}{\nu-1}-\frac{\mu_3}{\mu_3-1}\frac{\mu_1}{\mu_1-1}\frac{1}{\nu-1} \\ &\quad +\frac{1}{\mu_1-1}\frac{\nu}{\nu-1}+\frac{1}{\nu-1}+1 \\ &= \frac{\mu_3}{\mu_3-1}\frac{\mu_1}{\mu_1-1}=\frac{1}{1-m_1}\frac{1}{1-m_3}.\end{aligned}$$

Let $t_4-t_3 = m_5/((1-m_1)(1-m_3))$. Then at t_4,

$$\bar{Q}(t_4) = (m_5/((1-m_1)(1-m_3)),0,0,0,0)'.$$

From t_4 the solution enters a new cycle with initial total fluid level

$$m_5/((1-m_1)(1-m_3)) \geq 1.$$

Hence the discipline is unstable for the fluid model. Therefore we have proved that

$$\bar{\mathfrak{G}} = \left\{m \in \mathbb{R}_+^5 : \rho_1 < 1, \rho_2 < 1, m_5 < (1-m_1)(1-m_3)\right\},$$

and the worst discipline is the buffer priority discipline given earlier.

In [11], Dai and Wang showed that when $m_1 = 0.1\alpha_1$, $m_2 = 0.05\alpha_1$, $m_3 = 0.9\alpha_1$, $m_4 = 0.05\alpha_1$ and $m_5 = 0.8\alpha_1$, the corresponding Brownian model proposed by Harrison and Nguyen [18] does not exist under the FIFO discipline. Specializing the stability region to this case, we have

$$\bar{\mathfrak{G}} = \left\{\alpha_1 \geq 0 : \alpha_1 < 10\left(1-2\sqrt{2}/3\right) \approx 0.57191\right\}.$$

By solving an LP problem numerically and performing computer simulations, Down and Meyn [14] were able to predict the critical value to be

0.57191. By using a different LP, quadratic LP, Kumar and Meyn [20] had shown earlier that

$$\{\alpha_1 \geq 0 : \alpha_1 < 0.55587\} \subset \bar{\mathfrak{G}}.$$

The quadratic LP solution captures the greater part of the stability region. However, it was demonstrated that the quadratic LP method of Kumar and Meyn cannot give a sharp region. It is interesting to note that the critical utilization rates at both stations are

$$\rho_1 = \rho_2 < 9.5 \left(1 - 2\sqrt{2}/3\right) \approx 0.543314,$$

which is far below one.

7.3. Re-entrant line without immediate feedback. When there is no immediate feedback in a two station re-entrant line, the station visitation sequence takes the simple form: 1, 2, 1, 2, ... It was shown in Dai and Weiss [12, Section 3] that for $K = 3$ when

$$\rho_1 = \alpha_1(m_1 + m_3) < 1 \quad \text{and} \quad \rho_2 = \alpha_1 m_2 < 1,$$

the conditions in Corollary 6.1 hold for any non-idling discipline. Therefore, we have for any non-idling discipline π,

$$\bar{\mathfrak{D}}_\pi = \bar{\mathfrak{G}} = \mathfrak{D}_0,$$

where $\mathfrak{D}_0$ is defined in (5.4). The argument was generalized in Dai and VandeVate [10] to networks with $K = 4$,

$$\bar{\mathfrak{G}} = \mathfrak{D}_0.$$

In both cases, we also have

$$\mathfrak{G} = \mathfrak{D}_0.$$

8. Moments. So far we have shown that the stability of a fluid model implies the positive Harris recurrence for the Markov process in the corresponding queueing network. In this section, we show that under some stronger moment conditions on interarrival and service times, we can obtain some stronger stability results for the queueing network. In particular, we can bound the moments of queue lengths, which are the primary performance measures of a network. Assume that

(A2') For some integer $p \geq 1$, $E[\xi_\ell(1)^{p+1}] < \infty$ for $\ell \in \mathcal{E}$ and $E[\eta_k(1)^{p+1}] < \infty$ for $k = 1, \ldots, K$.

Recently, Dai and Meyn [9] proved the following result.

THEOREM 8.1. *Assume that the fluid model for a service discipline is stable, and that (A1) and (A2') hold. Then*

(i) *For some constant κ_p, and for each initial condition $x \in X$,*

$$\limsup_{t \to \infty} \frac{1}{t} \int_0^t \mathsf{E}_x[|Q(t)|^p] \, ds \leq \kappa_p,$$

where p is the integer used in (A2').
Assume further that (A3) holds. Then, the service discipline is stable with stationary distribution π, and moreover, for each initial condition,
(ii) *The transient moments converge to their steady state values: for $r = 1, \ldots, p$, $k = 1, \ldots, K$,*

$$\lim_{t \to \infty} \mathsf{E}_x[Q_k(t)^r] = \mathsf{E}_\pi[Q_k(0)^r] \leq \kappa_r.$$

(iii) *The first moment converges at rate t^{p-1}:*

$$\lim_{t \to \infty} t^{(p-1)} |\mathsf{E}_x[Q(t)] - \mathsf{E}_\pi[Q(0)]| = 0.$$

(iv) *The strong law of large numbers holds: for $r = 1, \ldots, p$, $k = 1, \ldots, K$,*

$$\lim_{t \to \infty} \frac{1}{t} \int_0^t Q_k^r(s) \, ds = \mathsf{E}_\pi[Q_k(0)^r], \quad P_x\text{-}a.s.$$

9. Concluding remarks. For any given service discipline, it is a challenging problem to *characterize* its stability region. It appears that the piecewise linear Lyapunov function used in Botvitch and Zamyatin [1], Dai and Weiss [12] and Down and Meyn [14] is a powerful tool in determining a global stability region. For a buffer priority discipline, Down and Meyn [14] were also able to apply the piecewise linear Lyapunov function technique. For the FIFO discipline, we refer readers to Chen and Zhang [7] and Bramson [4] for the latest developments.

Acknowledgments. I am grateful to Mike Harrison for his numerous comments on an earlier version of this paper. I thank Gideon Weiss and Sean Meyn for allowing me to cite recent joint work with them in this paper. I also thank an anonymous referee for suggesting many helpful improvements.

REFERENCES

[1] D. D. BOTVITCH AND A. A. ZAMYATIN, *Ergodicity of conservative communication networks.* Rapport de recherche 1772, INRIA, October 1992.
[2] M. BRAMSON, *Instability of FIFO queueing networks*, Annals of Applied Probability, (to appear).
[3] ———, *Instability of FIFO queueing networks with quick service times*, Annals of Applied Probability, (to appear).
[4] ———, *Private communications.* 1994.
[5] H. CHEN, *Fluid approximations and stability of multiclass queueing networks I: Work-conserving disciplines*, Annals of Applied Probability, (Submitted).

[6]   H. CHEN AND A. MANDELBAUM, *Hierarchical modeling of stochastic networks, Part I: fluid models*, In D.D. Yao (ed.), Applied Probability in Manufacturing Systems, (forthcoming).

[7]   H. CHEN AND H. ZHANG, *Fluid approximations and stability of multiclass queueing networks II: FIFO discipline*. in preparation.

[8]   J. G. DAI, *On positive Harris recurrence of multiclass queueing networks: A unified approach via fluid limit models*, Annals of Applied Probability, (to appear).

[9]   J. G. DAI AND S. P. MEYN, *Stability and convergence of moments for multiclass queueing networks via fluid limit models*, IEEE Transactions on Automatic Control, (submitted).

[10] J. G. DAI AND J. VANDEVATE, *Characterizing stability region for heterogeneous fluid models*. in preparation.

[11] J. G. DAI AND Y. WANG, *Nonexistence of Brownian models of certain multiclass queueing networks*, Queueing Systems: Theory and Applications, 13 (1993), pp. 41–46.

[12] J. G. DAI AND G. WEISS, *Stability and instability of fluid models for certain re-entrant lines*, Mathematics of Operations Research, (submitted).

[13] J. G. DAI AND R. J. WILLIAMS, *Existence and uniqueness of semimartingale reflecting Brownian motions in convex polyhedrons*, Theory of Probability and its Applications, (to appear).

[14] D. DOWN AND S. MEYN, *Piecewise linear test functions for stability of queueing networks*, Proceedings of the 33rd Conference on Decision and Control, (1994). submitted.

[15] P. DUPUIS AND R. J. WILLIAMS, *Lyapunov functions for semimartingale reflecting Brownian motions*, Annals of Probability, (to appear).

[16] J. M. GU, *Convergence and Performance for some Kelly-like Queueing Networks*, PhD thesis, University of Wisconsin, Madison, 1994.

[17] J. M. HARRISON, *Private communications*. 1994.

[18] J. M. HARRISON AND V. NGUYEN, *Brownian models of multiclass queueing networks: Current status and open problems*, Queueing Systems: Theory and Applications, 13 (1993), pp. 5–40.

[19] F. P. KELLY, *Networks of queues with customers of different types*, J. Appl. Probab., 12 (1975), pp. 542–554.

[20] P. R. KUMAR AND S. MEYN, *Duality and linear programs for stability and performance analysis of queueing networks and scheduling policies*. Preprint.

[21] S. H. LU AND P. R. KUMAR, *Distributed scheduling based on due dates and buffer priorities*, IEEE Transactions on Automatic Control, 36 (1991), pp. 1406–1416.

[22] S. P. MEYN AND D. DOWN, *Stability of generalized jackson networks*, Annals of Applied Probability, 4 (1994), pp. 124–148.

[23] S. P. MEYN AND R. L. TWEEDIE, *Stability of Markovian processes II: continuous time processes and sample chains*, Advances of Applied Probability, 25 (1993), pp. 487–517.

[24] A. N. RYBKO AND A. L. STOLYAR, *Ergodicity of stochastic processes describing the operation of open queueing networks*, Problems of Information Transmission, 28 (1992), pp. 199–220.

[25] T. I. SEIDMAN, *'First come, first served' can be unstable!*, IEEE Transactions on Automatic Control, (to appear).

[26] A. STOLYAR, *On the stability of multiclass queueing networks*. Submitted to the Proceeding of Second Conference on Telecommunication Systems—Modeling and Analysis, Nashville, March 22–27, 1994.

[27] W. WHITT, *Large fluctuations in a deterministic multiclass network of queues*, Management Sciences, 39 (1993), pp. 1020–1028.

ON OPTIMAL DRAINING OF RE-ENTRANT FLUID LINES

GIDEON WEISS*

Abstract. Recent results have shown a close connection between stability of stochastic networks and stability of their fluid models. A converse to the problem of stability is the question of optimal control of such networks. This motivates our search for optimal control of fluid models. In this paper we consider re-entrant lines in which parts move in a fixed route that may revisit machines several times. We pose several problems of optimal control for the fluid models of such lines.

Key words. queueing networks, re-entrant lines, fluid models, scheduling, optimal control, continuous linear programming.

1. Model. Consider a system in which a fluid is moving through a sequence of K buffers, initially entering buffer 1 from outside, moving in succession from buffer k to $k+1$, $k = 1, \ldots, K-1$, and finally exiting from buffer K; we will when convenient denote the outside source of the flow as buffer 0 and its sink as buffer $K+1$. Stations numbered $1, \ldots, I$ pump the fluid from one buffer to the next, where buffer k is pumped by station $i = \sigma(k)$; the constituency of station i is $C_i = \{k : \sigma(k) = i\}$ — we can imagine that buffer k is located at station $\sigma(k)$, and C_i is the collection of buffers located at station i. Note that while the fluid is following a fixed route of buffers (a flow-line) it can visit the same station more than once along the route. We assume that fluid is entering the system at a rate α, and that the pumping capacity of station i for buffer k (with $\sigma(k) = i$, and $k \in C_i$) is given by a pumping rate of μ_k: Pumping station i will move μ_k units of fluid from buffer k to buffer $k+1$, per unit time, if all the capacity of station i is devoted to buffer k. Let $m_k = 1/\mu_k$, then m_k is the effort necessary to pump one unit of fluid through buffer k. If pumping station i is dividing its pumping effort between the buffers of C_i, so that it pumps out of buffer k at a rate of u_k units of fluid per unit time then the total effort of station i is $\sum_{k \in C_i} m_k u_k$ and this will need to be less or equal to the normalized capacity 1.

This model has a queueing network analog in which discrete customers or parts move through a sequence of service steps or operations, some of which may require the same server or machine. Kumar [10] introduced the name re-entrant line for this model because the customers or parts may re-enter the same server or machine several times. Re-entrant lines occur in practice in manufacturing systems, in particular in semi-conductor wafer fabrication. Recent results of Rybko and Stolyar [17,18], Dai [6] and Chen

---

* School of Industrial and Systems Engineering, Georgia Tech, Atlanta, GA 30332-0205. Research supported by NSF grants DDM-8914863 and DDM-9215233, and the fund for the promotion of research at the Technion. Part of this work was done while the author was visiting the *Institute For Mathematics And Its Applications*, whose hospitality is gratefully acknowledged.

[4] show that such queueing networks are stable if their fluid limits are stable. This motivates us to look at the fluid model above.

Let $x_k(t)$ denote the fluid level in buffer k, $k = 1, \ldots, K$, at time t. The evolution of the fluid re-entrant line is described by the following equations and constraints:

$$\dot{x}_1 = \alpha - u_1$$

(1.1)
$$\dot{x}_k = u_{k-1} - u_k \qquad k = 2, \ldots, K$$

where:

$$x_k(0) = a_k, \qquad k = 1, \ldots, K, \text{ (initial state)}$$
$$x \geq 0, \qquad u \geq 0,$$

(1.2)
$$\sum_{k \in C_i} m_k u_k \leq 1, \qquad i = 1, \ldots, I$$

The following condition is necessary if this system is ever to reach state $x = 0$:

(1.3)
$$\alpha \sum_{k \in C_i} m_k \leq 1, \qquad i = 1, \ldots, I$$

It was shown by Dai and Weiss and by Kumar and Kumar [7,11] that if the fluid line is operated by the *LBFS* (last buffer first served) priority policy or by the *FBFS* (first buffer first served) priority policy, then under the condition (1.3) the system will indeed reach an empty state $x = 0$ in a finite time. On the other hand, one can demonstrate [13,7,3], networks for which FIFO (first in first out) policy, SPT (shortest processing time first) policy, or some other fixed buffer priority policy will fail to empty the system even when condition (1.3) holds.

In this paper we want to address the question of choosing the control flow rates u so as to optimize various cost functions in the process of draining the system.

We pose:

Problem 1: Empty the system in minimal time,

$$\min \int_0^\infty 1_{x \neq 0} dt.$$

Problem 2: Minimize the inventory at a target date T,

$$\min \sum_{k=1}^K x_k(T).$$

Problem 3: Minimize the holding costs at a target date T [5],

$$\min \sum_{k=1}^K c_k x_k(T).$$

Problem 4: Minimize average inventory (WIP) and flowtime,

$$\min \int_0^\infty \sum_{k=1}^K x_k(t)dt.$$

Problem 5: Empty the system in minimal holding costs,

$$\min \int_0^\infty \sum_{k=1}^K c_k x_k(t)dt,$$

where $c_k \geq 0$ is the holding cost rate for fluid in buffer k.

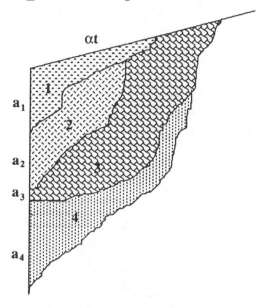

FIG. 1.1. *The Dynamics of a Re-Entrant Line*

Before discussing any of these problems we introduce the following graphical device (Figure 1.1), to describe the dynamics of a re-entrant line: We plot the $K + 1$ lines $\sum_{j=1}^K a_j + \alpha t$, $\sum_{j=k}^K a_j + \int_0^t u_{k-1}(s)ds$, $k = 2, \ldots, K$, and $\int_0^t u_K(s)ds$. Then the k'th line at time t will represent the total amounts of fluid which were initially present and which have flowed up to time t into the subsystem consisting of buffers $k, \ldots, K + 1$. The differences between these lines, or the height of the horizontal bands between these lines are the levels of fluid in each of the successive buffers, as given by the defining relation (1.1), and in particular the vertical intervals between the lines at time 0 are the initial levels $x_k(0), k = 1, \ldots, K$. The lines are monotone non-decreasing corresponding to $u \geq 0$, and do not cross corresponding to $x \geq 0$. The constraint that $\sum_{k \in C_i} m_k u_k \leq 1$ for $i = 1, \ldots, I$ imposes restrictions on the slopes of the lines.

Note that if buffer k is empty then the k'th and the $k + 1$'th lines coincide. This does not mean that $u_k = 0$. To keep the buffer empty it is necessary to pump it out at the same rate at which it is filling up, so *buffer k has positive flow without accumulation*. Zero flow corresponds to a horizontal line.

2. Draining in minimal time. If the system is empty at time t the total pumping effort of station i must have been $\sum_{k \in C_i} \left(\sum_{j=1}^{k} a_j + \alpha t \right) m_k$, for all $i = 1, \ldots, I$. Since each station has capacity 1, we obtain:

$$(2.1) \qquad t \geq \sum_{k \in C_i} \left(\sum_{j=1}^{k} a_j + \alpha t \right) m_k, \qquad i = 1, \ldots, I$$

Solving this as equality for each i we obtain:

$$(2.2) \qquad t_i = \frac{\sum_{k \in C_i} m_k \sum_{j=1}^{k} a_j}{1 - \alpha \sum_{k \in C_i} m_k}, \qquad i = 1, \ldots, I$$

and a lower bound on the earliest emptying time is:

$$(2.3) \qquad t^* = \max_{i=1,\ldots,I} t_i$$

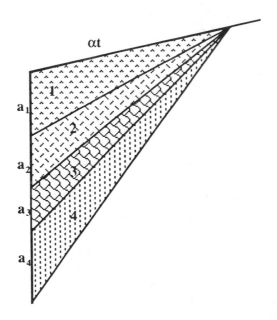

FIG. 2.1. *Draining in Minimal Time, by Constant Controls*

PROPOSITION 2.1. *It is possible to empty the system by time t^*.*

Proof. Fix the controls for all $0 \leq t \leq t^*$ at the constant levels (see Figure 2.1):

$$(2.4) \qquad u_k = \frac{\sum_{j=1}^{k} a_j + \alpha t^*}{t^*}, \qquad k = 1, \ldots, K$$

It is easy to verify that this defines a feasible control that drains all the buffers simultaneously precisely at t^*.   □

Replacing t^* by $t' \geq t^*$ in (2.4), will drain the system in time t'.

The above solution to draining the system in minimum time is obviously not unique. It is also not a buffer priority policy. Dai and Weiss as well as Lu and Kumar [7,13] show that FBFS and LBFS priority policies are stable in the sense that condition (1.3) is both necessary and sufficient for the system to drain in a bounded amount of time. In contrast they also show that in some situations other buffer priority policies may not drain the system at all, even when condition (1.3) holds. It is worth noting that, as the next two examples show, stable buffer priority policies such as FBFS and LBFS are not particularly adept at draining the system in a short time.

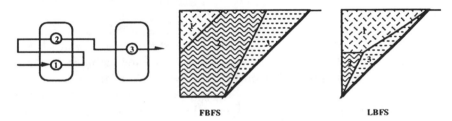

FIG. 2.2. *FBFS Does Not Drain in Minimal Time*

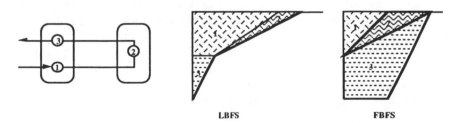

FIG. 2.3. *LBFS Does Not Drain in Minimal Time*

Example 2.2. Three buffers, two servers, $C_1 = \{1, 2\}$, $C_2 = \{3\}$, $\alpha = 0$, $m = (1, 0.5, 1)$, $a = (1, 1, 0)$ (see Figure 2.2):

$$t_{FBFS} = 3, \qquad t^* = t_{LBFS} = 2$$

Example 2.3. Three buffers, two servers, $C_1 = \{1,3\}$, $C_2 = \{2\}$, $\alpha = 0$, $m = (1,2,0.5)$, $a = (1,0,1)$ (see Figure 2.3):

$$t_{LBFS} = 2.5, \qquad t^* = t_{FBFS} = 2$$

3. The single server re-entrant line. The complexities of the re-entrant line manifest themselves only when there are more than one station pumping simultaneously. Before we tackle those complexities, we consider in this section the simpler case where there is a single common server for all the buffers, $I = 1$. We show:

PROPOSITION 3.1. *The LBFS buffer priority policy will minimize the inventory $\sum_{k=1}^{K} x_k(t)$ of the single station re-entrant line pointwise at every t.*

Proof. The LBFS policy will empty the buffers of the system one after the other, in the order of K first, $K-1$ second, and so on down to buffer 1. After buffers $k+1, \ldots, K$ have been emptied, buffer k is emptied by pushing the fluid out of buffer k and directly through buffers $k+1, \ldots, K$ out of the system; this is done at rate $u_j = \frac{1}{m_k + m_{k+1} + \cdots + m_K}$ for $j = k, \ldots, K$, while $u_j = 0$, $j = 1, \ldots, k-1$ (see Figure 3.1).

Let τ_k denote the time at which buffer k is emptied (and it remains empty thereafter), for $k = 2, \ldots, K$, let τ_1 equal the time at which buffer 1 is emptied of all the fluid that was in it initially, and let τ_0 be the time at which the whole system is empty. Then $\tau_k - \tau_{k-1} = a_k(m_k + m_{k+1} + \cdots + m_K)$, $k = 1, \ldots, K$, and $\tau_0 = \frac{\tau_1}{1-\alpha(m_1 + \cdots + m_K)}$. The total cumulative quantity of fluid pumped out of the system by time t is therefore a piecewise linear continuous function with slope $\frac{1}{m_k + m_{k+1} + \cdots + m_K}$ in the time interval (τ_{k+1}, τ_k), $k = K, \ldots, 1$, slope $\frac{1}{m_1 + \cdots + m_K}$ in the interval (τ_1, τ_0), and slope α thereafter. As we saw, τ_0 is the minimal time to empty the system. We will prove the theorem by showing that at any time t the cumulative quantity pumped out of the system under LBFS is at least as large as under any other policy.

The proof is by induction on the number of buffers. For $K = 0$ or $K = 1$ there is nothing to prove, so we now assume the theorem holds for less than K buffers and prove it for K. By the minimality of τ_0, the amount drained by LBFS is pointwise maximal for $t > \tau_0$. Consider the straight line:

$$l(t) = \sum_{k=1}^{K} a_k + \alpha\tau_0 + \frac{t - \tau_0}{m_1 + \cdots + m_K}$$

which coincides over the interval (τ_2, τ_0) with the amount drained from the system by LBFS. We claim that for any $t \leq \tau_0$ no policy can drain

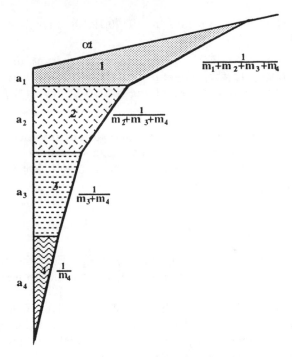

FIG. 3.1. *Draining of a Single Server Re-Entrant Line, by LBFS*

more than $l(t)$. Assume otherwise. Then the system could continue to be drained at a rate of at least $\frac{1}{m_1 + \cdots + m_K}$ (this is the rate if all the fluid is in buffer 1, otherwise a faster rate is possible), and so the system could be emptied before τ_0, which contradicts the minimality of τ_0.

Consider now the time up to τ_2. Consider a system in which buffer 1 has been removed, and its initial fluid placed in buffer 2 (which starts with $a_1 + a_2$).

The minimal inventory for this system is certainly no more than for the original system, since the new system requires less pumping from the server. However one can see that up to time τ_2 the amount of fluid pumped out of the new system under LBFS coincides with that of the original system. Furthermore, by the induction hypothesis this amount is pointwise maximal up to time τ_2. This completes the proof. □

Obviously, LBFS is optimal for Problems 1,2 and 4.

For general holding cost rates c_k, $k = 1, \ldots, K$ we define for $k < l$:

$$\nu_{k,l} = \frac{c_k - c_l}{m_k + \cdots + m_{l-1}}$$

$$\nu_k = \max_{l > k} \nu_{k,l}$$

This index is an example of a Gittins index [8], and the following proposition is a special case of the Klimov model [9,12,19] of single server queues with feedback:

PROPOSITION 3.2. *For a single station re-entrant line the policy which gives priority to the non-empty buffer with the highest index ν_k, $k = 1, \ldots, K$, will minimize the holding costs $\sum_{k=1}^{K} c_k x_k(t)$ pointwise at every t.*

4. A pointwise bound and optimization of inventory at a target date. Consider a general re-entrant line, and examine for each server (station) $i = 1, \ldots, I$ the single server re-entrant line consisting only of the buffers $k \in C_i$, with initial state

$$\tilde{a}_k = \sum_{j=\eta(k)+1}^{k} a_j, \qquad k \in C_i,$$

where $\eta(k) = \max(0, \max\{j : j \in C_{\sigma(k)}, j < k\})$. Equivalently, for each station i consider the original re-entrant line with infinite pumping rates at all the stations $\neq i$.

Define $L_i(t)$, $t \in [0, t^*]$ to be the line of cumulative out flow from the last buffer of station i, when LBFS is used. To that end, let $\tilde{u}_k(t)$ be the outflow rate from buffer $k \in C_i$, under LBFS, and let $K(i) = \max\{k : k \in C_i\}$ be the last buffer served by station i, then:

$$L_i(t) = \sum_{k=K(i)+1}^{K} a_k + \int_0^t \tilde{u}_{K(i)}(s)ds.$$

We now argue that each $L_i(t)$ is a pointwise upper bound on the cumulative out flow of the original system:

$$\sum_{k=1}^{K} x_k(t) \geq \sum_{k=1}^{K(i)} x_k(t) \geq \qquad \text{(by Proposition 3.1)}$$

$$\geq \sum_{k=1}^{K(i)} a_k + \alpha t - \int_0^t \tilde{u}_{K(i)}(s)ds = \sum_{k=1}^{K} a_k + \alpha t - L_i(t)$$

Hence (see Figure 4.1)

$$\overline{L}(t) = \min_i L_i(t)$$

provides a pointwise upper bound on the amount pumped out of the system, and $\sum_{k=1}^{K} a_k + \alpha t - \overline{L}(t)$ is a pointwise lower bound on the inventory, $\sum_{k=1}^{K} x_k(t)$ for every t.

PROPOSITION 4.1. *For every T it is possible to achieve the inventory:*

$$\sum_{k=1}^{K} x_k(T) = \sum_{k=1}^{K} a_k + \alpha T - \overline{L}(T)$$

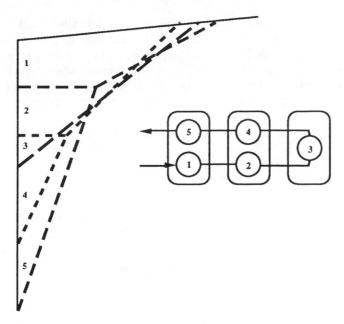

FIG. 4.1. *A Pointwise Bound on Inventory*

Proof. Let k^* be the buffer such that:

$$\sum_{k=k^*+1}^{K} a_k < \overline{L}(T) \le \sum_{k=k^*}^{K} a_k$$

if such exists, and $k^* = 1$ otherwise. Define, for $0 \le t \le T$

(4.1) $\qquad u_k(t) = \begin{cases} 0 & k < k^* \\ (\overline{L}(T) - \sum_{j=k+1}^{K} a_j)/T & k \ge k^* \end{cases}$

It is easily seen that these controls are feasible, and will reach the bound value at the target date T $\qquad\qquad\qquad\qquad\qquad$ □

This solves problem 2.

When there are holding cost rates c_k, one can construct similar point-wise bounds for the holding cost rates for the inventory at time t, for any t, and again it is possible to achieve these bounds for any target date T. This is carried out by Chen and Yao [5]. This solves problem 3.

It is unfortunately not possible in general to achieve these bounds simultaneously for all t.

5. Minimizing average inventory. We now consider the objective function:

(5.1) $\qquad\qquad\qquad V = \min \int_0^T \sum_{k=1}^{K} x_k(t)dt.$

i.e. we want to minimize the total costs on inventory, when each unit of fluid has unit cost, irrespective of the buffer. Equivalently this minimizes the average inventory (WIP — work in process) or, by Little's formula, the average flow time (time it takes fluid to move through the line). If (1.3) holds, then presumably for T large enough the optimal solution will drain the system by time T, and will remain optimal for all $t > T$ or when T is replaced by ∞.

Integrating by parts, and using the motion equations (1.1) one obtains:

$$(5.2) \qquad V = \sum_{k=1}^{K} a_k T + \frac{1}{2}\alpha T^2 - \max \int_0^T (T-t)u_K(t)dt.$$

We rewrite the constraints (1.2) to obtain the following optimization problem: Find control flow rates $u_1(t), \ldots, u_K(t)$, $0 \le t \le T$ so as to

$$(5.3) \qquad \max \int_0^T (T-t)u_K(t)dt$$

Subject, for all $t \in [0, T]$, to:

$$(5.4) \qquad \int_0^t u_1(s)\,ds \quad \le \quad a_1 + \alpha t,$$

$$(5.5) \qquad \int_0^t (u_k(s) - u_{k-1}(s))\,ds \quad \le \quad a_k, \qquad k = 2, \ldots, K$$

$$(5.6) \qquad \sum_{k \in C_i} m_k u_k(t) \quad \le \quad 1, \qquad i = 1, \ldots, I$$

$$(5.7) \qquad u_k(t) \quad \ge \quad 0 \qquad k = 1, \ldots, K.$$

We now present the solution for a 3 buffer example. From this it is seen that it is not always possible to achieve the pointwise bound.

Example 5.1. Three buffers, two servers, $C_1 = \{1, 3\}$, $C_2 = \{2\}$, and we assume that $m_2 > m_1 + m_3$. The optimal policy is to use $u_1 = 0$, $u_2 = \frac{1}{m_2}$, $u_3 = \frac{1}{m_3}$ while $x_2(t) > 0$. Once $x_2(t) = 0$, one uses $u_1 = u_2 = 0$, and $u_3 = \frac{1}{m_3}$ as long as

$$\frac{x_3(t)}{x_1(t)} > \frac{1}{1 - \alpha m_2} \frac{m_2 - m_1 - m_3}{m_1 + m_3}.$$

When the ratio $\frac{x_3(t)}{x_1(t)}$ falls lower than the above value, the optimal control is: $u_1 = u_2 = \frac{1}{m_2}$, $u_3 = \frac{1}{m_3}(1 - \frac{m_1}{m_2})$ as long as $x_3(t) > 0$. Once $x_2(t) = x_3(t) = 0$ use $u_1 = u_2 = u_3 = \frac{1}{m_2}$, until the system is empty (this last part is not unique).

The optimal solution is presented in Figure 5.1. Also shown in this figure is the effect of changing the time at which the control of station 1

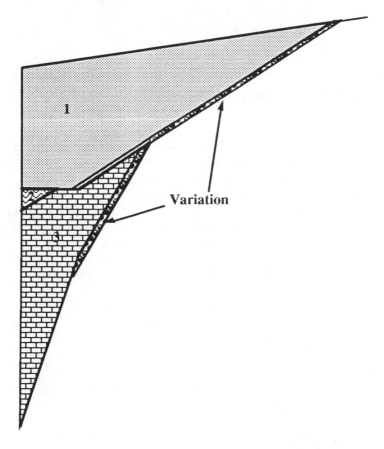

FIG. 5.1. *Optimal Control of Three Buffer System, Not Pointwise Optimal*

switches from $u_3 = \frac{1}{m_3}$, $u_1 = 0$, to $u_3 = \frac{1}{m_3}(1 - \frac{m_1}{m_2})$, $u_1 = \frac{1}{m_2}$. It is simple algebra to check that at the time at which $\frac{x_3(t)}{x_1(t)} = \frac{1}{1-\alpha m_2}\frac{m_2-m_1-m_3}{m_1+m_3}$, the variation introduced by a small change is zero. The optimality follows from this observation. An alternative proof, based on infinite dimensional LP duality theory, can also be given.

The above solution illustrates the tradeoff between myopic draining of the system at the steepest rate, and the necessity to prevent future starvation. It is a challenging problem to generalize the solution of this simple re-entrant line to more general lines.

6. General formulation as separated continuous linear program. The problem (5.3–5.7) is of the form:

$$\max \quad \int_0^T c^{\mathrm{T}}(s)u(s)\,ds$$

$$\text{s.t.} \quad \int_0^t G\, u(s)\, ds \le a(t)$$

$$Hu(t) \le b(t)$$

(6.1) $u(t) \ge 0, \qquad 0 \le t \le T$

Similarly, Problem 5, of minimizing average holding costs can be put in that form. Furthermore, this formulation is also obtained when one considers not just re-entrant lines but general fluid job-shop problems, as described by Anderson [1]. Note that in (6.1) G and H are fixed matrices, $c(t)$ and $a(t)$ are linear, and $b(t)$ is constant. Furthermore, it is easy to see that the feasible region for this problem is nonempty and bounded.

Anderson and Nash [2] call the problem (6.1) a separated continuous linear program. Recently Pullan [14,15,16] has made substantial progress towards solving such problems. A dual to (6.1) in reversed time (slightly modified from Pullan's formulation) is:

$$\min \quad \int_0^T a^{\mathrm{T}}(T-s)\, dP(s) + \int_0^T b^{\mathrm{T}}(T-s)\, q(s)\, d(s)$$

$$\text{s.t.} \quad G^{\mathrm{T}} P(t) + H^{\mathrm{T}} q(t) \ge c(T-t)$$

$$P(t) \text{ nondecreasing}, \quad P(0) = 0,$$

(6.2) $q(t) \ge 0 \text{ a.e.}, \qquad 0 \le t \le T.$

Pullan shows that strong duality holds between (6.1) and (6.2) [15], shows that the solution $u(t)$ to (6.1) is piecewise constant [16], and proposes an algorithm for its solution [14].

Acknowledgment. Florin Avram has contributed significantly to initial stages of this research. Malcolm Pullan provided a code for solving SCLP. I am grateful to Frank Kelly, Bruce Hajek, and Dimitri Bertsimas for useful discussions.

REFERENCES

[1] Anderson, E. J. A new continuous model for job-shop scheduling. *International J. Systems Science* **12**, 1469–1475 (1981).

[2] Anderson, E. J. and Nash, P. *Linear Programming in Infinite Dimensional Spaces.* Wiley-Interscience, Chichester (1987).

[3] Bramson, M. Instability of FIFO queueing networks. *Annals of Applied Probability.* To appear (1994).

[4] Chen, H. Fluid approximations and stability of multiclass queueing networks I: Work-conserving disciplines. (Preprint).

[5] Chen, H. and Yao, D. Dynamic scheduling of a multi class fluid network. *Operations Research* **41**, 1104–1115 (1993).

[6] Dai, J. G. On the positive Harris recurrence for multiclass queueing networks: A unified approach via fluid limit models. *Annals of Applied Probability* To appear (1994).

[7] Dai, J. G. and Weiss, G. Stability and instability of fluid models for certain re-entrant lines. *Mathematics of Operations Research.* To appear.

[8] Gittins, J. Bandit processes and dynamic allocation indices. *J. Royal Statistical Society Series B* **14**, 148–177 (1979).

[9] Klimov, G. P. Time sharing systems I. *Theory of Probability and Applications* **19**, 532–551 (1974).

[10] Kumar, P. R. Re-entrant lines. *Queueing Systems: Theory and Applications* **13**, 87–110 (1993).

[11] Kumar, S. and Kumar, P. R. The last buffer first serve priority policy is stable for stochastic re-entrant lines. (Preprint).

[12] Lai, T. L. and Ying, Z. Open bandit problems and optimal scheduling of queueing networks. *Advances Applied Probability* **20**, 447–472 (1988).

[13] Lu, S. H. and Kumar, P. R. Distributed scheduling based on due dates and buffer priorities. *IEEE Transactions on Automatic Control* **36**, 1406–1416 (1991).

[14] Pullan, M. C. An algorithm for a class of continuous linear programs. *SIAM J. Control and Optimization* **31**, 1558–1577 (1993).

[15] Pullan, M. C. A duality theory for separated continuous linear programs. Cambridge University Report CUED/E-MS/TR.2 (1993).

[16] Pullan, M. C. Forms of optimal solutions for separated continuous linear programs. Cambridge University Report CUED/E-MS/TR.3 (1993).

[17] Rybko, A. N. and Stolyar, A. L. Ergodicity of stochastic processes describing the operation of open queueing networks. *Problemy Peredachi Informatsii* **28**, 3–26 (1992).

[18] Stolyar, A. On the stability of multiclass queueing networks. Submitted to the Proceeding of Second Conference on Telecommunication Systems—Modeling and Analysis, Nashville, March 22–27, 1994.

[19] Weiss, G. Branching bandit processes. *Probability Engineering Information Sciences* **2**, 269–278 (1988).

TWO BADLY BEHAVED QUEUEING NETWORKS

MAURY BRAMSON*

Abstract. We consider FIFO queueing networks with customers arriving according to rate-1 Poisson processes. The service times are assumed to be exponentially distributed, with total mean service time at each queue strictly less than 1. Unlike Kelly networks, such networks can be unstable. We examine here the evolution of two classes of such systems, and how this leads to their instability.

Key words. queueing networks, equilibrium distribution, instability, first-in first-out

AMS(MOS) subject classifications. 60K25, 68M20

1. Introduction. There has been considerable interest recently in the existence/nonexistence of equilibria for queueing networks. It seems intuitively clear that equilibria will exist for systems whose customers are served substantially more quickly than the rate at which they enter. Despite various attempts, general results in this direction are lacking. We present here two classes of examples which contradict this intuition. Hopefully, they will shed some light on appropriate conditions for the existence of equilibria for general queueing networks.

As motivation, we first review the situation for the standard $M/M/1$ or *simple* queue. Customers are assumed to enter a queue according to a rate-1 Poisson process. Customers are served one at a time, leaving the queue according to independent rate-λ exponential holding times. Let N_t denote the number of customers waiting to be served at time t. The process N_t is a birth-death process on $\{0, 1, 2, \ldots\}$, and so any equilibrium distribution must be reversible. Using this, one obtains the equilibrium distributions

$$(1.1) \qquad \pi_\lambda(k) = (1 - \lambda^{-1})\lambda^{-k}, \; k = 0, 1, 2, \ldots,$$

for $\lambda > 1$. It is easy to check that for $\lambda \leq 1$, there are no equilibria, and in fact for $\lambda < 1$, the system is transient, with $N_t \to \infty$ as $t \to \infty$. Also, from elementary Markov chain theory, the above equilibria are, for given λ, unique.

A natural generalization of the above model is to open networks consisting of a series of m queues, $m \geq 1$. Customers enter the system, as before, according to a rate-1 Poisson process. Each customer proceeds along the same prescribed *route*, visiting all the queues, and then exiting from the system. The service times are again assumed to be independent and exponentially distributed. Individual queues may be visited more than

---

* Mathematics Department, University of Wisconsin, Madison, WI 53706. Partially supported by the Institute for Mathematics and its Applications, and by NSF Grant DMS-9300612.

once, and the rate a customer is served at a given queue may depend on his/her position, or *stage*, along the route; denote by λ_{ij} the rate for the jth visit to the ith queue. One can generalize the model further to H types of customers instead of just one type. In this setting, customers enter the system according to independent rate-ν_k Poisson processes, with $\sum_{h=1}^{H} \nu_h = 1$. Everything holds as before, although different customer types may proceed along different routes. The service rates are now denoted by λ_{hij}. For the first class of examples we will be considering, $H = 1$, and for the second class, $H = 2$. There is also the issue of priority for the order of service among customers at a given queue. Here, we assume the networks are first-in, first-out (FIFO) unless otherwise stated.

A fundamental question is under what conditions on λ_{hij} do equilibria exist for the above queueing networks. If an equilibrium exists, the network is *stable*, otherwise, it is *unstable*. For this, we introduce the notation $\mu_{hij} = \lambda_{hij}^{-1}$, $J(h, i) =$ the number of times the ith queue is visited along the route of a customer of type h, $\mu_{hi} = \sum_{j=1}^{J(h,i)} \mu_{hij}$ and $\mu_i = \sum_{h=1}^{H} \nu_h \mu_{hi}$. Here, μ_{hij} is the mean service time for a customer during a visit at a queue, and μ_i can be interpreted as the total mean service time at i. One can check that if $\mu_i \geq 1$ for some i, then the network is unstable.

A natural conjecture is that an equilibrium will exist under the condition

$$(1.2) \qquad\qquad \mu_i < 1 \ \text{ for all } \ i .$$

It is well known that this is in fact true if μ_{hij} does not depend on h and j, that is, the rate a customer is served depends only on the queue, not on the stage along the route (see Kelly [6] and [7]). We refer to these systems as *Kelly networks*. Inspired by the $M/M/1$ queue, one can, under this restriction, construct equilibria for which the probability of there being k_i customers at the ith queue, $i = 1, \ldots, m$, is given by

$$(1.3) \qquad\qquad \pi(k_1, \ldots, k_m) = \pi_{\lambda_1}(k_1) \cdots \pi_{\lambda_m}(k_m) ,$$

where π_λ is as in (1.1) and $\lambda_i = \mu_i^{-1}$. The different possible configurations of customers at different stages along the route with given $k_1, \ldots, k_m$ are equally likely here.

Until recently, it has been generally believed that the condition (1.2) should suffice even when μ_{hij} is dependent on h and j. Examples in Lu-Kumar [9] and Rybko-Stolyar [10] of unstable priority queues satisfying (1.2) did not shake this conviction. Bramson [1] has presented a class of examples which show that (1.2) does not suffice even in the FIFO context. In Bramson [2], a more complicated class of unstable examples has been constructed with

$$(1.4) \qquad\qquad \mu_i < \mu \ \text{ for all } \ i ,$$

where $\mu > 0$ is a constant which can be chosen as small as desired (for appropriate members of the class). Seidman [11] has also recently constructed deterministic systems which satisfy (1.2) and yet are unstable; these systems possess instantaneous stages.

The purpose of these notes is to provide a feel for the evolution of the examples constructed in [1] and [2] without getting bogged down in the technical details. In Section 2, we present the models. We then give our main results, and discuss some related material. The significant features in the evolution of the models from [1], and how these lead to instability, can be explained reasonable succinctly. This is done in Section 3. The models from [2] will be discussed in Section 4.

2. The models. The examples from [1] are assumed to possess two queues, labelled 1 and 2. Upon entering the system, customers move along the prescribed route

$$(2.1) \qquad\qquad 1 \to 2 \to 2 \to \cdots \to 2 \to 1 \,,$$

at which point they exit from the system. (Since all customers are of a single type, we can drop the subscript h in our notation.) The stage of a customer will be denoted by (i, j), with $j = 1, 2$ for $i = 1$, and $j = 1, \ldots, J$ for $i = 2$: the first coordinate denotes the queue, and the second coordinate the number of times that the queue has been visited up to then. The mean service time for a customer is given by

$$(2.2) \qquad \begin{array}{l} c \;\; \text{at} \;\; (1, 2) \; \text{and} \; (2, 1) \,, \\ \delta \;\; \text{at} \;\; (1, 1) \; \text{and} \; (2, j), \, j = 2, \ldots, J \,. \end{array}$$

We assume that

$$(2.3) \qquad \tfrac{399}{400} \leq c \lessdot 1, \quad c^J \leq \tfrac{1}{50}, \quad 0 \leq \delta \leq (1 - c)/50J^2 \,.$$

Under (2.3), (1.2) is clearly satisfied. For the sake of concreteness, the reader can set $c = \tfrac{399}{400}$, $J = 1,600$, and $\delta = 10^{-11}$. These restrictions can be weakened by a more parsimonious choice of bounds, but for our argument to work, c needs to be chosen fairly close to 1 and δ quite small. (One can choose c to be of order $\tfrac{9}{10}$, while applying essentially the same argument as in [1]; better bounds require more detailed estimates. One would expect the sharp bounds for c to be $\tfrac{1}{2} < c < 1$ under appropriate conditions on J and δ, in keeping with the discussion in Section 3. In Dai [3], the network with $J = 4$ and mean service times .897 at $(1, 2)$, .899 at $(2, 1)$, and .001 elsewhere is simulated, and exhibits the desired instability.)

The main result in [1] is the following.

THEOREM 2.1. *Any FIFO queueing network satisfying (2.1)–(2.3) is unstable, with the number of customers in the system approaching infinity as $t \to \infty$.*

Theorem 2.1 will follow easily from Theorem 2.2 below. For here and later on, we find it convenient to introduce the following notation. Let Ξ_t denote the state at time t of any of the above FIFO queueing networks satisfying (2.1)–(2.3). $\xi_t(i, j)$ will denote the number of customers at (i, j) at time t, and ξ_t the total number of customers in the system. By $(i, j)^+$ (resp., $(i, j)^-$), we will mean the set of stages in the system strictly beyond (resp., before) (i, j), and by $\xi_t(i, j)^+$ (resp., $\xi_t(i, j)^-$), the number of customers in $(i, j)^+$ (resp., $(i, j)^-$). For instance,

$$\xi_t(2, j)^+ = \xi_t(1, 2) + \sum_{\ell=j+1}^{J} \xi_t(2, \ell).$$

THEOREM 2.2. *Assume that*

$$(2.4) \qquad \xi_0(1, 1) = M, \quad \xi_0(1, 1)^+ \leq \tfrac{1}{50}M.$$

Then for some $\varepsilon > 0$, large enough M, and appropriate T (depending on M),

$$(2.5) \qquad P(\xi_T(1, 1) \geq 100M, \; \xi_T(1, 1)^+ \leq M) \geq 1 - e^{-\varepsilon M}$$

and

$$(2.6) \qquad P\left(\xi_t \geq \tfrac{1}{4}M, \; \forall t \in [0, T]\right) \geq 1 - e^{-\varepsilon M}.$$

As we will later see, T will be close to $2cM/(1 - c)$. Note that the factor 50 in (2.4) is not special, although we require the ratio $\xi_0(1, 1)^+/\xi_0(1, 1)$ to be small.

Suppose that Ξ_0 satisfies (2.4) for some large M. Repeated application of Theorem 2.2 yields

$$(2.7) \qquad P(\xi_t < \tfrac{1}{4}M \text{ for some } t \geq 0) \leq 2 \sum_{k=0}^{\infty} e^{-(100)^k \varepsilon M},$$

which $\to 0$ as $M \to \infty$. Since all states in the system are accessible from one another, (2.7) implies that $\xi_t \to \infty$ as $t \to \infty$ w.p. 1 for any Ξ_0. Theorem 2.1 thus follows from Theorem 2.2. In Section 3, we outline the proof of Theorem 2.2. The argument exhibits the typical evolution of Ξ_t over $[0, T]$ for the initial data given in (2.4).

The above iterative procedure, viewing time intervals over which the number of customers in the system grows geometrically while the system returns to a "multiple" of the original state, was also employed in [9] (in its deterministic form). Some thought should convince the reader that this is the most natural path to follow for analyzing the behavior of a large class of unstable networks. One question one can ask (which is not answered here) is whether there are essentially different ways in which Ξ_t can approach infinity. More generally, one can ask about the nature of the Martin boundary for this or other unstable queueing networks.

We now turn to the second class of models. In the examples from [2], there are two types of customers. Customers are assumed to move along the prescribed routes

$$(2.8) \quad \begin{array}{l} 1 \to 2 \to \cdots \to 2 \to 3 \to \cdots \to 3 \to \cdots \to m \to \cdots \to m \\ 1 \to 1 \to 2 \to \cdots \to 2 \to 3 \to \cdots \to 3 \to \cdots \to m \to \cdots \to m \to 1 \end{array}$$

at the end of which they exit from the system. Each portion $i \to \cdots \to i$ of the route consists of seven visits to the ith queue. We refer to customers employing the upper route as *upper* customers, and the others as *lower* customers. Note that the routes for both types of customers are similar, the only difference being that lower customers visit the 1st queue an extra time both at the beginning and at the end of the route. In keeping with the notation used for the first class of models (in (2.1)–(2.3)), we denote the stage of a customer by (h, i, j). Here, $h = u, \ell$, $i = 1, \ldots, m$, and $j = 1, \ldots, 7$ for $i = 2, \ldots, m$; $j = 1$ for $h = u$ and $i = 1$; and $j = 3$ for $h = \ell$ and $i = 1$. Each type of customer is assumed to enter the system at rate $\nu_h = \frac{1}{2}$. The mean service time for a customer is given by

$$(2.9) \quad \begin{array}{llll} c & \text{at} & (h, i, 1), & \text{for } h = u, \ell \text{ and } i = 2, \ldots, m, \\ c & \text{at} & (\ell, 1, 3), & \\ \delta & \text{at} & (h, i, j), & \text{for } h = u, \ell, i = 2, \ldots, m \text{ and } j = 2, \ldots, 7, \\ \delta & \text{at} & (u, 1, 1), & (\ell, 1, 1) \text{ and } (\ell, 1, 2). \end{array}$$

We assume that

$$(2.10) \qquad 0 < c \le \tfrac{1}{100}, \quad 0 \le \delta \le c^8, \quad m = [2c^{-1}\log(c^{-1})],$$

where $[\cdot]$ denotes the integer part. Each of the queues $2, \ldots, m$ therefore has one comparatively slow, and six very quick stages for each type of customer; the 1st queue has only the single quick stage for the upper customers, and two quick stages and one slow stage for lower customers. The choice of parameters is made for technical reasons. (The coefficient 2 in the definition of m has been chosen so that $(1-c)^{-m} \sim c^{-2}$; this will be indicated in Section 4. The bound $c \le \frac{1}{100}$ is somewhat arbitrary.) One can think of this class of models as being constructed by piecing together m copies of the type given in (2.1). (We only need $J = 7$ here.) As is later discussed, the models in (2.8)–(2.10) evolve in the same basic way as those in (2.1)–(2.3), although one must work considerably harder to show this. (One should focus on the lower customers in (2.8); the upper customers serve only to "guide" them.) We also note that if $m < (c + 6\delta)^{-1}$, then $\sum_i \mu_i < 1$, in which case the system in (2.8)–(2.9) must have an equilibrium.

Under (2.10), $\mu_i \leq c + 6\delta < 2c$. So the total mean service time can be chosen as small as desired at every i for these queueing networks. The main result in [2] corresponds to Theorem 2.1, but for the system given by (2.8)–(2.10).

THEOREM 2.3. *Any FIFO queueing network satisfying (2.8)–(2.10) is unstable, with the number of customers in the system approaching infinity as $t \to \infty$.*

To demonstrate Theorem 2.3, one employs an appropriate analog of Theorem 2.2 (see Section 4). One can then argue in exactly the same way as in (2.7) to finish the proof. Most likely, the analog of Theorem 2.3 also holds for a single-type network with route corresponding to that of the lower customers in (2.8), although the reasoning in [1] then no longer suffices.

Theorem 2.3 has the following consequence. One can compare any FIFO queueing network satisfying (2.8)–(2.10) with the network which is obtained from it by replacing (2.9) with the simple assumption that the mean service time $\mu_{hij} \equiv c$ at every stage of the route. The lengths of the mean service times for the new network are, of course, everywhere at least as great as for the original network. This new system is a Kelly network, and has an equilibrium distribution of the form given in (1.3). In fact, the equilibrium probability of k customers at any given queue is at most $(1 - 7c)(7c)^k$, $k \geq 1$, which means that the network is in fact "very stable" for small c. This observation shows that decreasing mean service times within a queueing network may have the effect of making it unstable.

Thinking along the lines of the previous paragraph, one can ask to what extent it is possible to create a stable queueing network from an unstable one satisfying (1.2) by increasing the mean service times. This corresponds to regulating the rate at which customers can move through the system. It is in fact not difficult to prove the following result for any unstable queueing network satisfying (1.2) (and not necessarily FIFO). Define a new queueing network by introducing single-stage queues between each of the already extant stages of the old network. Assume that the (exponential) mean service times for the new queues are greater than the total mean service times at each of the old queues, but less than 1. Then this new network is stable. The argument consists of first showing that the distribution of the number of customers at each of the old queues has an exponential tail. Using these bounds and the simple structure of the new queues, one then also bounds the number of customers at these new queues. Note that by perturbing the original network by adding in single-stage queues as before, but with small enough mean service times, the new network will continue to be unstable. By lengthening these times as before, the corresponding network becomes stable. So if one wishes, one can also create stable queueing networks from unstable ones without introducing new queues. The details of these results will appear elsewhere.

One can also ask (Kumar-Meyn [8]) whether the assumption (1.2) suffices for the stability of a queueing network if one drops the requirement that the network be FIFO, but requires as in Kelly networks that the rate a customer is served depend only on the queue, and not on the stage along the route. (1.2) does not suffice, as is shown in Dai [4] and Gu [5]. By perturbing such examples, it also follows that (1.2) does not suffice for networks which are "quickest serve, first serve", that is, networks for which customers with the smallest values of μ_{ij} are served first.

Theorem 2.1, Theorem 2.3, and the comparisons with stable networks given above provide an explanation for the lack of general criteria so far on queueing networks which ensure stability. It is perhaps risky to propose criteria at this point, but the following formulations suggest themselves, at least for FIFO networks. We set $\mu_i^{\min}(\mu_i^{\max})$ equal to the minimum (maximum) over all μ_{hij} where i is fixed, and $\mu_i^R = \mu_i^{\min}/\mu_i^{\max}$.

QUESTION 1. *For each given $\mu^R > 0$, does there exist a $\mu > 0$, so that any FIFO network satisfying $\mu_i^R \geq \mu^R$ and $\mu_i \leq \mu$ for all i is stable?*

QUESTION 2. *For each given $\mu < 1$, does there exist a $\mu^R < 1$, so that any FIFO network satisfying $\mu_i^R \geq \mu^R$ and $\mu_i \leq \mu$ for all i is stable?*

An affirmative answer to Question 1 would say that if the mean service times at a given queue are not too different, then small enough values of μ_i suffice for stability. An affirmative answer to Question 2 would say that the system is stable for $\mu_i < 1$ as long as μ_i^R are close enough to 1. The networks with μ_{hij} not dependent on h and j satisfy $\mu_i^R \equiv 1$, and so make up the limiting case in the latter setup.

3. Evolution of the first model. We go over in some detail here the evolution of the class of models specified in (2.1)–(2.3). For more detail, the reader should consult [1].

We begin by introducing a sequence of stopping times $S_1, S_2, \ldots,$ $S_\ell, \ldots$ for the process Ξ_t. Customers at any given queue at time t may be ordered according to the times at which they are next served, so we can talk about a "first" or "last" customer in this sense. (Due to the multiple stages at each queue, customers entering the network earlier on may be ordered behind more recent arrivals. For instance, a customer who has just been served at the 2nd queue will typically return to the end of this queue.) Let S_1 denote the time at which the last of the original customers (customers at $t = 0$) at the 1st queue is next served. Let $S_2, S_3, \ldots, S_\ell, \ldots$ denote the successive times at which the last customer at the 2nd queue is served, where the ordering is made at $t = S_{\ell-1}$. Set $S_\infty = \lim_{\ell \to \infty} S_\ell$, and let $S_\ell = S_{\ell+1} = \cdots = S_\infty$ when the 2nd queue becomes empty. ($S_\infty < \infty$ w.p.1, on account of (1.2).) We can think of the intervals $(S_\ell, S_{\ell+1}]$, $\ell = 1, 2, \ldots$, as "cycles" for which *each customer starting at $(2, j)$, $j < J$, ends up at $(2, j + 1)$.* Note that no customer can be served twice at the 2nd queue before every other customer there is served once. We also let T (as in Theorem 2.2) denote the time at which the last customer at $(1, 2)$ at time S_{2J} leaves the queue.

The queueing network can be analyzed at the times S_ℓ. First, assume as in (2.4) that $\xi_0(1, 1) = M$, with M large, and that there are few customers elsewhere in the system. Since $\delta \ll 1$, one has $S_1 \ll M$. Consequently, at S_1, nearly all of the original customers in the network are at $(2, 1)$. Also, comparatively few new customers have entered the network up to time S_1, and so there are few customers at $(1, 1)$. Of course, since we are working with random events here, this behavior will sometimes be violated. However, such exceptional events occur with probabilities that are exponentially small in M, and can therefore be ignored without affecting the basic nature of the evolution of Ξ_t. Here and later on, we can therefore treat the network as if it were evolving deterministically.

Over the time interval $(S_1, S_2]$, the (approximately) M customers at $(2, 1)$ all move to $(2, 2)$. On account of (2.2), the time it takes to serve these customers is (approximately) cM, so $S_2 - S_1 \approx cM$. So over this time (approximately) cM new customers enter the system, which quickly move to $(2, 1)$. Thus, at $t = S_2$, there are (comparatively) few customers in the system except at $(2, 2)$ and $(2, 1)$, where there are (approximately) M and cM customers, respectively.

Continuing our reasoning along the same lines, we observe that over $(S_2, S_3]$, the customers at $(2, 1)$ and $(2, 2)$ advance to $(2, 2)$ and $(2, 3)$, respectively. The time required to serve the customers at $(2, 2)$ is negligible; the time required for the customers at $(2, 1)$ is $c^2 M$, so $S_3 - S_2 \approx c^2 M$. Over this time, $c^2 M$ new customers enter the system, which quickly move to $(2, 1)$. (We are using the fact that $(1, 2)$ is nearly empty over $(S_2, S_3]$.) So, at $t = S_3$, there are few customers in the system except at $(2, 3)$, $(2, 2)$ and $(2, 1)$, where there are M, cM and $c^2 M$ customers, respectively. Proceeding inductively, we obtain that at $t = S_J$, there are M customers at $(2, J)$, cM customers at $(2, J-1)$, and so on down to $(2, 1)$, where there are $c^{J-1} M$ customers. On account of (2.3), there are

$$(3.1) \qquad \sum_{\ell=0}^{J-1} c^\ell M \approx M/(1-c)$$

customers in the system. Likewise, $S_J \approx cM/(1-c)$.

At this point, the evolution of the system changes. By $t = S_{J+1}$, there are M customers at $(1, 2)$. These customers require time cM to be served, until which new customers cannot pass from $(1, 1)$ to $(2, 1)$. The cycles $(S_\ell, S_{\ell+1}]$, $\ell = J, \ldots, 2J - 1$, are, however, of much shorter duration, as is $(S_J, S_{2J}]$. The customers already at the 2nd queue at $t = S_J$ have all already arrived at $(1, 2)$ by $t = S_{2J}$. By (3.1), there are $M/(1 - c)$ such customers. So at $t = S_{2J}$, there are essentially $M/(1-c)$ customers at $(1, 2)$ and no customers elsewhere. Of course, here and elsewhere, we are taking liberties in ignoring negligible quantities of customers and probabilities.

Over $(S_{2J}, T]$, the $M/(1 - c)$ customers at $(1, 2)$ exit the system. The time required to serve these customers is therefore $cM/(1 - c)$, over which time $cM/(1 - c)$ customers enter the system. These new customers are obliged to remain at $(1, 1)$ until time $T \approx 2cM/(1 - c)$. At this time, there are few customers elsewhere in the system. So at time T, the state of the system is a "multiple", by the factor $c/(1 - c)$, of the state at time 0. Since by (2.3), $c \geq \frac{399}{400}$, one has enough leeway to obtain the bound (2.5) in Theorem 2.2, also when the proof is done rigorously. (Presumably, the number of customers in the system grows for $c > \frac{1}{2}$.) We also note that, except on a set of small probability, ξ_t will not drop much below M over $[0, T]$. This is because, up to time S_J, most of the original M customers at $(1, 1)$ remain in the system, whereas by time S_J, there are approximately $M/(1 - c)$ new customers. Before most of these new customers have left the system, an additional $cM/(1 - c)$ customers enter the system, which are trapped at $(1, 1)$ until time T. This demonstrates (2.6), and completes the argument for Theorem 2.2, and hence Theorem 2.1.

4. Evolution of the second model. We now provide an account for the evolution of the class of models specified in (2.8)–(2.10). For more detail, the reader should consult [2].

We use notation which is analogous to that employed for the models given by (2.1)–(2.3), with Ξ_t denoting the state of the system at time t, $\xi_t(h, i, j)$ denoting the number of customers at (h, i, j), etc. Set

$$\xi_t(i, j) = \xi_t(u, i, j) + \xi_t(\ell, i, j), \quad \xi_t(i) = \sum_j \xi_t(i, j).$$

Also, let $\xi_0(u, 1, 1) = M_u$, $\xi_0(\ell, 1, 1) = M_\ell$ and $M = M_u + M_\ell$. The following result is the analog of Theorem 2.2.

THEOREM 4.1. *Assume that*

(4.1) $\frac{1}{4}M \leq M_u \leq \frac{3}{4}M$, $\quad \xi_0(1, 1)^+ \leq c^4 M$, $\quad \xi_0(1, 2) + \xi_0(2) \leq c^5 M$.

Then for some $\varepsilon > 0$, large enough M, and appropriate random T (depending on M),

(4.2)
$$P\left(\tfrac{1}{4}\xi_T(1, 1) \leq \xi_T(u, 1, 1) \leq \tfrac{3}{4}\xi_T(1, 1), \, \xi_T(1, 1) \geq \tfrac{1}{4}c^{-1}M,\right.$$
$$\left.\xi_T(1, 1)^+ \leq \tfrac{1}{8}c^3 M, \, \xi_T(1, 2) + \xi_T(2) \leq \tfrac{1}{8}c^4 M\right) \geq 1 - e^{-\varepsilon M}$$

and

(4.3) $$P\left(\xi_t \geq \tfrac{1}{5}M, \, \forall t \in [0, T]\right) \geq 1 - e^{-\varepsilon M}.$$

Theorem 2.3 follows easily from Theorem 4.1 by using the analog of (2.7). For this, one should note here that, since $c \leq \frac{1}{100}$, one has $\frac{1}{4}c^{-1}M \geq$

$25M$. Moreover, the stated bounds for $\xi_t(1, 1)$ relative to $\xi_t(1, 1)^+$ and $\xi_t(1, 2) + \xi_t(2)$ improve from (4.1) to (4.2), which we use for the iteration. The random time T above will be about $c^{-2}M$.

Here, we outline briefly the proof of Theorem 4.1. The argument exhibits the usual evolution of Ξ_t over $[0,T]$ for the initial data given in (4.1). As in Section 3, we introduce stopping times $S_{i,j}$, which correspond to "cycles". Again, customers at any given queue at time t are ordered according to the times at which they are next served, and one can talk about first and last customers in this sense. (Upper and lower customers are ordered without distinction.) Let $S_{1,1}$ denote the time at which the last of the original customers at the 1st queue is next served. Similarly, $S_{2,1}$ denotes the time when the last of the customers at the 1st queue is served, with the ordering being made at $t = S_{1,1}$. (By $S_{2,1}$, all of the customers originally at $(1, 1)$ and $(1, 2)$ have moved to at least the 2nd queue.) Denote by $S_{2,2}, \ldots, S_{2,7}$ the times at which the last customer at the 2nd queue is served, with the orderings being made at $t = S_{2,1}, \ldots, S_{2,6}$. Let $S_{3,1}, S_{3,1} \geq S_{2,7}$, denote the next time at which the 2nd queue is empty. (On account of (1.2), $S_{3,1} < \infty$ w.p.1.) Proceeding inductively, one denotes by $S_{i,2}, S_{i,3}, \ldots, S_{i,7}$ the times at which the last customer at the ith queue is served, with the orderings being made at $t = S_{i,1}, \ldots, S_{i,6}$, and by $S_{i+1,1}$ the next time at which the ith queue is empty. In this manner, one defines random times up through $S_{m,2}, \ldots, S_{m,7}$ and $S_{1,2}$, where $S_{1,2} \geq S_{m,7}$ is the next time at which the mth queue is empty. Lastly, let T (as in Theorem 4.1) denote the time when the last of the customers at the 1st queue is served, the ordering being made at $t = S_{1,2}$.

We can think of the intervals $(S_{i,j}, S_{i,j+1}]$, $j = 1, \ldots, 6$, as "cycles", over which each customer starting in the ith queue is served exactly once at this queue. Together with $(S_{i,7}, S_{i+1,1}]$, at the end of which the ith queue is empty, these intervals regulate the movement of customers. So, in principle, we have a situation similar to what was covered in Section 3 for the first model, where we were able to examine how customers moved throughout the network over $[0, T]$. There is here, however, the unfortunate fact that we cannot completely control the evolution of the second model over $(S_{i,7}, S_{i+1,1}]$. Since $S_{i+1,1}$ is determined by the behavior at the ith queue, customers already at $(i + 1, 1)$ before $S_{i+1,1}$ may possibly advance to other stages, rather than remaining at $(i + 1, 1)$ until $S_{i+1,1}$, where we want them. This can occur for each $i = 1, \ldots, m$, and makes bounds on Ξ_t more difficult. (For the first model, this type of behavior can only occur over the single interval $(0, S_1]$, for which good bounds are straightforward.)

One can nevertheless control the evolution of the process sufficiently so as to obtain the bounds in (4.2) of Theorem 4.1. One can show that at $t = S_{i,j}$, $i \geq 2$, most of the customers are at (i, j). There will be (approximately) $M/(1 - c)^{i-2}$ customers there, and (approximately) $c^{j-j'}M/(1 - c)^{i-2}$ customers at (i, j'), with $j' < j$. At $S_{i+1,1}$ (nearly all of) these customers will be at $(i+1, 1)$. In all cases, there will be (approx-

imately) equal numbers of upper and lower customers at (i, j). There will
also be few customers elsewhere in the system.

Assuming the behavior portrayed in the previous paragraph, there will
be $c^{-2}M$ customers at the mth queue at $t = S_{m, 7}$. (m was chosen as in
(2.10) for the factor c^{-2}.) Since there are few customers in $(m, 2)^-$ at this
time, the service times for the customers at the mth queue (by (2.9)) will
be short, and therefore $S_{1, 2} - S_{m, 7}$ will be small. Few new customers will
have entered the system over this time, and the customers previously in
the system will have either exited (upper customers) or be at $(1, 3)$, the
last stage of the route (lower customers). One can check that $S_{1, 2}$ is about
$c^{-2}M$.

As in Section 3, one next considers the network over $(S_{1, 2}, T]$, at the
end of which the customers in $(1, 3)$ have left the system. Since there are
$\frac{1}{2}c^{-2}M$ such lower customers, it takes time $\frac{1}{2}c^{-1}M$ to serve them. During
this time, $\frac{1}{4}c^{-1}M$ upper customers and $\frac{1}{4}c^{-1}M$ lower customers enter the
network, remaining at $(1, 1)$ until $t = T$. At this time, there are few
customers elsewhere in the network. So at $t = T$, the state of the system
is a "multiple", by the factor $\frac{1}{2}c^{-1}$, of the state at $t = 0$. This (when done
rigorously) gives the bound (4.2) in Theorem 4.1. The bound given in (4.3)
is arrived at in the same way as (2.6) of Theorem 2.2, by keeping track of
lower bounds for the number of customers in the system over different
subintervals of time.

We have so far not justified the claims three paragraphs up regarding
the location of customers in the network at $t = S_{i,j}$. This, not surprisingly,
constitutes the main part of the argument for Theorem 4.1. We need, in
effect, to establish that (a) few customers fall substantially behind the
"main body" of customers at (i, j) at $t = S_{i, j}$ and (b) few customers
"race ahead" of this main body of customers. One can show (a) by using
an induction argument. Since, up to $S_{m, 7}$, there will be few customers
at $(1, 3)$, new customers will be able to move freely through the system
until reaching the ith queue, where they must wait with the main body of
customers for further advancement. It will follow that at any time $S_{i,j}$,
there will necessarily be few customers at $(i, 1)^-$.

One uses the presence of upper customers to obtain (b). Note that from
(2.8), upper customers exit the system immediately after $(m, 7)$, without
visiting the 1st queue a final time. Upper customers, therefore, do not directly affect the movement of new customers when at the end of their route.
So the existence of a "moderately small" number of upper customers racing
through the system ahead of the main body of customers can be tolerated.
Such bounds are not too difficult. On the other hand, because of the extra stage $(1, 2)$ for lower customers, the M_ℓ lower customers originally at
$(1, 1)$ will fall behind the corresponding M_u upper customers. Specifically,
by $t = S_{1, 1}$, all of these upper customers will have advanced beyond $(1, 2)$,
whereas all of the lower customers will still be there. This relationship
will be maintained at all times, with all of these lower customers being

served at each (i, j), $i = 2, \ldots, m$, only after all the corresponding upper customers have been served there. In particular, none of the lower customers originally at $(1, 1)$ can arrive at $(1, 3)$ until at least M_u $(\geq M/4)$ upper customers have left the network. So, few lower customers (those originally in $(1, 1)^+$) can race ahead of the main body of customers, and cause trouble at $(1, 3)$. This is the spirit behind (b). The reader should observe that the above outlines for (a) and (b), while essentially correct, are simplified versions. In particular, the two parts are involved in the same induction argument. We also point out that in (4.1), different bounds were given for $\xi_0(1, 1)^+$ and $\xi_0(1, 2) + \xi_0(2)$, which then reappeared at $t = T$ in (4.2). Without going into detail here, we just note that this distinction is necessary for the induction argument.

REFERENCES

[1] M. BRAMSON, *Instability of FIFO queueing networks*, Ann. Appl. Prob. **4** (1994), pp. 414–431 (correction on p. 952).

[2] M. BRAMSON, *Instability of FIFO queueing networks with quick service times*, Ann. Appl. Prob. **4** (1994), pp. 693–718.

[3] J. DAI, *On positive Harris recurrence of multiclass queueing networks: A unified approach via fluid limit models*, Ann. Appl. Prob. (to appear).

[4] J. DAI, *Stability of open multiclass queueing networks via fluid models*, this volume.

[5] J. GU, *Convergence and Performance for some Kelly-like Queueing Networks*, University of Wisconsin (thesis).

[6] F.P. KELLY, *Networks of queues with customers of different types*, J. Appl. Prob. **12** (1975), pp. 542–554.

[7] F.P. KELLY, *Reversibility and Stochastic Networks*, Wiley & Sons, New York (1979).

[8] P.R. KUMAR AND S.P. MEYN, *Stability of queueing networks and scheduling policies* (preprint).

[9] S.H. LU AND P.R. KUMAR, *Distributed scheduling based on due dates and buffer priorities*, IEEE Transactions on Automatic Control **36** (1991), pp. 1406–1416.

[10] S. RYBKO AND A.L. STOLYAR, *Ergodicity of stochastic processes that describe the functioning of open queueing networks*, Problems of Information Transmission **28** (3) (1992), pp. 3–26 (in Russian).

[11] T.I. SEIDMAN, *'First Come, First Serve' can be unstable!*, IEEE Transactions on Automatic Control (to appear).

SOME BADLY BEHAVED CLOSED QUEUEING NETWORKS

J. MICHAEL HARRISON* AND VIÊN NGUYEN[†]

Abstract. We study several variations of a multiclass closed queueing network model, all closely related to the open priority network of Lu and Kumar. For each of our examples it is shown that, as the population size $n \to \infty$, the long-run utilization rate of each server approaches a limit strictly less than one. Thus no "bottleneck station" emerges in the heavy traffic limit. This is analogous to the "bad behavior" of open network examples developed by Bramson and by Lu and Kumar, among others, in which queue sizes grow without bound even though each service station has a traffic intensity parameter less than one. Two of the three examples considered here have preemptive-resume priority service disciplines at each station, as in the Lu-Kumar open network example, and one has FIFO service at each station, as in Bramson's open network examples.

Key words. Closed queueing networks, multiclass networks, priority networks, bottlenecks

AMS(MOS) subject classifications. 90B15,60K25,60K20,60K05

1. Introduction. Over the last few years queueing theorists have been shocked to discover that there exist open network models which are unstable even though each service station has more than enough capacity to handle the load imposed on it. Perhaps the simplest example is the two-station priority network pictured in Figure 1, which was originally studied

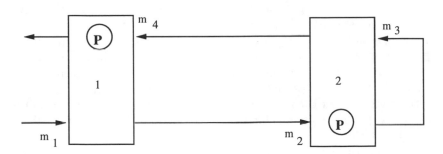

Classes 4 and 2 have priority at stations 1 and 2, respectively

FIG. 1.1. *The Lu-Kumar priority network*

by Lu and Kumar [3]. Here customers arrive according to a renewal process at average rate λ, and they follow a fixed deterministic route requiring

---
* Graduate School of Business, Stanford University, Stanford, CA 94305.
† Sloan School of Management, M.I.T., Cambridge, MA 02139.

four service operations. The four operations are performed at stations 1, 2, 2 and 1 successively. Customers that are waiting for or undergoing their k^{th} operation will be called *class* k customers. Each class is allowed to have its own service time distribution (thus there are four mutually independent service time sequences, also independent of the arrival process), and we denote by m_k the mean service time for class k. Finally, there is a preemptive-resume priority service discipline at each station, the priorities being as shown in Figure 1.

In their original paper, Lu and Kumar [3] treated a version of this model with deterministic interarrival and service times. Assuming specific numerical values for $\lambda, m_1, \ldots$ and m_4, they showed that queue lengths can grow without bound in the deterministic priority network even though

$$(1.1) \qquad \rho_1 \equiv \lambda(m_1 + m_4) < 1 \quad \text{and} \quad \rho_2 \equiv \lambda(m_2 + m_3) < 1.$$

Independently of this work, Rybko and Stolyar [4] analyzed the first legitimately stochastic network model shown to be unstable with traffic intensity parameters less than one. In an important recent paper, Dai and Weiss [2] studied the stochastic version of the Lu-Kumar model described above, and discovered the following additional condition:

$$(1.2) \qquad\qquad\qquad \lambda(m_2 + m_4) < 1.$$

Dai and Weiss [2] showed that (1.1) and (1.2) together are sufficient for stability (that is, positive Harris recurrence) of the stochastic Lu-Kumar network. They also showed that (1.2) is *necessary* for the network's deterministic fluid analog to be stable, which strongly suggests that (1.2) is necessary for stability of the queueing network as well.

It may be argued that "legitimate" networks can be made unstable by applying a "bad" priority scheme, such as those used in the Lu–Kumar and Rybko–Stolyar examples. Bramson [1] recently presented the surprising result of a multiclass network that is unstable even when its intensity parameters are less than one and stations serve customers on a FIFO basis. Whitt [7] and Seidman [6] also studied unstable deterministic networks with FIFO service.

Can a closed queueing network exhibit "bad behavior" analogous to the instability described above for open networks? In this paper we answer that question in the affirmative, as follows. We consider several closely related two-station closed networks, denoting by n the fixed population size, and show that for each of the examples considered, *the long-run utilization rates of both servers approach limits strictly less than one* as $n \to \infty$. Thus no "bottleneck station" emerges as the population grows in size, although the existence of such a bottleneck has often been implicitly assumed in theoretical analyses of closed networks [5]. That is, previous theoretical analyses have simply assumed that as $n \to \infty$, one observes a long-run utilization of 1 at some station(s) in the network, and this defines the

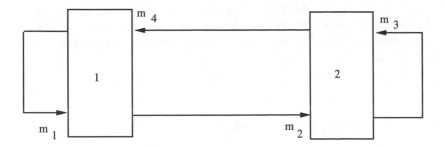

FIG. 1.2. *Structure of closed network examples to be studied*

bottleneck node(s). Readers should note that our "badly behaved" closed network examples are true multiclass networks, with different classes served at the same station having different mean service times, whereas most previous work on both open and closed networks has assumed a single service time distribution for all classes served at a station.

To be more specific, we will focus on the closed version of the Lu-Kumar model pictured in Figure 1.2, with two single-server stations, four customer classes, and the deterministic cyclic routing shown. Two very simple examples are treated in the next section, each characterized by extreme distributional assumptions (some service times are small and bounded above, others are large and bounded below), which eliminate the need for sophisticated analysis. The first of our elementary examples uses the Lu-Kumar preemptive-resume priority scheme (see Figure 1), and our second example shows that essentially the same system behavior may be observed with FIFO scheduling, at least if one allows some service times to vanish as $n \to \infty$.

In section 3 we study a precise closed analog of the Lu-Kumar open network, with preemptive-resume priority scheduling at each station, with all four service time distributions assumed to be exponential, and with mean service times assumed to satisfy

$$(1.3) \qquad m_2 + m_4 > \max(m_1 + m_4, m_2 + m_3).$$

Thus our final example, unlike those treated in section 2, is not characterized by exotic distributional forms or extreme parameter values, although one realizes after the fact that the priority rule assumed is a rather bad one. The limiting long-run average throughput rate (as $n \to \infty$) is explicitly calculated, and this corresponds to less-than-full utilization of each server.

2. Two simple examples.

Priority service. With the routing picture in Figure 1.2, first assume that classes 4 and 2 have preemptive-resume priority at stations 1 and 2, respectively. Let $\{v_k(i), i = 1, 2, \ldots\}$, $k = 1, 2, 3, 4$, be independent sequences of i.i.d non-negative random variables representing processing times of class k jobs. We suppose that for some $\delta > 0$ the following inequalities hold almost surely:

(2.1)
$$\begin{array}{cc} v_2(1) > \delta & v_4(1) > \delta \\ v_1(1) < \delta & v_3(1) < \delta. \end{array}$$

These bounds imply (recall $m_k \equiv \mathbf{E}[v_k(1)]$)

(2.2)
$$\max(m_1, m_3) < \min(m_2, m_4).$$

Suppose that all n customers are queued initially at station 1 as class 1 (i.e., $Q_1(0) = n$ and $Q_2(0) = Q_3(0) = Q_4(0) = 0$), so that station 1 contains only low priority work. Set

(2.3)
$$T_0 \equiv \inf\{t \geq 0 : Q_1(t) = n\} = 0 \quad \text{and}$$

(2.4)
$$T_{l+1} \equiv \inf\{t \geq T_l + \delta : Q_1(t) = n\}.$$

It follows from the bounds on service times (2.1) that while station 1 is processing low priority work, each class 1 customer will depart station 1 and enter station 2 before the previous customer can complete service at station 2. Consequently, assuming that station 1 is not interrupted by arrival of a high priority (class 4) customer, station 2 will be continuously busy with class 2 work after the initial service of a class 1 customer. Let τ be the sum of the first class 1 service at station 1 and the first n class 2 services at station 2. Because of the preemptive-resume priority policy at station 2, no class 4 customers are generated between T_0 and $T_0 + \tau$ and so at time $T_0 + \tau$ all n customers are queued at station 2 as low priority class 3 work.

The network now evolves in exactly a symmetrical manner as before. Services of high priority (class 4) customers at station 1 begin after the intial service of a class 3 customer, and station 1 continuously serves class 4 until all n customers have been served, at which time the entire population is again queued at station 1 as class 1.

It should be clear from this discussion that

$$\mathbf{E}[T_1] = n\,(m_2 + m_4) + (m_1 + m_3)\,.$$

Moreover, $T_1, T_2, \ldots$ form regeneration points for the queue length process $Q(t) = [Q_1(t), Q_2(t), Q_3(t), Q_4(t)]$. Denoting by λ be the long-run throughput rate of customers, it follows from the theory of regenerative processes that

$$\lambda = \frac{\mathbf{E}[\text{number of customers exiting station 1 between } T_0 \text{ and } T_1]}{\mathbf{E}[T_1]}$$

(2.5)
$$= \frac{n}{n\,(m_2 + m_4) + (m_1 + m_3)},$$

and we arrive at the main result of this subsection:

$$(2.6) \qquad \lambda \to \frac{1}{m_2 + m_4} \quad \text{as } n \to \infty.$$

Equations (2.2) and (2.6) together imply that the long-run utilization rate for station 1 is given by

$$\lambda(m_1 + m_4) \to \frac{m_1 + m_4}{m_2 + m_4} < 1 \quad \text{as } n \to \infty,$$

and similarly for station 2, the long-run utilization rate is

$$\lambda(m_2 + m_3) \to \frac{m_2 + m_3}{m_2 + m_4} < 1 \quad \text{as } n \to \infty.$$

FIFO service. We now consider a *FIFO* version of the network discussed in the previous subsection. The routing of customers remains exactly the same as shown in Figure 1.2 and customers are served on a FIFO basis at each station. Using the same notation as before, let us now impose the following bounds on service times:

$$(2.7) \qquad \begin{array}{llll} v_2(1) & > & \delta & \qquad v_4(1) & > & \delta \\ v_1(1) & < & \delta/n & \qquad v_3(1) & < & \delta/n, \end{array}$$

which of course implies $\max(m_1, m_3) < \min(m_2, m_4)$. We will find it convenient to refer to classes 1 and 3 as customers requiring "short" services and classes 2 and 4 as those requiring "long" services. Let us again assume initial conditions $Q_1(0) = n$ and $Q_2(0) = Q_3(0) = Q_4(0) = 0$; that is, all customers are initially queued at station 1 as class 1 awaiting a short service. From (2.7), one can verify that station 1 will finish serving all n class 1 customers (that is, all n customers will arrive to station 2) *before* station 2 completes its first long service. Thus all n class 2 services must be completed at station 2 before the first class 3 service can begin. Defining τ exactly as above, we find that at time τ all n customers are queued at station 2 as class 3. One may then argue exactly as above to conclude that (2.5) remains valid for a single FIFO system satisfying (2.7). In letting $n \to \infty$, we may keep the service time distributions for classes 2 and 4 (the long service operations) fixed, but (2.7) demands that service times for classes 1 and 3 vanish, implying that $m_1, m_3 \to 0$. Thus (2.6) remains valid in the current setting, and the long-run utilization rates at stations 1 and 2, respectively, converge as $n \to \infty$ to

$$\frac{m_4}{m_2 + m_4} < 1 \quad \text{and} \quad \frac{m_2}{m_2 + m_4} < 1.$$

3. A closed version of the Lu-Kumar priority network. With the routing portrayed in Figure 1.2, let us return to the assumption of preemptive-resume priority scheduling, with classes 4 and 2 having priority at stations 1 and 2, respectively, as in the Lu-Kumar open network (see Figure 1). Also, let all four service time distributions be exponential. With mean service times satisfying (1.3) we can (and do) choose units so that

(3.1) $m_2 + m_4 = 1$ but $m_1 + m_4 < 1$ and $m_2 + m_3 < 1$.

With exponentially distributed service times and preemptive resume priority scheduling the queue length process $Q(t) = [Q_1(t), Q_2(t), Q_3(t), Q_4(t)]$ is a Markov process with finite state space, but some of its states are transient, as explained below. Suppose that both $Q_2(0) > 0$ and $Q_4(0) > 0$. Then both servers will work on priority customers at their respective stations over an initial interval $[0, \tau]$, at the end of which either $Q_2(\tau) = 0$ or $Q_4(\tau) = 0$. Moreover, no new priority customers are created during this initial period, so it is immediate that $\mathbf{E}[\tau] < \infty$.

Suppose for the sake of concreteness that $Q_2(\tau) > 0$ and $Q_4(\tau) = 0$. Then server 2 will work on (priority) customers of class 2 over an interval (τ, τ'), during which period no new customers of class 4 can be created. Thus server 1 is either working on class 1 or else idle throughout the period, and at the end we have $Q_2(\tau') = Q_4(\tau') = 0$. Since no more than n customers can be served at station 2 in (τ, τ'), $\mathbf{E}[\tau' - \tau] < \infty$. The evolution of the system after time τ' can be divided into alternating "idle periods" and "busy periods" of length $I_1, B_1, I_2, B_2, \ldots$. Each idle period begins with $Q_2 = Q_4 = 0$, contains a single service completion, and ends with either $Q_2 = 1$ or $Q_4 = 1$ but not both. Using the memoryless property of the exponential distribution we have that

(3.2) $a_j \equiv \mathbf{E}[I_j] \leq \alpha \equiv \max(m_1, m_3)$ for all $j = 1, 2, \ldots$.

Of course, the expectation in (3.2) depends on the particular initial condition $Q(0)$ assumed, but we suppress that dependence to simplify notation, and the inequality in (3.2) is understood to hold *regardless* of initial state.

A "busy period" begins with one of the two servers having a single priority customer to work on (there may be non-priority customers present at one or both stations), and it ends as soon as that server runs out of priority customers. Because of the preemptive-resume priorities and the routing pictured in Figure 1.2, no priority customers are created for the *other* server during such a busy period, so it ends with $Q_2 = Q_4 = 0$ once again. The key to our analysis is the following bound, whose proof is postponed until the end of this section.

LEMMA 3.1. *There exists a constant $c > 0$, not depending on n, such that*

(3.3) $b_j \equiv \mathbf{E}[B_j] \geq cn$ *for all* $j = 1, 2, \ldots$.

The transient states $q = (q_1, q_2, q_3, q_4)$ referred to earlier are those with $q_2 > 0$ *and* $q_4 > 0$. As we have seen, such states cannot be revisited after the initial interval $[0, \tau]$, but all other states of the Markov chain $Q(t)$ communicate and hence there is a unique stationary distribution. Let us denote by π_2 the stationary probability of those states q with $q_2 > 0$, by π_4 the stationary probability of those states q with $q_4 > 0$, and by π_{13} the stationary probability of those q with $q_1 + q_3 = n$ (that is, $q_2 = q_4 = 0$). From the path decomposition described above (an initial period of finite expected duration, followed by alternating idle and busy periods, each with finite expected duration) it follows that

$$(3.4) \qquad\qquad \pi_2 + \pi_4 + \pi_{13} = 1.$$

Moreover, with $a_j \equiv \mathbf{E}[I_j]$ and $b_j \equiv \mathbf{E}[B_j]$ as in (3.2) and (3.3), we have that

$$(3.5) \qquad\qquad \pi_{13} = \lim_{k \to \infty} \left[\frac{\sum_{j=1}^{k} a_j}{\sum_{j=1}^{k} (a_j + b_j)} \right].$$

Two equivalent expressions for the long-run throughput rate λ are (here $\mu_k \equiv 1/m_k$)

$$(3.6) \qquad\qquad \lambda = \mu_2 \pi_2 \quad \text{and} \quad \lambda = \mu_4 \pi_4,$$

implying that $\lambda(m_2 + m_4) = \pi_2 + \pi_4$. With our convention that $m_2 + m_4 = 1$, this and (3.4) together imply

$$(3.7) \qquad\qquad \lambda = (1 - \pi_{13})/(m_2 + m_4) = 1 - \pi_{13}.$$

Together, (3.2), (3.3) and (3.5) imply that $\pi_{13} \to 0$ as $n \to \infty$, so we have the following major conclusion.

PROPOSITION 3.1. $\lambda \to 1$ as $n \to \infty$.

COROLLARY 3.1. *The long-run utilization rate for server 1 is* $\lambda(m_1 + m_4) \to m_1 + m_4 < 1$ *as* $n \to \infty$. *Similarly, the long-run utilization rate for server 2 is* $\lambda(m_2 + m_3) \to m_2 + m_3 < 1$ *as* $n \to \infty$.

It remains to prove Lemma 3.1. From (3.1) we have that

$$(3.8) \qquad\qquad \mu_1 > \mu_2 \quad \text{and} \quad \mu_3 > \mu_4,$$

Let us suppose that the jth busy period (its duration is B_j) begins with the creation of a class 2 customer, and that k customers of class 1 remain in existence after that creation ($k \le n - 1$). Until the supply of class 1 customers has been exhausted, the queue length process $Q_2(t)$ is precisely that of an $M/M/1$ queue with input rate μ_1, service rate μ_2, and hence traffic intensity parameter $\rho = \mu_1/\mu_2 > 1$. For such an $M/M/1$ queue there is a probability $p_{12} > 0$ that a busy period *never* ends. Using this and Wald's identity, we obtain a lower bound of $p_{12} m_2 k$ for the conditional expectation of B_j given k. Similarly, if the ith busy period begins with creation of a class 4 customer this creation leaving $n - k - 1$ class 3 customers behind, then the conditional expectation of B_j is bounded below by $p_{34}m_4(n - k)$, where $p_{34} > 0$. It follows that (3.3) holds with, for example,

$$(3.9) \qquad c = \min \left(\frac{\mu_1 p_{12} m_2}{\mu_1 + \mu_3}, \frac{\mu_3 p_{34} m_4}{\mu_1 + \mu_3} \right) > 0.$$

REFERENCES

[1] BRAMSON, M., *Instability of FIFO queueing networks*, Annals of Applied Probability, to appear 1994.

[2] DAI, J. G. AND G. WEISS, *Stability and instability of fluid models for certain multiclass networks*, preprint, 1994.

[3] LU, S. H. AND P. R. KUMAR, *Distributed scheduling based on due dates and buffer priorities*, IEEE Transactions on Automatic Control, 36 (1991), 1406–1416.

[4] RYBKO, S. AND A. L. STOLYAR, *Erogodicity of stochastic processes describing the operation of an open queueing network*, Problemy Peredachi Informatsil, 28 (1991), 2–26.

[5] SCHWEITZER, P. J., *Bottleneck determination in networks of queues*, in Applied Probability — Computer Science, The Interface, R. L. Disney and T. J. Ott (eds.), Vol. 10, Birkhauser, 1988.

[6] SEIDMAN, T. L., *'First Come, First Served' is unstable!*, preprint, 1994.

[7] WHITT, W., *Large fluctuations in a deterministic multiclass network of queues*, Management Science, 39 (1993), 1020–1028.

SEMIMARTINGALE REFLECTING BROWNIAN MOTIONS IN THE ORTHANT

R.J. WILLIAMS*

Abstract. This paper surveys recent work on semimartingale reflecting Brownian motions in the orthant. These diffusion processes have been proposed as approximate models of multiclass open queueing networks in heavy traffic. The topics covered by this survey are the problems of existence and uniqueness, recurrence classification and stationary distributions for these diffusions.

1. Introduction. Reflecting Brownian motions in the orthant have been proposed as approximate models of open queueing networks in heavy traffic (see Harrison-Nguyen [19]). Such networks are of current interest for studying congestion and delay in computer communication and manufacturing systems. For single class and feedforward multiclass networks with first-in-first-out service discipline, limit theorems to justify the diffusion approximation have been proved by Reiman [37] and Peterson [36], respectively. An outstanding open problem is to prove such a limit theorem for multiclass networks with feedback. Indeed, it has recently been discovered (see for example, Lu-Kumar [29], Rybko-Stolyar [39], Dai-Wang [11], Whitt [44], Bramson [3], [4] and Seidman [40]), that the conditions for stability of a multiclass network are not well understood. There is currently a good deal of activity directed towards resolving this problem which is a precursor to the development of a heavy traffic diffusion limit theorem for such networks. Despite the lack of a limit theorem in the multiclass case, it is still of interest and also of potential benefit for this approximation phase to develop a theory for the diffusions. This article focusses on the latter by surveying recent work on reflecting Brownian motions in the orthant. In fact, attention will be confined here to semimartingale reflecting Brownian motions in the orthant (SRBMs). The reasons for this are that (a) many of the queueing networks of interest lead to such processes (although there are some models of routing that lead to non-semimartingale reflecting Brownian motions — see [26] for a two-dimensional example), and (b) there currently is no theory for non-semimartingale reflecting Brownian motions in dimensions higher than two (see [6], [43], [45], [13], [46] for some results on the one and two dimensional cases).

This survey will concentrate on the following two topics for SRBMs:
(i) Synthesis — existence and uniqueness with given geometric data,
(ii) Analysis — recurrence classification and stationary distributions.

---

* Department of Mathematics, UCSD, 9500 Gilman Drive, La Jolla, CA 92093-0112, USA. The research reported in this paper was supported in part by NSF grant GER-9023335. In addition, part of this paper was written whilst the author was visiting the Institute for Mathematics and Its Applications, whose financial support is gratefully acknowledged.

2. Synthesis — existence and uniqueness. This section concerns
the problem of existence and uniqueness of semimartingale reflecting Brownian motions in the orthant (SRBMs). Loosely speaking, such a process
has a semimartingale decomposition such that in the interior of the orthant
it behaves like a Brownian motion with a constant drift and covariance
matrix, and at each of the $(d-1)$-dimensional boundary faces, the finite
variation part of the process increases in a given direction (constant for a
particular face), so as to confine the process to the orthant. For historical
reasons, this "pushing" at the boundary is called instantaneous reflection.

For a more precise description of an SRBM, the following notation is
needed. Let d be a positive integer, $S = \{x \in \mathbb{R}^d : x_i \geq 0 \text{ for } i = 1, \ldots, d\}$,
θ be a constant vector in $\mathbb{R}^d$, Γ be a $d \times d$ non-degenerate covariance
matrix (symmetric and positive definite), and R be a $d \times d$ matrix. A
triple $(\Omega, \mathcal{F}, \{\mathcal{F}_t\})$ is called a *filtered* space if Ω is a set, $\mathcal{F}$ is a σ-field of
subsets of Ω, and $\{\mathcal{F}_t\} \equiv \{\mathcal{F}_t, t \geq 0\}$ is an increasing family of sub-σ-fields
of $\mathcal{F}$, i.e., a filtration.

DEFINITION 2.1. *An SRBM associated with the data (S, θ, Γ, R) is a
continuous, $\{\mathcal{F}_t\}$-adapted, d-dimensional process Z together with a family of probability measures $\{P_x, x \in S\}$ defined on some filtered space
$(\Omega, \mathcal{F}, \{\mathcal{F}_t\})$ such that for each $x \in S$, under P_x,*

$$(2.1) \qquad\qquad Z(t) = X(t) + RY(t) \in S \quad \text{for all } t \geq 0,$$

where

(i) *X is a d-dimensional Brownian motion with drift vector θ, covariance
matrix Γ, $\{X(t) - \theta t, \mathcal{F}_t, t \geq 0\}$ is a martingale, and $X(0) = x$,
P_x-a.s.,*

(ii) *Y is an $\{\mathcal{F}_t\}$-adapted, d-dimensional process such that P_x-a.s. for
$i = 1, \ldots, d$,*
  (a) *$Y_i(0) = 0$,*
  (b) *Y_i is continuous and non-decreasing,*
  (c) *Y_i can increase only when Z is on the face $F_i \equiv \{x \in S : x_i =
0\}$, i.e., $\int_0^t 1_{F_i}(Z(s)) \, dY_i(s) = Y_i(t)$ for all $t \geq 0$.*

One approach to constructing an SRBM is to try to solve the following
deterministic Skorokhod problem for all continuous paths $x(\cdot)$ in $\mathbb{R}^d$ that
start in S. This approach is based on the hope that the solutions might
be given by a measurable, adapted path-to-path mapping, which could be
applied to the paths of a Brownian motion to yield a "strong solution" of
the equations defining an SRBM.

In the following, $\mathbf{C} \equiv C([0, \infty), \mathbb{R}^d) = \{x : [0, \infty) \to \mathbb{R}^d, x \text{ is continuous}\}$
and $\mathbf{C}_+ = \{x \in \mathbf{C} : x(0) \in S\}$. We endow $\mathbf{C}$ (and hence $\mathbf{C}_+$) with the
topology of uniform convergence on compact time intervals and define the
associated σ-fields $\mathcal{M} = \sigma\{x(s) : 0 \leq s < \infty\}$, $\mathcal{M}_t = \sigma\{x(s) : 0 \leq s \leq t\}$
for all $t \geq 0$.

DEFINITION 2.2. **(Skorokhod Problem)** *Let $x \in \mathbf{C}_+$. Then $(z, y) \in$*

$\mathbf{C} \times \mathbf{C}$ *solves the Skorokhod problem (SP) for x (with respect to S and R) if*

(i) $z(t) = x(t) + Ry(t) \in S$ *for all $t \geq 0$,*

(ii) *y is such that for $i = 1, \ldots, d$,*

    (a) $y_i(0) = 0$,

    (b) *y_i is non-decreasing,*

    (c) *y_i can increase only when z is on F_i.*

Harrison and Reiman [20] considered this problem in connection with single class open queueing networks. The type of R matrix that arises from such networks is of the form $R = I - P'$ where P has zeros on its diagonal and is a transition matrix for a transient Markov chain on d states (corresponding to the d nodes in the network). For such an R matrix, Harrison and Reiman showed the existence of a Lipschitz continuous path-to-path mapping $\Phi : \mathbf{C}_+ \rightarrow \mathbf{C} \times \mathbf{C}$ which for each $x \in \mathbf{C}_+$ yields an adapted solution $(z, y) = \Phi(x)$ of the Skorokhod problem for x. (Here adapted means that the mapping $x \rightarrow (z_t, y_t)$ is $\mathcal{M}_t$-measurable for each $t \geq 0$.) Moreover, this solution is the unique one for x. Scrutiny of the proof in [20] reveals that the result extends to the case where $R = I - Q$, Q has spectral radius strictly less than one and there are no restrictions on the signs of the entries in Q (see also Dupuis and Ishii [14]). If we apply this result to the paths of a Brownian motion then we obtain a strong solution of the SRBM equations. In summary we have the following.

THEOREM 2.1. *Suppose that $R = I - Q$ where Q has spectral radius strictly less than one. Then for each $x \in \mathbf{C}_+$, there is a unique adapted solution $\Phi(x) = (z, y)$ of the Skorokhod problem for x. Define X, Y, Z on $\mathbf{C}_+$ by $X(x) = x$, $(Y, Z)(x) = \Phi(x)$ for each $x \in \mathbf{C}_+$, and for each $x_0 \in S$, let P_{x_0} denote the probability measure on $(\mathbf{C}_+, \mathcal{M})$ under which the canonical path $X = x(\cdot)$ is a Brownian motion with drift θ, covariance matrix Γ, and starts from x_0. Then Z together with $\{P_{x_0}, x_0 \in S\}$ is an SRBM associated with (S, θ, Γ, R). Furthermore, it defines a Feller continuous strong Markov process.*

Beyond the result of Harrison and Reiman [20], it is relatively easy to see that a necessary condition for one to be able to solve the Skorokhod problem for each $x \in \mathbf{C}_+$ is that at any point on the boundary of S, there is a positive linear combination of the vectors of reflection that one is allowed to use there, which points into the interior of the state space. This can also be shown to be necessary for the existence of an SRBM starting from each point in S (see Theorem 2.3 below). This geometric condition can be expressed succinctly as the following algebraic *completely-S* condition on the reflection matrix R.

DEFINITION 2.3. *A principal submatrix of the $d \times d$ matrix R is any square matrix obtained from R by deleting all rows and columns from R with indices in some (possibly empty) subset of $\{1, \ldots, d\}$. The matrix R is completely-S if and only if for each principal submatrix $\tilde{R}$ of R there is $\tilde{x} \geq 0$ such that $\tilde{R}\tilde{x} > 0$.*

Completely-S matrices are known in the operations research literature. Alternative names for these matrices are strictly semimonotone or completely-Q matrices. A useful property is that R is completely-S if and only if its transpose R' is completely-S (see [38], Lemma 3, p. 91).

If we denote the columns of R by $v_1, \ldots, v_d$ and the inward unit normals to the faces F_i by n_i for $i = 1, \ldots, d$, then the completely-S condition can be written in coordinate form as

(S.a) for each $\mathbf{K} \subset \{1, \ldots, d\}$ there is a positive linear combination $v = \sum_{i \in \mathbf{K}} a_i v_i$ $(a_i > 0 \ \forall i \in \mathbf{K})$ of the $\{v_i, i \in \mathbf{K}\}$ such that $n_i \cdot v > 0$ for all $i \in \mathbf{K}$,

and its adjoint form is:

(S.b) for each $\mathbf{K} \subset \{1, \ldots, d\}$ there is a positive linear combination $n = \sum_{i \in \mathbf{K}} b_i n_i$ $(b_i > 0 \ \forall i \in \mathbf{K})$ of the $\{n_i, i \in \mathbf{K}\}$ such that $n \cdot v_i > 0$ for all $i \in \mathbf{K}$.

Two sets of authors, namely, Bernard and El Kharroubi [2] and Mandelbaum and Van der Heyden [34], independently showed that there is a solution of the Skorokhod problem for all $x \in \mathbf{C}_+$ when R is completely-S. Briefly their approach may be described as follows. Approximate the path of $x \in \mathbf{C}_+$ by a piecewise linear path $\tilde{x}$ and control this approximate path using a pushing path $\tilde{y}$ which keeps the resultant path $\tilde{z} = \tilde{x} + R\tilde{y}$ in S. Control of the affine segments of $\tilde{x}$ is achieved by solving linear complementarity problems of the following form.

DEFINITION 2.4. (**Linear Complementarity Problem**) *Let* $x \in \mathbb{R}^d$. *Then* $(z, y) \in \mathbb{R}^d \times \mathbb{R}^d$ *is a solution of the linear complementarity problem for x and R if*
(i) $z = x + Ry \in S$,
(ii) *for* $i = 1, \ldots, d$,
      (a) $y_i \geq 0$,
      (b) $z_i y_i = 0$.

One can pass to a limit in the approximation (possibly along a subsequence) using the following oscillation estimate (see Lemma 2.1) to obtain the necessary compactness. This then yields a solution of the Skorokhod problem, which is also known as the *dynamic complementarity problem*. For the statement of the oscillation estimate we need the following. For any continuous function f defined on $[t_1, t_2]$ into $\mathbb{R}^k$, some $k \geq 1$, let

$$\mathrm{Osc}(f, [t_1, t_2]) = \sup_{t_1 \leq s \leq t \leq t_2} |f(t) - f(s)|.$$

LEMMA 2.1. *Suppose that R is completely-S. Then there is a constant κ which depends only on R, such that for any $x \in \mathbf{C}_+$ and solution (z, y) of the Skorokhod problem for x, we have for all $0 \leq t_1 < t_2 < \infty$,*

$$\mathrm{Osc}(z, [t_1, t_2]) \leq \kappa \, \mathrm{Osc}(x, [t_1, t_2]), \quad \mathrm{Osc}(y, [t_1, t_2]) \leq \kappa \, \mathrm{Osc}(x, [t_1, t_2]).$$

Proof. See Bernard and El Kharroubi [2] or Dai and Williams [12], Lemma 4.3. □

Summarizing the above we have the following.

THEOREM 2.2. *There is a solution of the Skorokhod problem for each* $x \in \mathbf{C}_+$ *if and only if R is completely-$\mathcal{S}$.*

The linear complementarity problem has unique solutions whenever R is a P-matrix (all principal minors are positive) (see [7]). However, solutions of the Skorokhod problem need not be unique, even if R is a P-matrix, as shown by examples of Bernard and El Kharroubi [2] and Mandelbaum [33]. This non-uniqueness has the further consequence that the authors of [2] and [34] were not able to show that there is a measurable path-to-path mapping $\Phi : \mathbf{C}_+ \to \mathbf{C} \times \mathbf{C}$ which for each $x \in \mathbf{C}_+$ yields an adapted solution $(z, y) = \Phi(x)$ of the Skorokhod problem for x. The reason for this is that in taking the limit of the approximate solutions, one may need to pass to a subsequence that depends on x, and since there is no guarantee of uniqueness, one is not able to show that the solution can be chosen to be an adapted function of x. Consequently, Theorem 2.2 cannot be used to construct strong solutions for the SRBM equations. It is still an open problem to determine a necessary and sufficient condition for the existence and uniqueness of strong solutions to the SRBM equations.

Taking a different approach, in [41], Taylor and Williams sought weak solutions of the SRBM equations. That is, rather than trying to find (Z, Y) adapted to a given X, they sought (X, Y, Z) together which satisfied the SRBM equations. This approach yielded the following weak existence and uniqueness of an SRBM.

THEOREM 2.3. *There exists an SRBM with data (S, θ, Γ, R) if and only if R is completely-$\mathcal{S}$. In this case, the SRBM is unique in law and defines a Feller continuous strong Markov process.*

Proof of Theorem 2.3. The necessity of the completely-$\mathcal{S}$ condition follows easily by considering starting points that range over all of the facets of the boundary with dimensions between zero and $d - 1$ inclusive (see Reiman and Williams [38] for a proof). The sufficiency of the completely-$\mathcal{S}$ condition, as well as the uniqueness in law, is proved in Taylor and Williams [41]. An alternative method for proving existence that exploits Kurtz's [27] patchwork and constrained martingale problem methodology is used in Dai and Williams [12]. The uniqueness proof follows the general line of argument in Bass and Pardoux [1] or Kwon and Williams [28], but hinges on an ergodic property which is verified using the oscillation estimate of Lemma 2.1. The Feller continuity and strong Markov property follow by standard arguments once the existence, uniqueness and tightness of the associated probability measures are established (see [41], p. 316).   □

Remark. An extension of the results in this section to semimartingale reflecting Brownian motions in convex polyhedrons has been obtained by Dai and Williams [12]. Their result is sharpest for simple convex polyhedrons. In a simple polyhedron, precisely d faces meet at a vertex in d dimensions, and then a natural generalization of the completely-$\mathcal{S}$ condition (S.a) is necessary and sufficient for the weak existence and uniqueness of an SRBM.

For non-simple convex polyhedrons, a generalized completely-$\mathcal{S}$ condition together with an adjoint condition (cf. **(S.b)**) provides a sufficient condition for the weak existence and uniqueness of an SRBM. See [12] for more details. Reflecting Brownian motions in convex polyhedrons are of interest as approximate models of closed, capacitated and fork-join networks. In particular, the latter can lead to SRBMs in non-simple convex polyhedrons (see Nguyen [35]).

3. Analysis — recurrence and stationary distributions. Henceforth we assume that the matrix R is completely-$\mathcal{S}$ and without loss of generality we assume that it has ones on its diagonal (this can always be achieved by a renormalization of the vectors of reflection). We let Z together with $\{P_x, x \in S\}$, defined on some filtered space, denote a realization of the SRBM associated with (S, θ, Γ, R). Expectations under P_x will be denoted by E_x.

We first consider the problem of determining conditions for positive recurrence of Z. If Z is positive recurrence then there is a unique stationary distribution for Z. The second part of this section is devoted to the problem of characterizing (and computing) such a stationary distribution.

DEFINITION 3.1. *The SRBM Z is positive recurrent if for each closed set A in S having positive Lebesgue measure we have $E_x[\tau_A] < \infty$ for all $x \in S$, where $\tau_A = \inf\{t \geq 0 : Z(t) \in A\}$.*

Recently Dupuis and Williams [15] proved that a sufficient condition for positive recurrence of the SRBM Z is that all solutions of a related deterministic Skorokhod problem are attracted to the origin in the following sense.

DEFINITION 3.2. *A path $z \in \mathbf{C}$ is attracted to the origin if and only if for each $\epsilon > 0$ there is $T < \infty$ such that $|z(t)| \leq \epsilon$ for all $t \geq T$.*

THEOREM 3.1. *Suppose that all solutions z of the Skorokhod problem for drift paths x of the form $x(t) = x_0 + \theta t$, $t \geq 0$, $x_0 \in S$, are attracted to the origin. Then the SRBM Z is positive recurrent and it has a unique stationary distribution.*

Proof. The details of the proof can be found in Dupuis and Williams [15]. We briefly mention some of the key aspects here. Firstly, because of the Brownian motion diffusive aspect of Z, it suffices for the positive recurrence of Z to show that $E_x[\tau_r] < \infty$ for all $x \in S$ and all r sufficiently large, where $\tau_r = \inf\{t \geq 0 : |Z(t)| \leq r\}$ (see the proof of Theorem 2.6 in [15]). The mechanism used in [15] to establish the finiteness of these first moment hitting times is to construct a Lyapunov function f that has the following properties. The function f is twice continuously differentiable on $S \setminus \{0\}$ and

(i) given $N < \infty$, there is $M < \infty$ such that $x \in S$ and $|x| \geq M$ imply
    $f(x) \geq N$,

(ii) given $\epsilon > 0$ there is $M < \infty$ such that $x \in S$ and $|x| \geq M$ imply
    $\|D^2 f(x)\| \leq \epsilon$, where $\|D^2 f(x)\|$ denotes the matrix norm of the

Hessian $D^2 f$ of f at x,

(iii) there exists $c > 0$ such that

   (a) $\theta \cdot \nabla f(x) \leq -c$ for all $x \in S \setminus \{0\}$,

   (b) $v_i \cdot \nabla f(x) \leq -c$ for all $x \in F_i$, $i = 1, \ldots, d$.

Here ∇ denotes the gradient of f. It follows easily from Itô's formula and the semimartingale decomposition of Z that for such a function f there is $\eta > 0$ such that for all r sufficiently large,

$$E_x[\tau_r] \leq 2f(x)/\eta.$$

The main work of [15] is in constructing the Lyapunov function f. Besides its use in proving positive recurrence, this Lyapunov function can be used to obtain bounds on moments and path excursion estimates for the SRBM. It can also be used to prove that certain functionals of processes approximating an SRBM converge weakly to the same functionals of the limit SRBM.   □

Theorem 3.1 can be applied to verify the sufficiency of the following conditions for positive recurrence of an SRBM (these results were previously established by other means). It is an interesting open problem to obtain some new conditions for positive recurrence using Theorem 3.1. Another problem of interest is to determine whether the condition of Theorem 3.1 is *necessary* for positive recurrence of the SRBM.

In the following, v_{ij} denotes the jth component of the vector v_i, which in turn is the ith column of the reflection matrix R.

THEOREM 3.2. *Suppose $d = 2$. Then the SRBM Z is positive recurrent if and only if*

$$\theta_1 + v_{21}\theta_2^- < 0 \quad and \quad \theta_2 + v_{12}\theta_1^- < 0,$$

where the minus sign superscript denotes the negative part of a number and since R has ones on its diagonal, $v_{11} = 1, v_{22} = 1$.

Proof. See Hobson and Rogers [24] for the case $\theta \neq 0$, and Williams [46] for $\theta = 0$. The sufficiency of the above conditions can also be verified by hand using Theorem 3.1.   □

THEOREM 3.3. *Suppose $R = I - P'$ where P has zeros on its diagonal and is a transition matrix for a transient Markov chain on d states. Then the SRBM Z is positive recurrent if and only if $R^{-1}\theta < 0$, where the inequality is understood to hold component by component.*

Proof. See Harrison and Williams [21]. The necessity of the condition is easy to establish. An alternative proof of the sufficiency can be given using Theorem 3.1 and the stability of solutions of the Skorokhod problem for drift paths established by Chen and Mandelbaum [5], Theorem 5.2.   □

Remark. Stimulated by the result in Theorem 3.1, J. G. Dai [8] has proved an analogue for queueing networks, namely that stability of fluid limits associated with a queueing network implies positive recurrence for a Markov process that describes the dynamics of the network. Dai's method of proof

is different from that in [15] in the sense that he does not construct a
Lyapunov function. See other papers in this volume for discussion of how
Dai's result can be used to determine sufficient conditions for the stability
of multiclass networks with feedback.

Remark. Malyshev and his collaborators [30], [32], [31], [25], have been
working on problems of recurrence for reflected random walks in orthants.
They too use a related deterministic dynamical system and Lyapunov functions to obtain conditions for positive recurrence of their reflected random
walks. However, the details of how they obtain their deterministic dynamical system from the reflected random walk seem to be different from
how the Skorokhod problem with drift paths is related to the SRBM. It
would be interesting to further investigate the connections and differences
between these approaches.

Let us now turn to the problem of characterizing the stationary distribution of the SRBM Z.

DEFINITION 3.3. *A stationary distribution for Z is a probability measure π on the state space S, endowed with the Borel σ-algebra, such that
for each real-valued, bounded Borel measurable function f defined on S, we
have*

$$\int_S E_x[f(Z(t))]\,\pi(dx) = \int_S f(x)\,\pi(dx) \quad \text{for all } t \geq 0.$$

For such a π, we shall let $P_\pi = \int_S P_x \pi(dx)$ and E_π denote expectation
under P_π. Parts (i) and (ii) of the following result were proved in Harrison
and Williams [21] for the case where $R = I - P'$, P has zeros on the diagonal
and is the transition matrix for a transient Markov chain. However, as
described in Dai and Harrison [9], the proofs easily extend to the case
where R is completely-$\mathcal{S}$. A simple proof of part (iii) is given in [9].

LEMMA 3.1. *Suppose that Z has a stationary distribution π. Then the
following hold.*

(i) *The stationary distribution π is unique and it has a density p_0 relative
to Lebesgue measure on S.*

(ii) *For each $i \in \{1,\ldots,d\}$, there exists a finite Borel measure ν_i on F_i
that has a density p_i relative to surface measure σ_i on F_i, and*

$$E_\pi\left[\int_0^t f(Z(s))\,dY_i(s)\right] = t\int_{F_i} f(x)\nu_i(dx),$$

*for all real-valued, bounded Borel measurable functions f defined
on F_i.*

(iii) *The matrix R is invertible.*

From the abstract theory of Markov processes, we know that a stationary distribution for Z can be characterized by the property that it
annihilates all of the functions in the range of the infinitesimal generator
(see [16] for example), i.e., $\int_S f d\pi = 0$ for all f in the range of the infinitesimal generator. However, in general, for multidimensional diffusions

with boundary and especially for the case treated here where the boundary conditions are non-smooth, it is virtually impossible to characterize the range of the generator. Thus, it is natural to seek an alternative annihilation characterization. The one to be given here has the virtue that it incorporates boundary information into the annihilation relation, rather than incorporating it by restricting to a class of test functions f that satisfy certain boundary conditions. To obtain the characterization we need to make a connection between the process Z and its associated analytical theory. Since Z is a semimartingale, the natural mechanism for this is Itô's formula. Let $C_b^2(S)$ denote the space of continuous, bounded, real-valued functions defined on S that have bounded continuous first and second partial derivatives on S, and suppose that the semimartingale decomposition of Z is given by (2.1). Then, for each $f \in C_b^2(S)$ and $x \in S$, we have P_x-a.s. for all $t \geq 0$,

$$(3.1) \qquad f(Z(t)) - f(Z(0)) = \int_0^t \nabla f(Z(s)) \cdot dB(s)$$

$$+ \sum_{i=1}^d \int_0^t D_i f(Z(s)) \, dY_i(s)$$

$$+ \int_0^t Lf(Z(s)) \, ds,$$

where the first integral is a stochastic integral with respect to the driftless Brownian motion $B(t) = X(t) - \theta t$,

$$Lf = \frac{1}{2} \sum_{i,j=1}^d \Gamma_{ij} \frac{\partial^2 f}{\partial x_i \partial x_j} + \sum_{i=1}^d \theta_i \frac{\partial f}{\partial x_i},$$

and

$$D_i f = v_i \cdot \nabla f, \quad \text{for } i = 1, \ldots, d.$$

Suppose π is a stationary distribution for Z. Taking expectations in (3.1) under E_π and using Lemma 3.1 we see that $p = (p_0; p_1, \ldots, p_d)$ must satisfy the following *basic adjoint relation:*

$$(3.2) \qquad \int_S Lf \, p_0 \, dx + \sum_{i=1}^d \int_{F_i} D_i f \, p_i \, d\sigma_i = 0 \quad \text{for all } f \in C_b^2(S).$$

Thus, (3.2) is a necessary condition that must be satisfied by the density p associated with a stationary distribution π and the auxiliary measures ν_i, $i = 1, \ldots, d$. In fact, (3.2) characterizes the stationary distribution, as shown by the following theorem of Dai and Kurtz [10].

THEOREM 3.4. *Suppose that p_0 is a probability density on S (relative to Lebesgue measure) and for each $i \in \{1, \ldots, d\}$, p_i is a non-negative* *integrable (with respect to σ_i) Borel measurable function defined on F_i. If $p = (p_0; p_1, \ldots, p_d)$ satisfies the basic adjoint relation (3.2), then p_0 is the stationary density for Z and $d\nu_i = p_i d\sigma_i$ defines the boundary measures described in Lemma 3.1(ii).*

We may think of the integral relation (3.2) as the weak form of an elliptic partial differential equation with oblique derivative boundary conditions. One might be tempted to try to work with these differential equations directly, rather than with the integral relation (3.2). However, there is a primary difficulty with this. From the probabilistic derivation of (3.2), we do not obtain any smoothness properties of p other than the obvious application of Weyl's lemma which yields that p_0 is C^∞ in the interior of S. In particular, it is difficult to determine the regularity properties of p_0 near the boundary of S. Furthermore, experience with closed form solutions for two dimensional cases (see [17], [18], [42]), suggests that p_0 may have singularities at some of the non-smooth parts of the boundary. Direct analysis of (3.2) seems to be more fruitful. Indeed, we have the following result on product form solutions.

DEFINITION 3.4. *The SRBM Z is said to have a product form stationary distribution if it has a stationary density p_0 which can be written in the form*

$$p_0(x) = \prod_{i=1}^{d} p_0^i(x_i) \quad \text{for all } x \in S,$$

where for each $i \in \{1, \ldots, d\}$, p_0^i is a probability density relative to Lebesgue measure on $[0, \infty)$.

THEOREM 3.5. *The SRBM Z has a product form stationary distribution if and only if*

$$\gamma = -R^{-1}\theta > 0$$

and

$$2\Gamma = R\Lambda + \Lambda R'$$

where Λ is a diagonal matrix with the same diagonal entries as Γ. In this case, there is a positive constant C such that

$$p_0(x) = C \exp(-2\gamma \cdot x) \quad \text{for all } x \in S,$$

and the boundary density p_i is the restriction of $\frac{1}{2}p_0$ to F_i, $i = 1, \ldots, d$.

Proof. As observed by Dai and Harrison [9], the proof given in Harrison and Williams [21] for the case of matrices R that come from single class open queueing networks extends to all completely-$\mathcal{S}$ matrices R. □

For matrices R that are associated with Brownian models of feedforward (multiclass) networks, an interpretation of the above product form condition was given in Harrison and Williams [23], in terms of a notion of quasireversibility for the Brownian model. Furthermore, examples were given in [23] of queueing networks which are not known to be of product form, but for which the approximating Brownian model has a product form stationary distribution.

In general, it is unlikely that one will be able to find closed form solutions to (3.2). Thus one is naturally led to consider numerical methods. Dai and Harrison [9] have initiated work in this direction by using an L^2 projection scheme to find approximate solutions of this relation with $p_i = p_0/2$. One of the interesting problems in this area is that it is difficult to impossible to test numerically whether a function is positive. Thus, one would like to have a characterization of the stationary density as in Theorem 3.4, but without the positivity assumptions on $p_0, p_1, \ldots, p_d$.

4. Conclusion. As the preceding survey shows, a good deal of progress has been made on the theory of reflecting Brownian motions in polyhedral domains in recent years. However, a number of open problems remain. A selection of these is summarized below.

(i) To prove a heavy traffic limit theorem for multiclass open queueing networks which justifies the use of SRBMs as approximate models for such.

(ii) To determine necessary and sufficient conditions for the existence and uniqueness of semimartingale reflecting Brownian motions in non-simple convex polyhedrons (only sufficient conditions are given in Dai-Williams [12]).

(iii) Develop a theory of non-semimartingale reflecting Brownian motions in convex polyhedrons in three and more dimensions.

(iv) Use the sufficient condition of Theorem 3.1 to obtain concrete algebraic conditions for the positive recurrence of SRBMs. Also, determine whether this condition is necessary.

(v) Try to free the analytic characterization (cf. Theorem 3.4) of the stationary density for an SRBM from the a priori assumption of positivity of the densities $p_0, p_1, \ldots, p_d$. Also, try to develop new numerical schemes for computing such stationary densities or related moments.

REFERENCES

[1] R. F. BASS AND E. PARDOUX, *Uniqueness for diffusions with piecewise constant coefficients*, Prob. Theory Rel. Fields **76** (1987), 557–572.

[2] A. BERNARD AND A. EL KHARROUBI, *Régulation de processus dans le premier orthant de R^n*, Stochastics and Stochastics Reports **34** (1991), 149–167.

[3] M. BRAMSON, *Instability of FIFO queueing networks*, Ann. Appl. Prob. (to appear).

[4] ———, *Instability of FIFO queueing networks with quick service times*, Ann. Appl. Prob. (to appear).

[5] H. CHEN AND A. MANDELBAUM, *Discrete flow networks: Bottleneck analysis and fluid approximations*, Math. Oper. Res. **16** (1991), 408–446.

[6] K. L. CHUNG AND R. J. WILLIAMS, *Introduction to Stochastic Integration*, Birkhäuser, Boston, 2nd edition, 1990.

[7] R. W. COTTLE, J.-S. PANG AND R. E. STONE, *The Linear Complementarity Problem*, Academic Press, San Diego, 1992.

[8] J. G. DAI, *On positive Harris recurrence of multiclass queueing networks: A unified approach via fluid limit models*, Ann. Appl. Prob. (to appear).

[9] J. G. DAI AND J. M. HARRISON, *Reflected Brownian motion in an orthant: numerical methods for steady-state analysis*, Ann. Appl. Prob. **2** (1992), 65–86.

[10] J. G. DAI AND T. G. KURTZ, *Characterization of the stationary distribution for a semimartingale reflecting Brownian motion in a convex polyhedron*, preprint.

[11] J. G. DAI AND Y. WANG, *Nonexistence of Brownian models of certain multiclass queueing networks*, Queueing Systems: Theory and Applications **13** (1993), 41–46.

[12] J. G. DAI AND R. J. WILLIAMS, *Existence and uniqueness of semimartingale reflecting Brownian motions in convex polyhedrons*, Theory of Probability and its Applications (to appear).

[13] R. D. DE BLASSIE, *Explicit semimartingale representation of Brownian motion in a wedge*, Stochastic Processes and their Applications **34** (1990), 67–97.

[14] P. DUPUIS AND H. ISHII, *On the Lipschitz continuity of the solution mapping to the Skorokhod problem, with applications*, Stochastics **35** (1991), 31–62.

[15] P. DUPUIS AND R. J. WILLIAMS, *Lyapunov functions for semimartingale reflecting Brownian motions*, Ann. Prob. (to appear).

[16] S. N. ETHIER AND T. G. KURTZ, *Markov Processes: Characterization and Convergence*, Wiley, New York, 1986.

[17] J. M. HARRISON, *The diffusion approximation for tandem queues in heavy traffic*, Adv. Appl. Prob. **10** (1978), 886–905.

[18] J. M. HARRISON, H. LANDAU AND L. A. SHEPP, *The stationary distribution of reflected Brownian motion in a planar region*, Ann. Prob. **13** (1985), 744–757.

[19] J. M. HARRISON AND V. NGUYEN, *Brownian models of multiclass queueing networks: Current status and open problems*, Queueing Systems: Theory and Applications, **13** (1993), 5–40.

[20] J. M. HARRISON AND M. I. REIMAN, *Reflected Brownian motion on an orthant*, Ann. Prob. **9** (1981), 302–308.

[21] J. M. HARRISON, AND R. J. WILLIAMS, *Brownian models of open queueing networks with homogeneous customer populations*, Stochastics **22** (1987), 77–115.

[22] J. M. HARRISON AND R. J. WILLIAMS, *Multidimensional reflected Brownian motions having exponential stationary distributions*, Ann. Prob. **15** (1987), 115–137.

[23] J. M. HARRISON, AND R. J. WILLIAMS, *Brownian models of feedforward queueing networks: quasireversibility and product form solutions*, Ann. Appl. Prob. **2** (1992), 263–293.

[24] D. G. Hobson and L. C. G. Rogers, *Recurrence and transience of reflecting Brownian motion in the quadrant*, Math. Proc. Cambridge Philosophical Society **113** (1993), 387–399.

[25] I. Ignatyuk and V. A. Malyshev, *Classification of random walks in Z_+^4*, Selecta Mathematica Sovietica **12** (1993), 129–194.

[26] F. P. Kelly and C. N. Laws, *Dynamic routing in open queueing networks: Brownian models, cut constraints and resource pooling*, Queueing Systems **13** (1993), 47–86.

[27] T. G. Kurtz, *Martingale problems for constrained Markov processes*, in J. S. Baras and V. Mirelli (eds.), Recent Advances in Stochastic Calculus, Springer-Verlag, New York, 1990.

[28] Y. Kwon and R. J. Williams, *Reflected Brownian motion in a cone with radially homogeneous reflection field*, Trans. Amer. Math. Soc. **327** (1991), 739–780.

[29] S. H. Lu and P. R. Kumar, *Distributed scheduling based on due dates and buffer priorities*, IEEE Transactions on Automatic Control **36** (1991), 1406–1416.

[30] V. Malyshev, *Classification of two-dimensional random walks and almost linear semimartingales*, Dokl. Acad. Nauk. USSR **217** (1974), 755–758.

[31] V. Malyshev, *Networks and dynamical systems*, Adv. Appl. Prob. **25** (1993), 140–175.

[32] V. Malyshev and M. V. Menshikov, *Ergodicity, continuity, and analyticity of countable Markov chains*, Trans. Moscow Math. Soc. **1** (1981), 1–48.

[33] A. Mandelbaum, *The dynamic complementarity problem*, Math. Oper. Research (to appear).

[34] A. Mandelbaum and L. Van der Heyden, *Complementarity and reflection*, unpublished work, 1987.

[35] V. Nguyen, *Processing networks with parallel and sequential tasks: heavy traffic analysis and Brownian limits*, Ann. Appl. Prob. **3** (1993), 28–55.

[36] W. P. Peterson, *Diffusion approximations for networks of queues with multiple customer types*, Math. Oper. Res. **9** (1991), 90–118.

[37] M. I. Reiman, *Open queueing networks in heavy traffic*, Math. Oper. Res. **9** (1984), 441–458.

[38] M. I. Reiman and R. J. Williams, *A boundary property of semimartingale reflecting Brownian motions*, Prob. Theory Rel. Fields **77** (1988), 87–97, and **80** (1989), 633.

[39] S. Rybko and A. L. Stolyar, *Ergodicity of stochastic processes that describe functioning of open queueing networks*, Problems of Information Transmission **28** (1992), 3–26. (In Russian.)

[40] T. I. Seidman, *'First come, first served' can be unstable!*, IEEE Transactions on Automatic Control (to appear).

[41] L. M. Taylor and R. J. Williams, *Existence and uniqueness of semimartingale reflecting Brownian motions in an orthant*, Prob. Theory Rel. Fields **96** (1993), 283–317.

[42] L. N. Trefethen and R. J. Williams, *Conformal mapping solution of Laplace's equation on a polygon with oblique derivative boundary conditions*, J. Computational and Applied Math. **14** (1986), 227–249.

[43] S. R. S. Varadhan and R. J. Williams, *Brownian motion in a wedge with oblique reflection*, Comm. Pure Appl. Math. **38** (1985), 405–443.

[44] W. Whitt, *Large fluctuations in a deterministic multiclass network of queues*, Management Sciences **39** (1993), 1020–1028.

[45] R. J. Williams, *Reflected Brownian motion in a wedge: semimartingale property*, Z. Wahrsch. verw. Geb. **69** (1985), 161–176.

[46] R. J. Williams, *Recurrence classification and invariant measure for reflected Brownian motion in a wedge*, Ann. Prob. **13** (1985), 758–778.

[47] R. J. Williams, *Reflected Brownian motion with skew symmetric data in a polyhedral domain*, Prob. Th. Rel. Fields **75** (1987), 459–485.

A CONTROL PROBLEM FOR A NEW TYPE OF PUBLIC TRANSPORTATION SYSTEM, VIA HEAVY TRAFFIC ANALYSIS

HAROLD J. KUSHNER*

Abstract. To relieve the problem of highway congestion (both urban and suburban) the following system has been suggested. There are a number of nodes, where customers would come to rent or borrow a small vehicle at low cost, use it for some random length of time and then return it to one of the nodes. The system is closed in the sense that, although the vehicles circulate betwwen the nodes, the total number of vehicles is fixed. The system is open in the sense that the number of customers is not fixed. Major factors in the design as well as in the acceptability and affordability involve questions of controlled queueing. We derive the system equations. Then under the appropriate heavy traffic assumptions, it is shown that the system can be well approximated by a controlled reflected diffusion process. The heavy traffic limit theorms are proved for the system with a fixed (or no) control, and then for the optimally controlled case. The heavy traffic approach simplifies the model and gives the results in a form which depends on the basic structural properties of the system. Much inessential local detail is taken out. For certain symmetric cases, the dimensionality of the result is low enough so that numerical methods can be applied. The presented data show the qualitative properties of such systems as well as the role of control.

Key words. Heavy traffic approximations, controlled queues, numerical analysis of queues, public transportation problem

AMSMOS 60F17, 93E20, 90B22, 90B15

1. Introduction. The following promising approach to the problem of highway congestion (both urban and suburban) has been suggested and is being developed in France at this time. One would have a number of nodes, where customers would come to rent or borrow a small vehicle, use it for some length of time and possibly return it to another node. The vehicles would be small and efficient, possibly being powered electrically. The public acceptabiity of such a system remains to be seen. But, certainly major factors in the acceptability and affordability involve questions of queueing and control management. The system is closed in the sense that the number of vehicles is fixed and they circulate between the various nodes. But it is open in the sense that the number of customers is not fixed.

To date there seems to be little or no analysis of the queueing management aspects. One would expect that for such a system to be successful, it would have to have many vehicles and many customers. This suggests that some sort of heavy traffic analysis might be helpful in simplifying the basic physical models. We attempt such an analysis here. The main aim

---

* Division of Applied Mathematics, Brown University, Providence, RI, 02912. The work was partially supported by grants ECS-9302137, DAAH04-93-6-0070 and DAAL03-92-G-01157. The author would like to thank Dennis Jarvis for writing the very efficient codes, and Guy Fayolle for the introduction to the problem.

concerns an understanding of the control problem. The controls might be either reservation systems or the actual transfer of vehicles from less used nodes to more heavily used nodes. Many questions are of interest: for example, waiting times, dependence of performance on the system parameters, relative numbers of vehicles on route vs. the mean numbers in the parking areas of the nodes, etc.

The fundamental model for the general system is high dimensional, since one needs to keep track of the queues (whether of customers or vehicles) as well as of the numbers of vehicles in transit between any two nodes. We start with a general model, state the assumptions, write the state equations, and then prove a general heavy traffic limit theorem. This theorem shows that the general problem can be approximated by a much simpler limit system, a reflected diffusion process, if the traffic is heavy and the system is well designed. The meaning of "well designed" will be clear from the discussion below. But, loosely speaking, it means that (modulo moderate mean delays) the supply of vehicles is just enough to meet demand. The limit theorem tells us what the correct dimensioning of the system is, and indicates the economies of scale that the large numbers imply. For simplicity this first model will not be controlled. Controls are to be added later. Under various symmetries, the dimensionality can be reduced. This is done in Sections 3–6.

In Section 6, we look at perhaps the simplest model which can still yield much useful information, where the probability of a customer going to a particular destination depends only on the destination and not the source, and the transition times have the same distribution for all source-destination pairs. For this case, we introduce the control explicitly and prove that the heavy traffic limits can be used to get arbitarily good controls for the actual physical problem. The cost structure penalizes customer waiting, customer and vehicle buffer overflow and the cost of controlled transfers of vehicles between the nodes. We work with the average cost per unit time criterion. This is actually the hardest case. There are analogous results for the discounted cost model. The controls are robust in that various reasonable approximations to the optimal controls also yield good results. There are numerical methods for solving the optimization problems for the heavy traffic limits. Numerical data is presented in Section 7, where one can see the salient features and scaling requirements of well operating systems.

Although the numerical results are for small systems, the numerical and general analytical results give clues for larger systems. They tell us what the proper orders should be and what one can expect in terms of tradeoffs between vehicle numbers, customer waiting and vehicle control problems. A reasonable next step would be to experiment with other aggregation models to get useful low dimensional models for larger systems; e.g., from the point of view of a given node, can the rest of the network be aggregated in a simple way, with useful "disaggregation" models available to get details of good control policies? A start in this direction might be via a numerical investigation of three node systems.

Heavy traffic analysis has been used to model other types of problems; for example, Harrison and coworkers (a small sample is [5,6,4]) have examined many manufacturing type problems. Although little attention had been given to heavy traffic limits of physical control problems in these references, the basic motivation is the same; namely, to simplify the modelling and analysis of large physical problems by exploiting scale and flow balance.

Note: The heavy traffic limit equations are much simpler than the actual equations which represent the original physical system. One can have essentially arbitrary interarrival times, and can easily add features such as batch arrivals and services. It is easier to see the essential structural form of the problem since much unimportant detail is taken away. The basic scaling (that by $\sqrt{N}$) is more clearly exposed, so that appropriate "dimensioning" of the control problem becomes easier. Due to the scaling, each limit equation (or each numerical solution to a limit problem) represents a family of physical problems, each of different size. These comments are even more important for the control problem, where we wish to see the controls as functions of the basic structural parameters of the system with the appropriate scaling used to eliminate less important detail.

From the numerical point of view, we approximate the original problem by a heavy traffic limit and then approximate that limit by a discrete problem, which is the one actually solved. In some cases, one can also do the optimization or control problem for the original physical model. Generally, this is much harder than what is required by our two step process. The original problem is not necessarily Markov. It will be Markov if the input is a Poisson process. But even then the state space might be much larger. The nature of the discretization used to get the Markov chain approximation to the heavy traffic limit is somewhat equivalent to an aggregation of the state space of the original problem. The approximating discrete process has the basic structure of the original problem but is often simpler. If the interarrival times are constant or if there are batch arrivals or services, then the actual physical model is substantially harder. We also note that the numerical results in Section 7 are not too sensitive to the value of the discretization parameter which is used.

2. The general uncontrolled model. The system has K nodes or service centers. Customers arriving at a node take a vehicle if one is waiting. Otherwise, they queue, subject to a finite buffer. A customer at node i goes to node j with probability p_{ij}. The choices are mutually independent and independent of the sequence of arrival and service times. The excursion (service) times are mutually independent and independent of the arrival times. The excursion time to go to j from i is exponentially distributed with rate μ_{ij}. We make the appropriate assumptions so that the system is "well designed" and a heavy traffic type of approximation holds. The system is scaled by N, a measure of the total demand and number of vehicles, as specified further below. We suppose that the vehicles as well as the customers can be queued in finite buffers (parking lot, waiting room). For given B_i^+, B_i^-, the buffer sizes are $B_i^+ \sqrt{N}$ for the vehicles and $B_i^- \sqrt{N}$ for the customers. Excess customers are lost, although the waiting rooms can be made as large as desired. Normally we would be interested in systems where the waiting time and waiting number are not large. Vehicles that do not fit into the parking buffers can be handled in several ways. One can introduce a large temporary storage area into which the excess vehicles are put (at a cost per unit), and then taken out and used when needed later. Alternatively, one might simply bring them to a parking area associated with some other node, also at a cost per unit moved. In addition, in order to keep the system in proper balance there might be a controlled transfer of vehicles from one node to another, depending on the state values. The control problem will be dealt with in Section 6 and in the numerical data in Section 7. In order to simplify the formulation and the dimensionality of the resulting system in the rest of the paper, we will suppose that overflow vehicles are actually used immediately. There will be an extra cost to be paid per overflow vehicle. This overflow model corresponds to the situation where there is an extra source of potential users who use the overflow cars at a much reduced rate. The case where there is a temporary storage area is handled in a very similar way.

The arrival process. Assumptions. Let $\{\alpha_k^{i,N}, k \geq 1\}$ denote the sequence of mutually independent and identically distributed interarrival intervals for the customers coming to node i. These sequences are assumed to be mutually independent and also independent of the set of service intervals and routing choices. Define $\overline{\alpha}^{i,N} = E\alpha_k^{i,N}$. On arrival, a customer takes a vehicle if one is available. Otherwise the customer joins the queue. If the customer queue is full, the arriving customer is rejected and disappears from the system. We assume that the set of random variables

$$(2.1) \qquad \left\{ \left(\frac{\alpha_k^{i,N}}{\overline{\alpha}^{i,N}} \right)^2 ; k < \infty, N < \infty, i = 1, \dots, K \right\}$$

is uniformly integrable and that there is $\sigma_i^2 < \infty$ such that

$$(2.2) \qquad E\left[\left(1 - \alpha_k^{i,N}/\overline{\alpha}^{i,N} \right)^2 \right] = \frac{\text{var } \alpha_k^{i,N}}{(\overline{\alpha}^{i,N})^2} \equiv \sigma_{i,N}^2 \to \sigma_i^2.$$

$\sigma_{i,N}$ is the coefficient of variation, and equals unity if the input is a Poisson process.

     In the heavy traffic setup, the mean rate at which vehicles become available at a node needs to be close to the mean customer arrival rate.

We will suppose that there are $\lambda_i > 0$ and real b_i (positive or negative) such that the mean arrival rate at i is

(2.3) $$(\bar{a}^{i,N})^{-1} = \lambda_i N + b_i \sqrt{N}.$$

The appropriateness of the square root factor will be seen from the limit theorems, which shows that these orders give the desired flexibiity to the system. If there are always vehicles available, then the mean rate at which vehicles are picked up to go from i to j is $p_{ij}[\bar{a}^{i,N}]^{-1}$. It is necessary for good operation to keep the dominant parts of the arrival and departure rates in balance (modulo $O(\sqrt{N})$). We will require that

(2.4) $$\lambda_i = \sum_{j=1}^{K} p_{ji}\lambda_j$$

It will be seen that (2.4) does not restrict the generality of the model. If the arrival and departure rates are unequal (of the order of N), then the difference will need to be compensated for by substantial vehicle transfers and/or lost customers. A high imbalance is conceivable at certain times of the day. But we suppose a rough balance here.

Supposing that there are always vehicles available, the stationary mean number in transit from i to j is $p_{ij}[\lambda_i N + b_i\sqrt{N}]/\mu_{ij}$. Let us suppose that the number of vehicles V^N in the system is the sum over all i,j of the dominant part of this quantity, modulo $O(\sqrt{N})$; namely,

(2.5) $$V^N = \sum_{i,j} \frac{p_{ij}\lambda_i}{\mu_{ij}} N + V_0\sqrt{N},$$

for some constant V_0 (positive or negative). The V_0 is a design variable. One can let (2.4) also hold modulo $O(1/\sqrt{N})$, but it is enough to put that flexibility into the mean arrival rates and the numbers of vehicles via the parameters b_i and V_0. To simplify the notation, we assume that only one arrival or departure event can occur at a time. The general case requires that we define an "order" for the events which occur simultaneously and leads to a more complex notation, but with the same end results.

It will be seen that, subject to (2.4), the relative order in terms of N given in (2.5) is that for a well designed system, and that if (2.4) doesn't hold then it might have to be forced to hold via extra controls. For a well designed system, with a high probability nearly all vehicles are in use at each time. (This is somewhat analogous to the "just in time" approach to supplying a manufacturing system, or to the proper dimensioning of a trunk line system [13].) Under (2.4) and (2.5), with appropriate controls the system can be kept in balance, with the number of cars being moved by the control, vehicle buffer overflows and customer losses being of the order of $\sqrt{N}$, with small controllable constants. If any of the arrival rates differed from the heavy traffic value by a larger order, then either there would be a much too exessive number of vehicles or not nearly enough to avoid substantial overflowing of the customer buffer. Such results are a consequence of the heavy traffic analysis to follow. The general heavy traffic model allows an exploration of some of the main issues in design, as done in Section 7.

The input-output equation. The queue at i is defined as the [number of waiting vehicles minus the number of waiting customers] (only one of the two can be non zero). Thus the queue can be either positive or negative. All the other variables are to be scaled by $1/\sqrt{N}$. Define

$A^{i,N}(t) = $ [number of customer arrivals to i by t] $/\sqrt{N}$,

$V^{ij,N}(t) = $ [number of vehicle arrivals at j from i by t]$/\sqrt{N}$,

$X^{i,N}(t) = $ [queue at i at t]$/\sqrt{N}$,

$L^{i,N}(t) = $ [number of customers at i rejected by t]$/\sqrt{N}$,

$U^{i,N}(t) = $ [number of vehicles at i rejected by t (full parking)]$/\sqrt{N}$,

$\hat{R}^{ij,N}(t) = $ [number of vehicles in transit between i,j at t]$/\sqrt{N}$.

Define the centering and sums

$$R^{ij,N}(t) = \hat{R}^{ij,N}(t) - [p_{ij}\lambda_i\sqrt{N}/\mu_{ij}], \quad R^N(t) = \sum_{i,j} R^{ij,N}(t),$$

$$\hat{R}^N(t) = \sum_{i,j} \hat{R}^{ij,N}(t), \quad V^{i,N}(t) = \sum_j V^{ji,N}(t).$$

Now, we can write the input-output equations for the nodes as

$$(2.6) \quad X^{i,N}(t) = X^{i,N}(0) - A^{i,N}(t) + V^{i,N}(t) + L^{i,N}(t) - U^{i,N}(t).$$

Representation of $A^N(\cdot)$. Let $\mathcal{B}_t^N$ denote the σ-algebra induced by all the data on customer and vehicle arrivals and departures which is available up to and including time t. Define the auxiliary process

$$\tilde{A}^{i,N}(t) = \frac{1}{\sqrt{N}} \sum_{k=1}^{Nt} (1 - \frac{\alpha_k^{i,N}}{\alpha^{i,N}}).$$

In such sums, we always take the indices of summation to be the integer parts of the written indices (e.g., of the Nt above). Define

$$S^{i,N}(t) \quad = \frac{A^{i,N}(t)}{\sqrt{N}} = \frac{1}{N} \times \max\{n : \sum_{k=1}^n \alpha_k^{i,N} \le t\}$$

$$= \frac{1}{N} \times \text{[number of customer arrivals to } i \text{ by } t].$$

Using the fact that we can write

$$A^{i,N}(t) = \sum_{i=1}^{NS^{i,N}(t)} 1/\sqrt{N},$$

we have the representation

(2.7) $\qquad A^{i,N}(t) = \tilde{A}^{i,N}(S^{i,N}(t)) + \dfrac{1}{\sqrt{N}} \sum_{k=1}^{NS^{i,N}(t)} \alpha_k^{i,N}/\overline{\alpha}^{i,N}.$

The sampled process $\tilde{A}^{i,N}(k/N), k = 0, 1, \ldots$ is a martingale with re-spect to the natural $\sigma-$algebra. Also, the sequence $(\tilde{A}^{i,N}(S^{i,N}(t)), \mathcal{B}_t^N)$ sampled at the arrival times of customers to i is a martingale. It will hold that the $\tilde{A}^{i,N}(S^{i,N}())$ converge to mutually independent Wiener processes with quadratic variations $\lambda_i \sigma_i^2 t$. Using (2.3), the last term on the right of (2.7) is

(2.8) $\qquad \dfrac{1}{\sqrt{N}} \sum_{k=1}^{NS^{i,N}(t)} \alpha_k^{i,N}(\lambda_i N + b_i \sqrt{N}).$

Equation (2.8) can be simplified by using the relationship:

(2.9) $\qquad \sum_{k=1}^{NS^{i,N}(t)} \alpha_k^{i,N} = t + O_N^i(t),$

where $O_N^i(t) = -[t-$ time of last arrival to i before or at $t]$ and it satisfies $EO_N^2(t) = O(1/N^2)$.

The above arrival model is used because it describes common situa-tions. More generally, $A^{i,N}(\cdot)$ can be any sequence such that the process $\hat{A}^{i,N}(\cdot)$ defined by

$$\hat{A}^{i,N}(t) = A^{i,N}(t) - \lambda_i \sqrt{N} t$$

converges weakly to a Wiener process with constant drift, the set of incre-ments

$$\{\hat{A}^{i,N}(n+1) - \hat{A}^{i,N}(n) ; n, N\}$$

is uniformly integrable, and the sequence is independent in i.

Representation of the vehicle arrival process. Using the fact that the service intervals for travel from j to i are mutually independent and exponentially distributed with the rates μ_{ji}, we can decompose $V^{ji,N}(\cdot)$ in "martingale" form as

$$
\begin{aligned}
V^{ji,N}(t) &= \mu_{ji} \int_0^t \hat{R}^{ji,N}(s)ds + \tilde{V}^{ji,N}(t) \\
&= \mu_{ji} \int_0^t \left[R^{ji,N}(s) + \frac{p_{ji}\lambda_j \sqrt{N}}{\mu_{ji}} \right] ds + \tilde{V}^{ji,N}(t)
\end{aligned}
$$

(2.10)

where $\tilde{V}^{ji,N}(\cdot)$ is a $\mathcal{B}_t^N$-martingale. Using the fact that the processes are scaled by $1/\sqrt{N}$, we find that the Doob-Mayer compensator is

$$
\begin{aligned}
\left[\tilde{V}^{ji,N} \right](t) &= \mu_{ji} \frac{1}{\sqrt{N}} \int_0^t \hat{R}^{ji,N}(s)ds \\
&= \frac{1}{\sqrt{N}} \mu_{ji} \int_0^t \left[R^{ji,N}(s) + \frac{p_{ji}\lambda_j \sqrt{N}}{\mu_{ji}} \right] ds.
\end{aligned}
$$

(2.11)

The martingale property follows essentially from the fact that the total arrivals on the interval $[t, t+\delta)$ come from those in transit at t plus a quantity whose mean value is of order $O(\delta^2 N)$, and also the "no memory" property of the exponential distribution. The martingales $\tilde{V}^{ji,N}(\cdot)$, $i, j = 1, \ldots, K$, are mutually orthogonal, each with jumps $O(1/\sqrt{N})$. It will hold that the weak limits of the $\tilde{V}^{ji,N}(\cdot)$ are mutually independent Wiener processes with quadratic variations $\lambda_j p_{ji} t$.

Input-output equations for a node. With the above representations we can now rewrite (2.6) as

$$
\begin{aligned}
X^{i,N}(t) &= X^{i,N}(0) - \left[\tilde{A}^{i,N}(S^{i,N}(t)) + \lambda_i \sqrt{N} t + b_i t \right] + \rho^{i,N}(t) \\
&\quad + \sum_j \left\{ \mu_{ji} \int_0^t \left[R^{ji,N}(s) + \frac{p_{ji}\lambda_j \sqrt{N}}{\mu_{ji}} \right] ds + \tilde{V}^{ji,N}(t) \right\} \\
&\quad + L^{i,N}(t) - U^{i,N}(t),
\end{aligned}
$$

(2.12)

where $E|\rho^{i,N}(t)|^2 \to 0$ as $N \to \infty$, uniformly in all other variables. Using (2.4) and defining

$$
W^{i,N}(t) = \left(\sum_j \tilde{V}^{ji,N}(t) \right) - \tilde{A}^{i,N}(S^{i,N}(t)),
$$

we can write

$$
\begin{aligned}
X^{i,N}(t) &= X^{i,N}(0) + \sum_j \mu_{ji} \int_0^t R^{ji,N}(s)ds - b_i t + W^{i,N}(t) \\
&\quad + L^{i,N}(t) - U^{i,N}(t) + \rho^{i,N}(t) \equiv Z^{i,N}(t).
\end{aligned}
$$

(2.13)

An expression for the scaled number of vehicles $\hat{R}^{ji,N}(t)$ in transit from j to i at t is also easily obtained. The number of such vehicles at t equals the number at time zero plus the total number of customers sent out from j to i by t, minus the total number of vehicles from j which actually arrived at i by t. By similar reasoning, the number of vehicles (equiv, customers) divided by $\sqrt{N}$ leaving j for *some* destination by t is

$$(2.14) \qquad C^{j,N}(t) = [X^{j,N}(0)]^{-} + A^{j,N}(t) - [X^{j,N}(t)]^{-}.$$

For each source node j and each customer the decision concerning the destination can be done via a "coin toss," with probability p_{ji} of selecting the i-th outcome (and with the trials being independent in j and the customer). Thus, we can write the number of customers (divided by $\sqrt{N}$) leaving j with destination i by t as

$$(2.15) \qquad C^{ji,N}(t) = p_{ji}C^{j,N}(t) + \tilde{C}^{ji,N}(t),$$

where $\tilde{C}^{ji,N}(\cdot)$ is a $\mathcal{B}_t^N$-martingale (when it is sampled at the times of departure from j) with compensator

$$(2.16) \qquad \left[\tilde{C}^{ji,N}\right](t) = \frac{C^{j,N}(t)}{\sqrt{N}}p_{ji}(1 - p_{ji}).$$

It will hold that

$$(2.17) \qquad \begin{aligned} \lim_N \left[\tilde{C}^{ji,N}\right](t) &= \lambda_j p_{ji}(1 - p_{ji})t, \\ \lim_N \left[\tilde{C}^{ji,N}, \tilde{C}^{jk,N}\right](t) &= -\lambda_j p_{ji}p_{jk}t, \ k \neq i. \end{aligned}$$

Using (2.14), the expressions for the customer arrivals in (2.7)–(2.9), the characterization of $\hat{R}^{ji,N}(\cdot)$ above (2.14) and (2.15), we can write

$$\begin{aligned} \hat{R}^{ji,N}(t) = \hat{R}^{ji,N}(0) &- \left[\mu_{ji}\int_0^t \hat{R}^{ji,N}(s)ds + \tilde{V}^{ji,N}(t)\right] \\ &+ p_{ji}\left\{\left[\tilde{A}^{j,N}(S^{j,N}(t)) + \lambda_j\sqrt{N}t + b_j t\right] + [X^{j,N}(0)]^{-} - [X^{j,N}(t)]^{-}\right\} \\ &+ \tilde{C}^{ji,N}(t) + \rho^{ji,N}(t), \end{aligned}$$

(2.18)

where $E|\rho^{ji,N}(t)|^2 \to 0$ as $N \to 0$, uniformly in all other variables. Define

$$W^{ji,N}(t) = p_{ji}\tilde{A}^{j,N}(S^{j,N}(t)) - \tilde{V}^{ji,N}(t) + \tilde{C}^{ji,N}(t).$$

Then we can write

$$(2.19) \qquad \begin{aligned} R^{ji,N}(t) = R^{ji,N}(0) &- \mu_{ji}\int_0^t R^{ji,N}(s)ds + p_{ji}b_j t + W^{ji,N}(t) \\ &+ p_{ji}\left\{[X^{j,N}(0)]^{-} - [X^{j,N}(t)]^{-}\right\} + \rho^{ji,N}(t). \end{aligned}$$

3. A limit theorem. In the weak convergence analysis, the Skorohod topology [3] on $D^k[0, \infty)$ is used for appropriate integers k. In this section, we obtain the heavy traffic limit processes for the system of Section 2. Let $\Psi^N(\cdot)$ denote the process whose components are

$$\left(X^{i,N}(\cdot), R^{ij,N}(\cdot), L^{i,N}(\cdot), U^{i,N}(\cdot), \tilde{C}^{ij,N}(\cdot), \tilde{V}^{ij,N}(\cdot), \tilde{A}^{i,N}(S^{i,N}(\cdot)); i, j \right).$$

Theorem 3.1. *Let the set of initial conditions* $\{R^{ij,N}(0); i, j, N\}$ *be tight. Then, under the conditions of Section 2, the sequence* $\Psi^N(\cdot)$ *is tight. Let* $\Psi(\cdot)$ *denote a process which is the limit of a weakly convergent subsequence of* $\Psi^N(\cdot)$, *and let* $\mathcal{B}_t$ *denote the filtration which it defines. Let the components of* $\Psi(\cdot)$ *be denoted by*

$$\left(X^i(\cdot), R^{ij}(\cdot), L^i(\cdot), U^i(\cdot), \tilde{C}^{ij}(\cdot), \tilde{V}^{ij}(\cdot), \tilde{A}^i(\cdot); i, j \right)$$

The limit processes satisfy

$$(3.1) \qquad dX^i = \sum_j \mu_{ji} R^{ji}(t) dt - b_i dt + dW^i + dL^i - dU^i,$$

$$(3.2) \quad dR^{ji} = \left[-\mu_{ji} R^{ji} dt + p_{ji} \left(b_j dt - d[X^j]^- - dL^j + dU^j \right) \right] + dW^{ji}.$$

The $W^{ji}(\cdot)$, $W^i(\cdot)$, *and* $\tilde{C}^{ij}(\cdot), \tilde{V}^{ij}(\cdot), \tilde{A}^i(\cdot)$ *are* $\mathcal{B}_t$−*martingales and Wiener processes with*

$$W^i(t) = \sum_j \tilde{V}^{ji}(t) - \tilde{A}^i(t),$$

$$W^{ji}(t) = p_{ji} \tilde{A}^j(t) - \tilde{V}^{ji}(t) + \tilde{C}^{ji}(t),$$

and the quadratic variations of the $\tilde{C}^{ij}(\cdot), \tilde{V}^{ij}(\cdot), \tilde{A}^i(\cdot)$ *are as asserted in the previous section.*

Note. There is an extra variable since the total number of vehicles is fixed. We have

$$V^N = \sqrt{N} \sum_{i,j} \hat{R}^{ij,N}(t) + \sqrt{N} \sum_i \left[X^{i,N}(t) \right]^+.$$

Thus, using the definition (2.5),

$$\sum_{i,j} R^{ij,N}(t) + \sum_i [X^{i,N}(t)]^+ = V_0.$$

If this equality holds at $t = 0$, then it holds at all t. The situation is the same for the limit processes.

Outline of proof. The proof is by now more or less standard, and is similar to proofs in, say, [10]. The $R^{ij,N}(t)$ variables are not bounded, and this complicates the proof of tightness. It is more convenient to work with truncated $R^{ij,N}(t)$ processes and then let the trunction levels go to infinity [8]. For positive M, Let $q_M(r)$ be a bounded and smooth real valued function of a real variable which equals r for $|r| \leq M$. To get the desired truncation, we replace the $R^{ij,N}(t)$ variables on the right sides of (2.13) and (2.19) by $q_M(R^{ij,N}(t))$. Let $\Psi_M^N(\cdot)$ denote the process with components

$$\left(X_M^{i,N}(\cdot), R_M^{ij,N}(\cdot), L_M^{i,N}(\cdot), U_M^{i,N}(\cdot), \tilde{C}_M^{ij,N}(\cdot), \tilde{V}_M^{ij,N}(\cdot), \tilde{A}^{i,N}(\cdot); i,j \right)$$

which is the solution to (3.1), (3.2), together with the martingales in (2.13), (2.19), but with the truncation function used. The quadratic variations of the martingales are to be calculated with the "truncated" variables. One proves tightness and weak convergence for these new processes and then lets $M \to \infty$. Let $\Psi_M(\cdot)$ (with the components appropiately designated) denote a weak limit of the $\Psi_M^N(\cdot)$.

It is evident that the set $\Psi_M^N(\cdot)$ is tight for each M. Thus there are weakly convergent subsequences for each M. To simplify the notation, suppose that the entire sequence $\Psi_M^N(\cdot)$ converges weakly for each M. Given any $\rho > 0$ and time interval $[0, T]$, it will be seen from the form of the equation satisfied by the $\Psi_M(\cdot)$ that there is a $B_{\rho,T}$ such that the $|R_M^{ij}(t)|$ are bounded by $B_{\rho,T}$ on $[0, T]$ with probability at least $1 - \rho$, for all M. Since the weak limits of the original processes will equal those of the truncated processes up until at least the first time that some $|R_M^{ij}(t)|$ exceeds M and ρ is arbitrary, it is sufficient to work with the truncated processes.

The sequence $S^{i,N}(\cdot)$ converges weakly to the process with values $\lambda_i t$. The process $\tilde{A}^{i,N}(\cdot)$ is tight and any weak limit is a Wiener process with variance $\sigma_i^2 t$. Hence $\tilde{A}^{i,N}(S^{i,N}(\cdot))$ converges weakly to a Wiener process $\tilde{A}^i(\cdot)$ with variance $\lambda_i \sigma_i^2 t$. The $R^{ji,N}(\cdot)/\sqrt{N}$ factor in the quadratic variation in (2.11) is replaced by $q_M(R^{ji,N}(\cdot))/\sqrt{N}$ and disappears in the limit as $N \to \infty$. Thus, we get that the $\tilde{V}_M^{ji,N}(\cdot)$ converge to Wiener processes $\tilde{V}_M^{ji}(\cdot)$ with variances $p_{ji} \lambda_j t$. By the argument used in and below (6.8), it can be shown that all the terms on the right of (2.14) (except for the $A^{j,N}$ –term, and with the M –subscript used) go to zero in the mean when divided by $\sqrt{N}$ and $N \to \infty$. Thus $C_M^{j,N}(t)/\sqrt{N} \to \lambda_i t$ in the mean as $N \to \infty$. This and the martingale property imply that the $\tilde{C}^{ji,N}(\cdot)$ converge to Wiener processes $\tilde{C}_M^{ji}(\cdot)$ with variances given in (2.17). Apart from the correlation among the $\tilde{C}_M^{ji}(\cdot)$, all the limit Wiener processes are independent. We have

$$W_M^{ji}(t) = p_{ji} \tilde{A}^j(t) - \tilde{V}_M^{ji}(t) + \tilde{C}_M^{ji}(t),$$

$$W_M^i(t) = \sum_j \tilde{V}_M^{ji}(t) - \tilde{A}^i(t).$$

Also, by the weak convergence, (3.1) and (3.2) hold for the M−subscripted limit processes when $q_M(R_M^{ji})$ replaces R^{ji}. The $L_M^{i,N}(\cdot), U_M^{i,N}(\cdot)$ are tight for each M since the other processes in the right side of the (truncated) (2.13), (2.19) are tight. Let $\mathcal{B}_{t,M}$ denote the filtration generated by the $\Psi_M(\cdot)$.

It is only necessary to show that the Wiener processes are actually $\mathcal{B}_{t,M}$-Wiener processes. This will be shown via the usual martingale method. Let $H(\cdot)$ be a real valued, bounded and continuous function of its arguments. For an arbitrary integer m, let $t_i \leq \tau \leq \tau + s, i = i, \ldots, m$ be arbitrary positive numbers. Let $E_{t,M}^N$ denote expectation given $\mathcal{B}_{t,M}$. We have

$$EH(\Psi_M^N(t_i), i \leq m)\left[\tilde{V}_M^{ij,N}(\tau + s) - \tilde{V}_M^{ij,N}(\tau)\right]$$
$$= EH(\Psi_M^N(t_i), i \leq m)E_{t,M}^N\left[\tilde{V}_M^{ij,N}(\tau + s) - \tilde{V}_M^{ij,N}(\tau)\right] \to 0.$$

This and the weak convergence and uniform square integrability of the set $\tilde{V}_M^{ij,N}(t), t \leq T$, for any positive T yield

$$EH(\Psi_M(t_i), i \leq m)\left[\tilde{V}_M^{ij,N}(\tau + s) - \tilde{V}_M^{ij,N}(\tau)\right] = 0.$$

This last expression and the arbitrariness of the $h(\cdot), m, t_i, \tau, s$ imply that $\tilde{V}_M^{ij}(\cdot)$ is an $\mathcal{B}_{t,M}$-Wiener processes. A similar calculation works for the other Wiener processes. One can also verify that the given quadratic variations continue to hold under the filtration $\mathcal{B}_{M,t}$.   □

4. Simplified models for symmetric systems: $\mu_{ji} = \mu_j$. If the p_{ij} and μ_{ij} have restricted dependencies on i or j, then the forms (2.13), (2.19) and the heavy traffic limit equations can be simplified with a reduction in dimension. We first do the case where

(4.1) $$\mu_{ji} = \mu_j.$$

A simpler case will be dealt with in the next two sections. In view of the complexity of the equations, it seems essential to make the simplifying assumptions if any insights to the qualitative behavior are to be obtained. Under (4.1), we need only keep track of the total number of vehicles in transit from each source node j and not of the numbers in transit between each source-destination pair. Define $\hat{R}^{j,N}(t) = \sum_i \hat{R}^{ji,N}(t)$ and the centered quantity $R^{j,N}(t) = \hat{R}^{j,N}(t) - \sqrt{N}\lambda_j/\mu_j$. Since the distributions of the transit times from source j do not depend on the destination, we can get the destinations of the vehicles which are in transit from source j in two steps. First wait an "exponential length of time" with mean $1/\mu_j$, and then decide on the destination by a random trial with outcome probabilities $p_{ji}, i = 1, \ldots, K$, the trials being independent between customers and in j. This representation allows us to write

$$(4.2) \qquad V^{ji,N}(t) = \mu_j p_{ji} \int_0^t \hat{R}^{j,N}(s)ds + \tilde{R}^{ji,N}(t),$$

where the $\tilde{R}^{ji,N}(\cdot), i, j = 1, \ldots, K$ are orthogonal $\mathcal{B}_t^N$-martingales with quadratic variations

$$\left[\tilde{R}^{ji,N}\right](t) = \frac{\mu_j p_{ji}}{\sqrt{N}} \int_0^t \hat{R}^{j,N}(s)ds = \lambda_j p_{ji} t + \frac{1}{\sqrt{N}} \mu_j p_{ji} \int_0^t R^{j,N}(s)ds.$$
(4.3)

We will simply write the dynamical equations for the physical system and the weak limit. The proof is that of Theorem 3.1.

Dynamical equations for $R^{j,N}(\cdot), X^{i,N}(\cdot)$. We can write the number in transit from j at time t as the number in transit from j at time zero, plus the number entering transit from j by time t minus the number from j which actually arrived at a destination by t. This yields (neglecting the small error)

$$(4.4) \quad \hat{R}^{j,N}(t) = \hat{R}^{j,N}(0) + C^{j,N}(t) - \left[\mu_j \int_0^t \hat{R}^{j,N}(s)ds + \sum_i \tilde{R}^{ji,N}(t)\right].$$

Define

$$\tilde{W}^{i,N}(t) = \sum_j \tilde{R}^{ji,N}(t) - \tilde{A}^{i,N}(S^{i,N}(t)),$$

$$\hat{W}^{j,N}(t) = -\sum_i \tilde{R}^{ji,N}(t) + \tilde{A}^{j,N}(S^{j,N}(t)).$$

Upon simplifying, we can write the equation (2.6) as

$$(4.5) \quad \begin{aligned} X^{i,N}(t) &= X^{i,N}(0) + \sum_j \mu_j p_{ji} \int_0^t R^{j,N}(s)ds - b_i t + \tilde{W}^{i,N}(t) \\ &\quad + L^{i,N}(t) - U^{i,N}(t) + \rho^{i,N}(t). \end{aligned}$$

Similarly, (2.14) and (4.4) yield

$$(4.6) \quad \begin{aligned} R^{j,N}(t) &= R^{j,N}(0) - \mu_j \int_0^t R^{j,N}(s)ds + b_j t + \hat{W}^{j,N}(t) \\ &\quad + [X^{j,N}(0)]^- - [X^{j,N}(t)]^- - L^{j,N}(t) + U^{j,N}(t) \\ &\quad - \rho^{j,N}(t). \end{aligned}$$

The limit processes. By the method of Theorem 3.1, the weak limit processes can be shown to satisfy

$$(4.7) \qquad dX^i = \sum_j \mu_j p_{ji} R^j \, dt - b_i \, dt + d\tilde{W}^i + dL^i(t) - dU^i(t),$$

$$(4.8) \quad dR^j = -\mu_j R^j \, dt + b_j \, dt + d\hat{W}^j(t) - d[X^j(t)]^- - dL^j(t) + dU^j(t).$$

The $\hat{W}^i(\cdot), \tilde{W}^i(\cdot)$ are Wiener $\mathcal{B}_t$–Wiener processes with quadratic variations defined by:

$$< \tilde{W}^i > (t) = [\sum_j \lambda_j p_{ji} + \lambda_i \sigma_i^2] t = \lambda_i (1 + \sigma_i^2) t,$$

$$< \hat{W}^j > (t) = \lambda_j (1 + \sigma_j^2) t,$$

$$< \hat{W}^i, \tilde{W}^i > (t) = -\lambda_i (p_{ii} + \sigma_i^2) t,$$

$$< \hat{W}^k, \tilde{W}^l > (t) = -\lambda_k p_{kl} t, \quad k \neq l.$$

The non listed quadratic variations are zero. Note that

$$< W^i > (t) = < \tilde{W}^i > (t), \quad < \sum_i W^{ji} > (t) = < \hat{W}^j > (t).$$

5. The case $\mu_{ij} = \mu, p_{ij} = p_j$. Under the assumption $\mu_{ij} = \mu, p_{ij} = p_j$, the formulas in the previous section can be simplified even further, since we now need only keep track of the sum of the vehicles in transit irrespective of the origins or destinations. Define $\Lambda = \sum_i \lambda_i$ and $\hat{R}^N(t) = \sum_i \hat{R}^{i,N}$. Equation (2.4) yields $p_i = \lambda_i / \Lambda$. Also,

$$(5.1) \qquad\qquad R^N(t) = \hat{R}^N(t) - \frac{\Lambda}{\mu} \sqrt{N}.$$

Pursuing the basic simplifying idea in the previous section a little further, we aggregate the vehicles in transit from all sources. To get the destination for any vehicle entering transit, we wait a random time with an exponential distribution with mean $1/\mu$, and then do an experiment with outcomes having probabilities $p_j, j = 1, \ldots, K$, the experiments being independent among the customers. We can then write $V^{i,N}(t) = \sum_j V^{ji}(t)$ in the form

$$V^{i,N}(t) = \mu p_i \int_0^t \hat{R}^N(s) ds + \tilde{V}^{i,N}(t),$$

where $\tilde{V}^{i,N}(\cdot)$ is a $\mathcal{B}_t^N$−martingale with compensator

$$\left[\tilde{V}^{i,N}\right](t) = \frac{1}{\sqrt{N}}\mu p_i \int_0^t \hat{R}^N(s)ds = \lambda_i t + \frac{\mu p_i}{\sqrt{N}} \int_0^t R^N(s)ds.$$

The $\tilde{V}^{i,N}(\cdot)$ are mutually orthogonal.

Now rewrite (2.16) or (4.5) as

$$\begin{aligned} X^{i,N}(t) &= X^{i,N}(0) - \left[\tilde{A}^{i,N}(S^{i,N}(t)) + b_i t\right] + \mu p_i \int_0^t R^N(s)ds \\ &\quad + \tilde{V}^{i,N}(t) + \rho^{i,N}(t) + L^{i,N}(t) - U^{i,N}(t). \end{aligned}$$

With the definition $\overline{W}^{i,N}(t) = \tilde{V}^{i,N}(t) - \tilde{A}^{i,N}(S^{i,N}(t))$, we can rewrite the state equation as

$$\begin{aligned} X^{i,N}(t) &= X^{i,N}(0) - b_i t + \mu p_i \int_0^t R^N(s)ds + \overline{W}^{i,N}(t) + \rho^{i,N}(t) \\ &\quad + L^{i,N}(t) - U^{i,N}(t). \end{aligned}$$

(5.2)

Next note that (2.5) equals

$$V^N = \frac{\Lambda}{\mu}N + V_0\sqrt{N}.$$

Using this and the fact that

$$\sum_i [X^{i,N}(t)]^+ + \hat{R}^N(t) = V^N/\sqrt{N},$$

we get

(5.3) $$\sum_i [X^{i,N}(t)]^+ + R^N(t) = V_0, \quad \text{all } t.$$

Substituting this into (5.2) yields

(5.4) $$\begin{aligned} X^{i,N}(t) &= X^{i,N}(0) + [\mu p_i V_0 - b_i]t - \mu p_i \int_0^t \sum_j [X^{j,N}(s)]^+ ds \\ &\quad + \overline{W}^{i,N}(t) + \rho^{i,N}(t) + L^{i,N}(t) - U^{i,N}(t). \end{aligned}$$

Also,

$$\left[\overline{W}^{i,N}\right](t) = \lambda_i(\sigma_i^2 + 1)t + \frac{\mu p_i}{\sqrt{N}} \int_0^t \left[V_0 - \sum_j [X^{j,N}(s)]^+\right] ds,$$

(5.5) $$\left[\overline{W}^{i,N}, \overline{W}^{j,N}\right](t) = 0, \quad j \neq i$$

The limit equations. By the method of Theorem 3.1, we get

$$dX^i(t) = [\mu p_i V_0 - b_i]dt - \mu p_i \sum_j [X^j(t)]^+ dt + d\overline{W}^i(t) + dL^i(t) - dU^i(t),$$

where the $\overline{W}^i(\cdot)$ are mutually independent Wiener processes with variances

$$< \overline{W}^i > (t) = \lambda_i(\sigma_i^2 + 1)t.$$

6. Controls. In this section, we introduce the controls (i.e., the planned transfer of vehicles among the nodes) and prove the associated limit theorems. The results show that the optimal or nearly optimal control for the ergodic cost problem for the limit model provide nearly optimal controls for the physical problem if N is sufficiently large. *For notational simplicity we work with a two node form of the problem in the previous section.* The actual physical control process consists in the physical transfer of vehicles from one node to another. This will ordinarily be done by taking groups of vehicles at a time in a larger conveyance (truck, train), at particular discrete times. The actual problem is hard to deal with. We consider two reasonable approximations which yield the general properties of the true system. Most of the development will suppose that the vehicles can be moved continuously and instantaneously from one node to the other. The "instantaneous" part seems to be a reasonable approximation when the time required for the controlled transfer is much smaller than the average time that the vehicles are in use by the customers. The "continuous" transfer approximation seems justified when there are many vehicles and the transfers occur at frequent intervals. Numerical calculations show that the optimal controls can be approximated by relatively simple controls with little penalty, and suggest that the performance will not deteriorate much for large N if the transfers are (moderately) lumped. The heavy traffic limit results of Sections 3–5 tell us that under the conditions (2.3) to (2.5), the mean rate of transferring cars should be of the order $O(\sqrt{N})$. A smaller order will not have an effect in the limit. A larger order is not needed to attain good performance in such a well balanced system. With this in mind, let $\sqrt{N} \int_0^t u^{i,N}(s)ds$ denote the total number of vehicles transferred from i to $j \neq i$ by time t. We suppose that there are maximum transfer rates $\bar{u}^i < \infty$. Thus $0 \leq u^{i,N}(t) \leq \bar{u}^i, i = 1, 2$. Then the physical equations are the following modification of (5.4).

$$X^{i,N}(t) = X^{i,N}(0) + [\mu p_i V_0 - b_i]t - \mu p_i \int_0^t \left[[X^{1,N}(s)]^+ + [X^{2,N}(s)]^+ \right] ds$$

$$+ \int_0^t \left[u^{j,N}(s) - u^{i,N}(s) \right] ds + \overline{W}^{i,N}(t) + \rho^{i,N}(t) + L^{i,N}(t) - U^{i,N}(t).$$

(6.1)

The limit equations with controls are $(i \neq j)$

(6.2)
$$
\begin{aligned}
dX^i \;=\;& [\mu p_i V_0 - b_i]dt - \mu p_i \left[[X^{1,N}]^+ + [X^{2,N}]^+\right] dt \\
& + [u^j - u^i]dt + d\overline{W}^i + dL^i - dU^i,
\end{aligned}
$$

with $0 \leq u^i(t) \leq \bar{u}^i$. The Wiener processes are mutually independent and have quadratic variations $< \tilde{W}^i > (t) = \lambda_i(\sigma_i^2 + 1)$. We note that, no matter what the control sequence, we will have tightness and the existence of a subsequence which converges to a limit with the representation (6.2).

For non negative cost coefficients, an appropriate cost function for the physical system is defined by

(6.3)
$$
\begin{aligned}
\gamma_T^N(u, X^N(0)) =& \\
\frac{1}{T}E & \int_0^T \left[a_1[X^{1,N}(s)]^- + a_2[X^{2,N}(s)]^- + c_1 u^1(s) + c_2 u^2(s)\right] ds \\
+\frac{1}{T}E & \left[d_1 L^{1,N}(T) + d_2 L^{2,N}(T) + k_1 U^{1,N}(T) + k_2 U^{2,N}(T)\right].
\end{aligned}
$$

Define

(6.4)
$$
\begin{aligned}
\gamma^N(u, X^N(0)) &= \limsup{}_T \gamma_T^N(u, X^N(0)), \\
\bar{\gamma}_T^N(X^N(0)) &= \inf \gamma_T^N(X^N(0)), \quad \bar{\gamma}^N = \inf \gamma^N(u, X^N(0)),
\end{aligned}
$$

where the infs are over the admissible controls and initial conditions. The analogous cost for the limit system is

$$
\begin{aligned}
\gamma_T(u, X(0)) =& \frac{1}{T}E \int_0^T \left[a_1[X^1(s)]^- + a_2[X^2(s)]^- + c_1 u^1(s) + c_2 u^2(s)\right] ds \\
& + \frac{1}{T}E \left[d_1 L^1(T) + d_2 L^2(T) + k_1 U^1(T) + k_2 U^2(T)\right].
\end{aligned}
$$

(6.5)

Define

(6.6)
$$
\gamma(u, X(0)) = \limsup_T \gamma_T(u, X(0)), \quad \bar{\gamma} = \inf \gamma(u, X(0)),
$$

where the inf is over the admissible controls and associated processes and initial conditions. Strictly speaking, the limit costs $\gamma(u, X(0))$ might not be simply a function of the control and initial condition only. But the notation should not be misleading, since we will be concerned with inf's of the costs. With a feedback control $u(\cdot)$ used in (6.2), there is a unique invariant measure and the limit cost will depend only on $u(\cdot)$, and not on $X(0)$.

The limit model can be used for approximations of the physical problem as shown by the following theorem. There are analogous (and easier to prove) results for the discounted and finite time costs. Extend the definitions of $\mathcal{B}_t^N$ and $\mathcal{B}_t$ so that they also measure the control values up to time t.

Theorem 6.1.

(6.7) $$\lim_{N\to\infty} \bar{\gamma}^N = \bar{\gamma}, \quad \lim_{N,T\to\infty} \bar{\gamma}_T^N(X^N(0)) = \bar{\gamma}.$$

Proof. The controls introduced below are always bounded by $(\bar{u}^1, \bar{u}^2)$. We prove the second limit in (6.7). Most of the pieces of the proof are in place in existing works and only a detailed outline will be given. First, we state some properties of (6.2) and the associated optimal controls which will be needed later. For any feedback control, the system (6.2) has a unique invariant measure and a unique weak sense solution and $\gamma(u, X(0))$ does not depend on $X(0)$. In general the measure of the solution process is uniquely determined by the measure of the set $X(0), u^i(\cdot), \overline{W}^i(\cdot), i = 1, 2$. There is an optimal feedback control for the system (6.2) with cost $\gamma(u, X(0))$. The proof of these assertions is similar to that used in [7,10]. The proof in Section 10 of [10] shows that the optimal feedback control is optimal with respect to any non anticipative control for the ergodic cost problem for any initial condition. Furthermore, for any $\epsilon > 0$ there is a continuous feedback control $u^\epsilon(\cdot) = (u^{1,\epsilon}(\cdot), u^{2,\epsilon}(\cdot))$ such that $\gamma(u^\epsilon) \leq \bar{\gamma} + \epsilon$.

Rewrite (6.1) as

$$X^{i,N}(t) = Z^{i,N}(t) + L^{i,N}(t) - U^{i,N}(t).$$

Then, for any $\delta > 0$ and $\Delta > 0$, there is $\alpha > 0$ such that for all large N, all controls and each t we have w.p.1

(6.8) $$P\{ \sup_{\delta \geq s \geq 0} |Z^N(t+s)| \geq \Delta | \mathcal{B}_t^N \} \leq 1 - \alpha.$$

Then the proof in [12, Section 4] can be used to show that the set

(6.9) $$\{ L^{i,N}(n+1) - L^{i,N}(n), U^{i,N}(n+1) - U^{i,N}(n); i = 1, 2, n, N, \text{ all } u^N(\cdot) \}$$

is uniformly integrable. This uniform integrability will be needed for the convergence of the cost functions $\gamma^N(u^N, X^N(0))$ or $\bar{\gamma}^N$ to a cost function for some limit system, as $N \to \infty$.

The rest of the proof use a type of functional occupation measure argument, which is quite similar to that used in Section 5 of [10]. Define

$$G^{i,N}(t) = \int_0^t u^{i,N}(s)ds, \quad G^i(t) = \int_0^t u^i(s)ds.$$

We say that the solution to (6.2) is *stationary* if the distribution of

$$\Phi_t(\cdot) = \{X^i(t + \cdot), W^i(t + \cdot) - W^i(t), L^i(t + \cdot) - L^i(t),$$
$$U^i(t + \cdot) - U^i(t), G^i(t + \cdot) - G^i(t), i = 1, 2\}$$

does not depend on t. Define $\Phi_t^N(\cdot)$ for the physical process analogously to the definition of $\Phi_t(\cdot)$. Let the 10-tuple $\overline{\Phi}(\cdot) = \{\overline{X}_i(\cdot), \overline{W}_i(\cdot), \overline{L}_i(\cdot), \overline{U}_i(\cdot), \overline{G}_i(\cdot), i = 1, 2\}$ denote the canonical element of the path space $D^{10}[0, \infty)$ of any of the $\Phi_t(\cdot), \Phi_t^N(\cdot)$ processes. Let $P^{N,t}(\cdot)$ denote the measure which is induced by the process $\Phi_t^N(\cdot)$, and define the *occupation measure* $P_T^N(\cdot)$ by

$$P_T^N(\cdot) = \frac{1}{T} \int_0^T P^{N,t}(\cdot) dt.$$

This is a measure on $D^{10}[0, \infty)$. Such measures on the path space are a very useful tool for getting approximation and limit results for ergodic cost problems. See also [2,9,11].

The cost function (6.3) can be written in terms of the $P_T^N(\cdot)$. To do this, first note that

$$(6.10) \qquad \frac{1}{T} \int_0^T [U^{i,N}(t+1) - U^{i,N}(t)] dt = \frac{1}{T} U^{i,N}(T) + \delta_T^N,$$

where

$$(6.11) \quad \delta_T^N = \frac{1}{T} \left[\int_T^{T+1} [U^{i,N}(t) - U^{i,N}(T)] \, dt - \int_0^1 U^{i,N}(t) dt \right].$$

Using this for the $L^{i,N}(\cdot)$ and $G^{i,N}(\cdot)$ processes as well, for any admissible sequence $u^N(\cdot)$ we can write

$$\gamma_T^N(u^N, X^N(0)) = \frac{1}{T} E \int_0^T \sum_i \left[a_i [X^{i,N}(t)]^- + c_i (G^{i,N}(1+t) - G^i(t)) \right.$$
$$\left. + d_i (L^{i,N}(t+1) - L^{i,N}(t)) + k_i (U^{i,N}(t+1) - U^{i,N}(t)) \right] dt + \hat{\delta}_T^N,$$
(6.12)
where the "error term" $\hat{\delta}_T^N$ goes to zero uniformly in N as $T \to \infty$ by the uniform integrability properties of (6.9) and the forms (6.10), (6.11).

We can now write (6.3) in terms of the occupation measure as

$$\gamma_T^N(u^N, X^N(0)) =$$
$$\int \sum_i \left[a_i [\overline{X}_i(0)]^- + c_i \overline{G}_i(1) + d_i \overline{L}_i(1) + k_i \overline{U}_i(1) \right] P_T^N(d\overline{\Phi}) + \hat{\delta}_T^N.$$
(6.13)
Thus, the asymptotic behavior of the cost is determined by the asymptotic behavior of the occupation measure.

Note that the measure $P_T^N(\cdot)$ is that of a process $\hat{\Phi}_T^N(\cdot)$ which is constructed exactly as $\Phi^N(\cdot) \equiv \Phi_0^N(\cdot)$ but where one *randomizes* the initial time; i.e.,

$$P\{\text{initial time} \in [t, t+\Delta]\} = \Delta/T$$

for $0 \leq t \leq t + \Delta \leq T$, and the initial time is independent of all other random variables in the system. The set of random variables $\{\Phi_t^N(0); N < \infty, t < \infty\}$ is tight. It can then be readily shown that the set of processes $\{\Phi_t^N(\cdot); N < \infty, t < \infty\}$ is tight. I.e., the set of measures $\{P^{N,t}(\cdot); N < \infty, t < \infty\}$ is tight. This implies the tightness of the set $\{P_T^N(\cdot); N < \infty, T < \infty\}$.

Now, to simplify the terminology, let $N \to \infty$, $T \to \infty$ index a weakly convergent subsequence of the occupation measures $P_T^N(\cdot)$ with limit denoted by $\hat{P}(\cdot)$. Let

$$\hat{\Phi}(\cdot) = \{\hat{X}^i(\cdot), \hat{W}^i(\cdot), \hat{L}^i(\cdot), \hat{U}^i(\cdot), \hat{G}^i(\cdot), i = 1, 2\}$$

denote the process which induces the measure $\hat{P}(\cdot)$. Equivalently, $\hat{\Phi}(\cdot)$ is the weak limit of the $\hat{\Phi}_T^N(\cdot)$ defined below (6.13). Since $0 \leq \hat{G}^i(t + s) - \hat{G}^i(s) \leq \bar{u}^i s$, the processes $\hat{G}^i(\cdot)$ can be represented as $\int_0^t \hat{u}^i(s)ds$, where $0 \leq \hat{u}^i(t) \leq \bar{u}^i$. By the weak convergence and the representation of $\hat{\Phi}_T^N(\cdot)$ as the process (6.1) but with a randomized initial time, $\hat{\Phi}(\cdot)$ satisfies (6.2). The martingale proof in Theorem 3.1 can be used to show that the other processes are non anticipative with respect to the Wiener processes $\hat{W}^i(\cdot), i = 1, 2$, and that the $\hat{W}_i(\cdot)$ have the quadratic variation properties cited below (6.2).

We next prove the stationarity of $\hat{\Phi}(\cdot)$.

For a path $\overline{\Phi}(\cdot) \in D^{10}[0, \infty)$ and $c > 0$, define the "left shift" $\overline{\Phi}_c(\cdot)$ by

$$\overline{\Phi}_c(\cdot) = \left(\overline{X}_i(c + \cdot), \overline{W}_i(c + \cdot) - \overline{W}_i(c), \overline{L}_i(c + \cdot) - \overline{L}_i(c),\right.$$
$$\left. \overline{U}_i(c + \cdot) - \overline{U}_i(c), \overline{G}_i(c + \cdot) - \overline{G}_i(c), i = 1, 2\right).$$

For a Borel set $K \subset D^{10}[0, \infty)$, define the left shift $K_c = \{\overline{\Phi}(\cdot) : \overline{\Phi}_c(\cdot) \in K\}$. Then we can write

$$P_T^N\{K_c\} = \frac{1}{T} \int_0^T P^{N,t}(K_c)dt = \frac{1}{T} \int_0^T P\{\Phi_{t+c}^N(\cdot) \in K\}dt,$$

(6.14)
$$P_T^N(K_c) - P_T^N(K)$$
$$= \frac{1}{T} \int_T^{T+c} P\{\Phi_{t+c}^N(\cdot) \in K\}dt - \frac{1}{T} \int_0^c P\{\Phi_t^N(\cdot) \in K\}dt.$$

Let $f(\cdot)$ be a bounded continuous real valued function on $D^{10}[0, \infty)$. Then the "error" estimate (6.14) and the (a.a.) continuity of the limit process $\hat{\Phi}(\cdot)$ imply that

$$Ef(\hat{\Phi}(\cdot)) = \lim_{N,T} \int f(\overline{\Phi}(\cdot))P_T^N(d\overline{\Phi}) = \int f(\overline{\Phi}(\cdot))\hat{P}(d\overline{\Phi})$$

$$= \lim_{N,T} \int f(\overline{\Phi}_c(\cdot))P_T^N(d\overline{\Phi}) = Ef(\hat{\Phi}(\cdot)_c).$$

Since $c > 0$ is arbitrary, this last equation implies the stationarity.

Since the paths of $\hat{\Phi}(\cdot)$ are continuous w.p.1, the functions with values $\overline{L}_i(1), \overline{U}_i(1)$ and $\overline{G}_i(1), i = 1, 2$, are continuous in the Skorohod topology almost everywhere (ω, t) with respect to the measure of $\hat{\Phi}(\cdot)$. Thus ([1, Theorem 5.1]) the convergence

$$\lim_{N,T\to\infty} \gamma_T^N(u^N, X^N(0)) \to \gamma(\hat{u})$$

follows from the weak convergence, the continuity of the paths of $\hat{\Phi}(\cdot)$ and the uniform integrability of the set (6.9). The cost for the stationary limit system is not (strictly speaking) a function only of the control, but we still get the inequality

(6.15) $$\liminf_{N,T\to\infty} \bar{\gamma}_T^N(X^N(0)) \geq \bar{\gamma}$$

due to the definition of $\bar{\gamma}$ as the inf over all non anticipative controls and initial conditions.

Now, to get the reverse inequality

(6.16) $$\limsup_{N,T\to\infty} \bar{\gamma}_T^N(X^N(0)) \leq \bar{\gamma},$$

use the feedback control $u^\epsilon(\cdot)$ of the first paragraph of the proof. By the weak convergence argument used above and the uniqueness of the invariant measure of (6.2) under $u^\epsilon(\cdot)$, we get that $\gamma^N(u^\epsilon, X^N(0)) \to \gamma(u^\epsilon)$. This convergence, (6.15) and the ϵ−optimality of $u^\epsilon(\cdot)$ yield (6.16), hence (6.7). $\qquad\square$

Alternative models for the vehicle transfer. The model used for the controlled transfer of vehicles in (6.1) supposed that they were transferred immediately to their destination. The results seem to be a reasonable approximation to the actual physical problem if the time required for the controlled transfers is much less than the average user time. A main difficulty in a more precise modelling is the increase in dimension of the model. This causes no problems for the theory, but it complicates even further the numerical problems. A slight variation allows the controlled transfers to be done via the use of "auxiliary" customers. We would then be obliged to keep track of the number in transit between each pair of nodes.

The following provides another mechanism for modelling controlled transfers with non zero delays: Let $\beta_j > 0$. Let the system be (6.1) with $T^{j,N}$ replacing the $u^{j,N}$ there, where

(6.17) $$\dot{T}^{j,N} = -\beta_j T^{j,N} + \beta_j u^{j,N}.$$

The integral of $u^{j,N}(t)$ equals the integral of $T^{j,N}(t)$, modulo a bounded error. The limit equation would be (6.2) with u^j replaced by T^j and

$$\dot{T}^j = -\beta_j T^i + \beta_j u^j.$$

This model allows transfers which are "spread out, but which have the "time constants" $1/\beta_i$. This last model is similar to the "delay" in the response of a simple electrical circuit to a step input.

7. Data. The tables below summarize some of the salient facts concerning the model of Section 6. In all runs $\mu = 1$ and $\bar{u}^1 = \bar{u}^2$. The arrival processes will be either Poisson or have constant interarrival times. Unless mentioned otherwise, it will be Poisson. First we discuss the fully symmetric cases of Tables 1a and 1b. The system is symmetric in that $p_1 = p_2$ and $\lambda_1 = \lambda_2$. The cost criterion (6.5) is used, with coefficients $a_i = 1, c_i = 2 \; d_i = 5$ and $k_i = 10$. Thus we weigh vehicle buffer overflow heavily. The tables give the optimal cost for various parameter settings, as well as the basic (unweighted) components of the cost, under the optimal stationary control. In all cases, $B_i^+ = 2, B_i^- = 3$. Generally, one would wish to vary the weights in the cost function in order to study the various possible tradeoffs among the components of the cost. For example, one might want to find the minimum vehicle buffer size which guarantees that the averge vehicle buffer overflow is less than a given quantity, or the effect of the maximum transfer rates $\bar{u}^i$ on this size, or the tradeoffs among the customer queue lengths and control costs. To do the latter, one would take a sequence of runs with various weights on the customer queue length and control cost. Such numerical explorations are an important use of optimal control techniques. They enable us to test the effects of constraints and to compare various alternatives, each being good in a specific sense.

In the tables, the symbol $U_i = EU^i(1)$ (resp., $L_i = EL^i(1)$) denotes the vehicle buffer overflow per unit time (resp., customer buffer overflow). $u_i = Eu^i(t)$ denotes the mean controlled vehicle transfer per unit time, and $x_i = EX^i(t)$ the mean customer queue length. These are all for the limit problem. To get the equivalent for the physical problem, multiply by $\sqrt{N}$, where N is the scaling parameter. The numbers in the tables are rounded, with ~ 0 used for a quantity which is less than .01.

Refer to Table 1a. The table shows that the optimal number of vehicles is slightly in excess of the mean demand rate. To fix ideas, set the scale parameter $N = 100$ and let the unit of time be one hour. Then, there are 60 customers arriving per hour to each node on the average, and each customer has an equal chance of returning to either node. With $V_0 = 1.5$, there are $(\lambda_1 + \lambda_2)N + V_0\sqrt{N} = 135$ vehicles in the system, and the excess number of vehicles is $1.5 \times 10 = 15$. With $V_0 = 0$, there are 120 vehicles in the system, and with $V_0 = -1.5$, there are only 105 vehicles in the system. It seems true in general that it is better to have a slight excess number of vehicles. For $V_0 = 1.5$, the mean customer and vehicle buffer overflows are negligible. On the average there are $.19 \times \sqrt{N} = 1.9$ vehicles transferred per hour from each node via the control. Since $x_i = .13$, the mean customer queue at each node is just $.13 \times \sqrt{N} = 1.3$, which corresponds to a mean wait of just $1.3/60$ hours $=1.3$ minutes per customer. With $V_0 = -1.5$, we see a sharp drop in system quality: The control is essentially useless, and the mean customer queue at each node is 23, with a customer overflow (loss) rate of 7.5 per hour per node. There is little difference if $\bar{u}^i = 2$ is used. The reader can examine the effects of different values of N.

The values for no control are also tabulated. Indeed, for $V_0 = 1.5, N = 100$, the system still works well, although the mean customer queue has increased to $.46 \times 10 = 4.6$ customers/node. For $V_0 = 0$, the mean queue length is 12.5. We see that in this well balanced case, control decreases the cost by the highest percentage where the cost is already small. Keep in mind that the vehicle overflows are another form of (inadvertent) control, since they correspond to an (expensive) form of transfer, and we have supposed that this form of transfer is always possible even when there is no deliberately planned control.

Table 1b gives the data when the interarrival times are constant, as they might be for an ideal reservation system. The omitted data (for $V_0 = -1.5$) is essentially the same as in Table 1a. We see that the gratest improvement is where the system already worked well, with negligible improvement in other cases.

Tables 2a and 2b use the system of Tables 1a and 1b, except for the change $b = -(1, 0)$. Thus, for $N = 100$, there are $\lambda_1 N - \sqrt{N} = 50$ arrivals per hour at node 1 and, but the probability of returning to either node is still .5. Thus, there is a net surplus of vehicles at node 1. This is reflected in the data. It is still true that the best performance requires a slight excess of vehicles. Since the total customer arrival rate has been reduced, the value $V_0 = 0$ is now best, and corresponds to a vehicle excess. As expected, the main effect of the control is the transfer of vehicles from node 2 to node 1. Without control, the mean queue length at node 2 is quite long, and with $V_0 = 0$ the loss of customers due to buffer oveflow is large. Other (not tabulated) data shows that there is little advantage to using $u^i = 2$. Table 2b concerns the case where the interarrival times are constant. Again, the percentage improvement in performance is best where the system already works well.

The Tables 3–7 deal with an unbalanced system, where $\lambda_1 = .4, \lambda_2 = 1.2$. The probability of returning to node 1 (resp., node 2) is .25 (resp., .75). Thus for $N = 100$, the mean arrival rate per hour at node 1 is 40 and at node 2 it is 120. In all cases, a slight excess of vehicles is best. For the unbalanced cases control becomes even more important, as one would expect. The rapid increase in customer waiting and overflow as V_0 decreases is again evident. In Table 4 there are $\sqrt{N}$ fewer arrivals per hour at node 1. This increased assymetry again increases the value of control. Control is most effective in the cases which already work well, but the improvement is more pronounced than in Tables 2a, 2b. This is because the reduction in input rate (due to $b_1 < 0$) is concentrated at the node with the lowest input rate. In Table 5, the reduction in input occurs at the node with the highest input rate.

Tables 6 and 7 (resp.) repeat the cases in Tables 4 and 5, but for the case where the inputs are reduced by $2\sqrt{N}$ at node 1 (node 2, resp.). Here it becomes important to use a higher value of $\bar{u}^i$. In fact, we see from Table 6 that $Eu^1(t) \approx 1$, signifying that the control transferring from node 1 to node 2 is active essentially continuously. The difference between the rates at which vehicles collect at node 1 and the rate at which they can be used is greater for the case of Table 6 over that of Table 7. Thus, a higher maximum transfer rate is needed.

Table 1a. Optimal cost and components.

$\lambda = (.6, .6), b = (0, 0), \bar{u} = 1$

V_0	U_1	L_1	u_1	x_1	$\bar{\gamma}$
2.5	.18	~0	.22	.03	4.5
1.5	.06	~0	.19	.13	2.2
0	~0	.12	.12	1.1	3.8
-1.5	~0	.75	~0	2.3	12.1
no control					
1.5	.15	.04	0	.46	4.4
0	.02	.20	0	1.25	4.9
$\bar{u} = 2$					
1.5	.04	~0	.21	.12	1.8
0	~0	.11	.13	1.03	3.7

Table 1b. Optimal cost and components.

$\lambda = (.6, .6), b = (0, 0), \bar{u} = 1$, const. interarrival times

V_0	U_1	L_1	u_1	x_1	$\bar{\gamma}$
1.5	~0	~0	.11	.04	.71
0	~0	.07	.12	.94	3.1
no control					
1.5	.07	.02	0	.4	2.5
0	~0	.12	0	1.35	3.9

Table 2a. Optimal cost and components.

$\lambda = (..6, .6), b = -(1, 0), \bar{u} = 1$

V_0	U_1	U_2	L_1	L_2	u_1	u_2	x_1	x_2	$\bar{\gamma}$
1.5	.25	.14	~0	~0	.52	.08	.02	.04	5.1
0	.04	.02	~0	.01	.52	.04	.28	.34	2.4
-1.5	~0	~0	.09	.53	.29	~0	1.05	1.94	6.6
no control									
1.5	.89	.03	~0	.15	0	0	~0	1.1	11.1
0	.41	~0	~0	.59	0	0	.05	2.1	9.2
-1.5	.07	~0	.05	.98	0	0	.59	2.4	8.8

Table 2b. Optimal cost and components.									
$\lambda = (.6, .6), b = -(1, 0), \bar{u} = 1$, const interarrival times.									
V_0	U_1	U_2	L_1	L_2	u_1	u_2	x_1	x_2	$\bar{\gamma}$
1.5	.08	.04	~0	~0	.52	.38	~0	~0	2.37
0	~0	~0	~0	~0	.50	~0	.01	.11	1.2
-1.5	~0	~0	~0	.55	.27	.05	.66	1.23	6.4
no control									
1.5	.85	~0	~0	.15	0	0	~0	1.6	10.9
0	.30	~0	~0	.70	0	0	~0	2.6	9.1

Table 3. Optimal cost and components.									
$\lambda = (.4, 1.2), b = (0, 0), \bar{u} = 1$									
V_0	U_1	U_2	L_1	L_2	u_1	u_2	x_1	x_2	$\bar{\gamma}$
1.5	.05	.12	~0	.02	.20	.22	.07	.39	3.1
0	~0	.02	.04	.26	.13	.16	.83	1.1	4.2
-1.5	~0	~0	.17	1.3	.04	.25	1.3	2.1	11.2
no control									
1.5	.1	.31	.03	.01	0	0	.51	.50	5.7
0	.02	.06	.13	.38	0	0	1.2	1.2	5.7

Table 4. Optimal cost and components.									
$\lambda = (.4, 1.2), b = -(1, 0), \bar{u} = 1$									
V_0	U_1	U_2	L_1	L_2	u_1	u_2	x_1	x_2	$\bar{\gamma}$
1.5	.29	.21	~0	.02	.63	.05	~0	.25	6.7
0	.06	.06	~0	.12	.72	.03	.18	.68	4.1
-1.5	~0	~0	~0	.71	.58	~0	.71	1.68	7.4
no control									
1.5	.91	.06	~0	.33	0	0	~0	1.2	12.6
0	.62	~0	~0	1.14	0	0	~0	2.05	14.0
-1.5	.33	~0	~0	2.03	0	0	.05	2.34	15.9

Table 5. Optimal cost and components.									
$\lambda = (.4, 1.2), b = -(0, 1), \bar{u} = 1$									
V_0	U_1	U_2	L_1	L_2	u_1	u_2	x_1	x_2	$\bar{\gamma}$
1.5	.14	.32	~0	~0	.12	.40	~0	.16	5.9
0	.02	.08	~0	.05	.1	.35	.21	.54	2.9
-1.5	~0	~0	.14	.50	.04	.31	1.23	1.48	6.7
no control									
1.5	.05	.97	.06	~0	0	0	.82	.08	11.4
0	~0	.39	.22	.04	0	0	1.6	.3	7.2
-1.5	~0	.05	.43	.35	0	0	2.2	1.2	7.7
$\bar{u} = 2$									
1.5	.12	.27	~0	~0	.17	.44	~0	.13	5.3
0	.01	.05	.03	.03	.12	.37	.20	.48	2.4
-1.5	~0	~0	.14	.48	.05	.31	1.2	1.41	6.5

Table 6. Optimal cost and components.									
$\lambda = (.4, 1.2), b = -(2, 0), \bar{u} = 1$									
V_0	U_1	U_2	L_1	L_2	u_1	u_2	x_1	x_2	$\bar{\gamma}$
1	.80	.16	~0	.09	.93	~0	~0	.53	12.4
0	.60	.05	~0	.34	.98	~0	~0	1.0	11.3
-1	.41	.01	~0	.80	1.0	~0	.03	1.17	12
-2	.24	~0	~0	1.3	1.0	~0	.10	2.1	13.1
no control									
1	1.8	.02	~0	.69	0	0	~0	1.7	23.0
0	1.5	~0	~0	1.35	0	0	~0	2.2	24.5
-1	1.31	~0	~0	2.06	0	0	~0	2.5	26.0
-2	1.07	~0	~0	2.77	0	0	~0	2.60	27.0
$\bar{u} = 2$									
1	.38	.27	~0	~0	1.31	~0	~0	.11	9.3
0	.16	.11	~0	.02	1.43	.03	.02	.25	6.0
-1	.04	.04	~0	.06	1.48	.02	.19	.53	4.8
-2	~0	~0	~0	.35	1.41	~0	.49	1.12	6.3

| \multicolumn{10}{c}{Table 7. Optimal cost and components.} |
| \multicolumn{10}{c}{$\lambda = (.4, 1.2), b = -(0, 2), \bar{u} = 1$} |
V_0	U_1	U_2	L_1	L_2	u_1	u_2	x_1	x_2	$\bar{\gamma}$
1	.18	.54	~0	~0	.07	.57	.01	.09	8.6
0	.07	.25	~0	.01	.07	.56	.04	.26	4.8
-1	.02	.1	~0	.05	.05	.54	.24	.55	3.38
-2	~0	.03	.05	.27	.03	.55	.81	1.15	5.0
-3	~0	~0	.57	.51	~0	.20	2.21	1.53	9.59
\multicolumn{10}{c}{no control}
1	.01	1.6	.16	~0	0	0	1.4	.01	18.0
0	~0	1.0	.32	~0	0	0	1.9	.04	14.2
-1	~0	.56	.49	.01	0	0	2.22	.16	10.4
-2	~0	.20	.63	.09	0	0	2.40	.56	8.6
-3	~0	.04	.79	.41	0	0	2.53	1.31	10.3
\multicolumn{10}{c}{$\bar{u} = 2$}
1	.17	.43	~0	~0	.12	.64	~0	.07	7.7
0	.07	.17	~0	~0	.10	.60	.02	.20	4.0
-1	.01	.53	~0	.03	.06	.56	.21	.46	2.71
-2	~0	.01	.12	.17	.03	.47	1.0	1.0	4.6
-3	~0	~0	.53	.52	~0	.24	2.18	1.54	9.41

Refer to Figures 7.1–7.4. The dark areas are where the control is active, with $u^1(\cdot)$ being used in the lower right and $u^2(\cdot)$ on the upper left. Refer to Figure 7.1. The general shape is more or less as expected. Note that the control can be active even when there is a queue of customers and no vehicles on hand (i.e., when some $X^i(t) < 0$). This phenomenon will be even more pronounced in the other figures, and it is to be interpreted as follows. Both vehicles and customers arrive at the nodes at a high rate. When there is a queue of customers, some of the arriving vehicles might be transferred to the other node by the control, and not be available for use by the waiting customers. Owing to the scaling, this number will be at most of the order of $O(\sqrt{N})$ per unit time, a small fraction of the total arrival rate in heavy traffic. While it might seem puzzling to the waiting customers that not all arriving vehicles are available for their use, it does not increase their wait by much and reduces the overall cost of operating the system. For the case of Figure 7.1, the system will rarely be in the region where this phenomena occurs, and if we disallow the possibility, very little will be lost.

Now refer to Figure 7.2. In this case, at node 1 the rate of vehicle arrivals is much greater than the rate of customer arrivals. Owing to the imbalance and to the fact that $V_0 = 1.5$, there is a tendency for the vehicle buffer at node 1 to overflow. Indeed, $u^1(\cdot)$ is active about half the time. In Figure 7.3, there are fewer arrivals to node 1, but the vehicle return

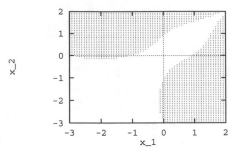

FIG. 7.1. *Control curve* $\lambda = (0.6, 0.6), b = (0,0), V_0 = 0$

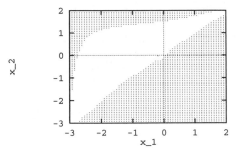

FIG. 7.2. *Control curve* $\lambda = (0.6, 0.6), b = (-1, 0), V_0 = 1.5$

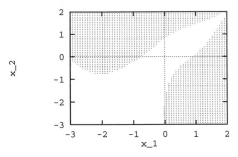

FIG. 7.3. *Control curve* $\lambda = (0.4, 1.2), b = (0, 0), V_0 = 0$

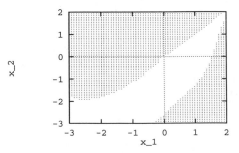

FIG. 7.4. *Control curve* $\lambda = (0.4, 1.2), b = (0, -1), V_0 = 0$

rate to each node is proportional to the customer arrival rate there. The control regions in the upper half plane and right half plane are more or less symmetric. But vehicles might be transferred from node 1 to node 2 even if there is a queue at node 1. The fact that the boundary of the upper control region first moves down as X^1 decreases and then moves up as X^1 approaches -3 is an artifact due to the values of the weights in the cost function. It seems that as the left boundary is approached, the likelihood increases that customers at node 1 are lost due to buffer overflow. But with our cost function weights, this loss is less expensive then the cost of a control effort to manage it. It is not an important phenomena, and little is lost if the boundaries of the control regions are made monotonic in a reasonable way. The example of Figure 7.4 is that of Figure 7.3, but the rate of customer arrival at node 2 is reduced. This shift explains the shift in the boundaries of the control regions.

REFERENCES

[1] P. Billingsley. *Convergence of Probability Measures*. John Wiley, New York, 1968.

[2] V.S. Borkar and M.K. Ghosh. Ergodic control of multidimensional diffusions: I, the existence results. *SIAM J. on Control and Optimization*, 26:112–126, 1988.

[3] S.N. Ethier and T.G. Kurtz. *Markov Processes: Characterization and Convergence*. Wiley, New York, 1986.

[4] J. M. Harrison and L. M. Wein. Scheduling networks of queues: heavy traffic analysis of a simple open network. *Queueing Systems*, 5:265–280, 1989.

[5] J.M. Harrison. Brownian models of queueing networks with heterogeneous customer populations. In *Proc. of the IMA*, volume 10, edited by P.-L. Lions and W. Fleming. Springer-Verlag, Berlin, 1988, pp. 147–186.

[6] J.M. Harrison and R.J. Williams. Brownian models of open queueing networks with homogeneous customer populations. *Stochastics*, 22:77–115, 1987.

[7] H.J. Kushner. Optimality conditions for the average cost per unit time problem with a difusion model. *SIAM J. on Control and Optimization*, 16:330–346, 1978.

[8] H.J. Kushner. *Approximation and Weak Convergence Methods for Random Processes with Applications to Stochastic System Theory*. MIT Press, Cambridge, MA, 1984.

[9] H.J. Kushner. *Weak Convergence Methods and Singularly Perturbed Stochastic Control and Filtering Problems*, volume 3 of *Systems and Control*. Birkhauser, Boston, 1990.

[10] H.J. Kushner. Control of trunk line systems in heavy traffic. Technical report, Brown University, Lefschetz Center for Dynamical Systems: Division of Applied Math., 1992. To appear SIAM J. on Control and Optimization.

[11] H.J. Kushner and L.F. Martins. Limit theorems for pathwise average cost per unit time problems for queues in heavy traffic. *Stochastics*, 42:25–51, 1993.

[12] H.J. Kushner and K.M. Ramachandran. Optimal and approximately optimal control policies for queues in heavy traffic. *SIAM J. on Control and Optimization*, 27:1293–1318, 1989.

[13] H.J. Kushner and J. Yang. Numerical methods for controlled routing in large trunk line systems via stochastic control theory. Technical report, Brown University, Lefschetz Center for Dynamical Systems, 1992. to appear in the ORSA Journal on Computing, 1994.

DYNAMIC ROUTING IN STOCHASTIC NETWORKS

F.P. KELLY*

Abstract. This paper reviews some current work on routing in loss and queueing networks. We describe two classes of bound on the performance of any dynamic routing scheme, together with some open questions concerning whether the bounds can be approached under certain limiting regimes. The first class of bound is particularly appropriate for networks in heavy traffic, where the key feature of a good routing scheme is its effective utilization of various pooled resources identified by a fluid version of the routing problem. The second class of bound is particularly appropriate for highly connected networks, with many alternative paths. Again, a network flow representation is central, but this time involving a collection of Markov decision processes, one for each resource of the network. Despite their simplicity, the bounds are able to identify the great variety of qualitatively distinct behaviour expected of a good dynamic routing scheme, depending on a network's size, connectivity, asymmetry and degree of overload.

1. Introduction. Modern telecommunications networks provide a variety of fascinating challenges for the mathematician interested in applications of probability and optimization. Dynamic routing schemes, for example, allow a network to respond to fluctuating demands and failures by rerouting traffic and reallocating resources. While the detailed design of a dynamic routing scheme is much influenced by the rapidly evolving technological environment, some important principles are apparent from an analysis of loss probabilities and queueing delays in simple network models.

In this paper we describe some current work on routing in loss and queueing networks. One of the promising approaches of the last few years has been the development of upper bounds on the performance of *any* dynamic routing scheme for a stochastic network ([1],[2],[19],[25]). We describe two classes of bound, each with an associated limiting regime under which it is conjectured that the bound is tight. Further open questions concern whether the bounds can be approached by dynamic routing schemes that are simple in some sense, perhaps depending only on local rules and thresholds. Despite their simplicity and potential coarseness, the bounds are able to identify the great variety of qualitatively distinct behaviour expected of a good dynamic routing scheme, depending on a network's size, connectivity, asymmetry and degree of overload.

In a queueing network an arriving demand makes use of different resources sequentially in time, and congestion causes delay. In a loss network an arriving demand requests simultaneous use of different resources, and congestion causes blocking. Despite these differences there is a remarkable similarity between the methods used to analyse loss and queueing networks:

---

* Statistical Laboratory, University of Cambridge, 16 Mill Lane, Cambridge CB2 1SB, England. Part of this paper was written whilst the author was visiting the Institute for Mathematics and its Applications, whose financial support is gratefully acknowledged.

see, for example, the discussion in [9] of product-form solutions, approximation procedures and optimization issues, and the study in [7] of least busy alternative routing. A subsidiary aim of this paper is to further indicate the potential for cross-fertilization of ideas between studies of the two forms of network.

In Section 2 we describe the multicommodity network flow results that apply to the deterministic, or fluid, version of the network routing problem. The flows that are achievable in the fluid version are defined by a set of generalized cut constraints, written in terms of the load upon, and the capacity of, various pooled resources. These constraints provide the starting point for the heavy traffic analysis of stochastic networks. In a queueing network heavy traffic is defined by a limiting regime where the load upon pooled resources approaches their capacity. In a loss network heavy traffic is defined within a limiting regime where the load upon, and capacity of, pooled resources increase together, with ratios held fixed. It is possible to bound the performance of a stochastic network in terms of the performance of its pooled resources, but there remain several open questions concerning whether these bounds can be approached and, if so, by what form of dynamic routing scheme. The analysis does, however, make clear that in heavy traffic the key determinant of network performance is the behaviour of the pooled resources identified by the fluid version of the routing problem.

It is also clear that, under a good routing scheme, the proportion of alternatively routed traffic within the network, routed along other than the first choice routes identified by the fluid version, should vanish in the heavy traffic limit. Rather different behaviour should be expected in highly connected networks with relatively small traffic and capacity at each resource. In Section 3 we outline a class of bound particularly appropriate for such networks, and discuss a limiting regime under which the proportion of alternatively routed traffic within the network does not vanish. Again a network flow representation is central, but this time involving a collection of Markov decision processes, one for each resource of the network. Once again, there remain several open questions concerning whether the bounds can be approached and, if so, by what form of dynamic routing scheme.

Finally in Section 4 we describe how, for any given finite network, the two classes of bound may be combined. In future numerical work with combined bounds it is hoped to get a clearer view of how the structural features of a given network affect the relative importance and accuracy of the various forms of bound.

2. Deterministic bounds, and heavy traffic. We begin by outlining the basic notation we use to describe loss and queueing networks.

2.1. Loss and queueing networks. The classical example of a loss network is a telephone network, and we shall phrase our description in terms of calls, circuits and routes. However the reader will observe that our model applies more widely to systems in which, before a request (which may be a call, or a task, or a customer) is accepted, it is first checked that sufficient resources are available to deal with the request.

Consider then a network with links labelled $i \in I$, for some finite set I, and suppose that link i comprises C_i circuits. A call on route $r \in R$ uses A_{ir} circuits from link i, where $A_{ir} \geq 0$. Thus if n_r calls are in progress on route r then

$$(2.1) \qquad\qquad\qquad An \leq C$$

where $A = (A_{ir}, i \in I, r \in R)$, $n = (n_r, r \in R)$ and $C = (C_i, i \in I)$.

Suppose that calls of type $j \in J$ arrive at the network at rate ν_j, where J is the set of call types. Set $D_{jr} = 1$ if a call of type j can be carried on route r, and set $D_{jr} = 0$ otherwise. This defines a $0 - 1$ matrix $D = (D_{jr}, j \in J, r \in R)$. When a call of type j arrives at the network, a dynamic routing scheme must decide which route in the set $R_j = \{r : D_{jr} = 1\}$ should be used to carry the call, or whether the call should be rejected. The call *must* be rejected if acceptance would lead to vector n violating the constraint (2.1). Once a call is accepted it has a holding period which, for definiteness, we shall assume has unit mean and is independent of earlier arrival times and holding periods.

We describe a queueing network in very similar terms. Suppose the network has stations labelled $i \in I$, for some finite set I, and that a customer on route $r = (r(1), r(2), \ldots, r(M_r))$ passes through the sequence of queues

$$r(1), r(2), \ldots, r(M_r)$$

before leaving the system. Thus the queue which a customer on route r visits at stage m $(= 1, 2, \ldots, M_r)$ of its route is queue $r(m)$, and M_r is the number of stages on route r. Let station i have a single server, capable of dispensing service effort at rate C_i, and, for definiteness, assume that all service requirements, even of the same customer at different stages of its route, are independent of each other and of earlier arrival times, and have unit mean. Let

$$A_{ir} = |\{m : r(m) = i\}|,$$

the number of times a customer on route r visits station i. Suppose that customers of type $j \in J$ arrive at the network at rate ν_j. As before set $D_{jr} = 1$ if a customer of type j can be carried on route r, and set $D_{jr} = 0$ otherwise. Since resources are used sequentially in time it may not be necessary for a dynamic routing scheme to choose the entire route r of a customer when the customer enters the network – it may, for example, be possible to select a first stage for the route, but leave open the choice of second stage until the first stage is completed. In a queueing network it is also necessary to make sequencing decisions at stations: a station must decide the order in which to serve the various customers on different routes that may be queueing there. We view sequencing choices as part of the overall dynamic routing scheme.

2.2. Deterministic bounds. Before developing bounds for a stochastic network it is useful to consider a deterministic or fluid version of the routing problem, using standard multicommodity network flow results.

Let

$$S = \{\nu \geq 0 : \exists \, x \geq 0 \text{ with } Dx = \nu, Ax \leq C\}.$$

We can view S as the achievable region for the fluid analogue: for any $\nu \in S$ there exists a deterministic flow pattern x which will satisfy the demand pattern ν (since $Dx = \nu$), but will not overload any resource (since $Ax \leq C$). Let

$$T = \{x \geq 0 : Ax \leq C\}.$$

Thus T is the intersection of the half-spaces $(x : (Ax)_i \leq C_i), i \in I$, and the non-negative orthant. With no loss of generality assume no column of A is null: hence the set T is bounded, and is the convex hull of a finite number of extreme points. Hence $S = DT$ is the convex hull of a finite number of extreme points or, equivalently, the bounded intersection of a finite set of half-spaces. Thus there exists a representation

(2.2) $$S = \{\nu \geq 0 : \overline{A}\nu \leq \overline{C}\}$$

for some choice of $\overline{A}, \overline{C}$. Moreover $\overline{A}, \overline{C}$ can be chosen to have non-negative element (since $0 \in S$; and $0 \leq \nu' \leq \nu \in S$ implies $\nu' \in S$).

This result, that the achievable region (2.2) for the fluid analogue is subject to the same form of linear constraint (2.1) as the original loss network, is a well known result [5] . The usual derivation is less direct, but achieves a valuable insight from its use of dual variables. In [5] it is shown that $\nu \in S$ if and only if for all $\beta \geq 0$

(2.3) $$\sum_j \alpha_j \nu_j \leq \sum_i \beta_i C_i$$

where

(2.4) $$\alpha_j = \min_{r:D_{jr}=1} \left\{ \sum_i \beta_i A_{ir} \right\} \quad j \in J.$$

Here the dual variable β_i may be viewed as a price charged for the use of one unit of capacity at resource i: if there exists a vector $\beta \geq 0$ such that (2.3) is violated then clearly the demand pattern ν cannot be achieved, since with these prices the maximum revenue from resources is exceeded by the amount demands must pay to cross the network. Laws [17] notes that (2.3) and (2.4) will follow for all $\beta \geq 0$ if they hold for a certain finite collection of vectors β: thus $\nu \in S$ if and only if

(2.5)
$$\sum_j \alpha_{kj} \nu_j \leq \sum_i \beta_{ki} C_i \qquad k \in K$$

for some collection $\{\{\alpha_{kj}, \beta_{ki}, i \in I, j \in J\}, k \in K\}$, where K is finite. The linear constraints (2.5) defining S are termed *generalized cut constraints*. Each such constraint describes the load on, and the capacity of, a single *pooled resource*.

2.3. Heavy traffic. What can we deduce about the behaviour of a stochastic network from the above fluid version? Consider first a loss network, and suppose that the arrival rates ν are such that at least one of the constraints (2.5) is violated: let us say that the constraint labelled $k = 0$ is violated. Then the performance of the loss network is clearly dominated by the performance of a single link, of capacity $\sum_i \beta_{oi} C_i$ circuits, at which calls of type j arrive, as in the network, at rate ν_j, each such call requesting α_{oj} circuits. More generally, if $(n_j, j \in J)$ describes the number of calls of type j in progress then necessarily

$$\sum_j \alpha_{kj} n_j \leq \sum_i \beta_{ki} C_i \qquad k \in K.$$

Thus the performance of a loss network under *any* dynamic routing scheme is dominated by the performance of a loss network, with no routing choices, where calls of type j arrive at rate ν_j, and require α_{kj} circuits from link k, a link of capacity $\sum_i \beta_{ki} C_i$, for $k \in K$.

When might the bounds[1] implied by the above comparison be good bounds, that can be approximately achieved? Consider a sequence of networks where $(\nu_j(N), j \in J)$, $(C_i(N), i \in I)$ label the arrival rates and capacities, respectively, in the Nth network, and suppose that as $N \to \infty$

(2.6)
$$\frac{\nu_j(N)}{N} \to \nu_j, \qquad \frac{C_i(N)}{N} \to C_i.$$

Then it is possible to show (under rather weak conditions on arrival processes: see [13],[14] for the case of Poisson arrivals) that the set of loss probabilities $(L_j(N), j \in J)$ achievable in network N approaches the set of loss probabilities achievable in the fluid version. Interest then focuses on whether simple decentralized schemes, using threshold rules like trunk reservation, are able to perform close to optimally (see [14],[18]: for networks with a small number of links it is possible to compute the optimal control [13], [16], but this seems impractical for a large network such as that

---

[1] When calls may be repacked the bound becomes a characterization: see [10] for a discussion of this interesting case, which here we consider no further.

studied in [3]). To give an example, suppose that arrivals of different types of call form independent Poisson processes, that the limit vectors ν, C from (2.6) satisfy with strict inequality all but one ($k = 0$) of the generalized cut constraints (2.5), and that

$$(2.7) \qquad \sum_j \alpha_{oj}\nu_j(N) - \sum_i \beta_{oi}C_i(N) = \gamma N^{\frac{1}{2}}$$

for some constant γ. Then for the bounding single link it seems plausible (cf. [18],[27]) that, by rejecting traffic of type $j = 1$ if the pooled resource is closer to saturation than, say, $\log N$ circuits, it is possible to achieve loss probabilities satisfying

$$L_j(N) = o(N^{\frac{1}{2}}) \qquad j \neq 1,$$

$$N^{\frac{1}{2}}L_1(N) \to \frac{\left(\sum_j \alpha_{oj}^2 \nu_j\right)^{\frac{1}{2}}}{\alpha_{01}\nu_1} \frac{\phi(\delta)}{1 - \Phi(\delta)}$$

where ϕ, Φ are respectively the normal density and distribution function, and

$$\delta = \gamma \left(\sum_j \alpha_{oj}^2 \nu_j\right)^{-\frac{1}{2}}.$$

(Note that if $\alpha_{oj} = 1, j \in J$, $\beta_{oi} = 1, i \in I$, and $\sum_j \nu_j = 1$, then $\delta = \gamma$ and these expressions become those established for a single link by Reiman [27]. For more general parameter choices the constant γ must be renormalized, and some insight into the expression for δ is obtained by observing that if n_j is a Poisson random variable with mean ν_j, then the variance of $\sum_j \alpha_{oj} n_j$ is $\sum_j \alpha_{oj}^2 \nu_j$.) It is a challenge to determine if a dynamic routing scheme for the loss network could approach the above putative bound, under the limiting regime (2.6) and (2.7).

Consider next a queueing network, and suppose the arrival rates ν are such that all of the constraints (2.5) are satisfied. Nevertheless choose one, say the constraint labelled $k = 0$. Then the performance of the queueing network is dominated by the performance of a single station constructed as follows. Let the single station have multiple servers, a server i replicating the server at station i for each station i with $\beta_{oi} > 0$. Let customers of type j arrive at this station in a Poisson stream of rate ν_j. Any customer may be served by any server, but a single service completion at server i corresponds to the accrual of β_{oi} credits by the customer, and a customer of type j cannot leave until it has accrued at least α_{oj} credits. (Note that, by relation (2.4), it is certainly possible for a customer of type j to acquire exactly α_{oj} credits.)

Again it is reasonable to consider when such bounds may be approximately achieved. Consider a sequence of networks, where $(\nu_j(N), j \in J)$ label the arrival rates in the Nth network, and where as $N \to \infty$

(2.8) $$\nu_j(N) \to \nu_j.$$

Suppose that all of the constraints (2.5) are satisfied with strict inequality, save one, labelled $k = 0$, for which

(2.9) $$\sum_j \alpha_{oj} \nu_j(N) = \sum_i \beta_{oi} C_i - \theta N^{-\frac{1}{2}}$$

where $\theta > 0$. Further suppose that arrivals of type j customers form a renewal process, where interarrival times have mean $\nu_j(N)^{-1}$ and variance a_j, and service times at station i have mean C_i^{-1} and variance s_i. Then for the bounding single station it seems plausible ([26],[28],[18]) that, by giving priority to all customers except those for which α_{oj} is maximal $(= \alpha$, say), it is possible to achieve a mean number of customers in the queue of order

$$\frac{N^{\frac{1}{2}}}{2\alpha\theta} \left(\sum_j \alpha_{oj}^2 \nu_j^3 a_j + \sum_i \beta_{oi}^2 C_i^3 s_i \right)$$

Again it is a challenge to determine if a dynamic routing scheme for the queueing network could approach this putative bound, under the limiting regime (2.8) and (2.9).

Laws and Louth [19] analyse a queueing network where *two* of the generalized cut constraints (2.5) are close to saturation: they show that a bound can be constructed using the pathwise minimum of two queues, each representing a pooled resource, and provide numerical evidence that this bound is approached under the limiting regime. See [12] for further discussion of a variety of examples, emphasizing the use of the Brownian network models of Harrison [6].

3. Diverse routing. In this Section we describe a further class of bound, more appropriate to the case of a highly connected network where the network contains many alternative routes and a typical resource lies, potentially, on a large number of alternative routes.

3.1. Loss networks. Let I label the set of possible demands on a network, and also the set of network resources. Assume that demands of type i arrive at the network in a Poisson stream of rate ν_i, for $i \in I$, and that as i varies it indexes independent Poisson streams. Let C_i be the *capacity* of resource i. Again, we shall refer to a resource as a *link*, interpret C_i as the number of *circuits* on link i, and refer to a demand labelled i as a *call* of type i. Suppose that an arriving call of type i may potentially be routed directly, on link i, or alternatively, on a route $r \in R(i)$. Here a route r identifies a subset of I, and $R(i)$ is the set of alternative routes potentially available to a call of type i. Label routes so that the sets $R(i), i \in I$, are disjoint: let $R = \bigcup_{i \in I} R(i)$, and let $i(r)$ be the unique element in I such that $r \in R(i(r))$. If an arriving call of type i is sent to route $r \in \{\{i\}\} \cup R(i)$ then it uses one circuit from each link $j \in r$ for the holding period of the call. A call may only be sent to a route r with at least one free circuit on each link $j \in r$. The call may also be discarded, and it *must* be discarded if there are no routes with spare capacity available. The holding period of a call is arbitrarily distributed with unit mean, is independent of earlier arrival times and holding periods, and is unaffected by the route used to carry the call.

For example, we could model a fully connected telephone network on n nodes by letting I be the set of $\frac{1}{2}n(n-1)$ unordered pairs of nodes. A call labelled $i \in I$ desires to connect the two nodes identified by i, and C_i is the number of circuits on the direct link between these two nodes. Note that we can interpret I as the set of edges of the complete graph on n nodes, and then a route connecting the node pair i is just a path in the graph between these two nodes. We might let $R(i)$ be the set of $(n-2)$ two-link paths connecting the node pair i. Or $R(i)$ might be the entire set of paths of length two or more connecting the node pair i, or any given subset of this. If the network is not fully connected, then we just set $C_i = 0$ for the missing links.

A dynamic routing scheme is a control policy which decides for each arriving call whether to discard the call, or, if not, by which route to carry the call. We allow the control policy to have full information on the history of call arrivals, routing decisions and expired holding times, but do not allow it to anticipate future events, and in particular do not allow it to have any information on the future holding time of an arriving call. Suppose that an accepted call of type i is worth w_i, and that our objective is to maximize the expected average reward per unit time. If holding times are exponentially distributed then we may formulate the control problem as a finite state Markov decision process, where the state is a complete list of the routes used by all calls currently in progress (although it may be difficult to make computational progress with this formulation: see [22]). With arbitrarily distributed holding times a Markov description requires additional information on the elapsed holding times of the calls currently in progress. Throughout we shall restrict attention to policies under which the number of calls of type i accepted over the period $[0, t]$ converges, in expectation and almost surely, as $t \to \infty$. We lose no essential generality by this restriction, since the event that the system is empty has bounded mean recurrence time under any policy, by the assumption of Poisson arrivals.

Our aim is to obtain an upper bound on the expected reward per unit time. Before describing the bound we shall find it helpful to analyze a much simpler dynamic optimization problem. Consider a single link, of capacity C circuits, offered two streams of traffic. Suppose that acceptance of a type 1 call generates a reward w_1, and acceptance of a type 2 call generates a reward w_2, where $w_1, w_2 \geq 0$. Suppose that the two streams of arriving traffic form independent Poisson processes of rates ν_1 and ν_2 respectively, and that accepted calls have holding times which are arbitrarily distributed with unit mean, independently of earlier arrival and holding times. When a call arrives a decision is made to accept or reject the call, and the decision can depend on the type of the call and the entire history of the link.

Let $W(\nu_1, \nu_2, C; w_1, w_2)$ be the maximal expected reward per unit time over all policies. Observe that W is a convex function of (w_1, w_2), since it can be expressed as a supremum of linear functions of (w_1, w_2),

$$W(\nu_1, \nu_2, C; w_1, w_2) = \sup_{\pi}\{w_1 x(\pi) + w_2 y(\pi)\},$$

where $x(\pi)$ and $y(\pi)$ are the mean acceptance rate of calls of types 1 and 2 respectively under a policy π. Let

$$M(\nu_1, \nu_2, C; x) = \sup_{\pi}\{y(\pi) : x(\pi) \geq x\},$$

the maximal mean acceptance rate of calls of type 2, subject to the requirement that the mean acceptance rate of calls of type 1 must be at least x. M is a concave function of x – indeed W and M are conjugates:

$$W(\nu_1, \nu_2, C; w_1, w_2) = \sup_{x \geq 0}\{w_1 x + w_2 M(\nu_1, \nu_2, C; x)\},$$
$$M(\nu_1, \nu_2, C; x) = \inf_{w \geq 0}\{W(\nu_1, \nu_2, C; w, 1) - wx\}.$$

If holding times are exponentially distributed with unit mean, then it is sufficient to consider policies which summarize the entire history of the link by a single integer, the number of calls currently in progress through the link. By a monotonicity argument [21] the optimal policy for the resulting Markov decision process is (if $w_1 \geq w_2 \geq 0$) to accept type 1 calls provided the link is not full, and to accept type 2 calls provided the number of spare circuits is above a certain integer threshold. Policies of this form are termed *trunk reservation* policies. If $w_2 > w_1 \geq 0$ then the optimal policy is to use trunk reservation against type 1 calls. To achieve the maximal flow M at arbitrary values of x we must extend the definition of a trunk reservation policy to allow randomized acceptance when the number of spare circuits is exactly equal to the threshold. If holding times are not exponentially distributed then the optimal policy may not be a trunk reservation policy, but such policies will generally be near optimal, especially for larger values of the capacity C.

Return now to the network model, and let

$$M_i(x) = M(\nu_i, \sum_{j : i \in R(j)} \nu_j, C_i; x),$$

$$W_i(w_1, w_2) = W(\nu_i, \sum_{j : i \in R(j)} \nu_j, C_i; w_1, w_2).$$

Then in [11] (see also [8],[15],[24] for earlier, related, work) it is shown that under any policy the expected reward per unit time is bounded above by the value attained in the maximum flow problem

(3.1a)   maximize $\sum_i w_i(x_i + \sum_{r \in R(i)} y_r)$

(3.1b)   subject to $x_i + \sum_{r \in R(i)} y_r \leq \nu_i \qquad i \in I$

(3.1c)   and $\sum_{r:i \in r} y_r \leq M_i(x_i) \qquad i \in I$

(3.1d)   over $x_i \geq 0, i \in I, \quad \text{and} \quad y_r \geq 0, r \in R$.

or, equivalently, the value attained in the dual minimum cost problem

(3.2a)   minimize $\sum_i [W_i(w_i - s_i, c_i) + s_i \nu_i]$

(3.2b)   subject to $w_{i(r)} - \sum_{j \in r} c_j \leq s_{i(r)} \qquad r \in R$

(3.2c)   over $c_i, s_i \geq 0 \qquad\qquad i \in I$.

The primal problem is perhaps the easier to interpret, with x_i, y_r representing mean flows along routes $\{i\}, r$, respectively: thus inequality (3.1c) captures the insight that link i cannot achieve higher acceptance rates as part of a network than it could if the rest of the network were transparent. A sketch of the proof via the dual problem is instructive. Consider a feasible choice of c, s; that is, a collection $c_i, s_i, i \in I$, satisfying (3.2b) and (3.2c). Suppose that a call of type i is charged an amount s_i by the network when it is offered to the network, and in addition is charged an amount $w_i - s_i$ by link i if it is carried directly on link i, or an amount c_j at each link $j \in r$ if it is carried on alternative route r. The maximum revenue that can possibly be collected thus cannot exceed the objective function (3.2a), for any feasible choice of s, c. Since s, c satisfy (3.2b) each routed call of type i pays at least w_i, whether the call is routed directly or alternatively, and the result follows.

When holding times are exponentially distributed, the function $M_i(x_i)$ in piecewise linear in x_i, for $i \in I$. In this case both the primal and dual problems (3.1) and (3.2) may be written as linear programming problems, a fact exploited in [4].

It is interesting to interpret the complementary slackness conditions that interlink the solutions to problems (3.1) and (3.2) in terms of the charges used in the proof sketch. The main condition shows that if the charges along route r are too high (i.e. $\sum_{j \in r} c_j > w_{i(r)} - s_{i(r)}$) then the route is not used (i.e. $y_r = 0$). The interpretation of the remaining complementary slackness conditions is more familiar: the charge c_i at link i is zero if there is spare capacity at this link, as indicated by slackness in the constraint (3.1c); while the charge s_i on an offered call of type i is zero if calls of type i are being lost, as indicated by slackness in the constraint

(3.1b). We might term s_i the *surplus value* of an additional call of type i, and c_j the *implied cost* of using a circuit from link j: if a call of type i is accepted on route r it will earn w_i directly but at an implied cost of c_j for each circuit used from link j, leaving a surplus value of $s_i = w_i - \sum_{j \in r} c_j$ (cf. [9]).

The above discussion is very suggestive of routing schemes where each link operates a trunk reservation policy and calls are routed by the cheapest route provided that route costs less than the worth of the call. For certain forms of symmetric network, for example a fully connected network on n nodes, it is known that such schemes achieve a reward per link that approaches the bound as the number of nodes increases ([23], [8]). This phenomenon has also been observed in simulations of asymmetric networks [3]. It seems reasonable to expect the bounds to be approached by trunk reservation schemes whenever calls have many alternative routes to choose from, and each link lies, potentially, on a large number of alternative routes. It seems difficult, however, to prove this under appropriately general conditions (some general results are available when calls may be repacked: see [11], Section 3).

Next we show that a parallel development is able to provide bounds on the performance of arbitrary dynamic routing schemes for queueing networks.

3.2. Queueing networks. Again let I label the set of possible demands on the network as well as the collection of network resources. Assume that arrivals at the network of demands labelled i form a stationary stochastic process of rate ν_i. Assume now that resources are queues, and that demands are customers. Suppose that a customer of type i may be routed directly, via queue i, or alternatively on a route $r \in R(i)$. A route r now identifies an ordered subset of I, and $R(i)$ is the set of alternative routes potentially available to a call of type i. If a customer is sent via route r then it must be served in succession at each queue $j \in r$. Assume that the service requirement of a customer arriving at the network is arbitrarily distributed with unit mean, and that a customer's service requirement is the same at each queue that it visits. Let queue i serve at rate C_i, so that the service time of a customer at queue i is that customer's service requirement divided by C_i.

A dynamic routing scheme is a control policy which decides the route of each arriving customer, and decides for each queue who the server should deal with next (pre-emption may be either allowed or disallowed). We suppose that it costs w_i per unit of time that a customer of type i remains in the system, and our objective is to minimize the expected cost per unit time. Again our aim is to bound the performance of any policy, and again we begin by considering a much simpler dynamic optimization problem.

Consider first a single queue at which customer arrivals form a stationary stochastic process of rate ν. On arrival a customer may be accepted or rejected. Accepted customers queue, and have service times that are arbitrarily distributed with mean C^{-1}. (For notational convenience, and with no loss of essential generality, we shall assume that ν fixes the distribution of the arrival process, as well as its rate, and that C fixes the distribution of service times, as well as their mean.) Suppose that an accepted customer costs w per unit of time that it remains in the queue, and generates a reward s on service completion. Let $W(\nu, C; w, s)$ be the maximal net rate of return over all admission and scheduling policies. Observe that W is a convex function of (w, s), since it can be expressed as a supremum of linear functions of (w, s),

$$W(\nu, C; w, s) = \sup_{\pi}\{sx(\pi) - wn(\pi)\},$$

where $x(\pi)$ is the mean acceptance rate of customers and $n(\pi)$ is the mean number of customers in the queue under a policy π. Let

$$N(\nu, C; x) = \inf_{\pi}\{n(\pi) : x(\pi) \geq x\},$$

the minimal mean number of customers in the queue, subject to the requirement that the mean acceptance rate of customers must be at least x. N is a convex function of x – indeed W and N are conjugates:

$$W(\nu, C; w, s) = \sup_{x}\{sx - wN(\nu, C; x)\}.$$
$$N(\nu, C; x) = \sup_{s}\{sx - W(\nu, C; 1, s)\}.$$

Under simplified assumptions concerning arrival and service times it is possible to determine explicitly the functions W and N. For example, suppose that arrivals of customers form a Poisson process of rate ν, and that service times are exponentially distributed with mean C^{-1} independently of earlier arrival and service times. Then the optimal policy for the resulting Markov decision process is to accept an arriving customer if the number of customers already in the queue is less than a certain integer, and to reject the customer otherwise [21].

Return now to the network model, and let

$$W_i(w, s) = W(\nu_i, C_i; w, s)$$

$$N_i(x) = N(\nu_i, C_i; x).$$

Then in [11] it is shown that under any policy the expected cost per unit time is bounded below by the value attained in the problem

(3.3a)   minimize $\displaystyle\sum_i w_i N_i(x_i)$

(3.3b)   subject to $\displaystyle x_i + \sum_{r \in R(i)} y_r = \nu_i \qquad i \in I$

(3.3c)   and $x_i + \sum_{r:i\in r} y_r \leq C_i \qquad i \in I$

(3.3d)   over $x_i \geq 0, i \in I, \quad \text{and} \quad y_r \geq 0, r \in R,$

or, equivalently, the value attained in the dual problem

(3.4a)   maximize $\sum_i [s_i\nu_i - c_iC_i - W_i(w_i, s_i - c_i)]$

(3.4b)   subject to $s_{i(r)} \leq \sum_{j\in r} c_j \qquad r \in R$

(3.4c)   over $c_i \geq 0, i \in I, \quad \text{and} \quad s_i, i \in I.$

A sketch of the proof is again instructive. Consider values s, c that satisfy (3.4b) and (3.4c). Suppose a network operator gives to each arriving customer of type i an amount s_i, but let queue i charge an amount s_i to directly routed customers and an amount c_i to alternatively routed customers. Note that $\sum_i s_i\nu_i$ is the rate of disbursement to arriving customers and that each customer pays to the queues at least the amount received from the network operator, by condition (3.4b). Suppose that queue i pays to the network operator an amount w_i per unit of time it holds a directly routed customer. Then

$$W_i(w_i, s_i - c_i) + c_iC_i$$

is an upper bound on the net rate of return that queue i can attain, and, since at least $\sum_i s_i\nu_i$ accrues to the queues from charges to customers,

$$\sum_i s_i\nu_i - \sum_i [W_i(w_i, s_i - c_i) + c_iC_i]$$

is a lower bound on the delay costs that are incurred in the system.

The complementary slackness conditions again have an interesting interpretation. The main condition shows that if the charges along route r are too high (i.e. $\sum_{j\in r} c_j > s_{i(r)}$) then the route is not used ($y_r = 0$). The other condition is more familiar: the charge c_i to alternatively routed customers at queue i is zero if there is spare capacity at this queue, as indicated by slackness in the constraint (3.3c).

Under exponential assumptions concerning arrival and service times, the function $N_i(x_i)$ is a piecewise linear convex function of x_i, for $i \in I$. In this case both the primal and dual problems (3.3) and (3.4) can be rewritten as linear programming problems.

When might we expect the above bound to be tight? From its derivation the bound will be tight if alternatively routed flows incur zero delay. This seems a condition unlikely to be satisfied, but we note a circumstance when the condition may be approached. Suppose that for each demand of type i there is a large number of alternative routes, that through each resource j there pass a large number of alternative routes, and that it is possible to break a demand into small sub-demands. The sub-demands have service requirements then add to the service requirement of the original demand, but they can be routed separately: the delay incurred by the original demand is then the maximum delay across the network incurred by any sub-demand. Under these circumstances we expect the bound to be approached (see [7],[8] for methods which are able to establish this for sufficiently symmetric networks, under Poisson arrival and exponential requirement assumptions).

If it is not possible to break a demand into smaller sub-demands, then the above bound will not be tight, since an alternatively routed demand must incur delay, and may cause additional delay to directly routed demands. In [11] there is developed a bound that takes these delays into account, but simplifies other aspects of structure.

3.3. Pipelined networks. We finish the Section with another variation which combines certain interesting aspects of queueing and loss networks. Suppose that an alternatively routed demand must be *pipelined*: that is it must be served simultaneously and without pre-emption at each queue on the alternative route, necessarily having the same service time at each of those queues. Suppose further that all service times are independent and exponentially distributed with unit mean, that it costs w_i per unit of time that a demand of type i remains in the system, and that arrival streams at the network form independent Poisson processes. (The exponential and Poisson assumptions are made for simplicity of exposition; the model and bounds could be developed under more general assumptions – see [11].)

Again consider a single queue with two types of customer. Customers of type 1 arrive in a Poisson stream of rate ν, and each customer may be accepted or rejected. There is an infinite reservoir of type 2 customers. The arrival of a type 2 customer may be triggered at any instant the server chooses, but after arrival type 2 customers have priority at the server. All service times are exponentially distributed with unit mean. Customers of type 1 cost an amount w per unit of time they are in the queue, and a customer of type i generates a reward s_i on service completion. Let $S(\nu; w, s_1, s_2)$ be the maximal net rate of return over all admission policies. Again the form of the optimal policy is clear: type 2 customers will be accepted, if at all, only when the queue is empty, and type 1 customers will be accepted if the queue size is below a threshold. Let $N(\nu; x, y)$ be the minimal mean number of type 1 customers in the queue for a given flow x of type 1 customers and y of type 2 customers. Then

$$S(\nu; w, s_1, s_2) = \sup_{x,y \geq 0} \{s_1 x + s_2 y - w N(\nu; x, y)\},$$
$$N(\nu; x, y) = \sup_{s_1, s_2 \geq 0} \{s_1 x + s_2 y - S(\nu; 1, s_1, s_2)\}.$$

Further, the expected cost per unit time for the network is bounded below by the value attained in the problem

$$\text{minimize} \quad \sum_i w_i [N(\nu_i; x_i, v_i) + \sum_{r \in R(i)} y_r]$$

$$\text{subject to} \quad x_i + \sum_{r \in R(i)} y_r = \nu_i \qquad\qquad i \in I$$

$$\text{and} \quad \sum_{r : i \in r} y_r = v_i \qquad\qquad i \in I$$

$$\text{over} \quad x_i, v_i \geq 0, i \in I, \quad \text{and} \quad y_r \geq 0, r \in R,$$

or, equivalently, the value attained in the dual problem

$$\text{maximize} \quad \sum_i [s_i \nu_i - S(\nu_i; w_i, s_i, c_i)]$$

$$\text{subject to} \quad s_{i(r)} \leq w_{i(r)} + \sum_{j \in r} c_j \qquad r \in R$$

$$\text{over} \quad c_i, s_i, \quad i \in I.$$

We expect this bound to be approached in large highly connected networks under a form of threshold routing, where a demand is sent to an idle alternative route whenever the queue for its direct route exceeds a threshold.

4. Combined bounds. The bounds of Sections 2 and 3 are expected to be tight under the different limiting regimes of heavy traffic and diverse routing respectively. For a given finite network it is often possible to combine the bounds in a manner that gives improved accuracy and further insight into the qualitative behaviour of the network. We illustrate this point with two loss network examples.

4.1. Vertex constraints. Consider a loss network, as described in Section 3.1. Make the additional and natural assumption that the set of resources I are the edges of a graph with vertex set V, and that a route $r \in R(i)$ is a path through this graph connecting the vertices at either end of the edge i. Write $v \in i$ if vertex v is at an end of the edge i, and write $v \in T(r)$ if the route r passes through vertex v, but v is not an end vertex of the path r. Thus $T(r)$ is the set of *tandem* vertices on route r.

Let

$$M_v(x) \quad = \quad M(\sum_{j : v \in j} \nu_j, \infty, \sum_{j : v \in j} C_j; x)$$

$$W_v(w_1, w_2) \quad = \quad W(\sum_{j : v \in j} \nu_j, \infty, \sum_{j : v \in j} C_j; w_1, w_2).$$

Consider the flow of traffic into and out of vertex v. The amount of tandem traffic is

$$2 \sum_{r : v \in T(r)} y_r,$$

since each call that uses vertex v as a tandem vertex occupies a circuit into, *and* a circuit out of, vertex v. Hence under any dynamic routing scheme

$$(4.1) \qquad 2 \sum_{r:v \in T(r)} y_r \leq M_v \left(\sum_{j:v \in j} (x_j + \sum_{r \in R(j)} y_r) \right),$$

providing a single additional constraint for each vertex $v \in V$. We can improve the bound provided by problem (3.1) by appending to that problem the additional constraints (4.1). The corresponding dual problem takes the form

$$\text{minimize} \quad \sum_i \left[W_i(e_i, c_i) + s_i \nu_i \right] + \sum_v W_v(g_v, d_v)$$

$$\text{subject to} \quad w_i \leq s_i + e_i + \sum_{v \in i} g_v \qquad i \in I$$

$$\text{and} \quad e_{i(r)} \leq \sum_{j \in r} c_j + 2 \sum_{v \in T(r)} d_v \qquad r \in R$$

$$\text{over} \quad c_i, s_i, e_i \geq 0, i \in I, \quad \text{and} \quad g_v, d_v \geq 0, v \in V.$$

Comparing this formulation with problem (3.4), we observe the introduction of additional implied costs g_v and d_v: the vertex v charges an amount g_v for the use of a single circuit out of vertex v by a call terminating at vertex v, and an amount $2d_v$ for the use of two circuits out of vertex v by a call using vertex v as a tandem vertex. The conceptual simplicity is notable: there is a charge for each vertex, as well as each link, used by a call.

4.2. Other cut constraints. Each of the constraints (4.1) corresponds to the cut formed between a single vertex and the other vertices. A network's size, connectivity and asymmetry may make some other key subset of resources important: we next describe an example illustrating how an additional cut constraint may be included.

Consider again the loss network described in Section 3.1. Let $J \subset I$ be such that

$$j \in J, r \in R(j) \Rightarrow |r \cap J| = 1,$$

where $|r \cap J|$ is the number of resources from J on route r. Thus J forms a cut, in that every demand from J requires to use a single resource from J however it is routed. Let

$$M_J(x) = M(\sum_{j \in J} \nu_j, \infty, \sum_{j \in J} C_j; x)$$

$$W_J(w_1, w_2) = W(\sum_{j \in J} \nu_j, \infty, \sum_{j \in J} C_j; w_1, w_2).$$

Consider the flow of traffic across the cut J. Under any dynamic routing scheme

$$(4.2) \qquad \sum_{k \notin J} \sum_{r \in R(k)} |r \cap J| y_r \leq M_J \left(\sum_{j \in J} (x_j + \sum_{r \in R(j)} y_r) \right).$$

We can improve the bound provided by problem (3.1) by appending to that problem that additional constraint (4.2). The corresponding dual problem takes the form

$$\text{minimize} \quad \sum_i [W_i(e_i, c_i) + s_i \nu_i] + W_J(g_J, d_J)$$

$$\text{subject to} \quad w_i \leq s_i + e_i + I[i \in J]g_J \qquad i \in I$$

$$\text{and} \quad e_{i(r)} \leq \sum_{j \in r} c_j + |r \cap J| d_J \qquad r \in R$$

$$\text{over} \quad c_i, s_i, e_i \geq 0, i \in I, \quad \text{and} \quad g_J, d_J \geq 0.$$

Again the conceptual simplicity is notable: each call serving a demand $i \in J$ must pay a charge g_J, while other calls must pay a charge d_J per circuit used from the cut J.

Work in progress is concerned with illustrating these various bounds and assessing their accuracy for a variety of realistic network architectures: for early results the reader is referred to [3] and [4].

REFERENCES

[1] D. BERTSIMAS, I. CH. PASCHALIDIS AND J.N. TSITSIKLIS, *Scheduling of multiclass queueing networks: bounds on achievable performance*, Ann. Appl. Prob.

[2] R.J. GIBBENS AND F.P. KELLY, *Dynamic routing in fully connected networks*, IMA J. Math. Cont. Inf. **7** (1990), pp. 77–111.

[3] R.J. GIBBENS, F.P. KELLY, AND S.R.E. TURNER, *Dynamic routing in multi-parented networks*, IEEE Trans. Networking **1** (1993), pp. 261–270.

[4] R.J. GIBBENS AND P. REICHL, *Performance bounds applied to loss networks*, in: Complex Stochastic Systems and Engineering, editor D.M. Titterington (Oxford University Press, 1994).

[5] M. GONDRAN AND M. MINOUX, Graphs and Algorithms (Wiley, New York, 1984).

[6] J.M. HARRISON, *Brownian models of queueing networks with heterogeneous customer populations*, in: Stochastic Differential Systems, Stochastic Control Theory and Applications, IMA Vol 10, editors W. Fleming and P.-L. Lions (Springer-Verlag, New York, 1988), pp. 147–186.

[7] P.J. HUNT AND C.N. LAWS, *Least busy alternative routing in queueing and loss networks*, Probab. Eng. Inf. Sci. **6** (1992), pp. 439–456.

[8] P.J. HUNT AND C.N. LAWS, *Asymptotically optimal loss network control*, Math. Oper. Res. **18** (1993), pp. 880–900.

[9] F.P. KELLY, *The optimization of queueing and loss networks*, in: Queueing Theory and its Applications, editors O.J. Boxma and R. Syski (Elsevier, Amsterdam, 1988), pp. 375–392.

[10] F.P. KELLY, *Loss networks*, Ann. Appl. Prob. **1** (1991), pp. 317–378.

[11] F.P. KELLY, *Bounds on the performance of dynamic routing schemes for highly connected networks*, Math. Oper. Res. **19** (1994), pp. 1–20.

[12] F.P. KELLY AND C.N. LAWS, *Dynamic routing in open queueing networks: Brownian models, cut constraints and resource pooling*, Queueing Systems **13**, (1993), pp. 47–86.

[13] P.B. KEY, *Optimal control and trunk reservation in loss networks*, Prob. Engineering and Information Sciences **4** (1990), pp. 203–242.

[14] P.B. KEY, *Some control issues in telecommunication networks*, in: Probability, Statistics and Optimization, editor F.P. Kelly (Wiley, Chichester, 1994), pp. 383–395.

[15] K.R. KRISHNAN AND T.J. OTT, *State-dependent routing for telephone traffic: theory and results*, Proc. 25th Conf. Decision and Control, Athens (1986).

[16] H.J. KUSHNER, *Control of trunk line systems in heavy traffic*, Division of Applied Mathematics, Brown University (1992).

[17] C.N. LAWS, *Resource pooling in queueing networks with dynamic routing*, Adv. Appl. Prob. **24** (1992), pp. 699–726.

[18] C.N. LAWS, *Paper in this volume.*

[19] C.N. LAWS AND G.M. LOUTH, *Dynamic scheduling of a four-station queueing network*, Prob. Eng. Inf. Sci. **4** (1990), pp. 131–156.

[20] V.G. LAZAREV AND S.M. STAROBINETS, *The use of dynamic programming for optimization of control in networks of commutation of channels*, Engineering Cybernet. **15** (1977), pp. 107–117.

[21] S.A. LIPPMAN, *Applying a new device in the optimization of exponential queueing systems*, Oper. Res. **23** (1975), pp. 687–710.

[22] G.M. LOUTH, M. MITZENMACHER AND F.P. KELLY, *Computational complexity of loss networks*, Theor. Comp. Sci. **125** (1994), pp. 45–59.

[23] V.V. MARBUKH *Asymptotic investigation of a complete communications network with a large number of points and bypass routes*, Problemy Peradači Informacii **16** (1981), pp. 89–95.

[24] D. MITRA AND R.J. GIBBENS, *State-dependent routing on symmetric loss networks with trunk reservations, II: asymptotics, optimal design*, Ann. Oper. Res. **35** (1992), pp. 3–30.

[25] J. OU AND L.M. WEIN, (1992) *Performance bounds for scheduling queueing networks*, Ann. Appl. Probab. **2** (1992), pp. 460–480.

[26] M.I. REIMAN, *A multiclass feedback queue in heavy traffic*, Adv. Appl. Prob. **20** (1988), pp. 179–207.

[27] M.I. REIMAN, *Optimal trunk reservation for a critically loaded network*, Proc. 13th ITC. In Teletraffic and Datatraffic, editors A. Jensen and V.B. Iverson (Elsevier, Amsterdam, 1991).

[28] M.I. REIMAN AND B. SIMON, *A network of priority queues in heavy traffic: one bottleneck station*, Queueing Systems **6** (1990), pp. 33–58.

ON TRUNK RESERVATION IN LOSS NETWORKS

C.N. LAWS*

Abstract. Trunk reservation is a simple and effective mechanism for controlling loss networks which allows priority to be given to chosen traffic streams. We review some current work on the asymptotic optimality of trunk reservation policies. The first system we consider is a fully-connected network in the limiting regime as the number of links of the network increases. Secondly we consider a heavy traffic limit where capacity and arrival rates increase together, their ratios being held fixed. In both cases we illustrate how the asymptotically optimal trunk reservation policies emerge from bounds on the performance of any control policy.

1. Introduction. Loss networks model systems at which demands arrive and request use of some of the system's capacity: an arriving demand may either be accepted by the system, or it may be rejected and lost from the system. The classical example of a loss network is a circuit-switched telephone network which can be thought of as follows. The network is made up of a set of links where each link consists of an integer number of circuits. Each circuit is capable of carrying a call but often a call will require one circuit from each of several links in order to be connected. This subset of links is the call's route. When a call arrives it is a control decision as to whether the call should be accepted. If accepted a call uses one circuit from each of the links on its route for its duration, after which these circuits are released for use by other calls. An arriving call will be blocked and lost if there is no spare capacity to carry it, and it is often desirable to reject calls in other situations too. For example, if some calls have higher rewards than others, it may be beneficial to reject some low revenue calls now in order to be able to accept higher revenue calls later. Kelly [14] provides an extensive review of much recent work on loss networks, including both optimization issues and many others.

Trunk reservation is a simple yet effective mechanism for making such control decisions and the near/approximate optimality, and in particular the asymptotic optimality, of trunk reservation policies is the subject of this paper. The form of a trunk reservation policy is: only accept a call on route r if the number of free (i.e. currently unused) circuits on link j is greater than s_{jr} for each link j on route r. (If this condition is not satisfied the arriving call is blocked and lost.) Here s_{jr} is known as the trunk reservation parameter used at link j against calls on route r.

Of course, the above outlined description of a loss network can be generalized. Calls may have several routes available to them and, on any given route, a call may require more than one circuit from some/all of the links on the route. When calls have multiple potential routes the control

---

* Department of Statistics, University of Oxford, 1 South Parks Road, Oxford OX1 3TG, UK.

decision when a call arrives is in two parts: should the call be accepted and, if so, on which route?

Motivation for studying trunk reservation policies is that they are simple to describe and implement, they are robust to network traffic fluctuations, and they are close to optimal in many situations. The simplicity of trunk reservation is apparent from its description above: when considering whether to accept a call on route r, the only information required is whether the unused capacity on each link of route r exceeds the appropriate trunk reservation parameter. By robustness we mean that good network performance is often observed when call arrival rates vary, provided only that trunk reservation parameters are chosen within some sensible range. That is, neither do arrival rates need to be known particularly accurately, nor do trunk reservation parameters need to be precisely chosen in order to obtain good performance (e.g. [4], [5], [17], [25]).

In the remainder of this Introduction we consider some exact optimality properties of trunk reservation in single-link systems and the motivation that this provides for using trunk reservation in networks. Then we discuss asymptotic optimality properties of trunk reservation-based policies in Sections 2 and 3. Section 2 covers some work on fully-connected networks. Section 3 describes current joint work with Phil Hunt and Nicolas Notebaert [12] on loss networks in heavy traffic.

1.1. Exact optimality. Consider a network consisting of a single link. Suppose the link comprises C circuits and that it is offered two streams of traffic. Calls of type i, $i = 1, 2$, arrive as independent Poisson processes of rate ν_i and each call requests use of one circuit for the holding time of the call. If accepted, a call of type i generates a reward of w_i, and suppose $w_1 \geq w_2 \geq 0$. Call holding times are assumed to be independent of earlier arrival times and holding times, and are exponentially distributed with unit mean. The aim is to maximize the expected reward per unit time.

The optimal policy [23] is to accept type 1 calls provided the link is not full, and to accept type 2 calls provided the number of free (i.e. currently unused) circuits is greater than some fixed integer, s say. This is a trunk reservation policy under which type 1 calls are treated as high priority, and are accepted whenever possible, and type 2 are treated as lower priority, and are subject to a trunk reservation parameter of s. The optimal choice of s can be determined from the equilibrium distribution of the birth-death process describing the number of occupied circuits, or via policy improvement. Optimal values of trunk reservation parameters are studied in detail in [17], [27] where various limiting regimes are considered.

In fact Lippman [21] has shown that the same form of policy is optimal under a variety of discount and finite-horizon optimality criteria. Further, if the single link is offered I types of call, a call of type i generating a reward of w_i if accepted ($i = 1, 2, \ldots, I$), the optimal policy is again a trunk reservation policy but now with multiple priority levels ([21], [23]). If types are labelled so that $w_1 \geq w_2 \geq \ldots \geq w_I \geq 0$ then type i calls are only accepted if the number of free circuits is greater than the trunk reservation parameter s_i where $0 = s_1 \leq s_2 \leq \ldots \leq s_I$. Nguyen [26] considers a closely related problem with two arrival streams, but where one stream is Poisson and the other is an overflow stream from an $M/M/m/m$ queue: again the optimal policy is a (generalized) trunk reservation policy.

The exact optimality of trunk reservation does not extend to systems of more than one link: see [17] for a two-link example. We can generalize the above single link model by allowing mean holding times and the number of circuits required by a call to depend on the call's type. Again the exact optimality of trunk reservation does not extend ([18], [28]). The use of dynamic programming to investigate call acceptance and routing issues in networks has been considered in several papers (e.g. [17], [19], [20]). However, here we concentrate on trunk reservation policies (see also [18]).

Trunk reservation-based policies perform well in a variety of situations (e.g. [4], [5], [24], [25]): in making this claim it is especially useful to consider bounds on the performance of a network under any control policy (cf. [4], [5], [6], [15], [16]). For the systems examined in Sections 2 and 3 performance bounds can play an important role in revealing asymptotically optimal policies. In this paper we focus on cases where bounds can be attained or approached by trunk reservation (i.e. simple threshold-type) policies (cf. [16]).

2. Fully-connected networks. Consider a fully-connected network as follows. The network has N nodes and there is a link comprising C circuits between each pair of nodes, giving a total of $K = N(N-1)/2$ links. Between each pair of nodes of the network, calls arrive according to independent Poisson processes of rate ν. When a call arrives, it may be routed on the direct link between its source and destination nodes; it may be routed, via an intermediate node, on one of the $N-2$ two-link paths connecting its source and destination nodes; or it may be rejected and lost from the system. A call may only be routed via a given link if the link has a spare circuit to carry the call. A successfully routed call uses one circuit from each link of its path for the holding time of the call, after which these circuits are released. Call holding times are exponentially distributed with unit mean. Given that the objective is to successfully route as many calls as possible, what is the optimal control policy?

Consider the expected number of calls rejected per unit time per link. In [11] it is shown that, in the limit $K \to \infty$, this quantity is bounded below by the optimal value of the following linear program.

$$LP: \text{minimize} \quad \sum_{j=0}^{C} z_{jb}$$

(2.1) subject to $(j+1)x_{j+1} = z_j + \sum_{i=0}^{C} z_{ij}$ $j = 0, 1, \ldots, C-1$

(2.2) $\nu x_j = z_j I_{\{j \neq C\}} + z_{jb} + \frac{1}{2} \sum_{k=0}^{C-1} z_{jk}$ $j = 0, 1, \ldots, C$

$$\sum_{j=0}^{C} x_j = 1$$

$$x_j, z_j, z_{ij}, z_{jb} \geq 0.$$

The interpretation of this problem is as follows: x_j is the proportion of links with j circuits in use; z_j is the rate per link at which calls are routed directly via links with j calls in progress; z_{ij} is the rate per link at which calls arrive at links with i calls in progress and are rerouted via links with j in progress; and z_{jb} is the rate per link at which calls arrive at links with j calls in progress and are blocked. The objective is to minimize the rate per link at which calls are blocked; constraints (2.1) require the rates at which links move between states j and $j+1$ to balance; constraints (2.2) express the fact that all arrivals are either routed directly, or alternatively, or are blocked; and also the x_j are proportions and all variables are non-negative.

In addition to providing a lower bound, the solution of problem LP leads directly to an asymptotically optimal policy. In fact, the optimal value also bounds the expected number of calls rejected per unit time per link for all finite values of K: this was first shown in [5] and is a special case of the general bounds of Kelly [15].

Problem LP is solved in [11] and its optimal solution describes the equilibrium behaviour of least busy alternative routing with trunk reservation (cf. [4], [14], [22], [24], [25]). The important features of this solution are as follows. Let ν^* be the unique solution of the equation

$$\frac{(\nu^*)^{C-[\nu^*]+1}}{C(C-1)\ldots[\nu^*]} = \frac{1}{2}$$

where $[x]$ denotes the integer part of x. Define u by

$$u = \min\left\{0 \leq n \leq C-1 : \frac{\nu^{C-n} n!}{C!} \leq \frac{1}{2}\right\} \qquad \text{if } \nu \leq \nu^*$$

$$u = \arg\min_{0 \leq n \leq C}\left(\frac{\nu^{C+1}}{C!} - \frac{n\nu^n}{2n!}\right)\left[\sum_{j=n}^{C} \frac{\nu^j}{j!}\right]^{-1} \qquad \text{if } \nu > \nu^*.$$

For all ν the optimal solution satisfies

$$x_j > 0 \quad \Leftrightarrow \quad j \in \{u, u+1, \ldots, C\}$$

(2.3) $z_j = \nu x_j \quad \text{for} \quad j = 0, 1, \ldots, C-1.$

Then for $\nu \leq \nu^*$

(2.4) $z_{jb} = 0$ for $j = 0, 1, \ldots, C$

(2.5) $z_{ij} > 0$ $\Leftrightarrow$ $(i, j) = (C, u-1)$ or (C, u),

while for $\nu > \nu^*$

(2.6) $z_{Cb}, \; z_{C,u-1} > 0$

                    $z_{jb} = z_{ij} = 0$ otherwise.

For all ν, (2.3) indicates that all calls should be directly routed if possible. Consider $\nu \leq \nu^*$. Relations (2.4) show that no calls need be blocked, and (2.5) indicates that calls are only rerouted (from fully-occupied links) to links with occupancy u or $u-1$. There is no contradiction in the rerouting of calls to links in state $u-1$: although $x_{u-1} = 0$, the actual number of links in state $u-1$ is $O(1)$ as $K \to \infty$. More precisely, this number of links is a process which operates on a fast time-scale (cf. [10], the following section, and see especially [8], [9] for similar behaviour in general networks): from the point of view of link proportions x_j, the jump process n_{u-1} describing the number of links in state $u-1$ is in equilibrium, and there is positive probability that a call can be rerouted via links in state $u-1$. There is also positive probability that $n_{u-1} = 0$, in which case a rerouted call uses two links in state u.

Now suppose $\nu > \nu^*$. Then (2.6) shows that some traffic is lost and that calls are only ever rerouted (from fully-occupied links) to links in state $u-1$. As above n_{u-1} is a process on a fast time-scale, with positive probability of being in states 0 and 1, and so some proportion of calls cannot be rerouted and are rejected. Although there are links in state u (since $x_u > 0$) the optimal solution is not to reroute calls to these links: that is, a trunk reservation parameter of $s = C - u$ applies against two-link calls to prevent them being routed via links with occupancy u or more. Similarly in the $\nu \leq \nu^*$ case above, we use a trunk reservation parameter of $s = C - u - 1$ against two-link calls: at first sight this appears unnecessary, but consideration of time-dependent behaviour or the case $K < \infty$ shows the importance of using trunk reservation in this case also ([10], [14]).

Following the above interpretations, an asymptotically optimal policy is as follows. When a call arrives, route it directly if possible. If not, route the call via the least busy two-link alternative route (to be defined below) if possible, using a trunk reservation parameter of s against the call on both links of this route. Otherwise reject the call.

In order to define the least busy alternative route, consider firstly a relaxed version of the original model in which the set of two-link routes available to a call consists of every pair of links of the system. That is, relax the condition that the two-link paths available to a call connect its source and destination nodes, and allow the call to use any pair of links instead. Then the least busy alternative route is simply the two links with most free circuits. A suitable definition for the original model is the route, of the available $N - 2$ alternatives, which could carry the highest number of additional two-link calls. Final justification of the asymptotic optimality of the least busy alternative/trunk reservation policy rests on convergence results for a sequence of networks operating under the policy. For the relaxed model this justification is simpler since one result ([11], Theorem 4) is more easily proved (the proof is in [8]). Asymptotic optimality then follows [11]. It is indicated in [11] that the techniques of [3] are expected to extend the result (i.e. [11], Theorem 4) from the relaxed model to the original model, and hence [11] to yield asymptotic optimality. For work concerning graph structure issues see [1], [7], [14]: in particular, Kelly [14] shows that a trunk reservation policy is asymptotically optimal in a fully-connected network when call repacking is permitted.

The least busy alternative strategy with trunk reservation parameter s is not the only asymptotically optimal policy for the problem considered here. The 'least busy alternative' part of the policy is not essential, though the trunk reservation part is. For example, consider a random routing strategy with multiple retries and trunk reservation, with parameter s defined as before. For this strategy an arriving call is routed directly if possible; otherwise it chooses a two-link path at random and is carried on that route provided both links have more than s free circuits. If this trunk reservation criterion fails the call is not immediately rejected: it attempts further two-link routes (chosen at random and subject to trunk reservation), a maximum of M attempts being permitted. If all M of these two-link attempts fail, the call is rejected and lost. This policy will be asymptotically optimal provided that M^K, the number of retries permitted in the Kth network, increases sufficiently fast as $K \to \infty$. More generally, the important features of an asymptotically optimal policy are that it respects trunk reservation and that, when considering two-link routes, it searches effectively for routes with link occupancies below the trunk reservation level. Simple dynamic schemes can be particularly effective in locating spare capacity ([4], [5]) and, when used with trunk reservation, give near optimal performance in finite-sized networks.

The above results can be generalized in certain directions [11]. Firstly, in the initial description we allowed calls to use only one- and two-link routes. However, consideration of the relaxed model shows that we actually have asymptotic optimality when alternative routes of three or more links are permitted as well. Next suppose the links of the fully-connected network are of type l with probability p_l ($\sum_l p_l = 1$), independently of each other. Suppose also that a type l link comprises C_l circuits and receives traffic at rate ν_l. A least busy alternative routing policy with trunk reservation again emerges from the solution of the suitably generalized version of problem LP. Using an asymptotic approach similar to that above, Key [18] obtains an optimal least busy alternative/trunk reservation scheme for a system of identical links each with a dedicated arrival stream, and where there is an additional stream which can be controlled.

3. Systems in heavy traffic. For most of this section we consider control of a system consisting of a single link.

Consider a single link comprising C circuits as follows. Let the finite set $\mathcal{I}$ index the arriving call-types. For $i \in \mathcal{I}$, let κ_i be the Poisson arrival rate of calls of type i, and assume that these Poisson streams are independent as i varies. Calls may be accepted or rejected, and if accepted a type i call uses $A_i \in \mathbb{Z}_+$ circuits from the link for its holding time. As usual a call may only be accepted if there is sufficient spare capacity to carry it, and the call may be rejected even when it could be carried. The holding time of a type i call is exponentially distributed with mean μ_i^{-1}, and holding times are independent of each other and of the arrival processes. An accepted call of type i earns a reward of w_i per unit time in progress. Which control policy maximizes the expected reward per unit time?

We consider an asymptotic approach to this problem but firstly we describe a bound on the expected reward rate. Let $\nu_i = \kappa_i/\mu_i$ and let summations over i range over $\mathcal{I}$. Then, as a special case of a result of [4], under any policy the expected reward per unit time is bounded above by the value attained in the following maximum flow problem.

$$MF: \quad \text{maximize} \quad \sum_i w_i x_i$$

(3.1)
$$\text{subject to} \quad \sum_i A_i x_i \leq C$$

$$0 \leq x_i \leq \nu_i.$$

This problem has a natural interpretation. Let x_i be the average number of calls of type i carried. Then constraint (3.1) requires that the average number of circuits in use does not exceed C. Also, x_i cannot exceed the average number of customers in an $M/M/\infty$ queue with arrival and service rates κ_i and μ_i, since this queue corresponds to the number of type i calls that would be in progress if all type i calls were accepted. Hence we have $x_i \leq \nu_i$.

We consider a sequence of systems indexed by N. In the Nth system replace C by $C(N)$, κ_i by $\kappa_i(N)$, and μ_i by $\mu_i(N)$, where

(3.2) $\quad C(N)/N \to C, \qquad \kappa_i(N)/N \to \kappa_i, \qquad \mu_i(N) \to \mu_i$

and all limits are as $N \to \infty$. We consider attainment of the bound from MF in the limit $N \to \infty$, a limit studied in detail by Kelly [13], [14] and termed the heavy traffic limit in [29]. Let $\bar{x}(N) = (\bar{x}_i(N), i \in \mathcal{I})$ be a solution of MF when the values of C, κ_i and μ_i are replaced by their values in the Nth system. We can choose $\bar{x}(N)$ such that $\bar{x}(N)/N \to \bar{x}$ as $N \to \infty$ and so $\sum_i w_i \bar{x}_i$ is a bound on the limiting normalized expected reward per unit time under regime (3.2), where the normalization is to divide the reward rate of the Nth system by N. Consider the following probabilistic routing strategy: when a call of type i arrives, attempt to carry it with probability $p_i = \bar{x}_i/\nu_i$, rejecting the call if this attempt fails, and automatically reject the call with probability $1 - p_i$. This strategy is asymptotically optimal in that its normalized reward rate converges to $\sum_i w_i \bar{x}_i$ as $N \to \infty$ [17].

Although probabilistic routing is asymptotically optimal, the limit here is a coarse one ([14], [17], [30]). To evaluate probabilities p_i, arrival rates κ_i must be known exactly—an unlikely situation in practice—and, further, other control policies which are also asymptotically optimal may well give higher reward rates for more realistic link capacities and arrival rates. Policies based on trunk reservation, which we now consider, are obvious candidates.

Consider problem MF with normalized limiting solution $\bar{x}$ as above. Since there is always an optimal basic solution to MF we can choose $\bar{x}$ with $|\{i : 0 < \bar{x}_i < \nu_i\}| \leq 1$. Fix a solution with this property: define type i to be *high priority* if $\bar{x}_i = \nu_i$; define type i to be *medium priority* if $\bar{x}_i \in (0, \nu_i)$; and define type i to be *low priority* if $\bar{x}_i = 0$. For ease of notation suppose there is exactly one call-type of each priority level. Suppose $\mathcal{I} = \{1, 2, 3\}$ where type 1 is high priority, type 2 is medium priority, and type 3 is low priority. Suppose also that each call requires use of a single circuit, i.e. $A_1 = A_2 = A_3 = 1$.

Suppose the Nth system operates as follows: calls of type 1 are accepted whenever possible; calls of type 2 are accepted subject to a trunk reservation parameter of $s_2(N)$; and calls of type 3 are accepted subject to a trunk reservation parameter of $s_3(N)$. If $s_2(N)$, $s_3(N)$ are fixed as N increases then we will not attain the asymptotic lower bound $\sum_i w_i \bar{x}_i$: a positive proportion of high, medium and low priority calls will be blocked (cf. [9]). Instead let $s_2(N) = \log N$, $s_3(N) = 2\log N$. (Trunk reservation parameters of $\log N$ and $2\log N$ are not the only suitable ones for our purpose ([9], [12]). We choose $\log N$ for concreteness and by comparison with [17] where the optimal trunk reservation parameter for a single link like ours is shown to be $O(\log N)$.) The behaviour of this trunk reservation policy can be examined by specializing results of [9] to a single link: see also [2] for asymptotic modelling of one member of the sequence with multiple trunk reservation levels. The limit results below from Hunt and Kurtz [9] rely on a generalization (which is described in [9]) of their initial results to the case where trunk reservation parameters vary with N.

Let $x^N(t) = N^{-1}(n_1^N(t), n_2^N(t), n_3^N(t))$ where $n_i^N(t)$ denotes the number of type i calls in progress at time t in the Nth system under the above trunk reservation policy. Then [9], assuming $x^N(0) \Rightarrow x(0)$ as $N \to \infty$ (where $\Rightarrow$ denotes weak convergence), $x^N(\cdot) \Rightarrow x(\cdot)$ where the limit process $x(t) = (x_1(t), x_2(t), x_3(t))$ satisfies

$$(3.3) \qquad x_i(t) = x_i(0) + \int_0^t \{\kappa_i \pi_i(u) - \mu_i x_i(u)\}\, du \qquad i \in \{1, 2, 3\}$$

where $\pi_i(t) = 1$ for all i if $\sum_i x_i(t) < C$, and otherwise

$$\pi_1(t) \;\; = \;\; \begin{cases} \sum \mu_i x_i(t)/\kappa_1 & \text{if } \sum \mu_i x_i(t) < \kappa_1 \\ 1 & \text{otherwise} \end{cases}$$

$$\pi_2(t) \;\; = \;\; \begin{cases} 0 & \text{if } \sum \mu_i x_i(t) < \kappa_1 \\ (\sum \mu_i x_i(t) - \kappa_1)/\kappa_2 & \text{if } \kappa_1 \leq \sum \mu_i x_i(t) < \kappa_1 + \kappa_2 \\ 1 & \text{if } \kappa_1 + \kappa_2 \leq \sum \mu_i x_i(t) \end{cases}$$

$$\pi_3(t) \;\; = \;\; \begin{cases} 0 & \text{if } \sum \mu_i x_i(t) < \kappa_1 + \kappa_2 \\ (\sum \mu_i x_i(t) - (\kappa_1 + \kappa_2))/\kappa_3 & \text{if } \kappa_1 + \kappa_2 \leq \sum \mu_i x_i(t) < \sum \kappa_i \\ 1 & \text{if } \sum \kappa_i \leq \sum \mu_i x_i(t). \end{cases}$$

We now give an intuitive description of this result (see also [2], [14] for the case of fixed trunk reservation parameters). A suitable interpretation of $\pi_i(t)$ is that it denotes the acceptance probability of a given type i call arriving at time t, in the limit $N \to \infty$. The dynamics of the limit process, given by equation (3.3), are what one might expect when $\sum_i x_i(t) < C$. For such $x(t)$ the evolution of each component is governed by a linear ODE: x_i increases at rate $\kappa_i - \mu_i x_i$, where the κ_i term is due to type i arrivals and the $-\mu_i x_i$ is due to type i departures. However the behaviour on the boundary $\sum_i x_i = C$ is more complicated. In order to consider this let $m^N(t) = C(N) - \sum_i n_i^N(t)$ denote the number of free circuits at time t in the Nth system. Suppose $\sum_i x_i(t) = C$ and $\sum_i \mu_i x_i(t) < \kappa_1$, and consider the short time interval $[t, t+\varepsilon]$. Owing to the normalization, $x^N(\cdot)$ is approximately constant over this interval (its components will change by $O(\varepsilon)$). Calls arrive and depart at rate $O(N)$ and so, in contrast, $m^N(\cdot)$ will evolve very rapidly over the interval. When $m^N(\cdot) \leq \log N$ it evolves, approximately, according to a birth-death process with constant birth and death rates of $N \sum_i \mu_i x_i(t)$ and $N\kappa_1$ respectively. Now, since $\sum_i \mu_i x_i(t) < \kappa_1$ and since the trunk reservation parameters are $O(\log N)$, $m^N(\cdot)$ will satisfy $m^N(\cdot) \leq \log N$ for almost the whole of the time interval $[t, t + \varepsilon]$. Hence the number of type 2 or 3 calls accepted over the interval will be approximately 0 and so, considering the limit $N \to \infty$, we have $\pi_2(t) = \pi_3(t) = 0$ as above. Further, since the trunk reservation parameters of $O(\log N)$ increase without bound under the limiting regime and since $m^N(\cdot)$ evolves so rapidly, $m^N(\cdot)$ can be approximated by a birth-death process on $\mathbb{Z}_+$ with constant rates. Hence the acceptance probability of a given type 1 call is approximately the equilibrium probability that the birth-death process on $\mathbb{Z}_+$, with normalized birth and death rates of $\sum_i \mu_i x_i(t)$ and

κ_1 respectively, is not at 0. This probability is $\sum_i \mu_i x_i(t)/\kappa_1$, as given by $\pi_1(t)$ above. Note that when $\sum_i \mu_i x_i(t) > \kappa_1$ the process $m^N(\cdot)$ has drift away from 0 and hence, in the limit $N \to \infty$, the probability of accepting a type 1 call is 1. In the limit $N \to \infty$ there is a separation of the time scales of $x^N(\cdot)$ and $m^N(\cdot)$ (or alternatively there is a time-scale decomposition, cf. Section 2) as shown rigorously by Hunt and Kurtz [9].

For interpretation in another case, again consider the time interval $[t, t + \varepsilon]$ and suppose that $\sum_i x_i(t) = C$ and $\kappa_1 < \sum \mu_i x_i(t) < \kappa_1 + \kappa_2$. As above, in this case $m^N(\cdot)$ has drift away from 0 and $\pi_1(t) = 1$. Let $\hat{m}^N(t) = m^N(t) - \log N$ denote the distance of the free circuit process above the level at which trunk reservation applies against the medium priority (type 2) calls. Then over $[t, t + \varepsilon]$, $\hat{m}^N(\cdot)$ will satisfy $|\hat{m}^N(\cdot)| < \log N$ for nearly all of the interval: this follows from the rapid evolution of $\hat{m}^N(\cdot)$ and since the conditions on $x(t)$ imply that $\hat{m}^N(\cdot)$ has drift toward 0. A type 2 call will be accepted provided that $\hat{m}^N(\cdot) > 0$. Since trunk reservation parameters are $O(\log N)$, as $N \to \infty$ the acceptance probability of a given type 2 call is approximately the equilibrium probability that a certain birth-death process on $\mathbb{Z}$ is greater than 0. The birth rates of this process are $\sum_i \mu_i x_i(t)$ in all states; the death rates are $\kappa_1 + \kappa_2$ in states $\{1, 2, \ldots\}$ and κ_1 in states $\{0, -1, \ldots, \}$ (type 2 calls are rejected when $\hat{m}^N(\cdot) \leq 0$). Hence the acceptance probability of a type 2 call can be determined from the double-geometric equilibrium distribution of this process, and is $(\sum \mu_i x_i(t) - \kappa_1)/\kappa_2$ as given by $\pi_2(t)$ above. When $\sum \mu_i x_i(t) > \kappa_1 + \kappa_2$ the process $\hat{m}^N(\cdot)$ has positive drift, $\hat{m}^N(\cdot)$ will be greater than 0 for almost all of the interval $[t, t + \varepsilon]$, and hence $\pi_2(t) = 1$.

To show that our trunk reservation policy is, in fact, asymptotically optimal the behaviour of the limit process $x(\cdot)$ must be investigated. Firstly recall that $\sum_i w_i \bar{x}_i$ is an upper bound on the normalized reward per unit time, in the limit $N \to \infty$. Secondly, if $x(0) = \bar{x}$ then $x(t) = \bar{x}$ for all $t \geq 0$ [12] (this can be checked using (3.3) and problem MF). That is, $\bar{x}$ is a fixed point of equations (3.3). Finally we must consider the trajectories of limit process $x(\cdot)$. For all possible $x(0)$, $x(t) \to \bar{x}$ as $t \to \infty$ ([12], proof due to Bruce Hajek). It follows that in the limit $N \to \infty$, the normalized reward rate of the trunk reservation policy attains the bound $\sum_i w_i \bar{x}_i$ [12]. So, in the sense of maximizing the normalized reward rate, the trunk reservation policy is asymptotically optimal.

The asymptotic optimality of trunk reservation extends to the general single link system introduced at the beginning of this section [12]. Call-types i are divided into high, medium and low priority levels according as $\bar{x}_i = \nu_i$, $\bar{x}_i \in (0, \nu_i)$ or $\bar{x}_i = 0$. As before, there need be at most one medium priority type. The asymptotically optimal policy uses trunk reservation parameters of $\log N$ and $2 \log N$ against medium and low priority calls respectively. Also, if $\mathcal{H}$ denotes the set of high priority types, it is convenient to use a trunk reservation parameter of $\max_{i \in \mathcal{H}} A_i$ against high priority types (so that all $\pi_i(\cdot)$, $i \in \mathcal{H}$, are equal). Observe that this additional trunk reservation parameter is fixed as N varies and its effect is to accept a high priority call of type i only when there is sufficient free capacity to accept a call of any type $i \in \mathcal{H}$.

Extending the results to networks appears more difficult. For the single link case the acceptance probabilities $\pi_i(\cdot)$ in equations (3.3) can be determined explicitly. For networks the equations governing the limit process are again (3.3) [9] though the corresponding $\pi_i(\cdot)$ are unknown in general. Here the $\pi_i(\cdot)$ can be obtained from one-dimensional calculations, while in general the comparable calculations involve multi-dimensional jump processes (the number of dimensions is determined by the number of links for which a capacity constraint of the form (3.1) is tight). Moreover, components of these processes can be in state $+\infty$, and some components can be in state $-\infty$. We have been able to avoid discussion of such states here, but they do play an important role in general: for details of the subtle issues involved see [8], [9]. However, suppose acceptance probabilities $\pi_i(\cdot)$ are approximated rather than calculated exactly. A common approximation procedure for calculating loss probabilities in networks is to assume that links block independently ([13], [29]). If we assume here that links are independently above/below the high/medium/low trunk reservation levels, then we can make further progress: see [12]. This approximation is known to fail when trunk reservation parameters are fixed [9]. It would be especially interesting to know whether the approximation is accurate when trunk reservation parameters vary with N (e.g. when they are $O(\log N)$) since this issue is intimately related to the asymptotic optimality (or not) of trunk reservation in the heavy traffic limit.

REFERENCES

[1] V. ANANTHARAM, *A mean field limit for a lattice caricature of dynamic routing in circuit switched networks*, Ann. Appl. Probab. **1** (1991), pp. 481–503.

[2] N.G. BEAN, R.J. GIBBENS AND S. ZACHARY, *Asymptotic analysis of single resource loss systems in heavy traffic, with applications to integrated networks*, Adv. Appl. Probab. **27** (1995), pp. 273–292.

[3] J.-P. CRAMETZ AND P.J. HUNT, *A limit result respecting graph structure for a fully connected loss network with alternative routing*, Ann. Appl. Probab. **1** (1991), pp. 436–444.

[4] R.J. GIBBENS AND F.P. KELLY, *Dynamic routing in fully connected networks*, IMA J. Math. Contr. Inf. **7** (1990), pp. 77–111.

[5] R.J. GIBBENS, F.P. KELLY AND S.R.E. TURNER, *Dynamic routing in multiparented networks*, IEEE/ACM Trans. Networking **1** (1993), pp. 261–270.

[6] R.J. GIBBENS AND P.C. REICHL, *Performance bounds applied to loss networks*, Research report 93-30, Statistical Lab., Univ. Cambridge (1993).

[7] C. GRAHAM AND S. MELEARD, *Propagation of chaos for a fully connected loss network with alternate routing*, Stoch. Proc. Appl. **44** (1993), pp. 159–180.

[8] P.J. HUNT, *Limit theorems for stochastic loss networks*, Ph.D. Thesis, Univ. Cambridge (1990).

[9] P.J. HUNT AND T.G. KURTZ, *Large loss networks*, Research report 92-16, Statistical Lab., Univ. Cambridge (1992).

[10] P.J. HUNT AND C.N. LAWS, *Least busy alternative routing in queueing and loss*

networks, Probab. Engrg. Inform. Sci. **6** (1992), pp. 439–456.

[11] P.J. HUNT AND C.N. LAWS, *Asymptotically optimal loss network control*, Math. Oper. Res. **18** (1993), pp. 880–900.

[12] P.J. HUNT, C.N. LAWS AND N. NOTEBAERT, In preparation.

[13] F.P. KELLY, *Blocking probabilities in large circuit-switched networks*, Adv. Appl. Probab. **18** (1986), pp. 473–505.

[14] F.P. KELLY, *Loss networks*, Ann. Appl. Probab. **1** (1991), pp. 319–378.

[15] F.P. KELLY, *Bounds on the performance of dynamic routing schemes for highly connected networks*, Math. Oper. Res. **19** (1994), pp. 1–20.

[16] F.P. KELLY, *Paper in this volume.*

[17] P.B. KEY, *Optimal control and trunk reservation in loss networks*, Probab. Engrg. Inform. Sci. **4** (1990), pp. 203–242.

[18] P.B. KEY, *Some control issues in telecommunications networks*, in: Probability, Statistics and Optimization, editor F.P. Kelly (Wiley, Chichester, 1994), pp. 383–395.

[19] K.R. KRISHNAN AND T.J. OTT, *State-dependent routing for telephone traffic: theory and results*, Proc. 25th IEEE Conf. Decision and Control (1986).

[20] V.G. LAZAREV AND S.M. STAROBINETS, *The use of dynamic programming for optimization of control in networks of commutation of channels*, Engrg. Cybernet. **15** (1977), pp. 107–117.

[21] S.A. LIPPMAN, *Applying a new device in the optimization of exponential queueing systems*, Oper. Res. **23** (1975), pp. 686–710.

[22] V.V. MARBUKH, *Asymptotic investigation of a complete communications network with a large number of points and bypass routes*, Problemy Peredači Informacii **16** (1981), pp. 89–95.

[23] B.L. MILLER, *A queueing reward system with several customer classes*, Mgmt. Sci. **16** (1969), pp. 234–245.

[24] D. MITRA AND R.J. GIBBENS, *State-dependent routing on symmetric loss networks with trunk reservations, II: asymptotics, optimal design*, Ann. Oper. Res. **35** (1992), pp. 3–30.

[25] D. MITRA, R.J. GIBBENS AND B.D. HUANG, *State-dependent routing on symmetric loss networks with trunk reservations, I*, IEEE Trans. Comm. **41** (1993), pp. 400–411.

[26] V. NGUYEN, *On the optimality of trunk reservation in overflow processes*, Probab. Engrg. Inform. Sci. **5** (1991), pp. 369–390.

[27] M.I. REIMAN, *Optimal trunk reservation for a critically loaded link*, Proc. 13th Internat. Teletraffic Congress (1991).

[28] K.W. ROSS AND D.H.K. TSANG, *Optimal circuit access policies in an ISDN environment: a Markov decision approach*, IEEE Trans. Comm. **37** (1989), pp. 934–939.

[29] W. WHITT, *Blocking when service is required from several facilities simultaneously*, A.T. & T. Tech. J. **64** (1985), pp. 1807–1856.

[30] P. WHITTLE, *Approximation in large-scale circuit-switched networks*, Probab. Engrg. Inform. Sci. **2** (1988), pp. 279–291.

FLUID MODELS OF SEQUENCING PROBLEMS IN OPEN QUEUEING NETWORKS; AN OPTIMAL CONTROL APPROACH

FLORIN AVRAM*, DIMITRIS BERTSIMAS†, AND MICHAEL RICARD‡

Abstract. We propose an optimal control approach to the optimization of fluid models of open multiclass queueing networks. Using Pontryagin's maximum principle we obtain insights on optimal policies and show that they are characterized by dynamic indices that lead to policies of the threshold type, the parameters of which we calculate explicitly in particular examples. We also propose a numerical approach to the problem that discretizes the problem and solves it as a linear programming problem - producing a solution that is nearly optimal. We finally propose a heuristic algorithm for the problem that captures interactions among various classes and has the attractive feature that it learns by solving smaller instances. The heuristic leads to a very fast and completely automatic process for solving the problem.

1. Introduction. A *multiclass queueing network* is one that services multiple types of customers which may differ in their arrival processes, service requirements, routes through the network as well as costs per unit of waiting time. The fundamental optimization problem that arises in open networks is to determine an optimal policy for sequencing and routing customers in the network that minimizes a linear combination of the expected sojourn times of each customer class. There are both *sequencing* and *routing* decisions involved in these optimization problems. A *sequencing policy* determines which type of customer to serve at each station of the network, while a *routing policy* determines the route of each customer. In the present paper we only consider sequencing problems.

There are several important applications of such problems: packet-switching communication networks with different types of packets and priorities, job shop manufacturing systems, scheduling of multi-processors and multi-programmed computer systems.

The control of multiclass queueing networks is a mathematically challenging problem. In order to achieve optimality, stations have to decide how to sequence competing customer types at each point in time, based on information about the load conditions of various other stations. These interactions between various stations create serious dependencies among them and prevent not only optimization but even performance analysis of a given policy. To indicate the difficulty of the problem, it is worth mentioning that even with Poisson arrivals and *class dependent* exponential service times, and for the simplest possible policy, FCFS, product form or

---

* Department of Mathematics, Northeastern University, Boston, MA 02115.

† Sloan School of Management and Operations Research Center, MIT, Cambridge, MA 02139. Research partially supported by a Presidential Young Investigator Award DDM-9158118 with matching funds from Draper Laboratory.

‡ Operations Research Center, MIT, Cambridge, MA 02139.

analytical solutions are not available. Naturally, optimizing a multiclass queueing network is an even harder problem. Thus, not surprisingly, simulation is the most common practice among researchers and practitioners as a tool of evaluating heuristic policies.

Optimizing a multiclass queueing network has three characteristics that add to the complexity of the problem: a) The problem is *dynamic*, i.e., sequencing decisions need to be made dynamically in time, b) the problem is *combinatorial*, i.e., there are many possible discrete solutions and c) the problem is *stochastic*, i.e., there are uncertainties regarding the arrival and service processes.

This paper takes the philosophical view that the most important components of the problem are the first two, i.e., the dynamic and combinatorial character of the problem. In other words, while the stochastic component of the problem is important, the structure of the optimal policies can be revealed by a deterministic and dynamic model. In particular, we study in this paper a deterministic fluid model for multiclass networks that has also been studied by Chen and Yao [6], who propose a suboptimal myopic solution using a sequence of linear programming problems (see also Atkins and Chen [3]). Our contributions in the present paper are as follows:

1. Since mathematically, the problem is a linear control problem over a polyhedral region, we use Pontryagin's maximum principle to obtain insights about the structure of the optimal policy. We show that the optimal policy is characterized by dynamic indices that lead to policies of the threshold type, the parameters of which we calculate explicitly in particular examples. In contrast with the work of Chen and Yao [6], our approach offers explicit threshold policies. Several of the examples we solve have not been solved previously in the literature. We also conjecture a strong relation among the optimal policy for the stochastic and the fluid network.

2. We also view the problem as an infinite-dimensional linear programming problem (see Anderson and Nash [2]). We then solve the problem based on a discretization scheme. Although this approach has dimensionality problems and only offers numerical insights, it provides us with a yardstick for evaluating the performance of heuristic policies.

3. Based on the exact solution of simpler cases we propose a family of heuristics that captures interactions among various classes in the network and has the attractive feature that it learns by solving smaller instances. The heuristic solves the problem both symbolically (thus offering qualitative insight on near optimal policies) and numerically very fast. Comparing the exact and approximate solutions we found that the current implementation of the heuristic provides very accurate policies that are close (and often identical) to the optimal ones even for large scale problems. An attractive feature of the heuristic is that it has a learning capability, in the

sense that the more problems one can solve optimally, the better the heuristic can be. Interestingly, when applied to feedforward networks, the heuristic yields a generalization of the $c\mu$ rule: serve at each station the class with the largest index $k_i = \mu_i[w_i - w_{n(i)}]$ where $n(i)$ denotes the class following i and $w_{n(i)}$, w_i are the indeces derived by the heuristic.

Literature Review

Optimal control approaches to fluid models in the context of routing in communication systems have been studied by several researchers. For an informative review of this work see Filipiak [7], Segal [19] and Segal and Moss [20]. Local optimal solutions can be found by gradient search techniques or global optimal solutions for specific problems may sometimes be found by exploiting the structure of the problems at hand. Solution techniques of both types can be found in [14], [21], [11] or [5].

Multiclass queueing networks operating in heavy traffic conditions have been successfully modeled via Brownian motions. Introduced by Harrison [8] and further explored by Wein, this approach proposes heuristic policies which typically outperform traditional ones. This approach has been more successful in closed networks [9] and networks with controllable input [22,23], but has not been as successful in scheduling open networks. In particular, Harrison and Wein show in [10] that a threshold policy is consistent with the optimality conditions for a Brownian two-station, three-class network which we also consider in this paper (we explicitly characterize the optimal policy, which is of the threshold type under some combination of input parameters). For a nice survey of the heavy-traffic approach for optimization of multiclass networks, the reader is referred to Kelly and Laws [12].

In studies that concern lower bounds for general networks, Ou and Wein [15] derive *pathwise* lower bounds for general open queueing networks with deterministic routing. Bertsimas, Paschalidis and Tsitsiklis [4] propose linear and nonlinear programming problems that provide lower bounds for the optimal solution values.

Structure of the paper

The rest of the paper is organized as follows: In Section 2, we formally define the sequencing problem for multiclass open fluid networks and describe the mathematical model. In Section 3 we review the maximum principle and use it to derive useful insights about properties of an optimal solution in a general network. In Section 4 we find explicit solutions for the optimal policy for particular multiclass fluid networks. In Section 5 we propose an optimal algorithm that solves the problem numerically. In Section 6 we propose a family of heuristic algorithms for the problem that solve the problem very fast, both analytically and numerically and report extensive computational results with the goal of obtaining insights into the structure of the optimal policy and the performance of the heuristic. The final section contains some concluding remarks and open problems.

2. Problem formulation. Before presenting the general problem formulation, we examine an example of a queueing network considered in [10]
composed of three classes and two stations; classes one and two are served
at stations one and two respectively, while class three competes with class
one for service at station one (see Figure 2.1). Jobs for classes one and three
arrive at rates λ_1 and λ_3 per time period and jobs of class i are processed
at rate μ_i per time period.

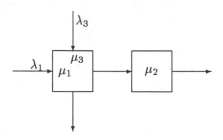

FIG. 2.1. *An example of a queueing network.*

Assuming that there are jobs in the system for all three classes, we
need to decide whether station one should service class one or three (it
is clear that neither station will be idle). We formulate this problem as
follows: For $i = 1$, 2, 3, let $x_i(t)$ denote the number of jobs of class i at
time t and $s(i)$ be the index of the station that processes jobs of class i.
The cost per unit time for holding a job of class i is c_i. Finally the control
variables, $u_i(t)$, will denote the fraction of effort that station $s(i)$ spends
processing jobs of class i at time t $(0 \le u_i(t) \le 1)$. Let $x(t)$, $u(t)$, c denote
the corresponding vectors. Let T be a large enough time so the system will
empty by time T. Let $\tilde{x}$ be the given vector of the number of customers at
time 0. The control problem is:

$$\min \int_0^T c'\, x(t)dt$$

$$
\begin{aligned}
\dot{x}_1(t) &= \lambda_1 - u_1(t)\mu_1 \\
\dot{x}_2(t) &= u_1(t)\mu_1 - u_2(t)\mu_2 \\
\dot{x}_3(t) &= \lambda_3 - u_3(t)\mu_3 \\
x(0) &= \tilde{x} \\
u_1(t) + u_3(t) &\le 1 \\
u_2(t) &\le 1 \\
x(t) &\ge 0 \\
u(t) &\ge 0.
\end{aligned}
$$

The primary goal in this paper is to provide insight to the question of what
the optimal policy $u(t)$ is as a function of the state vector $x(t)$.

In general, a queueing network has m stations and n job classes ($n \geq m$). A job class is specific to a station; as a particular job flows through the network, it defines a different class in each station. A single station may service multiple job classes, each with its own service rate and holding cost. The arrivals for each job class either come from another station or from outside the system. Each class i that has arrivals from another station has a unique previous class, $p(i)$. After the jobs are processed, they either become members of a new class or leave the system. If jobs of class i do not leave the system after processing, they have a unique next class, $n(i)$, otherwise $n(i) = 0$. The dynamics of the system then have the form

$$(2.1) \qquad \dot{x}_i(t) = \mu_{p(i)} u_{p(i)}(t) - \mu_i u_i(t).$$

if the class has internal arrivals and in the case that class i has external arrivals

$$(2.2) \qquad \dot{x}_i(t) = \lambda_i - \mu_i u_i(t).$$

Each class has a control variable associated with it specifying the fraction of time its station spends processing the jobs in this class. The data for the sequencing problem is as follows: a vector b of external arrivals, a vector μ of service rates and the routing data. The relationship between classes and stations can be represented by an $n \times n$ matrix A, such that $A_{ij} = \mu_j$, if $j = p(i)$ and $A_{ii} = -\mu_i$.

The dynamics of the fluid network are then expressed as:

$$\dot{x}(t) = Au(t) + b.$$

In the example above,

$$b = (\lambda_1, 0, \lambda_3)'$$

$$A = \begin{bmatrix} -\mu_1 & 0 & 0 \\ \mu_1 & -\mu_2 & 0 \\ 0 & 0 & -\mu_3 \end{bmatrix}$$

In addition, the sum of the control variables $u_i(t)$ for all the classes processed at the same station must be less than one. This can be represented with an $m \times n$ binary matrix, D, where

$$D_{ij} = \begin{cases} 1 & \text{if } s(j) = i \\ 0 & \text{otherwise.} \end{cases}$$

This constraint can then be written as

$$Du(t) \leq e,$$

where $e = (1, 1, \ldots, 1)'$. The general sequencing problem is as follows:

$$(2.3) \qquad (CONTROL) \min \int_0^T c' \, x(t) dt$$
$$(2.4) \qquad \dot{x}(t) \;=\; Au(t) + b$$
$$(2.5) \qquad Du(t) \;\leq\; e$$
$$(2.6) \qquad x(0) \;=\; \tilde{x}$$
$$(2.7) \qquad x(t) \;\geq\; 0$$
$$(2.8) \qquad u(t) \;\geq\; 0,$$

where $x(t)$ is in the space of continous functions and $u(t)$ is in the space of bounded functions. A special case of Problem *(CONTROL)*, which is of particular interest, is the case of reentrant lines (see Kumar [13]), in which only class one has external arrivals $(b = (\lambda, 0, \ldots, 0)')$.

Problem *(CONTROL)* can be viewed as a linear program in infinite dimensions. The number of variables is infinite since we have an $x(t)$ vector for each infinitesimal time period $[t, t + \Delta t]$. Such problems could be solved by using a discrete approximation and solving a linear program in a finite (albeit large) number of variables (see Pullan [18]); this approach will be discussed in Section 5.

Problem *(CONTROL)* can also be viewed as an optimal control problem. We can therefore use the Pontryagin maximum principle to yield necessary and sufficient conditions for an optimal solution and obtain insights on the structure of the optimal policies. We discuss this approach next.

3. The Pontryagin maximum principle. This section reviews necessary and sufficient optimality conditions arising from the Pontryagin Maximum Principle (see [17]) for Problem *(CONTROL)*, and examines their implications to insights regarding the optimal policy for Problem *(CONTROL)*.

3.1. Necessary and sufficient optimality conditions. We define the Hamiltonian function as

$$H(x, u, y, t) = -c' \, x(t) + y(t)' \, [Au(t) + b].$$

For any segment of the optimal trajectory that lies in the interior of the state space (i.e., $x(t) > 0$), there must exist multipliers, $y(t)$, that satisfy the "adjoint" system

$$\frac{dy_i}{dt} = -\frac{dH}{dx_i} = c_i, \; i = 1, \ldots, n.$$

Note that as long as $x_i(t) > 0$, the multipliers $y_i(t)$ are piecewise linear functions of time. For any segment of the optimal trajectory that lies entirely on a smooth surface of the boundary of the state space, i.e., $x_i(t) = 0$ for some i, the multipliers must satisfy

$$\frac{dy_i}{dt} = -\frac{dH}{dx_i} = c_i - p_i(t), \quad i = 1, \ldots, n.$$

where

$$p(t) \geq 0, \quad p(t)'x(t) = 0.$$

Moreover, along the optimal trajectory $\bar{x}$, $\bar{y}$, and $\bar{u}$ must satisfy

$$(3.1) \qquad \max_u H(\bar{x}, \bar{y}, u, t) = H(\bar{x}, \bar{y}, \bar{u}, t) = 0.$$

For problem *(CONTROL)* these conditions are also sufficient (see for example Seierstad and Sydsæter [21]). They are also equivalent to the strong duality theory for infinite-dimensional linear programming problems

Applying the maximum principle we obtain from (2.1) and (2.2) that

$$H = -c'x(t) + y(t)'b + \sum_i u_i(t)k_i(t)$$

where

$$k_i(t) = \mu_i[y_{n(i)}(t) - y_i(t)]$$

or if $n(i) = 0$

$$k_i(t) = -\mu_i y_i(t).$$

Since H must be maximized with respect to u, the optimal policy is characterized by the **dynamic indices** $k_i(t)$: serve at each station the class with the highest positive index $k_i(t)$.

Although the problem is in principle characterized by the presence of dynamic indices, which turn out to be piecewise linear functions of time the explicit calculation of these indices is nontrivial, because fitting together the various pieces, which is done using (3.1), can lead to quite complex computations. A general strategy will be outlined below and section 4 will illustrate its application in specific examples.

The optimal trajectory is characterized by changes of policy (switches), i.e., situations in which a certain fraction, d, of servicing effort is transferred from a class i to a competing class $c(i)$. These switches can be of two types: a) Forced switches, or *DEPLETIONS*, when a class i is depleted ($x_i(t) = 0$) and b) switches occurring in the interior of the state space, called from now on *INTERRUPTS*. Assuming that for both classes i and $c(i)$, $x_{n(i)}(\bar{t}) > 0$ and $x_{n(c(i))}(\bar{t}) > 0$, then after a switch at time $\bar{t}$, the Hamiltonian changes by $d\,[k_i(\bar{t}) - k_{c(i)}(\bar{t})]$, it follows from (3.1) that

$$k_i(\bar{t}) = k_{c(i)}(\bar{t}).$$

Notice that if there is no $c(i)$ (server $s(i)$ idles after time $\bar{t}$)

$$k_i(\bar{t}) = 0.$$

These "continuity" equations provide enough equations to solve for $y(t)$ and calculate the dynamic indices. This method is illustrated in section 4. Conceptually, the method can be applied for more general networks, but the computations involved are rather complex.

In order to contrast the insights gained via the maximum principle, notice that myopic policies (similar to the policies advocated by Chen and Yao [6]) lead naturally to static indices, while the true optimal solution via the maximum principle is characterized, as we have seen, by dynamic indices that lead to state dependent solutions (threshold curves).

3.2. Myopic policies and static indices for reentrant lines. Given the current state, $x(t) > 0$, a myopic policy attempts to optimize the immediate cost

$$(3.2) \qquad Z = c' \, x(t + \Delta t) = c' \, x(t) + c' \, \dot{x}(t)\Delta t + o(\Delta t).$$

For reentrant lines, let P be the set of classes for which $p(i)$ exists and N be the set of classes for which $n(i)$ exists $(n(i) > 0)$. Then substituting the dynamics (2.1) and (2.2) to (3.2), a myopic policy would optimize the expression:

$$\sum_{i \notin P} c_i[\lambda_i - \mu_i u_i(t)] + \sum_{i \in P} c_i[\mu_{p(i)}u_{p(i)}(t) - \mu_i u_i(t)] =$$

$$\sum_{i \notin P} c_i \lambda_i + \sum_{i \in N} u_i(t)\mu_i[c_{n(i)} - c_i] - \sum_{i \notin N} u_i(t)\mu_i c_i.$$

Therefore, under a myopic policy, the problem becomes:

$$Z_{myopic} = \min \sum_{i \in N} u_i(t)\mu_i[c_{n(i)} - c_i] - \sum_{i \notin N} u_i(t)\mu_i c_i$$
$$Du(t) \; \leq \; e$$
$$u(t) \; \geq \; 0.$$

Notice that as long as $x(t) > 0$, $x(t + \Delta t) \geq 0$ for sufficiently small Δt and therefore, the constraints $x(t + \Delta t) \geq 0$ are redundant. The myopic problem decomposes over stations, since the constraints $Du(t) \leq e$ decompose. Therefore, as long as $x(t) > 0$, the best myopic policy is simply to serve at each station among the classes with minimal index

$$f_i = \begin{cases} (c_{n(i)} - c_i)\mu_i & \text{if } i \in N \\ -c_i\mu_i & \text{if } i \notin N \end{cases}$$

The myopic solution turns out to be optimal in the case of one server networks, but in general it can miss important features of the optimal solution as we see in the next section.

4. Examples. This section illustrates the application of the maximum principle by solving examples with small networks. Our goal is to illustrate that our approach captures interesting phenemena that a myopic solution would miss. Moreover, the particular examples we solve are used in Section 6, where we develop a learning heuristic to solve larger networks based on solutions for smaller networks.

4.1. Competing queues. Consider the problem in Figure 4.1 with a single server and n classes. Jobs for class i arrive at a rate of λ_i per unit time and can be serviced at a rate μ_i per unit time. This problem is solved by the well known "$c\mu$-rule", but it will be illustrative to derive the "$c\mu$-rule" using the maximum principle.

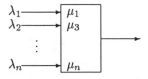

FIG. 4.1. *Multiple classes being served by one station.*

The problem is written as:

$$\min \int_0^T c'\, x(t)\, dt$$
$$\dot{x}_1(t) \;=\; \lambda_1 - u_1(t)\mu_1$$
$$\vdots$$
$$\dot{x}_n(t) \;=\; \lambda_n - u_n(t)\mu_n$$
$$x(0) \;=\; \tilde{x}$$
$$\sum_{i=0}^{n} u_i(t) \;\leq\; 1$$
$$x(t) \;\geq\; 0$$
$$u(t) \;\geq\; 0.$$

The Hamiltonian function is

$$H \;=\; -c'\, x(t) + \sum_{i=0}^{n} y_i(t)(\lambda_i - \mu_i u_i(t))$$
$$\;=\; -c'\, x(t) - \sum_{i=0}^{n} \mu_i y_i(t) u_i(t) + \sum_{i=0}^{n} y_i(t)\lambda_i$$

THEOREM 1. *The optimal policy is to use a static priority rule indexed by $c_i\mu_i$ (the class with the highest $c_i\mu_i$ has the highest priority). If $c_i\mu_i = c_j\mu_j$ for some i and j then the tie can be broken arbitrarily.*

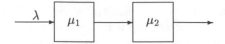

FIG. 4.2. *Two stations in tandem.*

Proof. Assuming that $x(0) > 0$, let class i be the one served first under an optimal policy. By the maximum principle we will continue to serve class i until time t_i, the switching time of class i. The index for class i in this case is $k_i(t) = -\mu_i y_i(t)$. By the maximum principle, for $t \leq t_i$, $k_i(t) = \max_{1 \leq j \leq n} -\mu_j y_j(t)$. At the epoch t_i a switch occurs and the server shifts effort to class j. By the continuity equations $\mu_i y_i(t_i) = \mu_j y_j(t_i)$. To ease the computation let us assume (without loss of generality) that $y_i(t_i) = 0$.

The indices $k_r(t) = -\mu_r y_r(t)$ for $t \leq t_i$, are linear functions with negative slopes of $-\mu_r c_r$. In particular, $k_i(t) = -\mu_i c_i(t - t_i)$ and $k_j(t) = -\mu_j c_j(t - t_i) + k_j(t_i)$. Since $k_i(0) \geq k_j(0)$, $\mu_i c_i \geq \mu_j c_j + k_j(t_i)$. Note that $k_j(t_i)$ must be non-negative or else $k_j(t) < 0$ for $t \geq t_i$ and class j will never be processed. Therefore, $\mu_i c_i \geq \mu_j c_j$. $\square$

This example supports our view that it is the dynamic and combinatorial character of the problem and not the stochastic character that is the most important. In this example the same policy is optimal for the stochastic model as well.

4.2. Tandem queues. Consider the queueing network in Figure 4.2 with two stations in tandem. Jobs enter the system with rate λ and are serviced at station one at rate μ_1. All jobs then pass to station two where they are served at rate μ_2 and then exit the system. The problem is written as:

$$\min \int_0^T c' \, x(t) dt$$

$$
\begin{aligned}
\dot{x}_1(t) &= \lambda - u_1(t)\mu_1 \\
\dot{x}_2(t) &= u_1(t)\mu_1 - u_2(t)\mu_2 \\
x(0) &= \tilde{x} \\
u_i(t) &\leq 1 \\
x(t) &\geq 0 \\
u_i(t) &\geq 0.
\end{aligned}
$$

It is clear that $\lambda < \mu_1$ and $\lambda < \mu_2$ are required for stability and that $u_2(t) = 1$ as long as $x_2(t) > 0$. The optimal policy is as follows.

THEOREM 2.   *While $x(t) > 0$, the optimal control for the tandem problem is given by the following cases.*

	Conditions	$u(t)$
Case 1	$c_1 \geq c_2$	(1,1)
Case 2	$c_1 < c_2$ $\mu_1 \geq \mu_2$	(0,1)
Case 3	$c_1 < c_2$ $\mu_1 < \mu_2$ $\frac{x_1(t)}{x_2(t)} < \frac{c_2}{c_1} \frac{\mu_1 - \lambda}{\mu_2 - \mu_1}$	(0,1)
Case 4	$c_1 < c_2$ $\mu_1 < \mu_2$ $\frac{x_1(t)}{x_2(t)} \geq \frac{c_2}{c_1} \frac{\mu_1 - \lambda}{\mu_2 - \mu_1}$	(1,1)

Proof. Cases 1 and 2 are intuitive (they can be proven by a formal argument based on the maximum principle). The focus will be on cases 3 and 4 which exhibit a threshold curve. The Hamiltonian function for this problem is

$$
\begin{aligned}
H &= -c_1 x_1(t) - c_2 x_2(t) + y_1(t)[\lambda - u_1(t)\mu_1] + y_2(t)[u_1(t)\mu_1 - u_2(t)\mu_2] \\
&= -c_1 x_1(t) - c_2 x_2(t) + y_1(t)\lambda + u_1(t)\mu_1[y_2(t) - y_1(t)] - u_2(t)[y_2(t)\mu_2]
\end{aligned}
$$

The assumptions for these cases are $c_1 < c_2$ and $\mu_1 < \mu_2$. In order to investigate whether a threshold curve exists, assume that at time 0 (without loss of generality) there is an INTERRUPT, i.e., station 1 changes from holding to serving. From then on both stations are serving until one of the two classes is depleted. Let t_i be the depletion times of the two classes.

We first examine the case $t_1 > t_2$, i.e., class 2 is depleted first. Since class 2 does not have a competitor class in the same station or a follower, the continuity equation implies $k_2(t) = -y_2(t)\mu_2 = 0$ for $t \geq t_2$. Moreover, since $\dot{y}_2 = c_2$ for $t \leq t_2$, we obtain

$$y_2(t) = c_2(t - t_2), \quad t \leq t_2.$$

Similarly, by the continuity equation $k_1(t) = \mu_1[y_2(t) - y_1(t)] = 0$ for $t \geq t_1$. Since $y_2(t_1) = 0$, $(t_1 > t_2)$, $y_1(t_1) = 0$ and since $\dot{y}_1 = c_1$ for $t \leq t_1$,

$$y_1(t) = c_1(t - t_1), \quad t \leq t_1.$$

Since $x_1(t) = (\lambda - \mu_1)t + x_1(0)$ and $x_2(t) = (\mu_1 - \mu_2)t + x_2(0)$ the depletion times are $t_1 = \frac{x_1(0)}{\mu_1 - \lambda}$, $t_2 = \frac{x_2(0)}{\mu_2 - \mu_1}$ (notice $\mu_1 < \mu_2$ is used here).

Since it was assumed that at time 0 there was an INTERRUPT, the continuity equation implies $k_1(0) = 0$, which implies that $y_1(0) = y_2(0)$, i.e., $c_1 t_1 = c_2 t_2$ leading to the threshold curve:

$$\frac{x_1(0)}{x_2(0)} = \frac{c_2}{c_1} \frac{\mu_1 - \lambda}{\mu_2 - \mu_1}.$$

Notice that the assumption $c_1 < c_2$ is used here to ensure that $t_1 > t_2$, since $c_1 t_1 = c_2 t_2$.

The case $t_1 < t_2$ leads to an infeasible set of equations. Therefore, both cases 3 and 4 follow. $\square$

When $x(t) \not> 0$, then the values of $u(t)$ must be modified to preserve the non-negativity constraints on $x(t)$. In case 1, both classes are always served, i.e., when $x_1(t) = 0$, $u_1(t) = \lambda/\mu_1$ and when $x_2(t) = 0$, $u_2(t) = \frac{\mu_1 u_1(t)}{\mu_2}$. In case 2, class one is held until $x_2(t) = 0$. At that time, the optimal policy is to set $u = (\mu_2/\mu_1, 1)$ (keeping $x_2(t)$ at zero). After $x_1(t)$ becomes zero, u becomes $(\lambda/\mu_1, \lambda/\mu_2)$.

For the remaining cases, the threshold does not change when $x(t) \not> 0$. When $x_2(t) = 0$, $\frac{x_1(t)}{x_2(t)}$ is infinity so class one is always served. Of course the control must be modified to account for this. For example, when $x_2(t) = 0$ and $x_1(t) > 0$, $u = (1, \mu_1/\mu_2)$.

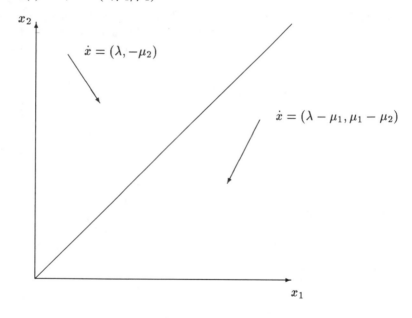

FIG. 4.3. *Switching curve for Cases 3 & 4.*

Geometrically, the state space is partitioned by a single line (Figure 4.3). It is interesting to contrast the optimal policy with the myopic solution. The myopic solution would be to serve at the first station if and only if $c_1 \geq c_2$, and to delay serving until $x_2 = 0$ if $c_1 < c_2$. However, the true optimal solution in this last case is to start serving before class 2 is depleted, i.e., a *HOLD* policy: keep station 1 idle until $\frac{x_1(t)}{x_2(t)} < \frac{c_2}{c_1} \frac{\mu_1 - \lambda}{\mu_2 - \mu_1}$. Chen and Yao [6] have already pointed out that the myopic policy should be changed adaptively, as soon as certain classes are depleted. In the above example we see that the correct optimal solution "forsees" future depletions on the route, and starts serving before the depletions actually occur.

4.3. Tandem queues with competition in the first node. Consider the queueing network in Figure 2.1 with two stations in tandem and two queues competing for service in the first node. It is clear that $\lambda_1/\mu_1 + \lambda_3/\mu_3 < 1$ and $\lambda_1 < \mu_2$ are required for stability, $u_2(t) = 1$ as long as $x_2(t) > 0$ and as long as $x_3(t) > 0$, $u_3(t) = 1 - u_1(t)$. The problem is to choose which class to serve at station 1.

THEOREM 3. *When $x(t) > 0$, the optimal control for the problem is given by the following cases.*

	Conditions	$u(t)$
Case 1	$c_1\mu_1 \leq c_3\mu_3$	*(0,1,1)*
Case 2	$c_1\mu_1 > c_3\mu_3$	*(1,1,0)*
	$c_1\mu_1 - c_3\mu_3 \geq c_2\mu_1$	
Case 3	$c_1\mu_1 > c_3\mu_3$	*(0,1,1)*
	$c_1\mu_1 - c_3\mu_3 < c_2\mu_1$	
	$\mu_1 \geq \mu_2$	
Case 4	$c_1\mu_1 > c_3\mu_3$	*(1,1,0)*
	$c_1\mu_1 - c_3\mu_3 < c_2\mu_1$	
	$\mu_1 < \mu_2$	
	$\dfrac{x_1(t)}{x_2(t)} \geq \dfrac{c_2\mu_1}{c_1\mu_1 - c_3\mu_3}\dfrac{\mu_1 - \lambda}{\mu_2 - \mu_1}$	
Case 5	$c_1\mu_1 > c_3\mu_3$	*(0,1,1)*
	$c_1\mu_1 - c_3\mu_3 < c_2\mu_1$	
	$\mu_1 < \mu_2$	
	$\dfrac{x_1(t)}{x_2(t)} < \dfrac{c_2\mu_1}{c_1\mu_1 - c_3\mu_3}\dfrac{\mu_1 - \lambda}{\mu_2 - \mu_1}$	

Proof. The Hamiltonian function in this case is

$$
\begin{aligned}
H \;=\; & -c_1 x_1(t) - c_2 x_2(t) - c_3\, x_3(t) + \\
& y_1(t)[\lambda_1 - u_1(t)\mu_1] + y_2(t)[u_1(t)\mu_1 - u_2(t)\mu_2] + y_3(t)[\lambda_3 - u_3(t)\mu_3] \\
=\; & -c_1 x_1(t) - c_2 x_2(t) - c_3\, x_3(t) + u_1(t)[y_2(t) - y_1(t)]\mu_1 \\
& -u_2(t)[y_2(t)\mu_2] - u_3(t)[y_3(t)\mu_3] \\
& +y_1(t)\lambda_1 + y_3(t)\lambda_3
\end{aligned}
$$

Again, the focus will be on the more interesting cases 4 and 5. Assume that at time 0 there is an INTERRUPT at station 1, i.e., the server switches from serving class 3 to serving class 1. Thereafter, classes 1 and 2 are served to depletion. Let t_1 and t_2 be the corresponding depletion times. At time t_1, station 1 switches to serve class 3. Let t_3 be the depletion time of class 3. Assume as with the previous problem that $t_2 < t_1$ (the case $t_2 > t_1$ does not lead to a consistent system of equations for $c_1\mu_1 > c_3\mu_3$ and $c_1\mu_1 - c_3\mu_3 < c_2\mu_1$).

Arguing as before, the continuity equation implies $k_2(t_2) = 0$, leading to $y_2(t) = c_2(t - t_2)$ for $t \leq t_2$ and $y_2(t) = 0$ for $t \geq t_2$. Similarly,

$y_3(t) = c_3(t - t_3)$ for $t \leq t_3$ and $y_3(t) = 0$ for $t \geq t_3$. Also, as long $x_1(t) > 0$, $\dot{y}_1(t) = c_1$, which implies that for $t \leq t_1$ $y_1(t) = c_1(t - t_1) + a$ for some constant a.

Since at time t_1 station 1 switches from class 1 to class 3 (a DEPLE-TION epoch), the continuity equation implies that $k_1(t_1) = k_3(t_1)$, leading to

$$[y_2(t_1) - y_1(t_1)]\mu_1 = -y_3(t_1)\mu_3.$$

Since $t_1 > t_2$, $y_2(t_1) = 0$ and therefore,

$$a = \frac{\mu_3}{\mu_1} c_3(t_1 - t_3).$$

Since at time 0 station 1 switches from class 3 to class 1 (an INTER-RUPT epoch), the continuity equation implies that $k_1(0) = k_3(0)$ leading to $[y_2(0) - y_1(0)]\mu_1 = -y_3(0)\mu_3$, which implies that

$$(-c_2 t_2 + z c_1 t_1 - a)\mu_1 = c_3 \mu_3 t_3,$$

which in turn simplifies to

(4.1) $$t_1(\mu_1 c_1 - \mu_3 c_3) = \mu_1 c_2 t_2.$$

Expressing the depletion times in terms of the initial conditions ($t_1 = \frac{x_1(t)}{\mu_1 - \lambda}$, $t_2 = \frac{x_2(t)}{\mu_2 - \mu_1}$) and substituting to (4.1) we obtain the threshold curve in Cases 4, 5 of the theorem. $\square$

Once again, the theorem is naturally extended when $x(t) \not> 0$. Whenever $x_2(t) = 0$, $u_2(t) = \min(1, \frac{\mu_1 u_1(t)}{\mu_2})$. In case 1, class three always has priority so when $x_3(t) = 0$, $u_3(t) = \lambda_3/\mu_3$; the value of $u_1(t)$ (either 0 or $1 - u_3(t)$) is determined from the second through last conditions of cases 2-5. In case 2, class 1 always has priority over three so once $x_1(t) = 0$, $u_1(t) = \lambda/\mu_1$ and $u_3(t) = 1 - u_1(t)$. In case 3, class one gets serviced after $x_2(t) = 0$ (i.e. at that point the "μc" rule prevails) so $u = (\mu_1/\mu_2, 1, 1 - \mu_1/\mu_2)$. In cases 4 and 5, $u_1(t)$ is determined by the threshold with $u_3(t) = 1 - u_1(t)$.

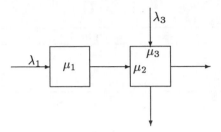

FIG. 4.4. *A tandem system with competition in the second node.*

4.4. Tandem queues with competition at the Second Node.
Consider the queueing network in Figure 4.4 with two stations in tandem and two queues competing for service in the second node. The problem is written as:

$$\min \qquad \int_0^T c'\, x(t)\, dt$$

$$\dot{x}_1(t) = \lambda_1 - u_1(t)\mu_1$$
$$\dot{x}_2(t) = u_1(t)\mu_1 - u_2(t)\mu_2$$
$$\dot{x}_3(t) = \lambda_3 - u_3(t)\mu_3$$
$$x(0) = \tilde{x}$$
$$u_1(t) \le 1$$
$$u_2(t) + u_3(t) \le 1$$
$$x(t) \ge 0$$
$$u(t) \ge 0.$$

It is clear that $\lambda_1 < \mu_1$ and $\lambda_1/\mu_2 + \lambda_3/\mu_3 < 1$ are required for stability and as long as $x_3(t) > 0$, $u_3(t) = 1 - u_2(t)$. The problem is to decide if class one should be held or not and if station two should process class two or three. Specifically, while $x(t) > 0$, the four possible choices for $u(t)$ are $(1,1,0)$, $(1,0,1)$, $(0,1,0)$ and $(0,0,1)$. In the following theorem, the optimal policy is presented. The proof is omitted as it is similar to, but somewhat more algebraically involved than, the previous cases. Once again, the theorem considers the case $x(t) > 0$ but if $x(t) \not> 0$, it is naturally extended.

THEOREM 4. *While $x(t) > 0$, the optimal control for the problem is given by the following cases.*

4.5. Three queues in tandem.
Consider the queueing network in Figure 4.5 with three stations in tandem.

	Conditions	$u(t)$
Case 1	$c_1 \geq c_2$ $c_2\mu_2 \geq c_3\mu_3$	$(1,1,0)$
Case 2	$c_1 \geq c_2$ $c_3\mu_3 > c_2\mu_2$	$(1,0,1)$
Case 3	$c_1 < c_2$ $\mu_1 \geq \mu_2$ $c_3\mu_3 > c_2\mu_2$	$(0,0,1)$
Case 4	$c_1 < c_2$ $\mu_1 \geq \mu_2$ $c_2\mu_2 \geq c_3\mu_3$	$(0,1,0)$
Case 5	$c_1 < c_2$ $\mu_1 < \mu_2$ $c_2\mu_2 \geq c_1\mu_2 \geq c_3\mu_3$ $\dfrac{x_3(t)+\frac{\mu_3}{\mu_2}x_2(t)}{\mu_3(1-\frac{\mu_1}{\mu_2})-\lambda_3} \geq \dfrac{x_1(t)}{\mu_1-\lambda_1}$ $\left(c_1 - \frac{\mu_3}{\mu_2}c_3\right)\dfrac{x_1(t)}{\mu_1-\lambda_1} \geq \left(c_2 - \frac{\mu_3}{\mu_2}c_3\right)\dfrac{x_2(t)}{\mu_2-\mu_1}$	$(1,1,0)$
Case 6	$c_1 < c_2$ $\mu_1 < \mu_2$ $c_2\mu_2 \geq c_1\mu_2 \geq c_3\mu_3$ $\dfrac{x_3(t)+\frac{\mu_3}{\mu_2}x_2(t)}{\mu_3(1-\frac{\mu_1}{\mu_2})-\lambda_3} \geq \dfrac{x_1(t)}{\mu_1-\lambda_1}$ $\left(c_1 - \frac{\mu_3}{\mu_2}c_3\right)\dfrac{x_1(t)}{\mu_1-\lambda_1} < \left(c_2 - \frac{\mu_3}{\mu_2}c_3\right)\dfrac{x_2(t)}{\mu_2-\mu_1}$	$(0,1,0)$
Case 7	$c_1 < c_2$ $\mu_1 < \mu_2$ $c_2\mu_2 \geq c_1\mu_2 \geq c_3\mu_3$ $\dfrac{x_3(t)+\frac{\mu_3}{\mu_2}x_2(t)}{\mu_3(1-\frac{\mu_1}{\mu_2})-\lambda_3} < \dfrac{x_1(t)}{\mu_1-\lambda_1}$ $\dfrac{c_1 x_1(t)}{\mu_1-\lambda_1} \geq \dfrac{c_2 x_2(t)}{\mu_2-\mu_1} + c_3\frac{\mu_3}{\mu_2}\left(\dfrac{x_3(t)+\lambda_3\frac{x_2(t)}{\mu_2-\mu_1}}{\mu_3(1-\frac{\mu_1}{\mu_2})-\lambda_3}\right)$	$(1,1,0)$
Case 8	$c_1 < c_2$ $\mu_1 < \mu_2$ $c_2\mu_2 \geq c_1\mu_2 \geq c_3\mu_3$ $\dfrac{x_3(t)+\frac{\mu_3}{\mu_2}x_2(t)}{\mu_3(1-\frac{\mu_1}{\mu_2})-\lambda_3} < \dfrac{x_1(t)}{\mu_1-\lambda_1}$ $\dfrac{c_1 x_1(t)}{\mu_1-\lambda_1} < \dfrac{c_2 x_2(t)}{\mu_2-\mu_1} + c_3\frac{\mu_3}{\mu_2}\left(\dfrac{x_3(t)+\lambda_3\frac{x_2(t)}{\mu_2-\mu_1}}{\mu_3(1-\frac{\mu_1}{\mu_2})-\lambda_3}\right)$	$(0,1,0)$
Case 9	$c_1 < c_2$ $\mu_1 < \mu_2$ $c_2\mu_2 \geq c_3\mu_3 > c_1\mu_2$ $\dfrac{c_1 x_1(t)}{\mu_1-\lambda_1} \geq \dfrac{c_2 x_2(t)}{\mu_2-\mu_1} + c_3\frac{\mu_3}{\mu_2}\left(\dfrac{x_3(t)+\lambda_3\frac{x_2(t)}{\mu_2-\mu_1}}{\mu_3(1-\frac{\mu_1}{\mu_2})-\lambda_3}\right)$	$(1,1,0)$

	Conditions	$u(t)$
Case 10	$c_1 < c_2$ $\mu_1 < \mu_2$ $c_2\mu_2 \geq c_3\mu_3 > c_1\mu_2$ $\dfrac{c_1 x_1(t)}{\mu_1 - \lambda_1} < \dfrac{c_2 x_2(t)}{\mu_2 - \mu_1} + c_3 \dfrac{\mu_3}{\mu_2} \left(\dfrac{x_3(t) + \lambda_3 \frac{x_2(t)}{\mu_2 - \mu_1}}{\mu_3(1 - \frac{\mu_1}{\mu_2}) - \lambda_3} \right)$	$(0,1,0)$
Case 11	$c_1 < c_2$ $\mu_1 < \mu_2$ $c_3\mu_3 > c_2\mu_2$ $\dfrac{c_2 x_2(t) + \frac{c_2 x_3(t)\mu_2}{\mu_3}}{(1 - \frac{\lambda_3}{\mu_3})\mu_2 - \mu_1} < \dfrac{c_1 x_1(t)}{\mu_1 - \lambda_1}$	$(1,0,1)$
Case 12	$c_1 < c_2$ $\mu_1 < \mu_2$ $c_3\mu_3 > c_2\mu_2$ $\dfrac{c_2 x_2(t) + \frac{c_2 x_3(t)\mu_2}{\mu_3}}{(1 - \frac{\lambda_3}{\mu_3})\mu_2 - \mu_1} \geq \dfrac{c_1 x_1(t)}{\mu_1 - \lambda_1}$	$(0,0,1)$

FIG. 4.5. *Three queues in tandem.*

The problem is written as:

$$\min \int_0^T c'\, x(t)\, dt$$

$$
\begin{aligned}
\dot{x}_1(t) &= \lambda - u_1(t)\mu_1 \\
\dot{x}_2(t) &= u_1(t)\mu_1 - u_2(t)\mu_2 \\
\dot{x}_3(t) &= u_2(t)\mu_2 - u_3(t)\mu_3 \\
x(0) &= \tilde{x} \\
u_i(t) &\leq 1, \ i = 1, \ldots, 3 \\
x(t) &\geq 0 \\
u(t) &\geq 0.
\end{aligned}
$$

It is clear that $\lambda < \mu_i$, $i = 1, 2, 3$ is required for stability and that station 3 will not be idle at time 0. This problem does exhibit a threshold curve for both station 1 and 2, however. The following theorem once again presents the optimal policy but omits the proof because of its algebraic complexity.

THEOREM 5. *While $x(t) > 0$, the optimal control for the problem is* *given by the following cases.*

	Conditions	$u(t)$
Case 1	$c_1 \geq c_2 \geq c_3$	$(1,1,1)$
Case 2	$c_1 \geq c_2$	$(1,0,1)$
	$c_2 < c_3$	
	$\mu_2 \geq \mu_3$	
Case 3	$c_1 \geq c_2$	$(1,1,1)$
	$c_2 < c_3$	
	$\mu_1 < \mu_2 < \mu_3$	
	$\frac{x_1(t)}{\mu_1 - \lambda} > \frac{x_2(t)}{\mu_2 - \mu_1}$	
	$\frac{x_3(t)}{x_2(t)} < \frac{c_2}{c_3} \frac{\mu_3 - \mu_2}{\mu_2 - \mu_1}$	
Case 4	$c_1 \geq c_2$	$(1,0,1)$
	$c_2 < c_3$	
	$\mu_1 < \mu_2 < \mu_3$	
	$\frac{x_1(t)}{\mu_1 - \lambda} \geq \frac{x_2(t)}{\mu_2 - \mu_1}$	
	$\frac{x_3(t)}{x_2(t)} \geq \frac{c_2}{c_3} \frac{\mu_3 - \mu_2}{\mu_2 - \mu_1}$	
Case 5	$c_1 \geq c_2$	$(1,1,1)$
	$c_2 < c_3$	
	$\mu_1 < \mu_2 < \mu_3$	
	$\frac{x_1(t)}{\mu_1 - \lambda} < \frac{x_2(t)}{\mu_2 - \mu_1}$	
	$\frac{x_3(t)}{x_1(t) + x_2(t)} < \frac{c_2}{c_3} \frac{\mu_3 - \mu_2}{\mu_2 - \lambda}$	
Case 6	$c_1 \geq c_2$	$(1,0,1)$
	$c_2 < c_3$	
	$\mu_1 < \mu_2 < \mu_3$	
	$\frac{x_1(t)}{\mu_1 - \lambda} < \frac{x_2(t)}{\mu_2 - \mu_1}$	
	$\frac{x_3(t)}{x_1(t) + x_2(t)} \geq \frac{c_2}{c_3} \frac{\mu_3 - \mu_2}{\mu_2 - \lambda}$	
Case 7	$c_1 \geq c_2$	$(1,1,1)$
	$c_2 < c_3$	
	$\mu_1 \geq \mu_2$	
	$\mu_2 < \mu_3$	
	$\frac{x_3(t)}{x_1(t) + x_2(t)} < \frac{c_2}{c_3} \frac{\mu_3 - \mu_2}{\mu_2 - \lambda}$	
Case 8	$c_1 \geq c_2$	$(1,0,1)$
	$c_2 < c_3$	
	$\mu_1 \geq \mu_2$	
	$\mu_2 < \mu_3$	
	$\frac{x_3(t)}{x_1(t) + x_2(t)} \geq \frac{c_2}{c_3} \frac{\mu_3 - \mu_2}{\mu_2 - \lambda}$	

	Conditions	$u(t)$
Case 9	$c_1 < c_2$ $c_2 \geq c_3$ $\mu_1 < \mu_2$ $\dfrac{x_3(t)}{\mu_2-\mu_2} < \dfrac{x_2(t)}{\mu_2-\mu_1}$ $\dfrac{x_1(t)}{x_2(t)} \geq \dfrac{c_2}{c_1}\dfrac{\mu_1-\lambda}{\mu_2-\mu_1}$	$(1,1,1)$
Case 10	$c_1 < c_2$ $c_2 \geq c_3$ $\mu_1 < \mu_2$ $\dfrac{x_3(t)}{\mu_3-\mu_2} < \dfrac{x_2(t)}{\mu_2-\mu_1}$ $\dfrac{x_1(t)}{x_2(t)} < \dfrac{c_2}{c_1}\dfrac{\mu_1-\lambda}{\mu_2-\mu_1}$	$(0,1,1)$
Case 11	$c_1 < c_2$ $c_2 \geq c_3$ $\mu_1 < \mu_2$ $\dfrac{x_3(t)}{\mu_3-\mu_2} > \dfrac{x_2(t)}{\mu_2-\mu_1}$ $\dfrac{x_3(t)+x_2(t)}{\mu_3-\mu_1} < \dfrac{x_1(t)}{\mu_1-\lambda}$ $\dfrac{c_3(x_3(t)+x_2(t))}{\mu_3-\mu_1} + \dfrac{x_2(t)(c_2-c_3)}{\mu_2-\mu_1} < \dfrac{c_1 x_1(t)}{\mu_1-\lambda}$	$(1,1,1)$
Case 12	$c_1 < c_2$ $c_2 \geq c_3$ $\mu_1 < \mu_2$ $\dfrac{x_3(t)}{\mu_3-\mu_2} \geq \dfrac{x_2(t)}{\mu_2-\mu_1}$ $\dfrac{x_3(t)+x_2(t)}{\mu_3-\mu_1} < \dfrac{x_1(t)}{\mu_1-\lambda}$ $\dfrac{c_3(x_3(t)+x_2(t))}{\mu_3-\mu_1} + \dfrac{x_2(t)(c_2-c_3)}{\mu_2-\mu_1} \geq \dfrac{c_1 x_1(t)}{\mu_1-\lambda}$	$(0,1,1)$
Case 13	$c_1 < c_2$ $c_2 \geq c_3$ $\mu_1 < \mu_2$ $\dfrac{x_3(t)}{\mu_3-\mu_2} \geq \dfrac{x_2(t)}{\mu_2-\mu_1}$ $\dfrac{x_3(t)+x_2(t)}{\mu_3-\mu_1} > \dfrac{x_1(t)}{\mu_1-\lambda}$ $\dfrac{x_2(t)(c_2-c_3)}{\mu_2-\mu_1} < \dfrac{x_1(t)(c_1-c_3)}{\mu_1-\lambda}$	$(1,1,1)$
Case 14	$c_1 < c_2$ $c_2 \geq c_3$ $\mu_1 < \mu_2$ $\dfrac{x_3(t)}{\mu_3-\mu_2} \geq \dfrac{x_2(t)}{\mu_2-\mu_1}$ $\dfrac{x_3(t)+x_2(t)}{\mu_3-\mu_1} \geq \dfrac{x_1(t)}{\mu_1-\lambda}$ $\dfrac{x_2(t)(c_2-c_3)}{\mu_2-\mu_1} \geq \dfrac{x_1(t)(c_1-c_3)}{\mu_1-\lambda}$	$(0,1,1)$

	Conditions	$u(t)$
Case 15	$c_1 < c_2$ $c_2 \geq c_3$ $\mu_1 \geq \mu_2$	$(0,1,1)$
Case 16	$c_1 < c_2 < c_3$ $\mu_1 < \mu_2$ $\mu_2 \geq \mu_3$ $\frac{c_2\mu_2 x_3(t)+c_2\mu_3 x_2(t)}{\mu_3(\mu_2-\mu_1)} < \frac{c_1 x_1(t)}{\mu_1-\lambda}$	$(1,0,1)$
Case 17	$c_1 < c_2 < c_3$ $\mu_1 < \mu_2$ $\mu_2 \geq \mu_3$ $\frac{c_2\mu_2 x_3(t)+c_2\mu_3 x_2(t)}{\mu_3(\mu_2-\mu_1)} \geq \frac{c_1 x_1(t)}{\mu_1-\lambda}$	$(0,0,1)$
Case 18	$c_1 < c_2 < c_3$ $\mu_1 \geq \mu_2 \geq \mu_3$	$(0,0,1)$
Case 19	$c_1 < c_2 < c_3$ $\mu_1 \geq \mu_2$ $\mu_2 < \mu_3$ $\frac{x_3(t)}{x_2(t)} \geq \frac{c_3\mu_2}{c_2(\mu_3-\mu_2)}$	$(0,1,1)$
Case 20	$c_1 < c_2 < c_3$ $\mu_1 \geq \mu_2$ $\mu_2 < \mu_3$ $\frac{x_3(t)}{x_2(t)} < \frac{c_3\mu_2}{c_2(\mu_3-\mu_2)}$	$(0,0,1)$
Case 21	$c_1 < c_2 < c_3$ $\mu_1 < \mu_2 < \mu_3$ $\frac{x_1(t)}{x_2(t)} \geq \frac{c_2}{c_1}\frac{\mu_1-\lambda}{\mu_2-\mu_1}$ $\frac{x_2(t)}{x_3(t)} \geq \frac{c_3}{c_2}\frac{\mu_2-\mu_1}{\mu_3-\mu_2}$	$(1,1,1)$
Case 22	$c_1 < c_2 < c_3$ $\mu_1 < \mu_2 < \mu_3$ $\frac{x_1(t)}{x_2(t)} < \frac{c_2}{c_1}\frac{\mu_1-\lambda}{\mu_2-\mu_1}$ $\frac{x_2(t)}{x_3(t)} \geq \frac{c_3}{c_2}\frac{\mu_2-\mu_1}{\mu_3-\mu_2}$	$(0,1,1)$
Case 23	$c_1 < c_2 < c_3$ $\mu_1 < \mu_2 < \mu_3$ $\frac{x_1(t)}{x_2(t)} \geq \frac{c_2}{c_1}\frac{\mu_1-\lambda}{\mu_2-\mu_1}$ $\frac{x_2(t)}{x_3(t)} < \frac{c_3}{c_2}\frac{\mu_2-\mu_1}{\mu_3-\mu_2}$	$(1,0,1)$
Case 24	$c_1 < c_2 < c_3$ $\mu_1 < \mu_2 < \mu_3$ $\frac{x_1(t)}{x_2(t)} < \frac{c_2}{c_1}\frac{\mu_1-\lambda}{\mu_2-\mu_1}$ $\frac{x_2(t)}{x_3(t)} < \frac{c_3}{c_2}\frac{\mu_2-\mu_1}{\mu_3-\mu_2}$	$(0,0,1)$

4.6. Stochastic versus deterministic models: a critical view.
In this subsection we take a critical view of the insights gained from the

maximum principle and conjecture that there is a strong connection between fluid and stochastic models.

In the previous subsections we demonstrated that the maximum principle leads to explicit forms for the optimal policy, including closed forms for the threshold curves. More importantly, although algebraically involved and combinatorially explosive (in order to describe the optimal policy in the last 3-class network we had to distinguish 24 cases), the derivation of the optimal policy in fluid models is conceptually simple, almost mechanical.

The optimal policy in fluid networks has the following structure. The parameters of the problem $(\lambda,\ c, \mu)$ are partitioned in several cases. In some of these cases the optimal policy is a static priority rule, while in some cases a dynamic priority rule (a threshold curve) needs to be followed. In addition, by its nature, the threshold curves are piecewise linear surfaces.

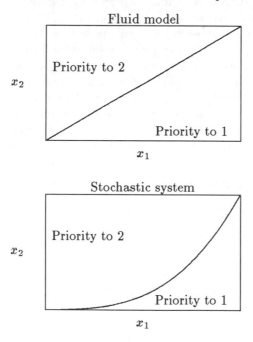

FIG. 4.6. *Conjectured relation between fluid and stochastic models.*

Naturally, a very interesting question is the relation of the optimal policy for the fluid models to the optimal policy for the stochastic queueing network models. We have seen that the optimal policy for the fluid model for the one station system is identical for the stochastic model. Notice that the policy was a static priority policy. We believe that this is an instance of a more general principle:

1) The partition of the parameter space $\lambda,\ c, \mu$ that characterizes an optimal policy for the fluid model is identical with the stochastic model.

2) Whenever the fluid model predicts that there is a static priority rule is

optimal, the same policy is optimal for the stochastic model.

3) Whenever the fluid model predicts that a linear threshold curve is optimal, the optimal policy for the stochastic model is also of the threshold type, but with a nonlinear threshold curve (see Figure 4.6).

5. A numerical approach to the general control problem. The goal of this section is to derive a numerical algorithm for Problem *(CONTROL)*, which can be used to assess the quality of heuristic solutions, by viewing Problem *(CONTROL)* as an infinite-dimensional linear program. Problem *(CONTROL)* is known in the literature as a separated continuous linear program *(SCLP)*. Anderson et. al. [1] discuss basic solutions for *(SCLP)*, while duality theorems are presented in [1], [2] and [18]. Pullan [18] presents an iterative algorithm to solve *(SCLP)*. The algorithm uses discrete approximations to find an upper bound and then generates an improvement step. As our goal was to understand the effectiveness of the learning heuristic that we present in the next section, we present in this section a numerical method to generate solutions for the problem that also uses discrete approximation, which is continuously refined until the solution no longer improves beyond a given tolerance.

Problem *(CONTROL)* can be approximated by discretizing the time interval and solving the resulting finite linear program as follows: Divide the time interval $[0, T]$ into T/h intervals of length h and only allow the controls to change value at the end of the intervals. This leads to a linear program with $2nT/h$ variables:

$(DISCRETE)$ $\min \sum_{t=0}^{T/h} c' \, x(th)$

$$x(th) = x((t-1)h) + [Au((t-1)h) + b]h \quad \text{for } t = 1 \ldots T/h$$

$$Du(th) \le e \quad \text{for } t = 0 \ldots T/h - 1$$

$$(th) \ge 0 \quad \text{for } t = 1 \ldots T/h$$

$$u(th) \ge 0 \quad \text{for } t = 0 \ldots T/h - 1$$

$$x(0) = \tilde{x}$$

Pullan [18] has shown that the optimal solution of *(DISCRETE)* can be used to derive a feasible solution to *(CONTROL)* and that the smaller the stepsize, h, the better the approximation will be. Of course, decreasing h has a sharp tradeoff in problem size. Our implementation of the algorithm begins with a moderate stepsize and gets an initial solution. It then keeps dividing the stepsize in half and solving the new problem. The process continues until the optimal objective value decreases less than 5% of that of the previous iteration.

Given the optimal solution with stepsize h, a feasible solution to the problem with stepsize $h/2$ is generated as follows:

$$x(t + h/2) = x(t) + [Au(t) + b]h/2$$

$$u(t + h/2) = u(t).$$

In this way the simplex method can start the new problem with a feasible solution, although not necessarily basic.

In order to obtain insight on the optimal policy from the numerical results (in particular generate threshold curves), we found the optimal policy for multiple values of $x(0)$ (notice that the same policy will be followed at time t if $x(t) = x(0)$). As an example, we solved the two queue tandem problem numerically for multiple values of $x(0)$. The solution given by the LP is compared with the threshold found in Theorem 2 in Figure 5.1. The numerical approach correctly reproduces the threshold curve of the optimal policy that we have derived analytically in the previous section.

$$c = (3,5), \lambda = 1, \mu = (2,3)$$

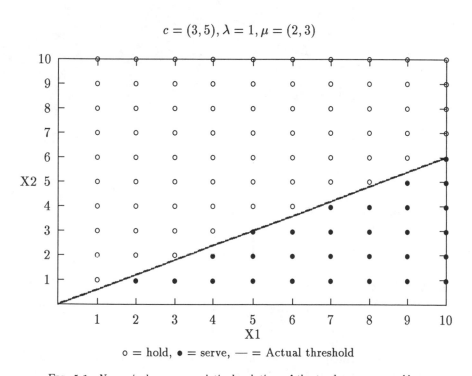

o = hold, • = serve, — = Actual threshold

FIG. 5.1. *Numerical versus analytical solution of the tandem queue problem.*

6. A learning heuristic that captures interactions among classes.
Given the complicated character of the optimal policy in the examples we have examined and given the dimensionality difficulties that the exact approach of the previous section has, it would be desirable to design an efficient approximate algorithm for the problem. This section develops such an algorithm.

We rewrite the dynamics of the system $\dot{x}(t) = Au(t) + b$ as (A is invertible)

(6.1) $$u(t) = A^{-1}(\dot{x}(t) - b).$$

Problem *(CONTROL)* becomes

$$\min \int_0^T c'\, x(t)dt$$

$$
\begin{aligned}
DA^{-1}(\dot{x}(t) - b) &\leq e \\
A^{-1}(\dot{x}(t) - b) &\geq 0 \\
x(0) &= \tilde{x} \\
x(t) &\geq 0
\end{aligned}
$$

which can be written compactly as

$$\min \int_0^T c'\, x(t)dt$$

$$
\begin{aligned}
\bar{A}\dot{x}(t) &\leq \bar{b} \\
x(0) &= \tilde{x} \\
x(t) &\geq 0
\end{aligned}
$$

where $\bar{A}$ is the $(m+n) \times n$ matrix

$$\bar{A} = \begin{bmatrix} DA^{-1} \\ -A^{-1} \end{bmatrix}$$

and $\bar{b}$ is the $m+n$ vector

$$\bar{b} = \begin{bmatrix} e + DA^{-1}b \\ -A^{-1}b \end{bmatrix}.$$

The maximum principle implies that while $x(t) > 0$, the optimal control will be an extreme point of $\{u : Du(t) \leq e, u \geq 0\}$. By definition, for any extreme point z of $\bar{A}\dot{x} \leq \bar{b}$ there is a cost w such that z is the unique optimal solution of

(6.2) $$\begin{aligned} \min w'\, \dot{x} \\ \bar{A}\dot{x} &\leq \bar{b} \end{aligned}$$

Our goal is to construct such a w from our knowledge of the optimal policy in smaller parts of the network. We will construct an approximate

w as follows:

We select subnetworks for which we know the optimal policy. Assume the subnetwork involves classes $i \in S$. When we consider the subnetwork in isolation, we will know the optimal policy and hence the optimal extreme point of $\bar{A}\dot{x} \leq \bar{b}$ and therefore, we can construct an optimal $w_i = \alpha_i$, $i \in S$. We will then try to produce a global w, which when restricted to the set of classes S is proportional to the vector α. In other words, we ask that w satisfies

$$\frac{w_i}{w_j} = \frac{\alpha_i}{\alpha_j},$$

which can be rewritten as $\alpha_j w_i - \alpha_i w_j = 0$, for any pair of interacting classes in the subnetwork S. In this way we form a linear system

$$\begin{aligned} Mw &= 0 \\ w'e &= 1, \end{aligned}$$

which can be potentially overdetermined. In such a case the actual w that is used is the least squares solution of the above system.

Overall, our strategy for obtaining w is to take into account limited interactions among various classes in the network and then use the exact solution we have obtained through the maximum principle. The level of interactions that are taken into account effect both the performance and the complexity of the heuristic. We will first examine the easiest case that only looks at pairwise interactions.

A pairwise interaction heuristic

In a general network, two classes i and j can interact in one of three ways:

   1. compete in the same station;

   2. be in tandem in consecutive stations;

   3. not be involved in the same or consecutive stations.

Having solved the problems of competing classes and two classes in tandem exactly in Sections 4.1 and 4.2 respectively, we can calculate the corresponding w, so that the reformulation (6.2) gives the exact optimal policy for each of these two class problems as follows: for two classes competing at the same node,

$$w = c$$

and for two classes in tandem,

$$w = \begin{cases} c & \text{if } \mu_1 > \mu_2 \\ \left(c_1, \min\left(\frac{x_2}{x_1} \frac{\mu_1 - \lambda}{\mu_2 - \mu_1} c_2, c_2 \right) \right) & \text{otherwise.} \end{cases}$$

When the pairwise interaction heuristic is applied to acyclic feed forward networks, the linear system will not be overdetermined and the heuristic yields a generalized $c\mu$ rule. Each station will serve the class with the highest index

$$k_i = \begin{cases} \mu_i[w_i - w_{n(i)}] & \text{if class } i \text{ has a next class} \\ \mu_i w_i & \text{otherwise} \end{cases}$$

where $w_i = c_i$ and $w_{n(i)} = \min\left(\frac{x_2}{x_1}\frac{\mu_1 - \lambda}{\mu_2 - \mu_1}c_2, c_2\right)$.

A three-way interaction heuristic

The power of the heuristic can be extended by considering three-way interactions. Three classes can compete in the same station, they can be involved in two stations with two classes competing in the first (Section 4.3) or in the second station (Section 4.4), or finally they can be involved in three stations in tandem (Section 4.5). Each relationship that involves the three classes i, j and k contributes two ratios w_i/w_j and w_j/w_k, or two rows, to the matrix M. The heuristic has been implemented as both a pairwise interaction heuristic and a three-way interaction heuristic.

6.1. An example. Having specified the building blocks of the pairwise interaction heuristic, we demonstrate its effectiveness by showing that the pairwise interaction heuristic can be used to solve the the 3-class network of Section 4.3 (see Figure 2.1). The mathematical formulation of the problem is presented in Section 4.3. We will first analyze the case $\mu_1 < \mu_2$. Classes one and three compete at station one; so the first row of the matrix M is

$$\frac{w_1}{w_3} = \frac{c_1\mu_1}{c_3\mu_3}.$$

Classes one and two are in tandem, so the second row is

$$\frac{w_1}{w_2} = \frac{(c_2 - c_1)\mu_1}{-c_2\mu_2}$$

if $\frac{x_2}{\mu_2 - \mu_1} > \frac{x_1}{\mu_1 - \lambda}$ and

$$\frac{w_1}{w_2} = \frac{[c_2 x_2(\mu_1 - \lambda) - c_1 x_1(\mu_2 - \mu_1)]\mu_1}{-c_2 x_2(\mu_1 - \lambda)\mu_2}$$

otherwise. Classes two and three are not related, so that pair does not contribute to the system of equations. If $x_2(\mu_1 - \lambda)/x_1(\mu_2 - \mu_1) < 1$, then w would be

$$w = ([c_2 x_2(\mu_1 - \lambda) - c_1 x_1(\mu_2 - \mu_1)]c_1\mu_1, -c_2 x_2(\mu_1 - \lambda)c_1\mu_2, [c_2 x_2(\mu_1 - \lambda) - c_1 x_1(\mu_2 - \mu_1)]c_3\mu_3)'.$$

Note that w can be scaled so $w'e = 1$ does not need to be enforced – it is only used to ensure a nontrivial solution for w. As mentioned in Section 4.3, $u_2(0) = 1$ and $u_3(0) = 1 - u_1(0)$ so the only extreme points to be considered correspond to $u = (1,1,0)$ and $u = (0,1,1)$. Direct calculation reveals that the control (1,1,0) will be chosen if

$$c_2 x_2 (\mu_1 - \lambda_1) \mu_1 < (c_1 \mu_1 - c_3 \mu_3)(\mu_2 - \mu_1) \mu_1 x_1$$

or

$$\frac{\mu_1 - \lambda}{\mu_2 - \mu_1} \frac{c_2 \mu_1}{c_1 \mu_1 - c_3 \mu_3} < \frac{x_1}{x_2}$$

if $c_1 \mu_1 > c_3 \mu_3$. If $c_1 \mu_1 < c_3 \mu_3$, the above condition will never be satisfied. This, along with the initial assumption $\mu_1 < \mu_2$ agree with Cases 2 and 4 of Theorem 3. If $x_2(\mu_1 - \lambda)/x_1(\mu_2 - \mu_1) > 1$, then

$$w = ((c_2 - c_1)c_1\mu_1, -c_2 c_1 \mu_2, (c_2 - c_1)c_3\mu_3)'.$$

By computing the objective function values, it is easy to show that control (0,1,1) will be used when $c_1 \mu_1 - c_3 \mu_3 < c_2 \mu_1$ (Case 5 of Theorem 3). This condition will always be satisfied when $c_3 \mu_3 > c_1 \mu_1$ so the control will always be (0,1,1) in this case (Case 1 of Theorem 3). When $\mu_1 > \mu_2$, then

$$w = ((c_2 - c_1)c_1\mu_1, -c_2 c_1 \mu_2, (c_2 - c_1)c_3\mu_3)'.$$

once again and the result agrees with (Case 2 of Theorem 3).

6.2. Implementation of the heuristic. Both the pairwise and three-way interaction heuristic are implemented in C using the symbolic manipulation program Maple. The user inputs the topology of a network, and specifies the parameters λ, μ and c symbolically. The algorithm operates in two modes: numerical or symbolic to either generate a policy for a given instance or to investigate the qualitative behavior of the problem. For every pair (pairwise) or collection of three interacting classes (three way) the algorithm derives rows of the matrix M that will be used to compute w with $Mw = 0$, $w' e = 1$. Some rows of M are only used if the parameters satisfy certain conditions.

A Maple program is then generated that computes, for each set of conditions, the algebraic solution for w, as well as the extreme points of $A\dot{x}(t) \leq \bar{b}$. This Maple program creates a LaTeX file that prints, for each set of conditions, the conditions in effect, the objective value for each extreme point and the control value associated with that extreme point. This information can then be used to derive the complete policy; after determining which conditions are being used, the correct policy is the one with the lowest objective value.

As an example, the output for two queues in tandem with competition in the front (processed with the pairwise interaction heuristic) is shown in Figures 6.1-6.3. It is easy to verify that the output agrees with the policy in Theorem 3.

Case 1
Conditions

$$\frac{\mu_0}{\mu_1} < 1$$

$$\frac{x_1 (\mu_0 - \lambda_0)}{x_0 (\mu_1 - \mu_0)} < 1$$

Objective Values (use minimum)
   Control (1 1 0)

$$\frac{-c_0 x_0 \mu_0 \mu_1 + c_0 x_0 \mu_0^2 + x_1 c_1 \mu_0^2 - x_1 c_1 \mu_0 \lambda_0}{c_2 \mu_1 x_0 - c_2 x_0 \mu_0 + \mu_1 c_0 x_0 - c_0 x_0 \mu_0 + x_1 c_1 \mu_0 - x_1 c_1 \lambda_0}$$

   Control (0 1 1)

$$\frac{-c_2 x_0 \mu_1 \mu_2 + c_2 x_0 \mu_0 \mu_2}{c_2 \mu_1 x_0 - c_2 x_0 \mu_0 + \mu_1 c_0 x_0 - c_0 x_0 \mu_0 + x_1 c_1 \mu_0 - x_1 c_1 \lambda_0}$$

FIG. 6.1. *Case 1 of heuristic output.*

A shortcoming of the current implementation is that the expressions for the objective function often require "manual" simplification. It is important to note, however, that the policy generation is completely automated and that the user is left with a Maple program that can be used as a base for further simplification that is more easily undertaken by a human. Under the numerical mode, the intermediate file is read along with instance data that specifies the numerical parameters for the network. This information is used to determine which conditions are in effect and then the exact M matrix is calculated. The value of w is then found numerically and the linear program is solved to produce the policy to be used for this instance of the network. Finally the heuristic policy is simulated to produce an estimate of the cost. In this case the software package outputs a policy and its cost automatically without user intervention. We have already seen that the pairwise interaction heuristic gives the correct optimal policy in one example. Naturally we would like to compare the performance of the heuristic with the optimal policy (found using the discretization technique of Section 5). The next subsection compares the three-way interaction heuristic with the optimal solution.

6.3. Performance of the heuristic. To test the performance of the heuristic several problems were processed both symbolically and numerically and the numerical results were compared with the optimal solution from the discrete approximation. Besides finding the heuristic policy as a

Case 2
Conditions

$$\frac{\mu_0}{\mu_1} < 1$$

$$\frac{x_0 (\mu_1 - \mu_0)}{x_1 (\mu_0 - \lambda_0)} < 1$$

Objective Values (use minimum)
Control (1 1 0)

$$-c_0 \mu_0 + c_1 \mu_0$$

Control (0 1 1)

$$-c_2 \mu_2$$

FIG. 6.2. *Case 2 of heuristic output.*

Case 3
Conditions

$$\frac{\mu_1}{\mu_0} < 1$$

Objective Values (use minimum)
Control (1 1 0)

$$-c_0 \mu_0 + c_1 \mu_0$$

Control (0 1 1)

$$-c_2 \mu_2$$

FIG. 6.3. *Case 3 of heuristic output.*

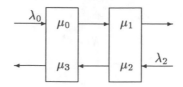

$\lambda_0 = 1, \lambda_2 = 2, \mu = (4, 8, 3, 4), c = (4, 3, 2, 3), x_0 = 4, x_3 = 2$

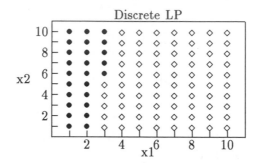

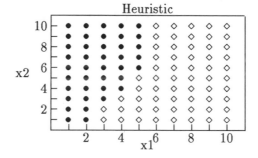

$\bullet = (1,1,0,0) \diamond = (0,1,0,1)$

FIG. 6.4. *Example with two competing tandem queues.*

function of the state, the heuristic was run in simulation mode (using the same step-size as that used in the discrete LP). The total cost of emptying the network under each policy is then compared. We present in this section the results of some sample networks.

A 4 class-2 station network

We first consider the network in Figure 6.4 that has four classes and two stations operating as two tandem queues. Class zero jobs arrive at the first station with rate λ_0 and then become class one jobs at the second station before leaving the system. Similarly, class two jobs arrive at the second station with rate λ_2 and become class three jobs at the first station before leaving the system. The parameters for λ, μ and c are shown in the figure. The values of x_0 and x_3 were fixed at 4 and 2 respectively while

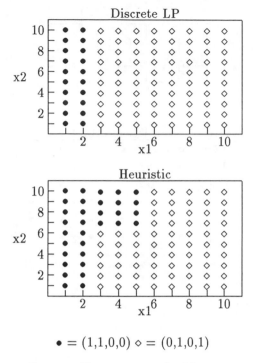

$\bullet = (1,1,0,0) \diamond = (0,1,0,1)$

FIG. 6.5. *The same network with* $x_3 = 7$.

x_1 and x_2 varied from 1 to 10. In order to obtain insight to the difference of the optimal policy and heuristic policy we have calculated both policies for all combinations of initial conditions for x_1 and x_2. We observe that the heuristic policy correctly identifies the existence of a threshold curve. The calculation of the slope of the curve is not exact but is very close to the optimal one. The average difference between the cost to empty the system for the heuristic and the discrete LP was 4.22% using the three-way interaction heuristic. The same network was also processed when x_3 was fixed at 7. Once again, a threshold curve was identified, but it does not match exactly the optimal threshold. The difference in total cost to empty the network was much closer however, the three-way interaction heuristic had an average difference of 1.58%.

A 4 class-3 station network

The next network tested is shown in Figure 6.6. Class zero jobs arrive at station one with rate λ_0 and then become class two jobs at the second station before leaving. Class one jobs arrive at the first station with rate λ_1 and are then processed at the third station before leaving. The values of x_2 and x_3 were fixed at 2 and 6 respectively while x_0 and x_1 took on the values $2, 4, \ldots, 10$. The control for classes two and three will always be to serve these classes, so the only decision is which class to process at

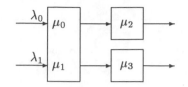

$$\lambda_0 = 1, \lambda_1 = 1, \mu = (4, 2, 6, 3), c = (3, 2, 4, 3)$$

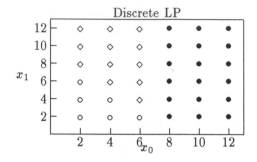

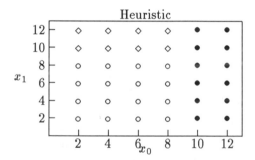

$$\circ = (0,0,1,1) \quad \bullet = (1,0,1,1) \quad \diamond = (0,1,1,1)$$

FIG. 6.6. *Results for the fork network.*

the first station. The control policy is shown in the graphs. As before the heuristic is very close to the optimal policy. The average difference in cost was 2.51% with the three-way heuristic.

Reentrant lines

We next consider the reentrant network shown in Figure 6.7. The values of x_0, x_1, x_3 and x_4 were fixed at 2,4,3 and 5 respectively. The values of x_2 and x_5 were in the set $2, 4, \ldots, 10$. In this example, the control chosen by the heuristic and the optimal control was always $u(0) = (0, 0, 0, 1, 1, 1)$.

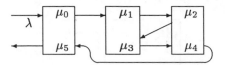

$$\lambda = 1, \mu = (5, 3, 4, 3, 4, 3), c = (4, 5, 6, 7, 6, 4)$$

$$x_0 = 2, x_1 = 4, x_3 = 3, x_4 = 5$$

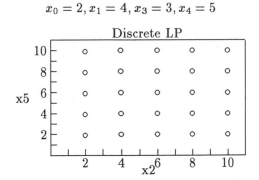

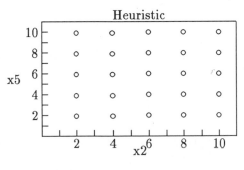

$$\circ = (0,0,0,1,1,1)$$

FIG. 6.7. *Results for a reentrant network.*

As a final example, consider the network in Figure 6.8 that was studied by Perkins and Kumar [16]. Because of the difficulty to solve the discrete LP to optimality only two instances of this network were processed. In the first instance, the parameters were

$$\lambda = 1, \mu = (2, 3, 2, 3, 2, 5, 4, 5, 4)$$

$$c = (2, 3, 3, 2, 4, 3, 4, 5, 3), x = (5, 6, 4, 7, 6, 5, 8, 5, 2)$$

and the control $u(0)$ was

$$u(0) = (0, 0, 0, 0, 0, 0, 0, 1, 1)$$

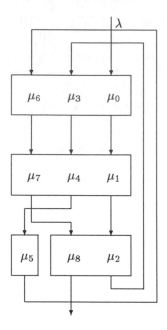

FIG. 6.8. *Network from Kumar and Perkins.*

for both the discrete LP and the heuristic. In the second instance, the parameters were

$$\lambda = 1, \mu = (2, 3, 2, 3, 2, 5, 4, 5, 4)$$

$$c = (2, 3, 6, 2, 10, 3, 4, 5, 3), x = (5, 6, 6, 7, 6, 5, 8, 5, 2)$$

Once again $u(0)$ was the same for both the heuristic and the LP; the value was

$$u(0) = (0, 0, 0, 0, 1, 0, 0, 0, 1).$$

7. Concluding remarks and open problems. We presented near optimal and heuristic approaches for a fluid model of a multiclass queueing network that offers, we believe, insights to the optimal policy (dynamic indices and threshold curves).

While the discretized linear programming algorithm solves the problem close to optimally, it has dimensionality problems so that we can solve problems with a very small number of stations. Obviously deriving an efficient optimal algorithm for the problem is a challenging open problem.

We proposed a hierarchy of heuristics that capture interactions between various classes and is able to obtain solutions for networks with a large number of classes of the type encountered in applications very efficiently. We have seen that the three-way interaction heuristic produces near optimal policies. It would be desirable to determine what the benefits of even higher interactions would be.

Finally, perhaps the most challenging problem is to determine the relation of fluid and stochastic networks. We have already discussed our belief that qualitatively the optimal policies for fluid and stochastic networks are the same. Refining our understanding of their relation, is a major research question.

Acknowledgements

We thank Arie Leizerovitch for useful discussions, Michael Harrison for insightful comments and an anonymous reviewer for constructive remarks.

REFERENCES

[1] E.J. ANDERSON, P. NASH, AND A.F. PEROLD (1984). *Some properties of a class of continuous linear programs*, SIAM J. Control and Optimization, 21, 758–765.

[2] E.J. ANDERSON AND P. NASH (1987). *Linear Programming in Infinite-Dimensional Spaces*, John Wiley & Sons, New York.

[3] D. ATKINS AND H. CHEN (1994). *Dynamic Scheduling Control for a Network of Queues*, The Proceedings of the 12th World Congress International Federation of Automatic Control, Sydney.

[4] D. BERTSIMAS, I. PASCHALIDIS, AND J. TSITSIKLIS (1994). *Optimization of Multiclass Queueing Networks: Polyhedral and Nonlinear Characterizations of Achievable Performance*, Annals of Applied Probability, 4(1), 43–75.

[5] A.E. BRYSON, JR., AND Y.C. HO (1969). *Applied Optimal Control*, Blaisdell, Waltham, Massachusetts.

[6] H. CHEN AND D. YAO (1989). *Optimal Scheduling of a Multi-Class Fluid Network*, working paper.

[7] J. FILIPIAK (1988). *Modelling and Control of Dynamic Flows in Communication Networks*, Springer–Verlag, Berlin.

[8] J.M. HARRISON (1986). *Brownian Models of Queueing Networks with Heterogeneous Customers*, Proceedings of IMA Workshop on Stochastic Differential Systems.

[9] J.M. HARRISON AND L.M. WEIN (1990). *Scheduling Networks of Queues; Heavy Traffic Analysis of a Two-Station Closed Network*, Operations Research, 38 (2), 1052–1064.

[10] J.M. HARRISON AND L.M. WEIN (1989). *Scheduling Networks of Queues; Heavy Traffic of a Simple Open Network*, Queueing Systems Theory and Applications, 5, 265–280.

[11] M. KAMIEN AND N. SCHWARTZ (1991). *Dynamic Optimization: The Calculus of Variations and Optimal Control in Economics and Management*, North-Holland, Amsterdam.

[12] F.P. KELLY AND C.N. LAWS (1993). *Dynamic routing in open queueing networks: Brownian models, cut constraints and resource pooling*, Queueing Systems 13, 47–86.

[13] P.R. KUMAR (1994). *Re-entrant lines, Queueing Systems and Applications*.

[14] I.H. MUFTI (1970). *Computational Methods in Optimal Control Problems*, Lecture Notes in Operations Research and Mathematical Systems, Springer–Verlag, Berlin.

[15] J. OU AND L.M. WEIN (1992). *Performance Bounds for Scheduling Queueing Networks*, The Annals of Applied Probability, 2 (2), 460–480.

[16] J.R. PERKINS AND P.R. KUMAR (1994). *Optimal Control of Pull Manufacturing Systems*, University of Illinois at Urbana, working paper.
[17] L.S. PONTRYAGIN, V.G. BOLTYANSKII, R.V. GAMKRELIDZE, AND E.F. MISHCHENKO (1962). *The Mathematical Theory of Optimal Processes*, Interscience Publishers, New York.
[18] M.C. PULLAN (1993). *An Algorithm for a Class of Continuous Linear Programs*, SIAM J. Control and Optimization, **31** (**6**), 1558–1577.
[19] A. SEGALL (1977). *Modelling of Adaptive Routing in Data Communication Networks*, IEEE Transactions of Comm., **25**, 85–95.
[20] A. SEGALL AND F. MOSS (1982). *An Optimal Control Approach to Dynamic Routing in Networks*, IEEE Transactions of Aut. Cont., **27**, 329–339.
[21] A. SEIERSTAD AND K. SYDSÆTER (1977). *Sufficient Conditions in Optimal Control Theory*, International Economic Review, **18** (**2**), 367–391.
[22] L.M. WEIN (1990). *Optimal Control of a Two-Station Brownian Network*, Mathematics of Operations Research, **15**, 215–242.
[23] L.M. WEIN (1992). *Scheduling Networks of Queues; Heavy Traffic Analysis of a Two Station Network with Controllable Inputs*, Operations Research, **40**, S312–S334.

CONVERGENCE OF DEPARTURES FROM AN INFINITE SEQUENCE OF QUEUES

T.S. MOUNTFORD* AND B. PRABHAKAR[†]

In this note we wish to discuss our recent paper (Mountford and Prabhakar [9]) which was presented at the workshop by the second author.

Given a stationary and ergodic arrival process $\mathbf{A}$ (representing customers) of rate $\alpha < 1$, arriving at an independent rate one exponential-server queue, one obtains the departure process $\mathbf{A}^1$ via Loynes' construction (see Loynes [8]). The process $\mathbf{A}^1$ is also stationary and ergodic of the same rate α. If this departure process is now regarded as an arrival process and fed into a second, independent, rate one exponential-server queue, one obtains in like manner a new departure process $\mathbf{A}^2$ which in turn may be regarded as an arrival process and fed into another exponential-server queue. Continuing thus, we obtain a sequence of point processes of customers $\mathbf{A}^n$, each stationary and ergodic of rate α.

The paper of Mountford and Prabhakar [9] proves

Theorem: The point processes $\mathbf{A}^n$ converge in distribution as n tends to infinity to the Poisson process of rate α.

This result answers affirmatively the unpublished Reiman-Simon [10] conjecture. The result stands in contrast to the result of Suhov [13] (see also Suhov and Vvedenskaya [12]) which showed that if the service times of each individual customer were fixed for all the queues and these service times were distributed among the customers as i.i.d. exponential mean one random variables, then the Palm measure for the $\mathbf{A}^n$ process converges to that of a rate one Poisson process. A similar result is obtained by Vere-Jones [15] who showed that if a stationary and ergodic arrival process of rate α is fed through a series of independent $\cdot/GI/\infty$ queues then the departures from the nth queue tend in distribution to a rate α Poisson process.

The paper is a strengthening of the following recent result of Anantharam [1]

Theorem: If a stationary and ergodic arrival process $\mathbf{A}$ (of rate < 1) is served by an independent rate one exponential server and the departure process $\mathbf{A}^1$ is equal to $\mathbf{A}$ in distribution then $\mathbf{A}$ must be Poisson.

Anantharam's proof introduced a natural metric between ergodic arrival processes of a fixed rate $\alpha < 1$. Then clever coupling arguments were employed to show that the transformation $\mathbf{A} \to \mathbf{A}^1$ regarded as a mapping of distributions, was a contraction. This implied the desired uniqueness.

---

* Department of Mathematics, University of California at Los Angeles, Los Angeles, CA 90024. Research supported by a grant from the Sloan Foundation.
† Department of Electrical Engineering, University of California at Los Angeles, Los Angeles, CA 90024.

Unfortunately point processes which are alternations of very long "bursts" of Poisson processes of rate β_1 and Poisson processes of rate β_2 $(\beta_1 \neq \beta_2)$, i.e. the so-called Markov-modulated Poisson processes, are almost preserved by the transformation. Therefore the transformation is not a strict contraction and our result does not follow immediately from this approach.

Our approach follows that of Liggett and Shiga [7] who recognized that the evolution of $\mathbf{A}^n$ as n varies had the common feature of mass preservation with exclusion processes as they varied over time. They saw that many of the classical exclusion process arguments of Liggett [5] (particularly the coupling constructions) could be adapted to yield information about the processes $\mathbf{A}^n$. That there was a connection between the asymmetric simple exclusion process and infinite series of queues had been long known; see, for example, Liggett [6], chapter eight, Kipnis [4] and recently Srinivasan [11]. But, in the above cases, an explicit isomorphism between the infinite queue process over time and the exclusion process over time (that is, an isomorphism between the process representing the evolution of the *whole* infinite sequence of queues over time, and the exclusion process over time) is used. Liggett and Shiga [7] were the first to see that one could adapt arguments developed for the exclusion process to understand the evolution of $\mathbf{A}^n$ as n, the queue number, varied. The original sequence of processes $\mathbf{A}^n$ was coupled with Poisson processes $\mathbf{P}_\gamma^n$ whose density γ varied over $(0, 1)$. One was able to conclude as in Liggett [6], chapter eight, that any limit measure $(\mathbf{A}, \mathbf{P})$ of the sequence $(\mathbf{A}^n, \mathbf{P}_\gamma^n)$ must be such that, pathwise, either all points of $\mathbf{A}$ are in $\mathbf{P}$ or all points of $\mathbf{P}$ are in $\mathbf{A}$. That is a.s. one of the two processes must dominate the other. From this (again as in Liggett [6], chapter eight) one could conclude that the limit of $\mathbf{A}^n$ existed and was a mixture of Poisson processes. This is the situation with general (non-symmetric) exclusion processes whose initial distribution is ergodic, stationary, apart from the nearest neighbour case where Andjel [2] proved convergence to a product measure.

The problem essentially was to rule out the possibility that the processes $\mathbf{A}^n$ could converge to a non-trivial mixture of Poisson processes in distribution. It is in addressing this point that the clever argument of Ekhaus and Gray [3] is invaluable. This argument is developed for the "marching soldiers" problem. For this problem if one looks at the underlying "slope process" then one has the important conservation of mass property common to exclusion processes and to our queuing problem. This argument is applied to the coupling of Liggett and Shiga [7] for $\mathbf{A}^n$ and $\mathbf{P}^n$ where the $\mathbf{P}^n$ are Poisson processes of density equal to that of the $\mathbf{A}^n$. The argument of Ekhaus and Gray [3] extends to show that every customer of $\mathbf{A}$ eventually coalesces with a customer of $\mathbf{P}$ and vice-versa. This gives the desired result.

While it was always natural to believe that any limit of the $\mathbf{A}^n$'s would be ergodic, Suresh and Whitt [14] describe long term subtle effects arising from feeding an arrival process through a series of queues. Consider feeding the arrival process to the $N + 1^{th}$ queue, $\mathbf{A}^N$, to an arbitrary network of queues. Given the weak convergence of the sequence of processes $\mathbf{A}^n$ to a Poisson process, an issue of practical importance is the size of the error involved in approximating $\mathbf{A}^N$ by a Poisson process of rate α for purposes of calculating quantities like average congestion, mean waiting, etc., at the various nodes of the network. Indeed, such considerations led to the original Reiman and Simon conjecture [10].

The paper of Suresh and Whitt [14] describes a simulation experiment which shows that such an approximation can be quite bad if there are "bottleneck queues". Essentially, as Suresh and Whitt [14] point out, this is due to the following reason. Although the process $\mathbf{A}^N$ may look locally Poisson (because weak convergence of point processes is the convergence of finite-dimensional distributions), congestion measures (like mean waiting, mean queue-size) at bottleneck queues depend on long-term statistics of the arrival process to the queue. Thus, one needs to be careful when approximating $\mathbf{A}^N$ by a Poisson process.

REFERENCES

[1] V. ANANTHARAM, Uniqueness of stationary ergodic fixed point for a ·/M/k node. *Annals of Applied Probability*, **3** (1) (1993), pp. 154–173.

[2] E. ANDJEL, The asymmetric simple exclusion process on Z^d. *Z. Wahrsch. Verw. Gebiete*, **58** (1981), pp. 423–432.

[3] M. EKHAUS AND L. GRAY, A Strong Law for the Motion of Interfaces in Particle Systems. *In preparation* (1993).

[4] C. KIPNIS, Central Limit Theorems for Infinite Series of Queues and Applications to Simple Exclusion. *Annals of Probability*, **14** (1985), pp. 397–408.

[5] T. M. LIGGETT, Coupling the simple exclusion process. *Annals of Probability*, **4** (1976), pp. 339–356.

[6] T. M. LIGGETT, *Interacting Particle Systems*. Springer-Verlag, New York (1985).

[7] T. M. LIGGETT AND T. SHIGA, A Note on the Departure Process of an Infinite Series of ·/M/1 Nodes. *Unpublished manuscript*.

[8] R. M. LOYNES, The stability of a queue with non-independent inter-arrival and service times. *Proc. Camb. Philos. Soc.*, **58** (1962), pp. 497–520.

[9] T. S. MOUNTFORD AND B. PRABHAKAR, On the weak convergence of departures from an infinite sequence of ·/M/1 queues. To appear in *Annals of Applied Probability* (1993).

[10] M. REIMAN AND B. SIMON, Private Communication (1981).

[11] R. SRINIVASAN, Queues in Series via Interacting Particle Systems. *Mathematics of Operation Research*, **18** (1993), pp. 39–50.

[12] Y. SUHOV AND N. D. VVEDENSKAYA, The limiting departure flow of an infinite series of queues. *Preprint* (1993).

[13] Y. SUHOV, The limiting departure flow of an infinite series of queues, II. *Preprint* (1993).

[14] S. SURESH AND W. WHITT, The Heavy-traffic Bottleneck Phenomenon in Open Queueing Networks, *Operations Research Letters*, **9** (1990), pp. 355–362.

[15] D. VERE-JONES, Some applications of probability generating functionals to the study of input-output streams. *Journal of the Royal Statistical Society, Series B*, **30** (1968), pp. 321–333.

STATE-DEPENDENT QUEUES: APPROXIMATIONS AND APPLICATIONS

AVI MANDELBAUM* AND GENNADY PATS*

Abstract. A state-dependent queue is an exponential service system, where arrival and service rates depend on queue length. For properly normalized queueing processes, we derive functional strong laws of large numbers and functional central limit theorems. The former support fluid approximations and the latter diffusion refinements. Our analysis is based on strong approximations, which provide a unified framework for most existing approximations of state-dependent queues.

Key words. State–Dependent Queues, Strong Approximations, Fluid and Diffusion Approximations, Many–Server and Finite–Populations Queues.

1. Introduction. State–dependent exponential $M_\xi/M_\xi/1$ queues are models in which arrival and service rates depend on the state ξ—the queue length. For properly normalized queueing processes, we derive functional strong law of large numbers (FSLLN, Theorem 4.1) and functional central limit theorems (FCLT, Theorems 4.2 and 4.3). The former support fluid approximations and the latter diffusion refinements. The current analysis is a first step in an ongoing effort to cover queueing networks.

The strong limit in FSLLN (henceforth called the *fluid limit*) is the unique solution to an autonomous first–order ordinary differential equation with reflection. In such an equation, the derivative depends explicitly only on state. Consequently, the fluid limit of a single queue is a monotone continuous function, which absorbs at zero if it ever reaches it.

The weak limit given in FCLT (henceforth called the *diffusion limit*) is the unique strong solution to a stochastic differential equation with a certain type of reflection. The diffusion limits are Markov processes with upper semi-continuous sample-paths. Weak convergence is with respect to Skorokhod's M_1-*topology* (see Appendix B and the discussion in Subsection 4.5).

Our technique for obtaining limit theorems is based on strong approximations. It is similar to Kurtz [31], who considers density-dependent population processes, for which limits do not involve the reflection phenomenon (see also Ethier and Kurtz [14, Chapter 11 §2,3]). It differs from Kurtz [32],[33] and [34], that relies on multiparameter time transformations.

In Section 5, derivations of many available fluid and diffusion approximations for state-dependent queues are unified. Examples covered are models with reneging, finite population and finite or infinite number of servers. We are not assuming boundedness of arrival rates, service rates or

---

* Technion Institute, Haifa, 32000 Israel. Avi Mandelbaum was partially supported by the Fund for the Promotion of Research at the Technion.

populations, at the expense of some additional technicalities in our proofs.

Pioneering works on fluid and diffusion approximations for queues are Oliver and Samuel [39], Newell (e.g., [38]), Kingman (e.g., [24],[25]), Borovkov [5],[6], and Iglehart and Whitt [20],[21]. For later advances, readers are referred to the following survey papers and references therein: Whitt [46], covering the period up to 1974, Lemoine [35], up to 1978, Coffman and Reiman [12], through 1984, Glynn [15], through 1990 and finally Chen and Mandelbaum [9],[10], up to 1992. Related recent research is Anulova [2] and Krichagina [30], who take a martingale approach to cover, as far as the single queue is concerned, special cases of our models. Additional representative examples of martingale-based fluid and diffusion approximations are Kogan *et al.* [28] and Kogan and Liptser [27], where certain types of closed exponential networks with state–dependent service are treated. Fluid approximations for state- and time-dependent queueing networks are described in [9]. Our analysis resembles Mandelbaum and Massey [37], who establish strong approximations, FSLLN and FCLT for the time-dependent $M_t/M_t/1$ system. The similarity is mainly a consequence of the fact that, in both models, diffusion approximations enjoy time-dependent drifts and variances (see (4.9)).

We use the model of an $M_\xi/M_\xi/1$ queue with the so-called autonomous server (Borovkov [5]). This means, roughly speaking, that the server is working permanently while actual departures are generated only when the system is not empty. (For an additional discussion see, e.g., Iglehart and Whitt [20]). Our mathematical formulation (equations (2.1)–(2.4)) are as in Prabhu [41] and Bremaud [7], but we focus on approximations rather than exact analysis. Fluid and diffusion approximations for state-*independent* systems have been commonly analyzed within the framework of "non-autonomous" server models (see the survey papers mentioned above). Difficulties, however, arise even in the mere interpretation of state-*dependent* non-autonomous queues, so this will not be pursued any further.

The $M_\xi/M_\xi/1$ queue is, in fact, a one-dimensional birth and death process (see Subsection 2.2). As such, it has been amply covered and a broad spectrum of (mainly elementary) tools is available and sufficient for its analysis. Here, however, we are concerned with transient evolution, and this seems challenging enough to deserve the analysis that follows. Also provided is a framework for most existing approximations of state-dependent queues, including stationary distributions when they exist. (Extensions to queueing networks, namely multi-dimensional birth and death processes, are currently being developed.)

The remainder of the paper is organized as follows. In Section 2 we present our model of the $M_\xi/M_\xi/1$ queue and discuss different representations of its queueing process. Section 3 deals with reflection maps, that characterize subsequent fluid and diffusion limits. In Section 4 we outline FSLLN and FCLT. Section 5 is devoted to applications of our results. Proofs of the main theorems are provided in Sections 6 and 7. In sec-

tion 8 we outline directions for future research. Technical background on
Skorokhod's reflection problem and on M_1-convergence is presented in Ap-
pendices A and B, while the main notation is summarized in Appendix C.

2. The model of the $M_\xi/M_\xi/1$ queue. The subject of our study
is the state-dependent $M_\xi/M_\xi/1$ queue. We analyze its *queueing process*
$Q = \{Q_t, t \geq 0\}$, whose value at time t, Q_t, describes the total number
of customers, waiting or being served at that time. A formal pathwise
construction of Q is an outcome of the observation that there exists a
unique stochastic process Q, satisfying the following relations at all $t \geq 0$:

$$(2.1) \qquad Q_t = Q_0 + A_t - D_t,$$

$$(2.2) \qquad A_t = N_+ \left(\int_0^t \lambda(Q_u) \, du \right),$$

$$(2.3) \qquad D_t = \int_0^t 1[Q_{u-} > 0] \, dS_u,$$

$$(2.4) \qquad S_t = N_- \left(\int_0^t \mu(Q_u) \, du \right).$$

Here Q is constructed in terms of the following primitives:

Q_0 is a nonnegative random variable,

λ, μ are nonnegative locally Lipschitz functions on $[0, \infty)$,

N_+, N_- are standard (rate 1) Poisson processes.

(The path construction is straightforward and of no significance to later
development, hence it is omitted). The random Q_0, N_+, N_- are defined
on a common probability space and are assumed to be independent. The
entities involved in the construction have the following interpretation: Q_0
is an initial queue; $A = \{A_t, t \geq 0\}$ and $D = \{D_t, t \geq 0\}$ are RCLL point
processes— A_t and D_t represent the cumulative number of arrivals and departures during $(0, t]$ respectively; finally, $\lambda(Q)$ and $\mu(Q)$ are, respectively,
instantaneous arrival and service rates while at state Q. Equations (2.3)
and (2.4) indicate that there are no departures when customers are absent.
Thus, $S = \{S_t, t \geq 0\}$ represents a *potential* for departures, which is fully
realized only when $Q > 0$.

Remark 2.1. The sample–paths of Q are piecewise constant RCLL
functions. If Q is non–explosive, that is $P\{Q_t < \infty, \forall t \geq 0\} = 1$, then
$D[0, \infty)$ is a suitable space for sample–paths. Otherwise, a one–point com-
pactification of $\mathbb{R}$ can be used, with λ and μ appropriately modified. A
simple sufficient condition, that ensures non-explosion of Q, is a linear
growth constraint on λ:

$$(2.5) \qquad \lambda(\xi) \leq K(1 + \xi), \quad \xi \geq 0,$$

for some constant $K > 0$, The limit theorems in this paper are stated
for non-explosive processes. (Generalizations are only commented on; see
Proposition 3.1 and Remark 4.1.)   □

2.1. Representation in terms of reflection. We recast equations (2.1)–(2.4) in a form that is convenient for analysis, namely the reflection problem described in Appendix A:

(2.6)
$$
\begin{cases}
Q_t = X_t + Y_t \geq 0, \quad t \geq 0, \\
Y \text{ nondecreasing}, \ Y_0 = 0, \\
\int_0^\infty 1[Q_t>0] \, dY_t = 0,
\end{cases}
$$

where

$$(2.7) \qquad X_t \;=\; Q_0 + N_+\left(\int_0^t \lambda(Q_u)\,du\right) - N_-\left(\int_0^t \mu(Q_u)\,du\right),$$

$$(2.8) \qquad Y_t \;=\; \int_0^t 1[Q_{u-}=0]\,dS_u\,.$$

The process Y represents cumulative losses of potential departures, due to server idleness.

Substituting X and Y into (2.6) and comparing the result with definition (2.1) of Q reveals that only the last equation of (2.6) requires verification. This equation is a complementarity relation between Y and Q: Y_t increases at time t only if $Q_t = 0$. By (2.8), it is equivalent to

$$(2.9) \qquad\qquad \int_0^\infty 1[Q_t>0]\,1[Q_{t-}=0]\,dS_t = 0,$$

whose verification we now outline. Assume, to the contrary, that (2.9) is not satisfied or, equivalently, that for some $t > 0$: $Q_{t-} = 0$, $Q_t > 0$, and $S_{t-} \neq S_t$, that is $S_t = S_{t-} + 1$. In words, the following two events occur at time t: first, a customer arrives to an *empty* system—A jumps; second, a potential service is completed—S jumps. However, as long as $Q = 0$, A and S evolve like independent Poisson processes with intensities $\lambda(0)$ and $\mu(0)$ respectively (see (2.2) and (2.4)). Such processes a.s. do not jump simultaneously, hence (2.9) a.s. prevails.

Remark 2.2. Equations (2.6) differ from the standard Skorokhod's reflection problem in that here, X itself depends on Q. Nevertheless, it turns out useful that

$$Q = \Phi(X), \quad Y = \Psi(X),$$

where Φ and Ψ are the Lipschitz operators in Appendix A, and X is given by (2.7).   □

2.2. Representation as a birth and death process. The distribution of Q is the same as that of a birth and death process on the integers, starting at Q_0 and evolving according to the following transition rates:

$$\begin{cases} q_{k,k+1} = \lambda(k), & k = 0, 1, \ldots, \\ q_{k,k-1} = \mu(k), & k = 1, 2, \ldots, \\ q_{0,-1} = 0; \end{cases}$$

(Theorem 4.1 of Chapter 6 in Ethier and Kurtz [14]). In particular, the effective service rate at a time $t \geq 0$ is $\mu_{eff}(Q_t) = 1[Q_t > 0]\mu(Q_t)$.

3. Reflection problems.
In this section, we introduce two reflection problems that provide the mathematical framework for our main theorems.

3.1. A differential equation with reflection.
Consider the following problem: given q_0—a nonnegative number, and θ—a locally Lipschitz function on $[0, \infty)$, find a pair (q, y) of absolutely continuous functions such that

(3.1)
$$\begin{cases} q_t = q_0 + \displaystyle\int_0^t \theta(q_s)\,ds + y_t \geq 0, & t \geq 0, \\ y \text{ nondecreasing}, \quad y_0 = 0, \\ \displaystyle\int_0^\infty 1[q_t > 0]\,dy_t = 0. \end{cases}$$

Remark 3.1. Analogously to Remark 2.2, (3.1) can be rewritten as

$$q = \Phi(x), \quad y = \Psi(x),$$

where

$$x_\bullet = q_0 + \int_0^\bullet \theta(q_s)\,ds,$$

and Φ, Ψ are the reflection operators from Appendix A. $\qquad \square$

Existence, uniqueness and some properties of the solution to (3.1) are given by

PROPOSITION 3.1. *If θ is locally Lipschitz then there exists a unique solution (q, y) to (3.1). For this solution, q is a monotone function and it is non–explosive if and only if at least one of the following two conditions is satisfied:*

$$\theta(\xi_1) \leq 0, \quad \text{for some } \xi_1 \geq q_0,$$

or

$$\int_{q_0}^\infty \frac{1}{\theta(s)}\,ds = \infty.$$

Remark 3.2. If $\theta(\xi) > 0$ for all $\xi \geq q_0$, then a linear growth of θ over $[q_0, \infty)$ (as in (2.5)) suffices for the second (integral) condition. $\qquad \square$

Outline of the Proof. Uniqueness follows from the Lipschitz properties of θ, Φ, and Ψ. To prove existence, associate with (3.1) the following ordinary differential equation:

$$(3.2) \qquad\qquad \dot{z}_t = \theta(z_t), \quad z_0 = q_0 .$$

When θ is locally Lipschitz, such an equation has a unique solution up to (a possible) *explosion time* [16]. This solution must be either strictly monotone or a constant function [16, page 40]. It gives rise to the unique solution of (3.1) in the following manner: q coincides with z up to the time $t > 0$ when z intersects zero, past which q vanishes. (If z is non-negative on $(0, \infty)$ then $q \equiv z$.) The first (non-positivity) condition of the theorem ensures that q remains bounded, and the second (integral)—that q approaches infinity only at infinite time.   □

To support later analysis, we now elaborate on the explicit forms that solutions to (3.1) can take. They are described in the following four Cases:

1. Strictly positive

1.1 *Strictly increasing.*
   If $\theta(\xi) > 0$ for all $\xi \geq q_0$, then

$$\dot{q}_t = \theta(q_t),\ t > 0; \quad q_t \uparrow\uparrow \infty; \quad y \equiv 0.$$

1.2 *Strictly increasing with horizontal asymptote.*
   If there exists $\xi_1 > q_0$ such that

$$\theta(\xi_1) = 0 \ \text{and}\ \theta(\xi) > 0,\ \xi \in [q_0, \xi_1),$$

   then

$$\dot{q}_t = \theta(q_t),\ t > 0; \quad q_t \uparrow\uparrow \xi_1; \quad y \equiv 0.$$

1.3 *Strictly decreasing with horizontal asymptote.*
   If $q_0 > 0$, and there exists $\xi_1 \in [0, q_0)$, such that

$$\theta(\xi_1) = 0 \ \text{and}\ \theta(\xi) < 0,\ \xi \in (\xi_1, q_0],$$

   then

$$\dot{q}_t = \theta(q_t),\ t > 0; \quad q_t \downarrow\downarrow \xi_1; \quad y \equiv 0.$$

1.4 *Non-zero constant.*
   If $q_0 > 0$ and $\theta(q_0) = 0$ then

$$q \equiv q_0 \quad y \equiv 0.$$

2. Vanishing, without reflection
   If $q_0 = 0$ and $\theta(0) = 0$ then

$$q \equiv 0, \quad y \equiv 0.$$

3. Vanishing, with reflection

If $q_0 = 0$ and $\theta(0) < 0$ then

$$q \equiv 0, \quad y_t = -\theta(0)t, \quad t \geq 0.$$

4. Strictly decreasing and absorbing at zero

If $q_0 > 0$ and $\theta(\xi) < 0$ for all $\xi \in [0, q_0]$ then

$$\begin{cases} \dot{q}_t = \theta(q_t), & t \in [0, t_0], \\ q_t = 0, & t \geq t_0 ; \end{cases}$$

$$\begin{cases} y_t = 0, & t \in [0, t_0], \\ y_t = -\theta(0)(t - t_0), & t \geq t_0 , \end{cases}$$

where $t_0 = \inf\{t \geq 0 : q_t = 0\}$, $0 < t_0 < \infty$.

Remark. Explosion can occur only in Case 1.1, otherwise q is bounded. Furthermore, q does not leave zero after reaching it. □

3.2. Derivatives of reflection operators. For a background and references on M_1-convergence see Appendix B. Notations are summarized in Appendix C. All functions below are defined on $[0, \infty)$. The following lemma plays a key role in our later formulation and proof of FCLT.

LEMMA 3.2. *Let b and x be continuous functions. Assume that $x_0 \geq 0$ and that x is either strictly monotone or constant. Suppose further that $b_0 \geq 0$ if $x_0 = 0$. Then, the sequence of continuous functions, given by*

$$\Psi(nx + b) - \Psi(nx) = \overline{(nx + b)^-} - \overline{nx^-}, \quad n = 1, 2, \ldots,$$

decreases monotonically, as $n \uparrow \infty$, to an upper semi-continuous function $\tilde{b}$. This convergence holds in the M_1-topology.

The proof is omitted as it resembles that of Lemma 4.2 in [37].

For each x satisfying the conditions of Lemma 3.2, denote by $C^x[0, \infty)$ the set of continuous functions

$$(3.3) \qquad C^x[0, \infty) \triangleq \begin{cases} C[0, \infty), & x_0 > 0, \\ C_0[0, \infty), & x_0 = 0. \end{cases}$$

Introduce the operators Ψ^x and Φ^x, with domain $C^x[0, \infty)$, by

$$(3.4) \qquad \Psi^x(b) \triangleq \tilde{b}, \quad \Phi^x(b) \triangleq b + \Psi^x(b).$$

The notation $\Phi^x(b)$ is justified in view of the M_1-convergence

$$(3.5) \quad \Phi(nx + b) - \Phi(nx) = b + \Psi(nx + b) - \Psi(nx) \longrightarrow b + \Psi^x(b),$$

which prevails by the continuity of b and the continuity of addition in the M_1-topology (See Appendix B).

To justify the title of the current subsection, note that Lemma 3.2 can be stated as follows:

$$\lim_{\varepsilon \downarrow 0} \frac{1}{\varepsilon} [\Psi(x + \varepsilon b) - \Psi(x)] = \Psi^x(b),$$

in the M_1-topology. Thus, $\Psi^x(b)$ can be interpreted as some form of a directional derivative of the operator $x \longrightarrow \Psi(x)$, at the point x in the direction b. Analogously, $\Phi^x(b)$ is a directional derivative of the operator $x \longrightarrow \Phi(x)$.

The transformation Φ^x is central for our results. We now elaborate on its explicit forms, recalling that its domain is $C^x[0, \infty)$ for those x that satisfy the conditions of Lemma 3.2. The following four Cases arise:

1. Identity operator

If x is strictly positive over $(0, \infty)$, then

$$\Phi^x(b) \equiv b.$$

2. Ordinary reflection operator

If x is identically zero, then

$$\Phi^x(b) = \Phi(b).$$

3. Delayed zero operator

If x is strictly decreasing with $x_0 = 0$, then

$$\Phi_t^x(b) = \begin{cases} b_0, & t = 0, \\ 0, & t > 0. \end{cases}$$

4. Restricted identity operator

If x is strictly decreasing with $x_0 > 0$, and if $x(t_0) = 0$ for some $t_0 \in (0, \infty)$, then

$$\Phi_t^x(b) = \begin{cases} b_t, & t < t_0, \\ 0, & t > t_0, \\ 0 \vee b_{t_0}, & t = t_0. \end{cases}$$

Remark 3.3. If $\Psi^x(b)$ is a continuous function at some point b, then the M_1-convergence in Lemma 3.2 reduces to U-convergence. A similar assertion holds with respect to Φ^x and the convergence in (3.5). Consequently, in Cases 1 and 2 the convergence in (3.5) is uniform on compact subsets of $[0, \infty)$. In Case 3, the convergence is uniform on compact subsets of $(0, \infty)$ if $b_0 \neq 0$, and of $[0, \infty)$ otherwise. In Case 4, the convergence is in $(\widetilde{D}[0, \infty), M_1)$, and the values of Φ^x are upper semi-continuous functions. Furthermore, in Case 4 when $b_{t_0} < 0$ (respectively $b_{t_0} > 0$), the convergence is uniform on compact subsets of $[0, t_0)$ and $[t_0, \infty)$ (respectively $[0, t_0]$ and (t_0, ∞)). If $b_{t_0} = 0$, the convergence is uniform on compact subsets of $[0, \infty)$.   □

The following explicit expression applies to Φ^x (by analogy to (4.5),(4.6) in [37]):

$$\Phi^x(b) = \sup_{s \in \widehat{H}_t} (-\widehat{b}_s), \quad t \geq 0,$$

where

$$\widehat{H}_t \equiv \left\{ 0 \leq s \leq t \mid x_s^- = \sup_{0 \leq u \leq t} x_u^- \right\},$$

$$\widehat{b}_t = \begin{cases} b_t, & x_t < 0, \\ b_t \wedge 0, & x_t = 0, \\ 0, & x_t > 0. \end{cases}$$

Such a representation is expected to be useful for the analysis of queues that are both time and state dependent.

4. Main theorems. FSLLN and FCLT are presented in Subsection 4.1 and 4.2 respectively. A refinement of FCLT, useful in applications, is formulated in Subsection 4.3. In Subsection 4.4 we analyze the rescaling procedure that lead to our limit theorems. The subject of Section 4.5 is an interpretation of discontinuous diffusion limits. We conclude the section with alternative types of rescaling. This motivates a later discussion, in Subsection 4.6, of models that are not covered in the current paper.

4.1. Fluid approximations (FSLLN). Consider a sequence $M_\xi^n/M_\xi^n/1$, $n = 1, 2, \ldots$, of queueing systems, each as in (2.6)–(2.8). The n-th system is described in terms of the following primitives: a random variable Q_0^n representing the initial queue, and non-negative locally Lipschitz functions λ^n and μ^n defining, respectively, the dependence of the arrival and service rates on the queue length Q^n. The queueing process Q^n can be realized as the unique solution to the following reflection problem (see Remark 2.2):

$$(4.1) \quad \begin{cases} Q^n = \Phi(X^n), \\ X_\bullet^n = Q_0^n + N_+ \left(\int_0^\bullet \lambda^n (Q_s^n)\, ds \right) - N_- \left(\int_0^\bullet \mu^n (Q_s^n)\, ds \right). \end{cases}$$

Introduce the rescaled processes $q^n = \{q_t^n, t \geq 0\}$ given by

$$(4.2) \quad q_t^n = \frac{1}{n} Q_t^n.$$

Then, due to the homogeneity of Φ and Ψ (Appendix A),

$$(4.3) \quad \begin{cases} q^n = \Phi(x^n), \\ x_\bullet^n = q_0^n + \frac{1}{n} N_+ \left(\int_0^\bullet \lambda^n (n q_s^n)\, ds \right) - \frac{1}{n} N_- \left(\int_0^\bullet \mu^n (n q_s^n)\, ds \right). \end{cases}$$

The asymptotic behavior of $\{q^n\}$ emerges from the following theorem, the proof of which is postponed to Section 6.

THEOREM 4.1 (FSLLN). *Suppose that*

$$(4.4) \qquad \frac{1}{n}\lambda^n(n\xi) \longrightarrow \lambda(\xi) \quad and \quad \frac{1}{n}\mu^n(n\xi) \longrightarrow \mu(\xi), \quad u.o.c.,$$

as $n \uparrow \infty$, where λ and μ are given locally Lipschitz functions, as well as

- $\frac{1}{n}\lambda^n(n\xi) \leq K(1 + \xi), \ \xi \geq 0$, *where K is a given positive constant;*
- $\lim_{n \uparrow \infty} q_0^n = q_0$ *a.s., where q_0 is a given non–negative scalar, and the sequence $\{Eq_0^n\}$ of expectations is uniformly bounded.*

Then, as $n \uparrow \infty$, the sequence $\{q^n\}$ of solutions to (4.3) converges u.o.c. over $[0, \infty)$, a.s., to a deterministic function q, given by

$$(4.5) \qquad \begin{cases} q = \Phi(x), \\ x_\bullet = q_0 + \int_0^\bullet (\lambda(q_s) - \mu(q_s))\, ds. \end{cases}$$

That is, q is the unique solution to the differential equation with reflection (3.1), with

$$(4.6) \qquad \theta(\xi) = \lambda(\xi) - \mu(\xi), \ \xi \geq 0.$$

In what follows, q will be referred to as the *fluid limit* associated with the queueing sequence under consideration. An analogous result holds for the sequence $\{y^n\}$, that is associated with losses of potential departures due to idleness. Specifically, for

$$y^n = \frac{1}{n}\Psi(X^n),$$

with X^n as in (4.1), we have $y^n \longrightarrow y = \Psi(x)$, a.s., u.o.c., where x is as in (4.5).

Remark 4.1. The growth condition imposed on λ^n ensures non–explosion of q^n and q. We believe, however, that FSLLN can be generalized to cover cases when q^n and/or q are explosive (see Remarks 2.1, 3.2). The theorem ought then to remain valid over the domain of existence of q. In particular, Theorem 4.1 ought to hold over $[0, \infty)$ when the linear growth constraint on λ^n is replaced by any condition that ensures non-explosion of q. Necessary and sufficient conditions for q to be non-explosive are given by Proposition 3.1. An example of a limit theorem that gives rise to explosive processes is Barbour [3]. □

The forms of the solutions to (3.1), listed at the end of Subsection 3.1, characterize possible fluid limits which, in turn, identify four modes of operation for the $M_\xi/M_\xi/1$ queue. They are depicted in Figure 1 and described by the following four Cases (based on (4.6)):

1. Permanent large queues

1.1 *Overloaded:* $\lambda(\xi) > \mu(\xi)$ for all $\xi \geq q_0$.

1.2 *Overloaded, with asymptotic transition to critically loaded:* there exists $\xi_1 > q_0$ such that

$$\lambda(\xi_1) = \mu(\xi_1); \quad \lambda(\xi) > \mu(\xi), \quad \xi \in [q_0, \xi_1).$$

1.3 *Underloaded, with large initial queue and asymptotic transition to critically loaded:* $q_0 > 0$, and there exists $\xi_1 \in [0, q_0)$, such that

$$\lambda(\xi_1) = \mu(\xi_1); \quad \lambda(\xi) < \mu(\xi), \quad \xi \in (\xi_1, q_0].$$

1.4 *Critically loaded with large initial queue:* $q_0 > 0$, $\lambda(q_0) = \mu(q_0)$.

2. Critically loaded: $q_0 = 0$ and $\lambda(0) = \mu(0)$.

3. Underloaded: $q_0 = 0$ and $\lambda(0) < \mu(0)$.

4. Underloaded with large initial queue: $q_0 > 0$ and $\lambda(\xi) < \mu(\xi)$, $\xi \in [0, q_0]$.

4.2. Diffusion approximations (FCLT). Introduce the sequences of stochastic processes $V^n = \{V_t^n, t \geq 0\}$, $n = 1, 2, \ldots$, given by

$$(4.7) \qquad V_t^n = \sqrt{n}\,(q_t^n - q_t), \quad t \geq 0.$$

This sequence amplifies deviations of the rescaled queueing processes q^n from their fluid limit q. The asymptotic behavior of $\{V^n\}$ is the subject of the next theorem, the proof of which is presented in Section 7.

THEOREM 4.2 (FCLT). *Let the conditions of Theorem 4.1 (FSLLN) be satisfied. Assume further that λ, μ in (4.4) are continuously differentiable with locally Lipschitz derivatives,*

$$(4.8) \qquad \begin{cases} \sqrt{n}\left[\frac{\lambda^n(n\xi)}{n} - \lambda(\xi)\right] \longrightarrow f_\lambda(\xi), & \text{u.o.c.,} \\ \sqrt{n}\left[\frac{\mu^n(n\xi)}{n} - \mu(\xi)\right] \longrightarrow f_\mu(\xi), & \text{u.o.c.,} \end{cases}$$

where f_λ, f_μ are locally Lipschitz functions, and that $V_0^n \xrightarrow{\ d\ } V_0$, as $n \uparrow \infty$, where V_0 is a given random variable.
Then the sequence $\{V^n\}$ converges weakly in $(\widetilde{D}[0, \infty), M_1)$ to a Markov process V with upper semi-continuous sample-paths. The process V is the unique (strong) solution to the following stochastic differential equation with reflection:

$$(4.9) \qquad \begin{cases} V = \Phi^x(X), \\ X_\bullet = V_0 + \displaystyle\int_0^\bullet (f_\lambda(q_s) - f_\mu(q_s))\,ds + \int_0^\bullet (\lambda'(q_s) - \mu'(q_s))V_s\,ds \\ \qquad + \displaystyle\int_0^\bullet \sqrt{\lambda(q_s) + \mu(q_s)}\,dW_s\,. \end{cases}$$

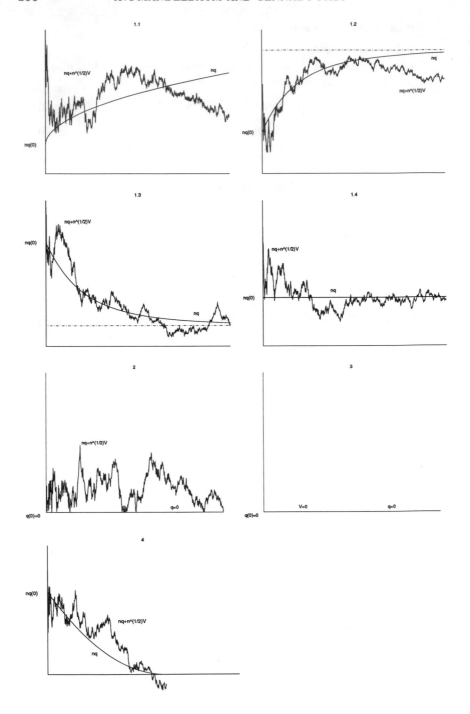

FIG. 1. *Fluid and Diffusion Limits*

Here x and q are given by (4.5), Φ^x is the operator defined by (3.4), and W is a standard Brownian motion.

In what follows, V will be referred to as the *diffusion limit* associated with the queueing sequence under consideration.

The possible forms of x in (4.5) (Cases 1–4 at the end of Subsection 4.1) reveal that x adheres to the conditions imposed in Lemma 3.2, and thus Φ^x is well-defined. In correspondence with the specific forms of Φ^x (Cases 1–4 at the end of Subsection 3.2), the relations (4.9) reduce to the following four Cases (see Figure 1 for suggestive sample paths of V):

1. Permanent large queues

(4.9) is the linear stochastic differential equation:

$$dV_t = [f_\lambda(q_t) - f_\mu(q_t) + (\lambda'(q_t) - \mu'(q_t)) V_t] \, dt + \sqrt{\lambda(q_t) + \mu(q_t)} \, dW_t$$

In particular, if V_0 is a Gaussian variable then V is a Gaussian process. With the notation

$$m_t \triangleq EV_t, \quad h_t \triangleq E[V_t - m_t]^2,$$

we have [22]

$$\begin{cases} \dot{m}_t = f_\lambda(q_t) - f_\mu(q_t) + (\lambda'(q_t) - \mu'(q_t))m_t \, , \\ \dot{h}_t = 2(\lambda'(q_t) - \mu'(q_t))h_t + \lambda(q_t) + \mu(q_t), \\ m_0 = EV_0, \\ h_0 = \mathrm{Var}V_0. \end{cases}$$

2. Critically loaded

(4.9) is a stochastic differential equation with reflection (see Remark 2.2 and Subsection 3.1):

$$V = \Phi(X),$$

where

$$X_\bullet = V_0 + (\lambda'(0) - \mu'(0)) \int_0^\bullet V_s \, ds + \sqrt{2\lambda(0)} \, W_\bullet \, .$$

Equivalently,

$$dV_t = (\lambda'(0) - \mu'(0))V_t \, dt + \sqrt{2\lambda(0)} \, dW_t + dY_t \, , \quad Y = \Psi(X).$$

3. Underloaded

(4.9) degenerates to

$$V_t = 0, \quad t > 0.$$

4. Underloaded with large initial queue

For $t < t_0$, (4.9) coincides with the linear stochastic differential equation of Case 1, and for $t \geq t_0$, it reduces to

$$V_t = \begin{cases} 0 \vee V_{t_0-}, & t = t_0, \\ 0, & t > t_0. \end{cases}$$

Further applications of (4.9) to specific λ, μ, q_0 and V_0, are the subject of Section 5.

4.3. Generalizations. We now present an extension of FCLT, covering λ and μ with piecewise continuous derivatives. This version will be used in Subsection 5.6, in the analysis of finite server queues.

THEOREM 4.3. *Assume that all the conditions of Theorem 4.2 are satisfied, but allow the derivatives λ' and μ' to be piecewise continuous functions with a finite number of discontinuities in each compact subinterval of $(0, \infty)$.*
If, in addition,

$$(4.10) \qquad \begin{cases} q_t \equiv q_0 > 0, \\ \lambda'(q_0-) \neq \lambda'(q_0+) \ \text{ or } \ \mu'(q_0+) \neq \mu'(q_0+), \end{cases}$$

then the sequence $\{V^n\}$ converges weakly in $(D[0, \infty), J_1)$ to the unique (strong) solution of the following stochastic differential equation (without reflection):

$$dV_t = [f_\lambda(q_0) - f_\mu(q_0) + f(V_t)] \, dt + \sqrt{\lambda(q_0) + \mu(q_0)} \, dW_t \,,$$

where

$$f(v) = \begin{cases} (\lambda'(q_0+) - \mu'(q_0+))v, & v \geq 0, \\ (\lambda'(q_0-) - \mu'(q_0-))v, & v < 0. \end{cases}$$

If (4.10) does not prevail, then Theorem 4.2 applies without any changes.

Comments on the proof of the theorem will be given in Subsection 7.7. Note that under (4.10), the diffusion limit has continuous sample-paths. Furthermore, (4.10) describes the only case that renders in doubt the existence of the second integral on the right-hand side of (4.9).

4.4. Time acceleration. The FSLLN rescaling (4.2) and (4.4) is equivalent to a procedure of accelerating time and aggregating space units, both by a factor of n. Indeed, consider the simplest, yet illuminating, example of $\{\lambda^n\}$ and $\{\mu^n\}$ that satisfy (4.4):

$$(4.11) \qquad \lambda^n(\xi) = n\lambda\left(\frac{\xi}{n}\right), \quad \mu^n(\xi) = n\mu\left(\frac{\xi}{n}\right),$$

for some given λ and μ. Equations (4.11) arise naturally for systems with linear, or piecewise–linear, dependence of arrival and service rates on the queue length. (See Subsections 5.4–5.8.) Alternatively to (4.11), consider a sequence $\bar{M}_\xi^n/\bar{M}_\xi^n/1$, $n = 1, 2, \ldots$, of queueing systems, with rates

$$(4.12) \qquad \bar{\lambda}^n(\xi) = \lambda\left(\frac{\xi}{n}\right), \quad \bar{\mu}^n(\xi) = \mu\left(\frac{\xi}{n}\right),$$

and queueing processes $\bar{Q}^n$. Introduce the processes $\bar{q}^n = \{\bar{q}_t^n, t \geq 0\}$, given by

$$(4.13) \qquad \bar{q}_t^n = \frac{1}{n}\bar{Q}_{nt}^n,$$

Then, rewriting equations (4.1) in terms of $\bar{\lambda}^n$, $\bar{\mu}^n$, and changing variables yields:

$$(4.14) \qquad \bar{q}^n = q^n, \quad n = 1, 2, \ldots.$$

4.5. An interpretation of discontinuous diffusion limits.
In this subsection we attempt a qualitative explanation of some phenomena that are amplified by our analysis.

As apparent from Subsection 4.2 and Subsection 4.1 (see Case 4), the diffusion limit has a discontinuity in light traffic (underloaded) with large initial queues: $\lambda(\xi) < \mu(\xi)$, for all $\xi \in [0, q_0]$, and

$$\frac{1}{n}Q_0^n \longrightarrow q_0 > 0, \quad \text{a.s.,} \quad n \uparrow \infty.$$

Discontinuity arises at time $t_0 > 0$, given by $t_0 = \inf\{t \geq 0 : q_t = 0\}$.

For simplicity of presentation, let us assume that $Q_0^n = nq_0$, for some $q_0 \in \mathcal{Z}^+$, and the rates in the n-th system are given by (4.11). We suppose also that the V^n converges a.s. to a process V with upper semi-continuous sample paths.

Consider first the case $V_{t_0-} > 0$ and thus $Q_{t_0}^n/\sqrt{n} \longrightarrow V_{t_0}$. To expose the causes of discontinuity, consider Q^n, which is a birth and death process on the integers with the following transitions rates:

$$(4.15) \qquad \begin{cases} q_{k,k+1}^n = n\lambda(k/n), & k = 0, 1, \ldots, \\ q_{k,k-1}^n = n\mu(k/n), & k = 1, 2, \ldots, \\ q_{0,-1}^n = 0; \end{cases}$$

One distinguishes three phases in the evolution of Q^n:

1. First relaxation phase of duration t_0: at the beginning of this phase, the queue length is nq_0, reducing to $\sim \sqrt{n}V_{t_0}$ at the end.

2. Second relaxation phase of duration $\sim 1/\sqrt{n}$, starting at t_0. This phase arises from the fact that the queue length at the outset is $\sim \sqrt{n}V_{t_0}$, while the rates in (4.15) are $\sim n\lambda(0)$ and $\sim n\mu(0)$ over this phase ($\lambda(0) < \mu(0)$). This phase shrinks, as $n \uparrow \infty$ ultimately resulting in a discontinuity of V at t_0. At the end of the phase, the queue is $o(\sqrt{n})$.

3. Light traffic phase. Fluid and diffusion limits vanish, as for the underloaded state-independent queues with constant rates $\lambda(0) < \mu(0)$ and with small, $o(\sqrt{n})$, initial queues. (See Subsection 5.1 for the explicit expressions of fluid and diffusion limits in state-independent systems.)

When $V_{t_0} < 0$, similar conclusions apply. The only difference is that the first phase ends $\sim 1/\sqrt{n}$ prior to t_0, the second phase terminates and the third phase starts at time t_0.

When V has a discontinuity at time zero (Case 3, Subsection 4.2 and Subsection 4.1), the first phase is skipped in the evolution of Q^n.

Note that the fluid and diffusion limits both vanish beyond t_0. These are simple examples of state-space collapse, when the limiting process is of a lower dimension than the process it approximates (see Reiman [42], Mandelbaum and Chen [8]). State-space collapse occurs here, when systems operate under light-traffic conditions. To obtain nonzero, more informative limits, one must formulate other limit theorems. Light traffic limit theorems usually involve various cumulative processes such as sums and integrals of the original process (see the survey by Glynn [15]). An alternative limit theorem for the distribution of Q^n in light traffic (in a particular closed network) is presented by Kogan and Liptser [27]. It is possible to obtain a non-degenerate diffusion limit for the second phase via a slower rescaling, namely, considering locally a process $Q^n(t_0 \pm \tau/\sqrt{n})$, for some $\tau > 0$.

Remark. An explanation for a discontinuity of V can be given also in terms of the transient behavior of $\bar{Q}^n$, introduced in the preceding subsection. Again, three successive phases arise in the evolution of $\bar{Q}^n$: relaxation of duration $\sim nt_0$, relaxation of duration $\sim \sqrt{n}$ and, finally, light traffic phase. The rescaling used (acceleration of time by a factor n) shrinks the duration of the second phase, resulting in a discontinuity of V.   □

4.6. Alternative rescaling. We now describe rescaling procedures other than (4.4),(4.8), which lead to different approximations of state-dependent queues. Specifically, assume that

$$(4.16) \quad \frac{1}{n}\lambda^n(n^\alpha\xi) \longrightarrow \lambda(\xi), \quad \sqrt{n}\left[\frac{\lambda^n(n^\alpha\xi)}{n} - \lambda(\xi)\right] \longrightarrow f_\lambda(\xi), \quad u.o.c.,$$

and

$$(4.17) \quad \frac{1}{n}\mu^n(n^\alpha\xi) \longrightarrow \mu(\xi), \quad \sqrt{n}\left[\frac{\mu^n(n^\alpha\xi)}{n} - \mu(\xi)\right] \longrightarrow f_\mu(\xi), \quad u.o.c.,$$

as $n \uparrow \infty$, for some specific $\alpha \geq 0$. Evidently, our limits correspond to the case $\alpha = 1$ (see (4.4) and (4.8)). Alternative rescaling procedures were considered by Yamada in [49] and [50]: the case $\alpha = 0$ is treated in [49], where the diffusion limit is of a Bessel type with a negative drift; the case $\alpha = 1/2$ is considered in [50], where the diffusion limit is a solution to a stochastic differential equations with *state*-dependent coefficients (while in our case the coefficients are *time*-dependent). The fluid limits vanish in both [49] and [50].

A comparison of the different approaches is summarized in Table 1 (with preference to clarity of presentation over precision). The expressions for the rates on the first upper part of Table 1 are based on (4.16). Combining these expressions with the fluid and diffusions limits from the third and forth parts of the table yields the last part.

Remark. The reflection $\widetilde{Y}$ in the expression for V, when $\alpha = 0$, is characterized by the condition:

$$\int_0^t V_s \, d\widetilde{Y}_s = \gamma \cdot t, \quad t \geq 0,$$

for some $\gamma > 0$. Such reflection gives rise to a Bessel–type distribution for V.   □

To recapitulate, our approach leads to a second order approximations for queueing processes: fluid limits provide approximations for actual values of queues, while diffusions limits—for their fluctuations. When fluid limits vanish, the three approaches provide approximations for systems in which arrival and service rates are sensitive to small $(\alpha = 1, \mathcal{O}(n^{-1/2}V))$, medium $(\alpha = 1/2, \mathcal{O}(V))$ and large $(\alpha = 0, \mathcal{O}(\sqrt{n}V))$ fluctuations of queues.

In Subsection 5.9, we compare the three types of rescaling, $\alpha = 0, 1/2, 1$, by applying them to a single queueing system.

5. Examples and applications. This section is devoted to some applications of the limit theorems presented in Section 4. In Subsections 5.1 and 5.3 we characterize the conditions under which diffusion limits are Brownian or Ornstein-Uhlenbeck processes. In Subsection 5.2 we consider asymptotically small initial queues.

A unified approach is offered to obtain fluid and diffusion limits for state dependent queueing systems. We demonstrate this through application and simplification of some completely or partially known results (see Subsections 5.4–5.8). Different types of rescaling are applied in Subsection 5.9 to a single model, thus highlighting their differences. In Subsection 5.10 we outline guidelines for implementing some of our approximations.

TABLE 1. *Rescaling State-Dependent Queues*

$$\lambda^n(Q^n) \approx n\lambda\left(\tfrac{1}{n^\alpha}Q^n\right) + \sqrt{n}\,f_\lambda\left(\tfrac{1}{n^\alpha}Q^n\right)$$

$$\mu^n(Q^n) \approx n\mu\left(\tfrac{1}{n^\alpha}Q^n\right) + \sqrt{n}\,f_\mu\left(\tfrac{1}{n^\alpha}Q^n\right)$$

Approaches	$\alpha = 1$	$\alpha = 1/2$	$\alpha = 0$
		Yamada 93	Yamada 86
Fluid	$\dfrac{Q^n}{n} \xrightarrow{a.s.} q$		$\dfrac{Q^n}{n} \xrightarrow{P} 0$
FSLLN	$\dot{q} = \lambda(q) - \mu(q) + dy,\ q \perp dy$	Critical Loading	
Diffusion	$\sqrt{n}\left(\dfrac{Q^n}{n} - q\right) \xrightarrow{d} V,\ M_1$	$\dfrac{Q^n}{\sqrt{n}} \xrightarrow{d} V,\ J_1$	
FCLT	$dV_t = b(q_t)V_t\,dt + c(q_t)\,dW_t + dY_t$	$dV_t = \beta(V_t)\,dt + \gamma(V_t)\,dW_t + dY_t$	$V_t = -bt + cW_t + \widetilde{Y}_t$
Effective	$\lambda^n(Q^n) \approx n\lambda(q) + \sqrt{n}\left[\lambda'(q)V + f_\lambda(q)\right]$	$\lambda^n(Q^n) \approx n\lambda(V) + \sqrt{n}\,f_\lambda(V)$	$\lambda^n(Q^n) \approx n\lambda(\sqrt{n}\,V) + \sqrt{n}\,f_\lambda(\sqrt{n}\,V)$
Rates	$\mu^n(Q^n) \approx\ -''-$	$\mu^n(Q^n) \approx\ -''-$	$\mu^n(Q^n) \approx\ -''-$

5.1. State–independent models: linear fluid and Brownian diffusion limits. Consider a sequence of state-independent queues $M^n/M^n/1$ with constant rates $\lambda^n = n\lambda$ and $\mu^n = n\mu$ $(f_\lambda = f_\mu \equiv 0$ has been chosen in (4.8) for simplicity).

Theorems 4.1 and 4.2 yield the following expressions for the fluid and diffusion limits:

$$x_t = q_0 + (\lambda - \mu)t, \quad q = x + y; \quad V = \Phi^x\left(V_0 + \sqrt{\lambda + \mu}\,W_\bullet\right).$$

Three modes of evolution arise

1. *Overloaded:* $\lambda > \mu$. Here $q_t = q_0 + (\lambda - \mu)t$ and

$$V_\bullet = V_0 + W((\lambda + \mu)\bullet) \stackrel{d}{\sim} BM_{V_0}(0, \lambda + \mu).$$

2. *Critically loaded:* $\lambda = \mu$. Here $q \equiv q_0$;
   If $q_0 > 0$, then $V_\bullet = V_0 + W((\lambda + \mu)\bullet) \stackrel{d}{\sim} BM_{V_0}(0, \lambda + \mu)$;
   If $q_0 = 0$, then $V = \Phi[V_0 + W((\lambda + \mu)\bullet)] \stackrel{d}{\sim} RBM_{V_0}(0, \lambda + \mu)$.

3. *Underloaded:* $\lambda < \mu$.
   If $q_0 > 0$ (large initial queues):

$$q_t = \begin{cases} q_0 + (\lambda - \mu)t, & t \le t_0 = \dfrac{q_0}{\mu - \lambda}, \\ 0, & t > t_0 \end{cases} ;$$

$$V_t = \begin{cases} V_0 + W((\lambda + \mu)t), & t < t_0, \\ 0, & t > t_0, \\ \max[V_0 + W((\lambda + \mu)t), 0], & t = t_0. \end{cases}$$

If $q_0 = 0, V_0 \ne 0$ (moderate initial queues): $q \equiv 0$ and $V_t = 0$, $t > 0$.
If $q_0 = V_0 = 0$ (small initial queues): $q \equiv 0$ and $V \equiv 0$.

5.2. Small initial queues. In this subsection, our limit theorems are applied with asymptotically small initial queues, that is $q_0 = V_0 = 0$. This case is highlighted for its simplicity: diffusion limits are continuous on $[0, \infty)$ and the behavior of the fluid (positive, vanishing) and diffusion limits (with/without reflection, vanishing) depends solely on $\lambda(0)$, $\mu(0)$. Specifically:

1. *Overloaded:* $\lambda(0) > \mu(0)$. Here q is strictly positive over $(0, \infty)$ (Cases 1.1 or 1.2 of Subsection 4.1) and V is a diffusion as in Case 1 at the end of Subsection 4.2.

2. *Critically loaded:* $\lambda(0) = \mu(0)$. Here $q \equiv 0$, and V is a reflected diffusion (Cases 2 in Subsections 4.1 and 4.2).

3. *Underloaded:* $\lambda(0) < \mu(0)$. Here $q \equiv 0$ and $V \equiv 0$ (Cases 3 in Subsections 4.1 and 4.2).

5.3. Constant fluid and Ornstein-Uhlenbeck diffusion limits.

We continue with queues whose diffusion limits are Ornstein-Uhlenbeck or reflected Ornstein-Uhlenbeck processes. As previously, $f_\lambda = f_\mu \equiv 0$ in (4.8) is chosen for simplicity.

Let λ, μ and $\xi_1 \geq 0$ satisfy:

$$(5.1) \qquad \lambda(\xi_1) = \mu(\xi_1) \quad \text{and} \quad \lambda'(\xi_1) < \mu'(\xi_1).$$

Ornstein-Uhlenbeck diffusion limit. Add to (5.1) the assumption $\xi_1 > 0$. Two cases arise:

1. $q_0 = \xi_1$

   Theorems 4.1 and 4.2 yield $q \equiv \xi_1$ and

   $$dV_t = (\lambda'(\xi_1) - \mu'(\xi_1))V_t \, dt + \sqrt{2\lambda(\xi_1)} \, dW_t \, .$$

   Thus V is an Ornstein-Uhlenbeck process with

   $$m_t \overset{\Delta}{=} EV_t = m_0 e^{-2\alpha t} \, ,$$

   $$h_t \overset{\Delta}{=} \mathrm{Var}V_t = \frac{\sigma^2}{2\alpha} + \left(h_0 - \frac{\sigma^2}{2\alpha} \right) e^{-2\alpha t} \, ,$$

   $$h_{s,t} \overset{\Delta}{=} \mathrm{Cov}(V_s \, , V_t) = \left[h_0 + \frac{\sigma^2}{2\alpha}(e^{2\alpha(t \wedge s)} - 1)]e^{-\alpha(t+s)} \right] \, ,$$

   where $\sigma = \sqrt{2\lambda(\xi_1)}$, and $\alpha = \mu'(\xi_1) - \lambda'(\xi_1)$. Taking $V_0 \overset{d}{\sim} \mathcal{N}\left(0, \frac{\sigma^2}{2\alpha}\right)$, V is the stationary Ornstein-Uhlenbeck process with

   $$h_{s,t} = \frac{\sigma^2}{2\alpha}e^{-\alpha|t-s|} \, .$$

2. $q_0 \neq \xi_1$

   Assume, in addition, that

   $$\lambda(\xi) > \mu(\xi), \ \ \xi < \xi_1 \, ,$$
   $$\lambda(\xi) < \mu(\xi), \ \ \xi > \xi_1 \, .$$

   Then, it follows from Subsection 4.1 (see Cases 1.2, 1.3) that, $q_t \downarrow\downarrow \xi_1$ or $q_t \uparrow\uparrow \xi_1$ as $t \uparrow \infty$. For V we have

   $$dV_t = (\lambda'(q_t) - \mu'(q_t))V_t \, dt + \sqrt{\lambda(q_t) + \mu(q_t)} \, dW_t \, .$$

   The random variable V_t converges weakly, as $t \uparrow \infty$, to $V(\infty) \overset{d}{\sim} \mathcal{N}\left(0, \frac{\sigma^2}{2\alpha}\right)$, where σ and α are as defined above.

Reflected Ornstein-Uhlenbeck diffusion limit. Assume $\xi_1 = 0$ in (5.1). The diffusion limit V is then

$$dV_t = (\lambda'(0) - \mu'(0))V_t \, dt + \sqrt{2\lambda(0)} \, dW_t + dY_t \, ,$$

which is a reflected Ornstein-Uhlenbeck process. This example was described by Liptser and Shiryayev [36, pages 753,754].

5.4. Finite population and general service (Liptser, Shiryayev [36]).
Consider a sequence of $M/M_\xi^n/1/\infty/n$ systems, as in [36, pages 638–636]. The parameters of the n-th system are given by

$$\lambda^n (Q^n) = \lambda \cdot (n - Q^n), \quad \mu^n (Q^n) = n\mu \left(\frac{Q^n}{n} \right),$$

for some $\lambda \geq 0$ and a function μ. We identify the parameters of the fluid and diffusion limits via (4.4),(4.8): $\lambda(\xi) = \lambda \cdot (1 - \xi)$, $\mu(\xi)$, and $f_\lambda = f_\mu \equiv 0$. Let Q_0^n and μ satisfy the conditions of the FSLLN and FCLT (Theorems 4.1 and 4.2). Then the fluid limit is given by

$$\dot{q}_t = \lambda(1 - q_t) - \mu(q_t),$$

and if, for example, $q_t > 0$ for all $t > 0$, the diffusion is

$$V_\bullet = V_0 - \int_0^\bullet (\lambda + \mu'(q_s)) V_s \, ds + \int_0^\bullet \sqrt{\lambda(1 - q_s) + \mu(q_s)} \, dW_s .$$

Assume, in addition, that $\lambda (1 - \xi_1) = \mu(\xi_1)$ and $\mu'(\xi_1) > -\lambda$, for some $\xi_1 \in (0, 1]$, and consider separately two cases:

1. $q_0 = \xi_1$. The expressions above for q and V yield $q_t \equiv \xi_1$, and

$$dV_t = -(\lambda + \mu'(\xi_1))V_t \, dt + \sqrt{2\lambda(1 - \xi_1)} \, dW_t .$$

   With $V_0 \overset{d}{\sim} \mathcal{N} (0, \sigma^2)$, $\sigma^2 = [\lambda(1 - \xi_1)]/[\lambda + \mu'(\xi_1)]$, V becomes a stationary Ornstein-Uhlenbeck process.

2. $q_0 \neq \xi_1$. Stipulating,

$$\lambda(\xi) > \mu(\xi), \ \ \xi < \xi_1 ; \quad \lambda(\xi) < \mu(\xi), \ \ \xi > \xi_1 ,$$

   we obtain that $V_t \overset{d}{\longrightarrow} V(\infty)$, where $V(\infty) \overset{d}{\sim} \mathcal{N} (0, \sigma^2)$, with σ^2 as in case 1. The random variable $V(\infty)$ can be used to approximate the long-run behavior of Q^n, for sufficiently large n (see [36, pages 653–656] and Remark 5.1).

5.5. Infinite number of servers (Whitt [48]).
Consider a sequence of $M^n/M/\infty$ systems, namely

$$\lambda^n(Q^n) \equiv n\lambda, \quad \mu^n (Q^n) = \mu Q^n,$$

for some $\lambda, \mu > 0$. This corresponds to an infinite–server queue under heavy traffic. By (4.4), $\lambda(\xi) \equiv \lambda$, $\mu(\xi) = \mu\xi$.

Assume that $q_0 = \rho \overset{\triangle}{=} \lambda/\mu$. Since $\mu(q_0) = \lambda$, we have $q_t \equiv \rho$, and

$$dV_t = -\mu V_t \, dt + \sqrt{2\lambda} \, dW_t ,$$

(see Case 1.4 in Subsection 4.1 and Case 1 in Subsection 4.2). That is, if $V_0 \overset{d}{\sim} \mathcal{N}(0, \rho)$, then V is a stationary Ornstein-Uhlenbeck process.

For the general case $q_0 \neq \rho$, we obtain

$$(5.2) \qquad q_t = \rho + (q_0 - \rho)\, e^{-\mu t},$$

and

$$(5.3) \qquad dV_t = -\mu V_t\, dt + \sqrt{2\lambda + (\mu q_0 - \lambda)\, e^{-\mu t}}\, dW_t$$

(Cases 1.2, 1.3 in Subsection 4.1 and Case 1 in Subsection 4.2). The process V has a steady-state distribution $\mathcal{N}(0, \rho)$, which can be used to approximate the distribution of $Q^n(\infty)$. (See [15], [23] and [36, pages 653–656].)

5.6. Finite number of servers (Iglehart [18], Halfin and Whitt [17]).

The limit procedure of Borovkov [6] and Iglehart [18]. Consider a sequence of $M^n/M/n$ systems such that

$$\lambda^n(Q^n) \equiv n\lambda, \qquad \mu^n(Q^n) = \mu \cdot (Q^n \wedge n),$$

for some $\lambda, \mu > 0$. By (4.4), $\lambda(\xi) \equiv \lambda$, $\mu(\xi) = \mu \cdot (\xi \wedge 1)$. The traffic intensity for the n-th system is given by

$$\rho^n = \frac{n\lambda}{n\mu} = \frac{\lambda}{\mu} \overset{\triangle}{=} \rho.$$

Three cases arise: $\rho < 1$, $\rho > 1$, $\rho = 1$.

1. $\rho < 1$

   Assume first that $q_0 \in [0, 1]$. Then q and V are the same as in the case of an infinite number of servers (see (5.2) and (5.3)). In this sense, the sequence of $M^n/M/n$ systems with $\rho < 1$ is asymptotically $(n \uparrow \infty)$ indistinguishable from the sequence $M^n/M/\infty$, $n = 1, 2, \dots$. In other words, due to (5.2),

   $$\frac{1}{n} Q^n \longrightarrow q, \text{ u.o.c, a.s.,} \quad \text{and} \quad q_t < 1, \text{ for all } t > 0,$$

   and thus the probability $\mathrm{P}\,[Q_t^n > n]$, that all servers are busy, converges to zero, as $n \uparrow \infty$.

   If $q_0 \in (1, \infty)$, then for $t \leq (q_0 - 1)/(\mu - \lambda) \overset{\triangle}{=} t_1$ the limits coincide with those of the state-independent system (Subsection 5.1, underloaded regime with large initial queues). For $t \geq t_1$ the limits are the same as for an infinite number of servers, given $q_0 = 1$ (see (5.2) and (5.3)).

2. $\rho > 1$

For the fluid limit,

$$q_t = \begin{cases} \rho + (q_0 - \rho)e^{-\mu t}, & t < t_2, \\ (\lambda - \mu)t, & t \geq t_2; \end{cases} \quad t_2 = \inf\{t : q_t \geq 1\}.$$

If $q_0 \geq 1$, then $t_2 = 0$ and only the second equation in the above expression is relevant. Therefore q is a combination of two limits: a model with an infinite number of servers, $t < t_2$ (Subsection 5.5) and a state-independent system, $t \geq t_2$ (Subsection 5.1, overloaded regime). The diffusion limit enjoys a similar structure.

3. $\rho = 1$

For $q_0 \in [0, 1)$, the limits are analogous to those of case $\rho < 1$. For $q_0 \in (1, \infty)$, the limits are the same as for the state-independent critically loaded case (Subsection 5.1).

For $q_0 = 1$, the fluid limit is trivial: $q_t \equiv q_0$. While μ is non-differentiable at $q_0 = 1$, the generalized FCLT (Theorem 4.3) is applicable:

$$dV_t = f(V_t)\,dt + \sqrt{2\mu}\,dW_t,$$

where

$$f(v) = \begin{cases} 0, & v \geq 0, \\ -\mu v, & v < 0. \end{cases}$$

Thus, V is a combination of a Brownian motion and an Ornstein-Uhlenbeck process. Such limits were proposed by Halfin and Whitt [17] and are described in what now follows.

The limit procedure of Halfin and Whitt [17]. Reconsider the limit procedure of Borovkov and Iglehart described above. For that case, if $\rho < 1$, the probability that all servers are busy converges to zero, as $n \uparrow \infty$. Halfin and Whitt [17] proposed another limit procedure for a sequence $M^n/M/\infty$ (with traffic intensity in the n-th system $\rho^n < 1$), such that the probability of delay converges to a non–degenerate limit. It was shown in [17] that

$$\lim_{n \uparrow \infty} P\{Q^n(\infty) \geq n\} = \alpha, \quad 0 < \alpha < 1,$$

if and only if

$$\lim_{n \uparrow \infty} (1 - \rho^n)\sqrt{n} = \beta, \quad 0 < \beta < \infty,$$

in which case $\alpha = \left[1 + \sqrt{2\pi}\beta\Phi(\beta)\exp(\beta^2/2)\right]^{-1}$. Here Φ is the standard normal distribution function. Note that $Q^n(\infty)$ exists since $\rho^n < 1$.

The limits described in [17] can also be deduced from our theorems. Indeed, let

$$\lambda^n(Q^n) \equiv \mu \cdot (n - \beta\sqrt{n}), \quad \mu^n(Q^n) = \mu \cdot (Q^n \wedge n), \quad \mu > 0.$$

The parameters of the fluid and diffusion limits are identified via (4.4) and (4.8): $\lambda(\xi) \equiv \mu$, $\mu(\xi) = \mu \cdot (\xi \wedge 1)$, and $f_\lambda \equiv 0$, $f_\mu \equiv -\beta\mu$. Moreover,

$$(5.4) \qquad\qquad \rho^n = \frac{\mu \cdot (n - \beta\sqrt{n})}{\mu n} = 1 - \frac{\beta}{\sqrt{n}},$$

that is $\rho^n \uparrow\uparrow 1$, and $(1 - \rho^n)\sqrt{n} \longrightarrow \beta$, as $n \uparrow \infty$.

Assuming $q_0 = 1$ and noting that μ is non-differentiable at this point, we obtain that $q \equiv 1$, and by the generalized FCLT (Theorem 4.3),

$$dV_t = f(V_t)\, dt + \sqrt{2\mu}\, dW_t,$$

where

$$f(v) = \begin{cases} -\beta\mu, & v \geq 0, \\ -\mu(\beta + v), & v \leq 0. \end{cases}$$

Here the diffusion limit is a combination of an Ornstein-Uhlenbeck process and a Brownian Motion, both with negative drifts.

Note that the drift $(-\beta\mu)$ appears in V due to the specific rates of convergence in (5.4), while the special choice of $q_0 = 1$ (at this point μ is non-differentiable) gives rise to the compound structure of the diffusion limit. The stationary distribution of V can be used in approximating the distribution of $Q^n(\infty)$, as shown in [17] (see also Remark 5.1).

5.7. The repairman problems (Iglehart and Lemoine [19], Iglehart [18]). Consider a sequence of $M/M/k^n/\infty/n$ systems. The n-th system is interpreted as follows [19]. There are n operating units subject to breakdowns and k^n repair facilities. The rates of breakdown and repair are λ and μ respectively. Thus Q_t^n here is the number of operating units which are being repaired or are awaiting repair at time t. Introduce a process $Y^n = n - Q^n$, which describes the number of units operating at time t. The fluid and diffusion limits for $\{Y^n\}$ can be obtained immediately from those for $\{Q^n\}$, therefore we focus on the latter only.

The arrival and service rates in the n-th system are given by

$$\lambda^n(Q^n) = \lambda \cdot (n - Q^n), \quad \mu^n(Q^n) = \mu \cdot (Q^n \wedge k^n).$$

Assume that

$$\frac{k^n}{n} \longrightarrow k, \quad \sqrt{n}\left(\frac{k^n}{n} - k\right) \longrightarrow 0; \quad 0 < k < 1,$$

as $n \uparrow \infty$. Then, by (4.4) and (4.8), $\lambda(\xi) = \lambda \cdot (1 - \xi)$, $\mu(\xi) = \mu \cdot (\xi \wedge k)$, and $f_\lambda = f_\mu \equiv 0$. There are three combinations of the parameters λ, μ and

k, each corresponding to an essentially different fluid and diffusion limits. For simplicity, we pursue the cases where the fluid limit $q \equiv q_0$, which are sufficient to demonstrate the main modes of behavior.

1. $\dfrac{\lambda}{\lambda + \mu} < k$, $q_0 = \dfrac{\lambda}{\lambda + \mu}$. Here $dV_t = -(\lambda + \mu)V_t\, dt + \sqrt{2\frac{\lambda\mu}{\lambda+\mu}}\, dW_t$.

2. $\dfrac{\lambda}{\lambda + \mu} > k$, $q_0 = 1 - \dfrac{\mu k}{\lambda}$. Here $dV_t = -\lambda V_t\, dt + \sqrt{2\mu k}\, dW_t$.

3. $\dfrac{\lambda}{\lambda + \mu} = k$, $q_0 = \dfrac{\lambda}{\lambda + \mu}$.

   Note that μ is not differentiable at q_0. Hence we have (by Theorem 4.3)

$$dV_t = f(V_t)\, dt + \sqrt{2\frac{\lambda\mu}{\lambda+\mu}}\, dW_t,$$

where

$$f(v) = \begin{cases} -\lambda v, & v \geq 0, \\ -(\mu + \lambda)v, & v < 0. \end{cases}$$

Remark. The FSLLN reveals that, for all t, the probability $\mathrm{P}\,[Q_t^n > k^n]$ of all repair facilities being busy, converges to zero in Case 1 and to unity in Case 2. Case 1 is therefore preferable over Case 2 for real repair systems. $\square$

Another repairman model is proposed in [18], which generalizes the one described above (see, also, Kurtz [33]). In this model, the n-th system has m^n spares, in addition to the elements described previously. These spares can immediately replace those operating units that have failed. In our terms, one can write

$$\lambda^n(Q^n) = \lambda n - \lambda \cdot (Q^n - m^n)^+, \quad \mu^n(Q^n) = \mu \cdot (Q^n \wedge k^n).$$

Following [18], assume that $m^n = nm$ and $k^n = nk$. In this case,

$$\lambda(\xi) = \lambda \cdot \left(1 - (\xi - m)^+\right), \quad \mu(\xi) = \mu \cdot (\xi \wedge k)$$

and $f_\lambda = f_\mu \equiv 0$. If, for example,

$$\mu k < \lambda, \quad k < m, \quad \text{and} \quad q_0 = 1 + m - \frac{\mu k}{\lambda},$$

then $q \equiv q_0$, and $dV_t = -\lambda V_t\, dt + \sqrt{2\mu k}\, dW_t$.

Note that our theorems apply to the more general case

$$\sqrt{n}\left(\frac{k^n}{n} - k\right) \longrightarrow \tilde{k}, \quad \sqrt{n}\left(\frac{m^n}{n} - m\right) \longrightarrow \tilde{m}, \quad \text{as } n \uparrow \infty; \quad q_0 \geq 0.$$

in which f_λ and f_μ need not vanish.

5.8. Queues with reneging (Coffman *et al.* **[11]).** Following [11], consider a sequence of queues with processor-shared service and reneging. Reneging means that a customer is lost when its sojourn time reaches an individual random deadline. Namely, we assume that, in the n-th system, the arrival and service rates are given by:

$$(5.5) \quad \lambda^n (Q^n) = n\lambda, \quad \mu^n (Q^n) = \frac{1}{Q^n} \cdot (n\lambda + \alpha\sqrt{n\lambda}) \cdot Q^n + \nu Q^n \,,$$

for arbitrary positive λ, α and ν. The quantity $n\lambda + \alpha\sqrt{\lambda n}$ is the service rate, shared among all customers in the system. The parameter ν is interpreted as the reneging rate. Assume that $q_0 = 0$. Then the fluid limit vanishes and the diffusion limit is the reflected Ornstein-Uhlenbeck process

$$dV_t = -\alpha\sqrt{\lambda} \, dt - \nu V_t \, dt + \sqrt{2\lambda} \, dW_t + dY_t \,,$$

which are limits for the critically loaded mode (Cases 2 in Subsections 4.1 and 4.2). With appropriate parameters in (5.5), one obtains limit theorems for other regimes, beyond [11].

5.9. A comparison of different rescaling procedures. The three types of rescaling, described in Section 4.6, are now applied to a single queueing system, operating in different modes.

Consider a sequence $M_\xi^n / M_\xi^n / 1$, $n = 1, 2, \ldots$, with arrival and service rates given by

$$(5.6) \quad \lambda^n(Q^n) = b^n + c^n \cdot (Q^n \wedge \delta^n), \quad \mu^n(Q^n) = \beta^n + \gamma^n \cdot (Q^n \wedge \delta^n),$$

where b^n, c^n, δ^n, β^n, γ^n are positive constants, $c^n \le \gamma^n$. Reviewing the examples from the previous subsections, one can offer various interpretations for the n-th system:

1. Service is provided simultaneously by δ^n servers (each at a rate γ^n) and by a processor-shared server (at a rate β^n). The arrival process consists of exogenous arrivals (rate b^n) and served customers that leave for a while, then return for rework with probability $c^n/\gamma^n \le 1$. (The time till their return is assumed short enough that the queue does not change much, and long enough that they are independent of exogenous arrivals.) This is a possible model to some human-service systems.

2. Service is provided by a single server, at a rate that increases with queue length, but only up to an exhaustion level $\beta^n + \gamma^n \cdot \delta^n$. Arrival rates, which increase with queue length, describe a possible scenario where a long queue attracts customers being a source of information on service value.

Assume that $Q_0^n = 0$. The three examples, presented below, exhibit different diffusion limits V, according the choice of parameters in (5.6).

These diffusion limits are obtained through the three types of rescaling discussed in Section 4.6.

1. $\alpha = 1$. Let $b^n = \beta^n = nb$, $c^n = \gamma^n = c$ and $\delta^n = n\delta$; then

$$V = \sqrt{2b}\,W + Y;$$

2. $\alpha = 1/2$. Let $b^n = nb$, $\beta^n = nb + \sqrt{n}$, $c^n = \sqrt{n}\,c$, $\gamma^n = \sqrt{n}\,c + 1$ and $\delta^n = \sqrt{n}\,\delta$; then

$$
\begin{aligned}
dV_t \;=\; & -[1 + (V_t \wedge \delta)]\,dt + \sqrt{b + c(V_t \wedge \delta)}\,dW_t^1 \\
& + \sqrt{b + c(V_t \wedge \delta)}\,dW_t^2 + dY_t;
\end{aligned}
$$

3. $\alpha = 0$. Let $b^n = \beta^n = nb$, $c^n = nc$, $\gamma^n = nc + \sqrt{n}$ and $\delta^n = \delta$; then

$$V_t = -\delta \cdot t + \sqrt{2c\delta + 2b}\,W_t + Y_t.$$

Here $b, c, \delta > 0$, W, W^1, W^2 are standard Brownian Motions (W^1 and W^2 are independent) and Y is a normal reflection term. For all three examples, the fluid limit $q \equiv 0$ and $Q^n/\sqrt{n} \xrightarrow{\mathrm{d}} V$. The examples mainly differ by the number of servers relative to the queue size, which are $n : \sqrt{n}$, $\sqrt{n} : \sqrt{n}$, $1 : \sqrt{n}$ in examples 1,2,3 respectively.

5.10. Approximating queueing systems. Our approximations typically apply when some natural parameters of the systems are taken to an extreme. For example, large number of servers, population size, initial queue or traffic intensity. The sequence $M_\xi^n / M_\xi^n / 1$, $n = 1, 2, \ldots$ is used to formalize the approximation, which always takes the form

$$(5.7) \qquad\qquad Q^n \stackrel{\mathrm{d}}{\approx} nq + \sqrt{n}\,V,$$

where q and V are the fluid and diffusion limits respectively.

Remark 5.1. A relation analogous to (5.7) can be written, at least formally, for the stationary distributions $Q(\infty)$ and $V(\infty)$, when they exist. Examples of theorems that support such approximations are Halfin and Whitt [17], Kaspi and Mandelbaum [23], Ethier and Kurtz [14, Chapter 4,§9], Liptser and Shiryayev [36].   □

6. Proof of FSLLN. This section is devoted to the proof of Theorem 4.1. To simplify the presentation, we consider (4.11) only. The general case requires minor notational changes.

The linear growth constraint on the function λ implies that both q^n, $n = 1, 2, \ldots$, and q are non-explosive (see Remarks 2.1 and 3.2).

Let T be an arbitrary positive constant. Subtracting the equation for q in (4.5) from the equation for q^n in (4.3) and using the Lipschitz property of Φ and Ψ (see Appendix A), we obtain

$$
\begin{aligned}
\| q^n - q \|_t \leq\ & C \,|\, q_0^n - q_0 \,| \\
& + C \left\| \frac{1}{n} N_+ \left(n \int_0^\bullet \lambda(q_s^n)\, ds \right) - \int_0^\bullet \lambda(q_s^n)\, ds \right\|_t \\
& + C \left\| \frac{1}{n} N_- \left(n \int_0^\bullet \mu(q_s^n)\, ds \right) - \int_0^\bullet \mu(q_s^n)\, ds \right\|_t \\
& + C \left\| \int_0^\bullet (\lambda(q_s^n) - \lambda(q_s))\, ds \right\|_t \\
& + C \left\| \int_0^\bullet (\mu(q_s^n) - \mu(q_s))\, ds \right\|_t,
\end{aligned}
$$

(6.1)

for all $t \leq T$, where C is the Lipschitz constant for Φ and Ψ. Note that the first term on the right-hand side of (6.1) converges to zero, by the conditions of the theorem.

It will be proved in Lemma 6.1 below that

$$
(6.2) \qquad \forall T > 0 \ \exists A_T < \infty \ \ni\ \varlimsup_n \| q^n \|_T \leq A_T \quad \text{a.s.}
$$

(A_T is a non-random scalar.) Consequently, the second and the third terms on the right-hand side of (6.1) converge to zero, by the continuity of λ and μ, combined with the FSLLN for any Poisson process N:

$$
\lim_{n \uparrow \infty} \left\| \frac{1}{n} N(nt) - t \right\|_T = 0, \quad \forall T \geq 0, \quad \text{a.s.}
$$

From (6.2) and the Lipschitz property of λ and μ, the last two terms in (6.1) satisfy a.s. the following inequality:

$$
\left\| \int_0^\bullet (\lambda(q_s^n) - \lambda(q_s))\, ds \right\|_t + \left\| \int_0^\bullet (\mu(q_s^n) - \mu(q_s))\, ds \right\|_t \leq C_T \int_0^t \| q^n - q \|_s\, ds,
$$

for all but finitely many values (in general, random) of n. Here

$$
C_T = C_T^\lambda + C_T^\mu,
$$

where C_T^λ and C_T^μ denote the Lipschitz constants for λ and μ respectively in $[0, (A_T \vee \| q \|_T) + 1]$. Now combining all of the above, we obtain

$$
(6.3) \qquad \| q^n - q \|_t \leq \epsilon^n(T) + B \int_0^t \| q^n - q \|_s\, ds, \quad 0 \leq t \leq T,
$$

where $\epsilon^n(T)$, which is the sum of the first three terms on the right-hand side of (6.1) (with $t = T$), converges to zero; and $B = C \cdot C_T$. Finally, applying Gronwall's inequality (e.g. [14, page 428]) to (6.3) completes the proof of the theorem. $\quad\square$

It remains to show that

LEMMA 6.1. *Assertion (6.2) holds under the conditions of Theorem 4.1.*

Proof. We prove the lemma by bounding $\{q^n\}$ from above, with a sequence of processes $\{g^n\}$, for which the assertion holds. To this end, consider the sequence of processes $G^n = \{G_t^n, t \geq 0\}$, $n = 1, 2, \ldots$, which are solutions to:

$$G_t^n = Q_0^n + N_+ \left(nK \int_0^t \left(1 + \frac{G_s^n}{n}\right) ds \right),$$

where K is the constant from the linear growth condition of the theorem. The process G^n is pure birth with parameters

$$G_0^n = Q_0^n, \quad q_{k,k+1}^n = nK \left(1 + \frac{k}{n}\right), \quad k \in \mathcal{Z}^+,$$

as apparent from the interpretation discussed in Subsection 2.2. A pathwise analysis can now be used to deduce that

(6.4) $Q_t^n \leq G_t^n, \quad t \geq 0, \quad n = 1, 2, \ldots,$ a.s.

In order to prove the lemma, it is sufficient to show that (6.2) holds with q_t^n replaced by

$$g_t^n = \frac{G_t^n}{n} = q_0^n + \frac{1}{n} N_+ \left(n \int_0^t (1 + g_s^n) ds \right).$$

However, by Theorem 2.2 of Kurtz [31], $g^n \longrightarrow g$, u.o.c., a.s., as $n \uparrow \infty$, where g is the unique solution to

$$g_\bullet = q_0 + K \int_0^\bullet (1 + g_s) ds, \quad t \geq 0.$$

Hence, assertion (6.2) for g^n is established. The proof is now complete. □

Remark. The domination argument was required, since the general theorems of [31] do not treat the reflection phenomenon. Note that reflection does not arise for g^n.   □

7. Proof of FCLT. This section is devoted to the proof of Theorem 4.2. As previously, we restrict the proof to the case (4.11) only (consequently, $f_\lambda = f_\mu \equiv 0$). Commentary on the general case is provided in Remark 7.1 at the end of Subsection 7.3.

7.1. Existence and uniqueness. First we confirm that (4.9) is well-defined and enjoys a unique strong solution.

The right-hand side of (4.9) is well-defined. To see that, according to the definition of Φ^x in (3.4) we must show, *first*, that x given by (4.5) satisfies the conditions of Lemma 3.2 and *second*, that the argument of Φ^x in (4.9) is always in $C^x[0, \infty)$ (see (3.3)). First, as explained in Subsection 3.1, x given by (4.5) is continuous with $x_0 = q_0 \geq 0$ and strictly monotone or constant, which is precisely what was imposed on x in Lemma 3.2. Second, it follows from (4.5) and (4.7) that if $x_0 = q_0 = 0$, then V_0^n, $n = 1, 2, \ldots$, are non-negative, as well as V_0. Observing that the argument of Φ^x is a continuous function therefore establishes that (4.9) is well-defined.

We now appeal to known results that support existence and uniqueness of the solution to (4.9). Review the four explicit forms of (4.9) listed at the end of Subsection 4.2. In Cases 1,3 and 4, a strong unique solution exists by the arguments, given in Section 5.6 of the book by Karatzas and Shreve [22]. In Case 2, $q \equiv 0$ and existence and uniqueness of (4.9) follows from Theorem 4.1 of Tanaka [44].

7.2. The set-up of strong approximations. We prove FCLT within the framework of strong approximations. For this, recall the strong approximation result presented in Ethier and Kurtz [14, Chapter 7, Corollary 5.5]. Adapted to our context, it guarantees the existence of a probability space on which a Poisson process N and a Brownian motion B can jointly be realized so that

$$\sup_{t \geq 0} \frac{|\, N(t) - t - B(t) \,|}{\log(2 \vee t)} < \infty, \quad \text{a.s.}$$

Thus, we may start with two *independent* Brownian motions W_+ and W_- such that, for all $t \geq 0$, the following inequalities hold a.s.:

$$(7.1) \quad \left| N_+ \left(n \int_0^t \lambda(q_s^n) \, ds \right) - n \int_0^t \lambda(q_s^n) \, ds - W_+ \left(n \int_0^t \lambda(q_s^n) \, ds \right) \right| \leq K_+ \log \left(2 \vee n \int_0^t \lambda(q_s^n) \, ds \right),$$

$$(7.2) \quad \left| N_- \left(n \int_0^t \mu(q_s^n) \, ds \right) - n \int_0^t \mu(q_s^n) \, ds - W_- \left(n \int_0^t \mu(q_s^n) \, ds \right) \right| \leq K_- \log \left(2 \vee n \int_0^t \lambda(q_s^n) \, ds \right),$$

for some random variables K_+ and K_-. Assume further that

$$(7.3) \qquad \lim_{n \uparrow \infty} V_0^n = V_0, \quad \text{a.s.,}$$

and V_0 is independent of W_+ and W_-. (See the assumptions on the primitives in Section 2.)

7.3. The main steps. As explained in Subsection 3.2, the limit process V is sometimes continuous over $[0, \infty)$ or $(0, \infty)$, in which case we have U-convergence. (Note that this depends only on x.) To prove the theorem for U-convergence, we construct a sequence $\{\tilde{V}^n\}$ of continuous path stochastic processes such that, for all $T > 0$,

$$
(7.4) \quad
\begin{cases}
\mathrm{P} - \lim_{n \uparrow \infty} \left\| V^n - \tilde{V}^n \right\|_T = 0, \\
\tilde{V}^n \overset{\mathrm{d}}{=} V.
\end{cases}
$$

The assertion of the theorem now follows from Theorem 4.1 of Billingsley [4], preparing the ground for M_1-convergence. We now outline the main steps. Technical details are given subsequently.

Start by rewriting the expression for V^n (defined by (4.7)) in a form that is amenable to further calculations. To this end, introduce the sequence of processes $\tilde{X}^n = \{\tilde{X}^n_t, t \geq 0\}$, by

$$
(7.5) \quad
\begin{aligned}
\tilde{X}^n_\bullet &= n q^n_0 + n \int_0^\bullet (\lambda(q^n_s) - \mu(q^n_s)) \, ds \\
&\quad + W_+ \left(n \int_0^\bullet \lambda(q^n_s) \, ds \right) - W_- \left(n \int_0^\bullet \mu(q^n_s) \, ds \right).
\end{aligned}
$$

Then, according to (4.3) and (4.5),

$$
(7.6) \quad
\begin{aligned}
V^n &= \sqrt{n} \left[\Phi(x^n) - \Phi(x) \right] \\
&= \left[\Phi(\sqrt{n}\, x^n) - \Phi\left(\frac{1}{\sqrt{n}} \tilde{X}^n \right) \right] + \left[\Phi\left(\frac{1}{\sqrt{n}} \tilde{X}^n \right) - \Phi(\sqrt{n}\, x) \right].
\end{aligned}
$$

Equations (7.5) and (7.6) imply that

$$
(7.7) \quad V^n = \Delta^n + \left[\Phi(\sqrt{n}\, x + H^n + \epsilon^n) - \Phi(\sqrt{n}\, x) \right].
$$

Here the processes $H^n = \{H^n_t, t \geq 0\}$, $n = 1, 2, \ldots$, are given by

$$
(7.8) \quad
\begin{aligned}
H^n_\bullet &= V^n_0 + \int_0^\bullet (\lambda'(q_s) - \mu'(q_s)) V^n_s \, ds \\
&\quad + \frac{1}{\sqrt{n}} W_+ \left(n \int_0^\bullet \lambda(q_s) \, ds \right) - \frac{1}{\sqrt{n}} W_- \left(n \int_0^\bullet \mu(q_s) \, ds \right),
\end{aligned}
$$

and the processes Δ^n and ϵ^n, by

$$
(7.9) \quad \Delta^n = \Phi(\sqrt{n}\, x^n) - \Phi\left(\frac{1}{\sqrt{n}} \tilde{X}^n \right),
$$

$$
(7.10) \quad \epsilon^n = \bar{\epsilon}^n + \epsilon^n_+ - \epsilon^n_-,
$$

$$
(7.11) \quad \bar{\epsilon}^n(\bullet) = \sqrt{n} \int_0^\bullet (\lambda(q^n_s) - \lambda(q_s)) \, ds - \sqrt{n} \int_0^\bullet (\mu(q^n_s) - \mu(q_s)) \, ds
$$
$$
\quad\quad\quad\quad - \int_0^\bullet (\lambda'(q_s) - \mu'(q_s)) V^n_s \, ds,
$$

$$
(7.12) \quad \epsilon^n_+(\bullet) = \frac{1}{\sqrt{n}} W_+ \left(n \int_0^\bullet \lambda(q^n_s) \, ds \right) - \frac{1}{\sqrt{n}} W_+ \left(n \int_0^\bullet \lambda(q_s) \, ds \right),
$$

$$
(7.13) \quad \epsilon^n_-(\bullet) = \frac{1}{\sqrt{n}} W_- \left(n \int_0^\bullet \mu(q^n_s) \, ds \right) - \frac{1}{\sqrt{n}} W_- \left(n \int_0^\bullet \mu(q_s) \, ds \right).
$$

We claim that there exist processes $\tilde{V}^n = \{\tilde{V}_t^n, t \geq 0\}$ and $\tilde{H}^n = \{\tilde{H}_t^n, t \geq 0\}$, such that

$$(7.14) \quad \begin{cases} \tilde{V}^n = \Phi^x(\tilde{H}^n), \\ \tilde{H}_\bullet^n = V_0 + \int_0^\bullet (\lambda'(q_s) - \mu'(q_s))\tilde{V}_s^n \, ds + \dfrac{1}{\sqrt{n}} W_+ \left(n \int_0^\bullet \lambda(q_s) \, ds \right) \\ \qquad - \dfrac{1}{\sqrt{n}} W_- \left(n \int_0^\bullet \mu(q_s) \, ds \right). \end{cases}$$

Note that the arguments, used to establish that (4.9) is well–defined and possesses a unique strong solution, apply equally to (7.14). For convenience, rewrite equation (4.9) in a form similar to (7.14):

$$(7.15) \quad \begin{cases} V = \Phi^x(H), \\ H_\bullet = V_0 + \int_0^\bullet (\lambda'(q_s) - \mu'(q_s))V_s \, ds \\ \qquad + \int_0^\bullet \sqrt{\lambda(q_s) + \mu(q_s)} \, dW_s. \end{cases}$$

In view of the scale invariance property of any Brownian motion B, $B(\bullet) \overset{d}{=} B(n\bullet)/\sqrt{n}$, and because

$$W_+ \left(\int_0^\bullet \lambda(q_s) \, ds \right) - W_- \left(\int_0^\bullet \mu(q_s) \, ds \right) \overset{d}{=} \int_0^\bullet \sqrt{\lambda(q_s) + \mu(q_s)} \, dW_s,$$

the relations (7.14) and (7.15) yield that for all n:

$$(7.16) \quad \begin{cases} \tilde{V}^n \overset{d}{=} V, \\ \tilde{H}^n \overset{d}{=} H. \end{cases}$$

Comparing now (7.16) with (7.4) reveals that only the first assertion of (7.4) remains to be proved. For this, rewrite (7.7) in the form

$$(7.17) \quad \begin{aligned} V^n = {} & \Delta^n + \left[\Phi(\sqrt{n}\,x + H^n + \epsilon^n) - \Phi(\sqrt{n}\,x + \tilde{H}^n) \right] \\ & + \left[\Phi(\sqrt{n}\,x + \tilde{H}^n) - \Phi(\sqrt{n}\,x) \right], \end{aligned}$$

that enables us to sketch the general idea behind the rest of the convergence proof. It will be shown that the first (Δ^n) and second terms in (7.17) (the expression within the first pair of brackets on the right-hand side) converge in probability to zero, as $n \uparrow \infty$, with respect to the U-topology. The last term in (7.17) is then shown to converge weakly in the M_1-topology to $\Phi^x(H) = V$. The proof of the theorem is thus complete, by the continuity property of addition (see Appendix B) and because the limits of the first and second terms in (7.17) are continuous and non-random. (See

Theorem 4.4 by Billingsley [4] and the paper by Whitt [47] for more details.) However, the convergence proof for the second term in (7.17) is not straightforward, since this term itself depends on V^n (see (7.8)). As a standard tool in such situations, Gronwall's inequality will be used.

The proof will be carried out in two steps—first, U-convergence, followed by M_1-convergence.

Remark 7.1. To prove the general case, given by (4.8) with nonzero f_λ, f_μ, one can replicate all the considerations, but with

$$
\begin{aligned}
\widetilde{X}^n_\bullet &= nq^n_0 + n\int_0^\bullet (\lambda(q^n_s) - \mu(q^n_s))\, ds + \sqrt{n}\int_0^\bullet (f_\lambda(q^n_s) - f_\mu(q^n_s))\, ds \\
&\quad + W_+\left(n\int_0^\bullet \lambda(q^n_s)\, ds\right) - W_-\left(n\int_0^\bullet \mu(q^n_s)\, ds\right),
\end{aligned}
$$

$$
\begin{aligned}
H^n_\bullet &= V^n_0 + \int_0^\bullet (\lambda'(q_s) - \mu'(q_s))V^n_s\, ds + \int_0^\bullet (f_\lambda(q_s) - f_\mu(q_s))\, ds \\
&\quad + \frac{1}{\sqrt{n}}W_+\left(n\int_0^\bullet \lambda(q_s)\, ds\right) - \frac{1}{\sqrt{n}}W_-\left(n\int_0^\bullet \mu(q_s)\, ds\right),
\end{aligned}
$$

instead of (7.5) and (7.8).

7.4. U-convergence. In this subsection we prove the theorem for those cases where, as $n \uparrow \infty$,

$$(7.18) \qquad \Phi(\sqrt{n}\, x + b) - \Phi(\sqrt{n}\, x) \longrightarrow \Phi^x(b), \quad \text{u.o.c.,}$$

for all $b \in C^x[0, \infty)$. (See Cases 1,2 and 3 in Subsection 3.2.)

We fix $T > 0$, restrict attention to the interval $[0, T]$ and verify the first assertion in (7.4). Subtracting the expression for $\widetilde{V}^n$, given by (7.14), from (7.17) and using the Lipschitz property of Φ (C being the Lipschitz constant), one can write for $t \le T$:

$$
(7.19) \quad
\begin{aligned}
\left\|V^n - \widetilde{V}^n\right\|_t &\le \|\Delta^n\|_T + C\,\|\epsilon^n\|_T + C\left\|H^n - \widetilde{H}^n\right\|_t \\
&\quad + \left\|\Phi(\sqrt{n}\, x + \widetilde{H}^n) - \Phi(\sqrt{n}\, x) - \Phi^x(\widetilde{H}^n)\right\|_T.
\end{aligned}
$$

For the third term on the right-hand side of (7.19) we have ($t \le T$):

$$(7.20) \qquad \left\|H^n - \widetilde{H}^n\right\|_t \le |V^n_0 - V_0| + C_T\int_0^t \left\|V^n - \widetilde{V}^n\right\|_s\, ds,$$

where $C_T = \|\lambda'(q_s) - \mu'(q_s)\|_T$ is finite by the continuity of λ', μ' and q. Combining (7.19) with (7.20) and applying Gronwall's inequality yields

$$
(7.21) \quad
\begin{aligned}
&\left\|V^n - \widetilde{V}^n\right\|_t \\
&\le (C\,|V^n_0 - V_0| + \|\Delta^n\|_T + C\,\|\epsilon^n\|_T + \|\epsilon^n_\Phi\|_T)e^{T \cdot C \cdot C_T},
\end{aligned}
$$

where the process ϵ_Φ^n is given by

$$\epsilon_\Phi^n = \Phi(\sqrt{n}\,x + \tilde{H}^n) - \Phi(\sqrt{n}\,x) - \Phi^x(\tilde{H}^n).$$

In view of (7.3), U-convergence will be established once it is shown that the second, third and fourth terms in the parentheses on the right-hand side of (7.21) converge in probability to zero. Each of these terms will now be analyzed separately.

Review the definition (7.9) of Δ^n. Denote the quantities between the absolute value signs on the left-hand side in (7.1) and in (7.2) by $\Delta_\lambda^n(t)$ and $\Delta_\mu^n(t)$ respectively. It follows, by the Lipschitz property of Φ (C being the Lipschitz constant), that

$$(7.22) \quad \|\Delta^n\|_T \le C\left\|\sqrt{n}\,x^n - \frac{1}{\sqrt{n}}\tilde{X}^n\right\|_T \le C\frac{1}{\sqrt{n}}(\|\Delta_\lambda^n\|_T + \|\Delta_\mu^n\|_T).$$

Due to the FSLLN (or by Lemma 6.1) and by the locally Lipschitz property of λ and μ, relations (7.1), (7.2) and (7.22) imply the convergence, as $n \uparrow \infty$:

$$(7.23) \qquad \frac{\Delta_\lambda^n}{\sqrt{n}}, \frac{\Delta_\mu^n}{\sqrt{n}}, \Delta^n \longrightarrow 0 \ \ u.o.c., \ a.s.$$

Review now the definition (7.10) of ϵ^n. We show that each term in the sum on the right-hand side of (7.10) converges in probability to zero with respect to the U-topology, and, hence, ϵ^n does so too.

It will be shown in Lemma 7.1 (see Subsection 7.6) that

$$(7.24) \qquad \lim_{n\uparrow\infty} \|\bar{\epsilon}^n\|_T = 0, \ \ \text{a.s.}$$

We now apply Lemma 7.2 presented in Subsection 7.6, with

$$g_\bullet^n = \int_0^\bullet \lambda(q_s^n)\,ds, \ \ g_\bullet = \int_0^\bullet \lambda(q_s)\,ds,$$

to get

$$(7.25) \qquad \text{P} - \lim_{n\uparrow\infty} \|\epsilon_+^n\|_T = 0.$$

(The conditions of Lemma 7.2 are satisfied by the FSLLN and because of the auxiliary assertion (7.30) obtained in Lemma 7.1.) Similarly, we obtain

$$(7.26) \qquad \text{P} - \lim_{n\uparrow\infty} \|\epsilon_-^n\|_T = 0.$$

Finally, (7.16) and (7.18) imply

$$(7.27) \qquad \text{P} - \lim_{n\uparrow\infty} \|\epsilon_\Phi^n\|_T = 0.$$

This completes the proof of U-convergence.

7.5. M_1-convergence. We now consider the case

$$\Phi(\sqrt{n}\,x + b) - \Phi(\sqrt{n}\,x) \longrightarrow \Phi^x(b), \quad \text{as } n \uparrow \infty,$$

in the M_1-topology, but not in the U-topology (see Cases 4 in Subsection 3.2). Then, combining (7.17) with (7.16) reveals that the third term in (7.17), namely the expression within the last pair of brackets, converges weakly in the M_1-topology to $\Phi^x(H) = V$. It will be shown further that the second term in (7.17) converges in probability to zero with respect to the U-topology. In view of (7.23), the proof is then complete. (See Theorem 4.4 by Billingsley [4] and the paper by Whitt [47] and recall the arguments at the end of Subsection 7.3.)

In order to show that the second term in (7.17) converges to zero in the U-topology, we apply the Lipschitz property of Φ^x to this term and obtain ($t \leq T$):

(7.28)
$$\left\| \Phi(\sqrt{n}\,x + H^n + \epsilon^n) - \Phi(\sqrt{n}\,x + \tilde{H}^n) \right\|_t$$
$$\leq C \left\| \epsilon^n \right\|_T + C \left\| H^n - \tilde{H}^n \right\|_T.$$

It will be shown further that the last term on the right–hand side of (7.28) converges to zero in probability. Then, combining (7.28) with definition (7.10) of ϵ^n and using (7.24)–(7.26) proves the desired assertion.

In order to deduce that the last term in (7.28) converges in probability to zero, review inequality (7.20). Since the first term on the right-hand side of (7.20) converges to zero a.s. (by (7.3)), it is sufficient to check that the second term in (7.20), for $t = T$, converges in probability to zero. But this follows from Remark 3.3 and (7.30), and the fact that theorem is already proved for the case of U-convergence.

7.6. Lemmata.

LEMMA 7.1. *The sequence $\{\bar{\epsilon}^n\}$ given by (7.11) satisfies (7.24).*

Proof. Our calculations resemble those in Chapter 8,§3 of the book by Liptser and Shiryayev [36, pages 635,636], where they are presented in the context of martingale theory.

From the definition (7.11) of $\bar{\epsilon}^n$ and by the locally Lipschitz continuity of λ' and μ' we obtain, in view of the definition (4.7) of V^n:

(7.29)
$$|\bar{\epsilon}^n(t)| \leq \int_0^t \left(|\lambda'(q_s + \varphi_s^1(q_s^n - q_s))\sqrt{n}(q_s^n - q_s) - \lambda'(q_s)V_s^n| \right.$$
$$\left. + |\mu'(q_s + \varphi_s^2(q_s^n - q_s))\sqrt{n}(q_s^n - q_s) - \mu'(q_s)V_s^n| \right) ds$$
$$\leq C_T \|V^n\|_t \|q^n - q\|_t\, t, \quad t \leq T,$$

where $\varphi_s^1, \varphi_s^2 \in [0,1]$, and C_T is a constant. By the FSLLN (Theorem 4.1), it follows from the last equation that to prove the lemma it is sufficient to show that the following holds a.s.:

(7.30) $$\overline{\lim_{n}} \|V^n\|_t < \infty, \quad t \leq T.$$

In order to prove (7.30) we continue as follows. From the definition of V^n (see (4.7)) one obtains

$$
\begin{aligned}
\|V^n\|_t \;\leq\;& C\sqrt{n}\,\|x^n - x\|_t \\
\leq\;& C\,|V_0^n| + C\left\| \sqrt{n}\left(\frac{1}{n}N_+\left(n\int_0^\bullet \lambda(q_s^n)\,ds \right) - \int_0^\bullet \lambda(q_s^n)\,ds \right) \right\|_t \\
& + C\left\| \sqrt{n}\left(\frac{1}{n}N_-\left(n\int_0^\bullet \mu(q_s^n)\,ds \right) - \int_0^\bullet \mu(q_s^n)\,ds \right) \right\|_t \\
& + C\left\| \sqrt{n}\int_0^t (\lambda(q_s^n) - \lambda(q_s))\,ds \right\|_t \\
& + C\left\| \sqrt{n}\int_0^t (\mu(q_s^n) - \mu(q_s))\,ds \right\|_t, \quad t \leq T.
\end{aligned}
$$

Using the FSLLN and the local Lipschitz continuity of λ and μ, we obtain the existence of a (possibly) random M, and positive non-random scalars F_T, L_T, such that for all $n \geq M$ the following inequality holds a.s. ($t \leq T$):

(7.31)
$$
\begin{aligned}
\|V^n\|_t \;\leq\;& C\,|V_0^n| + C\left\| \sqrt{n}\left(\frac{1}{n}N_+(ns) - s \right) \right\|_{F_T} \\
& + C\left\| \sqrt{n}\left(\frac{1}{n}N_-(ns) - s \right) \right\|_{F_T} + L_T\int_0^t \|V^n\|_s\,ds.
\end{aligned}
$$

Note that, as $n \uparrow \infty$,

$$\sqrt{n}\left(\frac{1}{n}N_+(n\bullet) - \bullet \right) \longrightarrow W_+, \quad \text{a.s.}$$

and analogously for N_-. This fact, the convergence (7.3) of $\{V_0^n\}$ and Gronwall's inequality applied to (7.31) complete the proof of the lemma. □

Condition (7.30) implies the so-called compact containment condition (see Ethier and Kurtz [14, page 129]):

$$\lim_{\ell \uparrow \infty} \overline{\lim_{n}} P\{\|V^n\|_t > \ell\} = 0, \quad t \leq T.$$

This condition is often involved in proving weak limit theorems and is used in the following

LEMMA 7.2. *Let $\{g^n\}$ be a sequence of stochastic processes with mono-tone increasing sample paths. Let g be a monotone increasing determinis-tic function, and let B denote a Brownian motion. Further, for all n, let $g_0^n = g_0 = 0$. Assume, in addition, that for all $T > 0$,*

(7.32) $$\lim_{n \uparrow \infty} \|g^n - g\|_T = 0, \quad \text{a.s.},$$

and

(7.33) $$\lim_{\ell\uparrow\infty}\overline{\lim_{n}}\,P\left\{\sqrt{n}\,\|g^n - g\|_T > \ell\right\} = 0.$$

Then

(7.34) $$P - \lim_{n\uparrow\infty}\frac{1}{\sqrt{n}}\,\|B(ng^n) - B(ng)\|_T = 0,$$

for all $T > 0$.

Proof. Introduce the random variables T^n and $\tilde{T}^n$ by

$$T^n = n\,\|g^n - g\|_T\,,$$

$$\tilde{T}^n = n\left(g^n(T) \vee g(T)\right).$$

Evidently, $0 \le T^n \le \tilde{T}^n$. Without loss of generality, consider the case $0 < T^n < \tilde{T}^n$.

Fix $\varepsilon > 0$. By (7.33),

$$\lim_{\ell\uparrow\infty}\overline{\lim_{n}}\,P\left\{\frac{T^n}{\sqrt{n}} > \ell\right\} = 0.$$

In view of (7.32), we can choose $\ell_\varepsilon > 0$, a natural number N_ε and a set B_ε such that for all $n > N_\varepsilon$

(7.35) $$P\left\{\frac{T^n}{\sqrt{n}} > \ell_\varepsilon\right\} < \varepsilon,$$

and

$$\begin{cases} \tilde{T}^n \le F^n \triangleq 2n(g(T) \vee 1) \text{ on } B_\varepsilon, \\ P\{B_\varepsilon^c\} \le \varepsilon. \end{cases}$$

Denoting

$$A^n = \frac{1}{\sqrt{n}}\,\|B(ng^n) - B(ng)\|_T\,,$$

$$S(\alpha,\beta,\gamma) = \{u, v : 0 \le u, v \le \alpha,\ \beta <\mid u - v \mid\le \gamma\}\,,$$

we obtain for $n > N_\varepsilon$:

$$P\{A^n > \varepsilon\} \leq P\left\{(A^n > \varepsilon) \bigcap B_\varepsilon\right\} + P\{B_\varepsilon^c\}$$

$$\leq P\left\{\sup_{S(F^n,0,T^n)} |B(u) - B(v)| > \varepsilon\sqrt{n}\right\} + \varepsilon$$

$$\leq P\left\{\sup_{S(F^n,0,\ell_\varepsilon\sqrt{n})} |B(u) - B(v)| > \varepsilon\sqrt{n}\right\}$$

$$+ P\left\{\sup_{S(F^n,\ell_\varepsilon\sqrt{n},T^n)} |B(u) - B(v)| > \varepsilon\sqrt{n}\right\} + \varepsilon.$$

By (7.35), the second term on the right-hand side of the last inequality is less than ε, and therefore we restrict our attention to the first term only.

Recall Lemma 1.2.1 in the book by Csörgő and Révész [13], which asserts the following. For any positive δ, there exists a constant $C = C(\delta)$ such that the inequality

$$P\left\{\sup_{S(F,0,\ell)} |B(u) - B(v)| > p\sqrt{\ell}\right\} \leq C(1 + \frac{F}{\ell})e^{-\frac{p^2}{2+\delta}}$$

holds for every positive p and $0 < \ell < F$. (This form of Lemma 1.2.1 in [13] is taken from [10].) Using this assertion and continuing our calculation, we obtain

$$P\{A^n > \varepsilon\}$$

$$\leq 2\varepsilon + P\left\{\sup_{S(F^n,0,\ell_\varepsilon\sqrt{n})} |B(u) - B(v)| > \frac{\varepsilon\sqrt{n}}{(\ell_\varepsilon\sqrt{n})^{1/2}}(\ell_\varepsilon\sqrt{n})^{1/2}\right\}$$

$$\leq 2\varepsilon + C\left(1 + \frac{2(g(T)\vee 1)\sqrt{n}}{\ell_\varepsilon}\right)e^{-\frac{\varepsilon^2\sqrt{n}}{\ell_\varepsilon(2+\delta)}},$$

which implies the assertion of the lemma.

7.7. Proof of Theorem 4.3. The proof of Theorem 4.3 is omitted, being similar to that of Theorem 4.2, except for the following comments. Recall that fluid limits q are strictly monotone or constant and reconsider (7.29) (the step in the proof where the Lipschitz properties of λ', μ' are used). By a simple modification of the arguments, one concludes that Theorem 4.2 holds without any changes, with the exception of the special situation (4.10). In that case, one must separate the analysis of (7.29) to the right and the left neighborhood of q_0.

8. Directions for future research. Of interest are extensions to the current model that cover time- and state-dependent rates, other performance measures such as waiting time and work-loads, and random or discontinuous λ and μ. The latter would enable, among other things, analysis of models with finite buffers, breakdowns and batch service.

Other possible extensions are to non-exponential models. The approach taken here should carry over, but the details would naturally depend on the particular model at hand. (See, for example, a steady-state analysis of state-dependent $M_\xi/G_\xi/1$ queues in Knessl *et al.* [26]; diffusion approximations of phase-type models in Whitt [48] and Krichagina [29]; fluid and diffusion approximations of various semi-Markovian models in Anisimov [1]).

Work is currently ongoing on approximating state–dependent networks, that includes state–dependent routing. Fluid limits for such networks are solutions to autonomous ordinary differential equations with *state–dependent* oblique reflection. Diffusion limits are solutions to stochastic differential equations with *time–dependent* oblique reflection. The diffusion limits are Markov processes with possibly discontinuous sample-paths. Weak convergence is with respect to Skorokhod's M_1-*topology*.

Fluid limits of networks (as solutions to a multi-dimensional differential equation) need not be monotone functions and can leave a boundary, after having reached it. As a consequence, the diffusion limits could have multiple points of discontinuity. Furthermore, the characterization of fluid and diffusion limits involves reflection problems with non-constant directions of reflections, varying with time and state. Such mappings are less well-behaved than the usual multi-dimensional Skorokhod maps (in particular, they need not be Lipschitz). All this suggests that new tools must be developed in order to establish convergence, existence and uniqueness of the limits.

A. Skorokhod's reflection problem. We use the following version of the one-dimensional Skorokhod's reflection problem (taken from [9]):

THEOREM. *For any $x \in D_0[0,\infty)$, there exist a unique pair $(q,y) \in D_0[0,\infty) \times D_0[0,\infty)$ satisfying*

$$\begin{cases} q_t = x_t + y_t \geq 0, \ \ t \geq 0, \\ y \ \text{nondecreasing, with} \ y_0 = 0, \\ \displaystyle\int_0^\infty 1[q_t>0] \, dy_t = 0. \end{cases}$$

The operators Φ and Ψ with domain $D_0[0,\infty)$, given by

$$q = \Phi(x), \quad y = \Psi(x),$$

are both Lipschitz continuous with respect to the uniform norm. Namely, there exists a constant $C > 0$, such that

$$\left\|\Phi(x^1) - \Phi(x^2)\right\|_T \leq C\left\|x^1 - x^2\right\|_T,$$
$$\left\|\Psi(x^1) - \Psi(x^2)\right\|_T \leq C\left\|x^1 - x^2\right\|_T,$$

for all $x^1, x^2 \in D_0[0, \infty)$ and $T > 0$. Furthermore, Φ and Ψ are both homogeneous of degree 1:

$$\Phi(\gamma x) = \gamma\Phi(x),$$
$$\Psi(\gamma x) = \gamma\Psi(x),$$

for all $x \in D_0[0, \infty)$ and $\gamma \geq 0$.

Note that both the theorem cited above and all properties of Φ and Ψ hold when we use, instead of $D_0[0, \infty)$, the space of the $\mathbb{R}^d$-valued RCLL functions with non-negative values at zero. However, only in the one-dimensional case do Φ and Ψ have the explicit forms:

$$\Psi_t(x) = \sup_{0 \leq s \leq t}(x_s^-), \quad t \geq 0,$$
$$\Phi(x) = x + \Psi(x) = x + \overline{x^-}.$$

B. Weak convergence. We use in this paper the set $\widetilde{D}[0, \infty)$ of all real-valued functions on $[0, \infty)$ with right and left limits at each point. Values of functions are assumed to be equal to either the left or the right limit. Note that discontinuities at zero are admissible.

Our weak convergence results are proved for the space $(\widetilde{D}[0, \infty), M_1)$, that is $\widetilde{D}[0, \infty)$ endowed with Skorokhod's M_1-topology, see [43]. The appropriate definitions of the M_1-topology and, respectively M_1-convergence, for $\widetilde{D}[0, \infty)$ (which slightly differs from the space used in [43]) can be given within the unified graph approach of Pomarede [40]. For the extension of Pomarede's definitions to the non-compact interval $[0, \infty)$, see, e.g., Whitt [45] and [47].

We use the following properties of the M_1-topology:

1. Let $\{x^n\}$ converge to x in the M_1-topology. If x is an element of $C[0, \infty)$, then the M_1-topology reduces to the topology of uniform convergence on compact sets (U-topology). Uniform convergence is referred to as U-convergence.

2. Theorem 3.1 by Pomarede [40], on M_1-convergence: Let $x^n \longrightarrow x$, $y^n \longrightarrow y$, as $n \uparrow \infty$. Then, $x^n + y^n \longrightarrow x + y$, if x and y have no common points of discontinuity.

In Theorem 4.3 we use the ordinary Skorokhod space $(D[0, \infty), J_1)$.

C. Notation.

RCLL	right-continuous with left limits
u.o.c	uniformly on compact
$1[S]$	indicator function of a set S
$f_t \uparrow\uparrow a$	f is strictly increasing and $\lim\limits_{t \uparrow \infty} f_t = a$
$f_t \downarrow\downarrow a$	f is strictly decreasing and $\lim\limits_{t \uparrow \infty} f_t = a$
$\vee$ and $\wedge$	maximum and minimum
$a^- = -(a \wedge 0)$	the negative part of a
$\overline{f}(t) = \sup\limits_{0 \leq s \leq t} f_s$	the upper envelope of f
$\|f\|_T = \sup\limits_{0 \leq s \leq T} \|f\|$	the uniform norm of f on the interval $[0, T]$
$\mathcal{Z}^+$ and $I\!\!R^+$	the sets of non-negative integer and real numbers
$C[0, \infty)$	the set of continuous real-valued functions on $[0, \infty)$
$C_0[0, \infty)$	$\{f \in C[0, \infty)\| f_0 \geq 0\}$
$D[0, \infty)$	the set of RCLL real-valued functions
$D_0[0, \infty)$	$\{f \in D[0, \infty)\| f_0 \geq 0\}$
$D_E[0, \infty)$	the set of RCLL E-valued functions
$\widetilde{D}[0, \infty)$	see Appendix B
$(\widetilde{D}[0, \infty), J_1)$	the space $\widetilde{D}[0, \infty)$ endowed with Skorokhod's J_1-topology
$(\widetilde{D}[0, \infty), M_1)$	the space $\widetilde{D}[0, \infty)$ endowed with Skorokhod's M_1-topology
$\overset{d}{\sim}$	is distributed as
$\overset{d}{\longrightarrow}$	convergence in distribution
$P - \lim$	limit in probability
$\mathcal{N}(\delta, \sigma^2)$	the normal distribution with mean δ and variance σ^2
$BM(\delta, \sigma^2)$	Brownian motion with drift δ and variance σ^2, starting at 0
$BM_x(\delta, \sigma^2)$	Brownian motion with drift δ and variance σ^2, starting at x
$RBM(\delta, \sigma^2)$	Reflected Brownian motion with drift δ and variance σ^2, starting at 0
$RBM_x(\delta, \sigma^2)$	Reflected Brownian motion with drift δ and variance σ^2, starting at x

REFERENCES

[1] V. V. Anisimov. Switching processes: Asymptotic theory and applications. In A. N. Shiryayev et. al., editors, New Trends in Probability and Statistics. To appear.

[2] S. V. Anulova. Functional limit theorems for network of queues. In IFAC Congress, Tallinn, 1990. Abstract.

[3] A. D. Barbour. On a functional central limit theorems for Markov population processes. Advances in Applied Probability, 6:21–39, 1974.

[4] P. Billingsley. Convergence of Probability Measures. John Wiley and Sons, New York, 1968.

[5] A. Borovkov. Some limit theorems in the theory of mass service, II. Theory of Probability and Its Applications, 10:375–400, 1965.

[6] A. Borovkov. On limit laws for service processes in multi-channel systems. Siberian Mathematical Journal, 8:746–763, 1967.

[7] P. Bremaud. Point Processes and Queues: Martingale Dynamics. Springer-Verlag, Berlin, 1981.

[8] H. Chen and A. Mandelbaum. Discrete flow networks: Diffusion approximations and bottlenecks. The Annals of Probability, 19:1463–1519, 1991.

[9] H. Chen and A. Mandelbaum. Hierarchical modelling of stochastic networks, part I: Fluid models. In D. Yao, editor, Stochastic Modelling and Analysis of Manufacturing Systems, pages 47–105. Springer-Verlag, New York, 1994.

[10] H. Chen and A. Mandelbaum. Hierarchical modelling of stochastic networks, part II: Strong Approximations. In D. Yao, editor, Stochastic Modelling and Analysis of Manufacturing Systems, pages 107–131. Springer-Verlag, New York, 1994.

[11] E. G. Coffman, Jr., A. A. Puhalskii, M. I. Reiman, and P. Wright. Processor shared buffers with reneging. Performance Evaluation, 19:25–46, 1994.

[12] E. G. Coffman, Jr. and M. I. Reiman. Diffusion approximation for computer communication systems. In G. Iazeolla, P. J. Courtois, and A. Hordijk, editors, Mathematical Computer Performance and Reliability, pages 33–53. North-Holland, Amsterdam, 1984.

[13] M. Csörgö and P. Révész. Strong Approximations in Probability and Statistics. Academic Press, New York, 1981.

[14] S. N. Ethier and T. G. Kurtz. Markov Process: Characterization and Convergence. John Wiley and Sons, New York, 1986.

[15] P. W. Glynn. Diffusion approximations. In D. P. Heyman and M. J. Sobel, editors, Handbooks in Operations Research and Management Science, Vol. 2, pages 145–198. North-Holland, Amsterdam, 1990.

[16] J. Hale. Ordinary Differential Equations. J. Wiley & Sons/Interscience, New York, 1969.

[17] S. Halfin and W. Whitt. Heavy-traffic limits theorem for queues with many exponential servers. Operations Research, 29:567–588, 1981.

[18] D. L. Iglehart. Limit diffusion approximations for the many-server queue and repairman problem. Journal of Applied Probability, 2:429–441, 1965.

[19] D. L. Iglehart and A. J. Lemoine. Approximations for the repairman problem with two repair facilities, I: No spares. Advances in Applied Probability, 5:595–613, 1973.

[20] D. L. Iglehart and W. Whitt. Multiple channel queues in heavy traffic, I. Advances in Applied Probability, 2:150–177, 1970.

[21] D. L. Iglehart and W. Whitt. Multiple channel queues in heavy traffic, II: Sequences, networks, and batches. Advances in Applied Probability, 2:355–364, 1970.

[22] I. Karatzas and S. E. Shreve. Brownian Motion and Stochastic Calculus. Springer-Verlag, New York, 1988.

[23] H. Kaspi and A. Mandelbaum. Regenerative closed queueing networks. Stochastics

and Stochastics Reports, 39:239–258, 1992.

[24] J. F. C. Kingman. The single server queue in heavy traffic. *Proceedings of the Cambridge Philosophical Society*, 57:902–904, 1961.

[25] J. F. C. Kingman. The heavy traffic approximations in the theory of queues. In W. Smith and W. Wilkinson, editors, *Proceedings of the Symposium on Congestion Theory*, pages 137–159. The University of North California Press, Chapel Hill, 1965.

[26] C. Knessl, B. J. Matkowsky, Z. Schuss, and C. Tier. On the performance of state-dependent single server queues. *SIAM Journal on Applied Mathematics*, 46(4):657–697, August 1986.

[27] Y. A. Kogan and R. S. Liptser. Limit non-stationary behavior of large closed queueing networks with bottlenecks. *Queueing Systems*, 14:33–55, 1993.

[28] Y. A. Kogan, R. S. Liptser, and A. V. Smorodinskii. Gaussian diffusion approximation of closed Markov models of computer networks. *Problems of Information Transmission*, 22(1):38–51, 1986.

[29] E. V. Krichagina. Diffusion approximation for a queue in a multiserver system with multistage service. *Automation and Remote Control*, 50(3):346–354, 1989.

[30] E. V. Krichagina. Asymptotic analysis of queueing networks (martingale approach). *Stochastics and Stochastics Report*, 40:43–76, 1992.

[31] T. G. Kurtz. Strong approximation theorems for density dependent Markov chains. *Stochastic Processes and Their Applications*, 6:223–240, 1978.

[32] T. G. Kurtz. Representation of markov processes as multiparameter time changes. *The Annals of Probability*, 8(4):682–715, 1980.

[33] T. G. Kurtz. Representation and approximation of counting processes. In W. H. Fleming and L. G. Gorostiza, editors, *Advances in Filtering and Optimal Stochastic Control, Lecture Notes in Control and Information Sci. 42*, pages 177–191. Springer-Verlag, Berlin, 1982.

[34] T. G. Kurtz. Gaussian approximations for markov chains and counting processes. In *44th Session of the International Statistical Institute*, Madrid, Spain, September 1983.

[35] A. J. Lemoine. Networks of queues—a survey of weak convergence results. *Management Science*, 24:1175–1193, 1978.

[36] R. S. Liptser and A. N. Shiryayev. *Theory of Martingales*. Kluwer Academic Publishers, 1989.

[37] A. Mandelbaum and W. A. Massey. Strong approximations for time–dependent queues. To be published in Mathematics of Operations Research, 1994.

[38] G. F. Newell. *Applications of Queueing Theory*. Chapman and Hall, 1982.

[39] R. M. Oliver and A. H. Samuel. Reducing letter delays in post offices. *Operations Research*, 10:839–892, 1962.

[40] J. L. Pomarede. *A Unified Approach via Graphs to Skorokhod's Topologies on the Function Space*. PhD thesis, Department of Statistics, Yale University, 1976.

[41] N. Prabhu. *Stochastic Storage Processes, Queues, Insurance Risk and Dams*. Springer-Verlag, New York, 1980.

[42] M. I. Reiman. Some diffusion approximations with state-space collapse. In F. Baccelli and G. Fayolle, editors, *Modelling and Performance Evaluation Methodology*. Springer-Verlag, 1984.

[43] A. V. Skorokhod. Limit theorems for stochastic processes. *Theory of Probability and Its Applications*, 1:261–290, 1956.

[44] H. Tanaka. Stochastic differential equations with reflected boundary conditions in convex region. *Hiroshima Mathematical Journal*, 9:163–174, 1974.

[45] W. Whitt. Weak convergence of first passage time processes. *Journal of Applied Probability*, 8:417–422, 1971.

[46] W. Whitt. Heavy traffic theorems for queues: A survey. In A. B. Clarke, editor, *Mathematical Methods in Queueing Theory*, pages 307–350. Springer-Verlag, Berlin, 1974.

[47] W. Whitt. Some useful functions for functional limit theorems. *Mathematics of*

Operations Research, 5(1):67–85, February 1980.

[48] W. Whitt. On the heavy traffic limit theorem for $GI/G/\infty$ queues. Advances in Applied Probability, 14:171–190, 1982.

[49] K. Yamada. Multi-dimensional Bessel processes as heavy traffic limits of certain tandem queues. Stochastic Processes and Their Applications, 23:35–56, 1986.

[50] K. Yamada. Diffusion approximations for open state-dependent queueing networks under heavy traffic situation. Technical report, Institute of Information Science and Electronics, University of Tsukuba, Tsukuba, Ibaraki 305, Japan, 1993.

A STATE-DEPENDENT POLLING MODEL
WITH MARKOVIAN ROUTING

GUY FAYOLLE* AND JEAN-MARC LASGOUTTES*

Abstract. A state-dependent 1-limited polling model with N queues is analyzed. The routing strategy generalizes the classical Markovian polling model, in the sense that two routing matrices are involved, the choice being made according to the state of the last visited queue. The stationary distribution of the position of the server is given. Ergodicity conditions are obtained by means of an associated dynamical system. Under rotational symmetry assumptions, average queue length and mean waiting times are computed.

1. Introduction. Consider a taxicab in a city in which there are N stations at which clients arrive and wait for the vehicle. When their turn has come to be served, they ask for a transit to a destination (one of the other stations) where they leave the system. Whenever the taxicab finds a station empty, it goes somewhere else to look for a client. The choice of the destinations by a client or by the empty taxicab are made via the two distinct routing matrices P and $\tilde{P}$.

This system can also be seen as a polling model with N queues at which customers arrive and one server which visits the queues according to the following rules: if the server finds a client at the current queue, it serves this client and chooses a new queue according to the routing matrix P; otherwise it selects the next queue according to $\tilde{P}$. This polling scheme is an extension of the classical Markovian polling model, with routing probabilities depending on the state of the last visited queue. As a taxicab only takes one client at a time, the service strategy is the 1-*limited* strategy, which is known to be more difficult to analyze than either the *gated* or *exhaustive* strategies—in which the server respectively serves the clients present on its arrival or serves clients until the queue is empty.

Since the bibliography on polling models is plethoric, we refer the reader to the references given in Takagi [16] and Levy and Sidi [10]. Among the studies related to our work, we can cite Kleinrock and Levy [9], who compute waiting times in random polling models in which the destination of the server is chosen independently of its provenance, and Boxma and West-strate [2] who give a pseudo-conservation law for a model with Markovian routing. Ferguson [6] and Bradlow and Byrd [3] study approximatively a model where the switching time is station-dependent. Srinivasan [15] analyzes a polling system with the same routing policy as in our model, and with different service policies at each queue. However, the waiting times are not computed in the case where the routing is state dependent. Another important issue concerns state-independent ergodicity conditions.

* Postal address: INRIA — Domaine de Voluceau, Rocquencourt, BP105 — 78153 Le Chesnay, France.

In this case, necessary and sufficient conditions were obtained by Fricker and Jaïbi [7,8] for deterministic and Markovian routing with various service disciplines. Borovkov and Schassberger [1] and Fayolle *et al.* [5] give necessary and sufficient conditions for a system with Markovian routing and 1-limited service. Finally, Schassberger [13] gives a necessary condition for the ergodicity of a polling model with 1-limited service and a routing that depends on the whole state of the system.

Here, we give a new method to get the stationary distribution of the position of the server and, in the case of a fully symmetrical system, a way to compute the mean waiting time of a customer. This extends known results on 1-limited polling models, especially for the symmetrical case. However, we do not provide a pseudo-conservation law as for example Boxma and Weststrate [2], since the computations are not as simple as in the state-independent routing case. Our method of proof allows to make only minimal assumptions on the arrival process and applies to either discrete or continuous time models. Moreover, in the symmetrical case the results allow to compare various polling strategies.

The paper is organized as follows: in Section 2, we present the model and give a functional equation describing its evolution. In Section 3, stationary probabilities of the position of the server are obtained and Section 4 is devoted to the general problem of ergodicity. Section 5 presents formulas to compute the first moment of the queue length at polling instants. These results are used in Section 6 to calculate the waiting time of an arbitrary customer. Applications to known models are also given.

2. Description of the model. The system consists of N stations at which clients arrive according to a stationary process. Let $S \overset{\text{def}}{=} \{1, \ldots, N\}$ be the set of stations. Assume that the server arrives after $n - 1$ moves at station $i \in S$ where a client is waiting. Then the server loads this client and goes to station $j \in S$ with probability $p_{i,j}$, such that $p_{i,1} + \cdots + p_{i,N} = 1$. Conversely, if station i is empty, the server polls station j with probability $\tilde{p}_{i,j}$. The service policy is known as *1-limited* policy, since at most 1 customer is served each time a station is visited. The two transition matrices will be denoted respectively by

$$P = (p_{i,j})_{i,j \in S}, \quad \tilde{P} = (\tilde{p}_{i,j})_{i,j \in S}.$$

The number of new customers arriving to station $q \in S$ between the polling of stations i and j when there have been a service (resp. no service) at station i is $B_{i,j;q}(n)$ (resp. $\tilde{B}_{i,j;q}(n)$). The vectors $B_{i,j}(n) = (B_{i,j;1}(n), \ldots, B_{i,j;N}(n))$ (resp. $\tilde{B}_{i,j}(n)$) are i.i.d. for different n, but their components may be dependents and not identically distributed. In the classical polling terminology, $B_{i,j}(n)$ is the number of arrivals during *both* the service and the switchover and $\tilde{B}_{i,j}(n)$ is the number of arrivals during a switchover. However, our model covers a wider class of applications. In particular, the service time may depend on both i and j. In addition, the switchover time distribution may depend on the state of the last visited queue (empty or not). Remark also that, unlike in Srinivasan [15], we can also have $\mathbb{E}\, B_{i,j}(n) < \mathbb{E}\, \widetilde{B}_{i,j}(n)$, which means that serving a customer can actually take less time than just moving. This would amount to *negative* service times in the usual polling formulation, which are not easy to handle!

It is worth noting that our setting is valid for both discrete-time and continuous-time models with or without batch arrivals. We do not need a separate analysis for each case, until the computation of waiting times, where we will have to give precise definitions of the vectors $B_{i,j}(n)$ and $\widetilde{B}_{i,j}(n)$. We do not describe yet the exact arrival process and the time taken to travel from one station to another, since we only need the distribution of the number of clients arriving during such a move. However, due to the assumption of independence between successive arrivals, the model applies mainly to the following situations:

- discrete-time evolution, when a batch of customers arrives at each station at the beginning of each time slot; batches are i.i.d. with respect to the time slots but they can be correlated between stations;
- continuous time evolution, when the arrival process is a compound Poisson process: customers arrive at the system in batches at the epochs of a Poisson process and the batches have the same properties as in the previous case.

Let $X_i(n)$ be the number of clients waiting at station i at polling instant n and $S(n)$ be the corresponding position of the server. If we define $X(n) \overset{\text{def}}{=} (X_1(n), \ldots, X_N(n))$, then $\mathcal{L} \overset{\text{def}}{=} (S, X)$ forms a Markov chain. Throughout this paper, we will assume that this chain is irreducible. For any vector $\vec{z} = (z_1, \ldots, z_N) \in \mathcal{D}^N$ (where $\mathcal{D}$ denotes the unit disc in the complex plane), we note $\vec{z}^B = z_1^{B_1} \cdot z_2^{B_2} \cdots z_N^{B_N}$ and define the following generating functions:

$$a_{i,j}(\vec{z}) \overset{\text{def}}{=} \mathbb{E}[\vec{z}^{B_{i,j}(n)}],$$

$$\tilde{a}_{i,j}(\vec{z}) \overset{\text{def}}{=} \mathbb{E}[\vec{z}^{\widetilde{B}_{i,j}(n)}],$$

$$F_i(\vec{z}; n) \overset{\text{def}}{=} \mathbb{E}[\vec{z}^{X(n)} \mathbf{1}_{\{S(n)=i\}}],$$

$$\widetilde{F}_i(\vec{z}; n) \overset{\text{def}}{=} \mathbb{E}[\vec{z}^{X(n)} \mathbf{1}_{\{S(n)=i, X_i(n)=0\}}],$$

$$= F_i(z_1, \ldots, z_{i-1}, 0, z_{i+1}, \ldots, z_N; n),$$

$$F_i(\vec{z}) \overset{\text{def}}{=} \lim_{n \to \infty} F_i(\vec{z}; n),$$

$$\widetilde{F}_i(\vec{z}) \overset{\text{def}}{=} \lim_{n \to \infty} \widetilde{F}_i(\vec{z}; n),$$

where $\mathbf{1}_{\mathcal{E}}$ is as usual the indicator function of the set $\mathcal{E}$. The following result holds:

THEOREM 2.1. *Let $A(\vec{z})$, $\widetilde{A}(\vec{z})$ and $\Delta(\vec{z})$ be matrices defined as follows:*

for all $i, j \in \mathcal{S}$, the elements of row i and column j are given by

$$\begin{aligned}
[A(\vec{z})]_{i,j} &= p_{j,i} a_{j,i}(\vec{z}), \\
[\widetilde{A}(\vec{z})]_{i,j} &= \tilde{p}_{j,i} \tilde{a}_{j,i}(\vec{z}), \\
[\Delta(\vec{z})]_{i,j} &= \frac{1}{z_i} \mathbf{1}_{\{i=j\}}.
\end{aligned}$$

Then the vectors $F(\vec{z}) = (F_1(\vec{z}), \ldots, F_N(\vec{z}))$ and $\widetilde{F}(\vec{z}) = (\widetilde{F}_1(\vec{z}), \ldots, \widetilde{F}_N(\vec{z}))$ are related by the functional equation

(2.1) $$[I - A\Delta(\vec{z})] F(\vec{z}) = [\widetilde{A}(\vec{z}) - A\Delta(\vec{z})] \widetilde{F}(\vec{z}),$$

where $A\Delta(\vec{z})$ stands for $A(\vec{z})\Delta(\vec{z})$.

Proof. We have, for all $i, j \in \mathcal{S}$ and $n > 0$,

$$\begin{aligned}
\mathbb{E}[\vec{z}^{X(n+1)} &\mathbf{1}_{\{S(n)=i,S(n+1)=j\}}] \\
&= \mathbb{E}[\vec{z}^{X(n)+B_{i,j}(n)-\vec{e}_i} \mathbf{1}_{\{S(n)=i,S(n+1)=j,X_i(n)>0\}}] \\
&\quad + \mathbb{E}[\vec{z}^{X(n)+\widetilde{B}_{i,j}(n)} \mathbf{1}_{\{S(n)=i,S(n+1)=j,X_i(n)=0\}}] \\
&= p_{i,j} a_{i,j}(\vec{z}) \mathbb{E}[\vec{z}^{X(n)-\vec{e}_i} \mathbf{1}_{\{S(n)=i,X_i(n)>0\}}] \\
&\quad + \tilde{p}_{i,j} \tilde{a}_{i,j}(\vec{z}) \mathbb{E}[\vec{z}^{X(n)} \mathbf{1}_{\{S(n)=i,X_i(n)=0\}}] \\
&= p_{i,j} a_{i,j}(\vec{z}) \frac{F_i(\vec{z}; n) - \widetilde{F}_i(\vec{z}; n)}{z_i} + \tilde{p}_{i,j} \tilde{a}_{i,j}(\vec{z}) \widetilde{F}_i(\vec{z}; n).
\end{aligned}$$

The second equality above uses the independence of the routing and the arrivals with respect to the past. Summing over all possible i, we get

$$F_j(\vec{z}; n+1) = \sum_{i=1}^{N} p_{i,j} a_{i,j}(\vec{z}) \frac{F_i(\vec{z}; n) - \widetilde{F}_i(\vec{z}; n)}{z_i} + \sum_{i=1}^{N} \tilde{p}_{i,j} \tilde{a}_{i,j}(\vec{z}) \widetilde{F}_i(\vec{z}; n),$$

so that, letting $n \to \infty$,

$$F_j(\vec{z}) - \sum_{i=1}^{N} \frac{p_{i,j} a_{i,j}(\vec{z})}{z_i} F_i(\vec{z}) = \sum_{i=1}^{N} \tilde{p}_{i,j} \tilde{a}_{i,j}(\vec{z}) \widetilde{F}_i(\vec{z}) - \sum_{i=1}^{N} \frac{p_{i,j} a_{i,j}(\vec{z})}{z_i} \widetilde{F}_i(\vec{z}),$$

which is equivalent to (2.1).   □

Defining, when it exists,

(2.2) $$D(\vec{z}) \stackrel{\text{def}}{=} [I - A\Delta(\vec{z})]^{-1} [\widetilde{A}(\vec{z}) - A\Delta(\vec{z})],$$

one sees that (2.1) can be rewritten as

(2.3) $$F(\vec{z}) = D(\vec{z}) \widetilde{F}(\vec{z}).$$

This functional equation contains all the information sufficient to characterize $F(\vec{z})$ and $\widetilde{F}(\vec{z})$. Although its solution for $N \geq 3$ is still an open question, partial results can be derived as shown in the following sections.

3. The stationary distribution of the position of the server.
What renders the polling model presented in the previous section difficult
to analyze is, among other things, the fact that the movements of the server
depend on the state of the visited stations. In particular, $\{S(n)\}_{n\geq 0}$ is not
a Markov process, as it would be when $p_{i,j} = \tilde{p}_{i,j}$ for all i and j. Some
computations are needed to get the stationary probabilities of the position
of the server, that is

$$
\begin{aligned}
F_i(\vec{e}) &= P(S = i) \\
\widetilde{F}_i(\vec{e}) &= P(S = i, X_i = 0).
\end{aligned}
$$

It appears that these stationary probabilities depend not only on the
transition probabilities, but also on the following mean values:

$$(3.1) \qquad \alpha_{i;q} \stackrel{\text{def}}{=} \sum_{j=1}^{N} p_{i,j} \, \mathbb{E} \, B_{i,j;q}(n),$$

$$(3.2) \qquad \tilde{\alpha}_{i;q} \stackrel{\text{def}}{=} \sum_{j=1}^{N} \tilde{p}_{i,j} \, \mathbb{E} \, \widetilde{B}_{i,j;q}(n).$$

Here $\alpha_{i;q}$ (resp. $\tilde{\alpha}_{i;q}$) is the mean number of new clients at station q
between the arrival of the server at station i which was non-empty (resp.
empty) and its arrival at the next (arbitrary) polled station. Since the
arrival process is stationary, it is possible to define for all $q \in \mathcal{S}$ the mean
number λ_q of arrivals at station q per unit of time and write, with obvious
definitions for τ_i and $\tilde{\tau}_i$,

$$(3.3) \qquad\qquad \alpha_{i;q} = \lambda_q \tau_i,$$
$$(3.4) \qquad\qquad \tilde{\alpha}_{i;q} = \lambda_q \tilde{\tau}_i.$$

Throughout this paper we will sometimes write F (resp. $\widetilde{F}$) instead of
$F(\vec{e})$ (resp. $\widetilde{F}(\vec{e})$) in order to shorten the notation. Moreover, $^*F$ denotes
the transpose of the vector F.

THEOREM 3.1. *Define the matrices* $A \stackrel{\text{def}}{=} (\alpha_{i;q})$ *and* $\widetilde{A} \stackrel{\text{def}}{=} (\tilde{\alpha}_{i;q})$. *When
the Markov chain* $\mathcal{L}$ *is ergodic,* F *and* $\widetilde{F}$ *satisfy the following linear system
of equations:*

$$(3.5) \qquad\qquad {}^*F\,\vec{e} = 1,$$
$$(3.6) \qquad\qquad {}^*F\,[I - P] = {}^*\widetilde{F}\,[\widetilde{P} - P],$$
$$(3.7) \qquad\qquad {}^*F\,[I - A] = {}^*\widetilde{F}\,[I - A + \widetilde{A}].$$

Define

$$\hat{\rho} \stackrel{\text{def}}{=} \sum_{i=1}^{N} \lambda_i(\tau_i - \tilde{\tau}_i).$$

When $\hat{\rho} \neq 1$, Equations (3.6) and (3.7) can be rewritten as

$$(3.8) \qquad F_j = \sum_{i \in \mathcal{S}} \tilde{p}_{i,j} F_i + \bar{\tau} \sum_{i \in \mathcal{S}} \lambda_i (p_{i,j} - \tilde{p}_{i,j}), \quad \text{for all } j \in \mathcal{S},$$

$$(3.9) \qquad \bar{\tau} = \frac{1}{1 - \hat{\rho}} \sum_{j \in \mathcal{S}} F_j \tilde{\tau}_j,$$

and $\widetilde{F}$ is given by

$$(3.10) \qquad\qquad\qquad \widetilde{F}_j = F_j - \lambda_j \bar{\tau}.$$

Moreover the mean time between two consecutive visits to queue j is

$$(3.11) \qquad\qquad\qquad \mathbb{E}\, T_{j,j} = \frac{\bar{\tau}}{F_j}$$

The quantity $\bar{\tau}$ defined in (3.9) can be seen as the mean time between two polling instants. Note that, with the above definitions, $\hat{\rho}$ is in general different from the classical ρ defined in queueing theory. However, they coincide if $P = \widetilde{P}$ and the time between two polling instants can be decomposed into a state-independent switchover time and a service time which only depends on the station where the customer is.

Proof of Theorem 3.1. Although this theorem could be proved analytically using the functional equation (2.1), we give here a simple probabilistic interpretation of (3.6) and (3.7). Indeed, when the system is ergodic, (3.6) follows directly from

$$P(S = j) = \sum_{i=1}^{N} \Big[P(S = i, X_i > 0) p_{i,j} + P(S = i, X_i = 0) \tilde{p}_{i,j} \Big].$$

Moreover, writing the equality of the outgoing and ingoing flows at station q gives

$$P(S = q, X_q > 0) = \sum_{i=1}^{N} \sum_{j=1}^{N} \Big[P(S = i, X_i > 0) p_{i,j} \, \mathbb{E}\, B_{i,j;q}(n)$$

$$+ P(S = i, X_i = 0) \tilde{p}_{i,j} \, \mathbb{E}\, \widetilde{B}_{i,j;q} \Big],$$

which is equivalent to (3.7) if we take into account (3.1) and (3.2). It is straightforward to see, using (3.3) and (3.4), that, when $\hat{\rho} \neq 1$,

$$[I - \mathcal{A}][I - \mathcal{A} + \widetilde{\mathcal{A}}]^{-1} = I - \frac{\widetilde{\mathcal{A}}}{1 - \hat{\rho}}.$$

With this relation and $\bar{\tau}$ as in (3.9), Equations (3.8) and (3.10) follow easily from (3.6) and (3.7). Let $N_{j,j;j}$ be the number of arrivals to a queue j between two consecutive visits of the server. The equality of flows reads

$$P(X_j > 0 \mid S = j) = \mathbb{E}\, N_{j,j;j}.$$

Since the last equation can be rewritten as

$$\frac{F_j - \widetilde{F}_j}{F_j} = \lambda_j \, \mathbb{E} \, T_{j,j} \, ,$$

we obtain (3.11) and the proof of the theorem is concluded.   □

The set of equations (3.5), (3.8), (3.9) and (3.10) provides a convenient way to compute F and $\widetilde{F}$. The following proposition gives simple conditions under which its rank is full.

PROPOSITION 3.2. *The linear system given by (3.5), (3.8) and (3.9) has a unique solution when $\widetilde{P}$ has exactly one essential class. When $\widetilde{P}$ has $K > 1$ essential classes $\mathcal{E}_1, \ldots, \mathcal{E}_K$, the Markov chain $\mathcal{L}$ is never ergodic unless*

$$(3.12) \qquad \sum_{j \in \mathcal{E}_m} \sum_{i \in \mathcal{S}} \lambda_i (p_{i,j} - \tilde{p}_{i,j}) = 0, \;\; \text{for all } m \leq K.$$

Proof. Define $\psi_i \overset{\text{def}}{=} F_i / \bar{\tau}$. Then the system (3.8)–(3.9) reads

$$(3.13) \qquad \psi_j - \sum_{i \in \mathcal{S}} \tilde{p}_{i,j} \psi_i \;\; = \;\; \sum_{i \in \mathcal{S}} \lambda_i (p_{i,j} - \tilde{p}_{i,j}), \;\; \text{for all } j \in \mathcal{S},$$

$$(3.14) \qquad \sum_{i \in \mathcal{S}} \psi_i \tilde{\tau}_i \;\; = \;\; 1 - \hat{\rho}.$$

The system (3.13) have rank $N - 1$ unless the stochastic matrix $\widetilde{P}$ has several essential classes. Then, for each $m \leq K$, $i \in \mathcal{E}_m$ and $j \notin \mathcal{E}_m$, $\tilde{p}_{i,j} = 0$. In this case, (3.13) has solutions only when (3.12) holds.   □

The conditions (3.12) represent in some sense "zero drift" relationships which, as usual, imply more involved derivations. However, when the arrivals form independent compound Poisson processes at each queue, the system is never ergodic! The proof is of analytic nature. It requires Taylor expansions of second order in equation (2.1), similar to those extensively used in Section 5. In general, when the batches are correlated, it is difficult to conclude and we suspect that all situations might occur (ergodicity, null recurrence or transience).

It is worth noting the special role played by $\widetilde{P}$. In particular, when the arrivals are Poisson, the Markov chain is *never* ergodic if $\widetilde{P}$ admits several essential classes (obviously, P has then to be chosen to ensure the irreducibility of $\mathcal{L}$). In our opinion, this result is not intuitive and we cannot be explained just by waving hands.

4. Conditions for ergodicity. The purpose of this section is to classify the process $\mathcal{L}$, viewed as a random walk on $\mathcal{S} \times \mathbb{Z}_+^N$, in terms of ergodicity and non-ergodicity.

4.1. Necessary condition. We give a necessary condition for ergodicity which is a simple consequence of the results of Theorem 3.1. When the system is ergodic, we have $\widetilde{F}_i > 0$ for all $i \in \mathcal{S}$ and Equation (3.10) implies the following.

THEOREM 4.1. *If the Markov chain $\mathcal{L}$ is ergodic, then*

$$(4.1) \qquad\qquad \hat{\rho} < 1,$$

$$(4.2) \qquad\qquad \lambda_i \bar{\tau} < F_i, \text{ for all } i \in \mathcal{S}.$$

These conditions extend in the state-dependent case the results obtained in [1]. When they hold, we prove another useful inequality. Indeed, instantiating (4.2) in (3.9) yields

$$\bar{\tau} > \frac{\sum_{j \in \mathcal{S}} \bar{\tau} \lambda_j \tilde{\tau}_j}{1 - \hat{\rho}},$$

or, equivalently,

$$(4.3) \qquad\qquad 1 - \sum_{j \in \mathcal{S}} \lambda_j \tau_j > 0.$$

4.2. Sufficient condition. Let us emphasize that this part could be skipped by readers not strongly interested (if any!) in ergodicity. The mathematical understanding requires the reading of Appendix A.1, which summarizes deep works of [12], [4] and [5]. The approach relies on the study of a dynamical system associated to $\mathcal{L}$. For the sake of readability, we recall hereafter two main notions.

- *Induced chain $\mathcal{L}^\wedge$.* It is a Markov chain corresponding to a polling system in which the queues belonging to the *face* $\wedge \subset \mathcal{S}$ are kept saturated. From a purely notational point of view, the original system would correspond to the case $\wedge = \emptyset$. The behaviours of $\mathcal{L}$ and $\mathcal{L}^\wedge$ are not *directly* connected to each other.
- *Second vector field $\vec{v}^\wedge$.* One can imagine that the random walk starts from a point which is close to $\wedge$, but sufficiently far from all other faces $\wedge'$, with $\wedge \not\subset \wedge'$. After some time—sufficiently long, but less than the minimal distance from $\wedge'$—, the stationary regime in the induced chain will be installed. In this regime, one can ask about the mean along $\wedge$: it is defined exactly by $\vec{v}^\wedge$.

As in Theorem 3.1, the stationary position of the server for any ergodic induced chain can be obtained as the solution of a linear system consisting of $N + 1$ equations. Indeed, for all $t \in \mathcal{S}$, $j \notin \wedge$, we have

$$(4.4) \qquad \pi^\wedge(t) = \sum_{s \in \wedge} \pi^\wedge(s) p_{s,t} + \sum_{s \notin \wedge} \pi^\wedge(s) \tilde{p}_{s,t}$$

$$+ \sum_{s \notin \wedge} \pi^{\wedge}(s, x_s > 0)(p_{s,t} - \tilde{p}_{s,t}),$$

$$(4.5) \qquad \pi^{\wedge}(j, x_j > 0) \; = \; \sum_{s \in \wedge} \pi^{\wedge}(s)\lambda_j \tau_s + \sum_{s \notin \wedge} \pi^{\wedge}(s)\lambda_j \tilde{\tau}_s$$

$$+ \sum_{s \notin \wedge} \pi^{\wedge}(s, x_s > 0)\lambda_j(\tau_s - \tilde{\tau}_s),$$

$$(4.6) \qquad \sum_{s \in \mathcal{S}} \pi^{\wedge}(s) \; = \; 1.$$

Let us introduce the quantities

$$\hat{\rho}^{\wedge} \; \overset{\text{def}}{=} \; \sum_{s \notin \wedge} \lambda_s(\tau_s - \tilde{\tau}_s),$$

$$\bar{\tau}^{\wedge} \; \overset{\text{def}}{=} \; \frac{1}{1 - \hat{\rho}^{\wedge}} \left[\sum_{s \in \wedge} \pi^{\wedge}(s)\tau_s + \sum_{s \notin \wedge} \pi^{\wedge}(s)\tilde{\tau}_s \right].$$

Note that when $\wedge$ is ergodic, $\pi^{\wedge}(j, x_j > 0) = \lambda_j \bar{\tau}^{\wedge}$, for all $j \notin \wedge$, and $\bar{\tau}^{\wedge}$ can be interpreted as the mean time between two polling instants of the induced chain. Then the system defined by (4.4) and (4.5) can be replaced by the forthcoming $N + 1$ equations: for all $t \in \mathcal{S}$,

$$(4.7) \; \pi^{\wedge}(t) \; = \; \sum_{s \in \wedge} \pi^{\wedge}(s)p_{s,t} + \sum_{s \notin \wedge} \pi^{\wedge}(s)\tilde{p}_{s,t} + \bar{\tau}^{\wedge} \sum_{s \notin \wedge} \lambda_s(p_{s,t} - \tilde{p}_{s,t}),$$

$$(4.8) \quad \bar{\tau}^{\wedge} \; = \; \frac{1}{1 - \hat{\rho}^{\wedge}} \left[\sum_{s \in \wedge} \pi^{\wedge}(s)\tau_s + \sum_{s \notin \wedge} \pi^{\wedge}(s)\tilde{\tau}_s \right].$$

When $\wedge = \emptyset$, the system (4.6), (4.7) and (4.8) coincides with (3.5), (3.8) and (3.9). In addition, under (4.3), we have

$$(4.9) \qquad\qquad \hat{\rho}^{\wedge} < 1, \text{ for all } \wedge \text{'s.}$$

An easy flow computation shows that the components of the drift vector $\vec{M}(s, \vec{x})$ can be expressed as

$$M_j(s, \vec{x}) \; = \; \lambda_j \tau_s \mathbf{1}_{\{x_s > 0\}} + \lambda_j \tilde{\tau}_s \mathbf{1}_{\{x_s = 0\}} - \mathbf{1}_{\{x_s > 0, s = j\}}.$$

Then the computation of the second vector field becomes easy. For $j \in \wedge$,

$$v_j^{\wedge} \; = \; \sum_{\substack{x \in C^{\wedge} \\ s \in \mathcal{S}}} \pi^{\wedge}(s, \vec{x})M_j(s, \vec{x})$$

$$= \; \sum_{\substack{x \in C^{\wedge} \\ s \notin \wedge}} \pi^{\wedge}(s, \vec{x})M_j(s, \vec{x}) + \sum_{\substack{x \in C^{\wedge} \\ s \in \wedge}} \pi^{\wedge}(s, \vec{x})M_j(s, \vec{x})$$

$$= \lambda_j \sum_{s \notin \wedge} \left[\pi^\wedge(s, x_s > 0)\tau_s + \pi^\wedge(s, x_s = 0)\tilde{\tau}_s \right]$$

$$+ \lambda_j \sum_{s \in \wedge} \pi^\wedge(s)\tau_s - \pi^\wedge(j)$$

$$= \lambda_j \bar{\tau}^\wedge - \pi^\wedge(j).$$

In order to apply Theorem A.1, it will be convenient to introduce

$$f_i(\vec{x}) \stackrel{\text{def}}{=} \left[{}^* \vec{x} [I - \mathcal{A} + \tilde{\mathcal{A}}]^{-1} \right]_i$$

(4.10)
$$= x_i + \frac{\lambda_i \sum_{j=1}^N x_j(\tau_j - \tilde{\tau}_j)}{1 - \hat{\rho}}.$$

This function is not as outlandish as it could seem at first sight. It is indeed directly related to flow conservation equations (see [5]). Let $\wedge$ be any ergodic face. For $i \in \wedge$, we get after a straightforward computation

$$f_i(\vec{v}^\wedge) = v_i^\wedge + \frac{\lambda_i}{1 - \hat{\rho}} \sum_{j=1}^N (\tau_j - \tilde{\tau}_j)v_j^\wedge$$

$$= \frac{\lambda_i \sum_{j=1}^N \pi^\wedge(j)\tilde{\tau}_j}{1 - \hat{\rho}} - \pi^\wedge(i).$$

For $i \notin \wedge$, and from the very definition of $f_i(\vec{x})$, we have

$$f_i(\vec{v}^\wedge) - \sum_{j=1}^N f_j(\vec{v}^\wedge)\lambda_i(\tau_j - \tilde{\tau}_j) = v_i^\wedge = 0,$$

or

$$f_i(\vec{v}^\wedge) = \frac{\lambda_i \sum_{j \in \wedge} f_j(\vec{v}^\wedge)(\tau_j - \tilde{\tau}_j)}{1 - \hat{\rho}^\wedge}.$$

THEOREM 4.2. *Assume that (4.1)–(4.2) hold and that, for any ergodic face $\wedge$,*

(4.11) $\quad f_i(\vec{v}^\wedge) \equiv \dfrac{\lambda_i \sum_{j=1}^N \pi^\wedge(j)\tilde{\tau}_j}{1 - \hat{\rho}} - \pi^\wedge(i) < 0, \quad \text{for all } i \in \wedge.$

Then the random walk $\mathcal{L}$ is ergodic. In particular, when $P = \tilde{P}$, the conditions (4.1) and (4.2) are necessary and sufficient for the random walk $\mathcal{L}$ to be ergodic.

Proof. We shall apply Theorem A.1 (quoted in Appendix A.1), which in itself contains a principle of gluing Lyapounov functions together. Most of the time, these functions are piecewise linear. Since the $f_i(\vec{v}^\wedge)$ enjoy "nice" properties, it is not unnatural to search for a linear combination like

$$f(\vec{x}) \stackrel{\text{def}}{=} \sum_{i=1}^{N} u_i f_i(\vec{x}),$$

where $\vec{u} = (u_1, \ldots, u_N)$ is a positive vector to be properly determined. Then, for any ergodic $\wedge$,

$$f(\vec{v}^\wedge) = \sum_{i \in \wedge} \left[u_i + \frac{\sum_{j \notin \wedge} \lambda_j u_j}{1 - \hat{\rho}^\wedge} (\tau_i - \tilde{\tau}_i) \right] f_i(\vec{v}^\wedge).$$

The basic constraint for $\vec{u}$ is to ensure the positivity of f. Using (4.3) and (4.9), it appears that a suitable choice is $u_i = \max(\tilde{\tau}_i - \tau_i, \varepsilon)$, for some ε, positive and sufficiently small. Then one can directly check that $f(\vec{x}) > 0$, for any $\vec{x} \in \mathbb{R}_+^N$, and $f(\vec{v}^\wedge) < 0$, for any ergodic face $\wedge$. $\qquad \square$

Although the conditions of Theorem 4.2 are sufficient for ergodicity, they might well be implied by conditions (4.1)–(4.2) only. We did not check this fact, since it is difficult to compare algebraically the solutions of system (4.7)–(4.8), for different $\wedge$'s. Let us simply formulate the conjecture that (4.1)–(4.2) are also sufficient for ergodicity when $P \neq \widetilde{P}$.

THEOREM 4.3. *Assume that $P = \widetilde{P}$ and that one of the following hold:*
1. *there exists $j \in S$, such that $\lambda_j \bar{\tau} - F_j > 0$;*
2. *$\hat{\rho} > 1$.*
Then the random walk $\mathcal{L}$ is transient.

Proof. We assume that the queues are numbered according to the following order:

$$(4.12) \qquad \frac{\lambda_1}{F_1} \leq \frac{\lambda_2}{F_2} \leq \ldots \leq \frac{\lambda_N}{F_N}.$$

Consider all faces $\wedge_i \stackrel{\text{def}}{=} \{i, \ldots, N\}$, for $i = 1, \ldots, N$. The idea is to show the transience of $\mathcal{L}$ by visiting the ordered set $\wedge_1, \wedge_2, \ldots \wedge_N$. In the usual terminology of dynamical systems, this amounts to finding a set of trajectories going to infinity with positive probability. To this end, the following algebraic relationship will be useful: for any $\wedge$ and $k \notin \wedge$, we have

$$(4.13) \qquad (\lambda_k \bar{\tau}^{\wedge + \{k\}} - F_k)(1 - \hat{\rho}^{\wedge + \{k\}}) = (\lambda_k \bar{\tau}^\wedge - F_k)(1 - \hat{\rho}^\wedge).$$

By definition, the face $\wedge_1 \equiv S$ is ergodic (see Appendix A.1). Assume now $\wedge_i$ is ergodic, for some fixed $i \in S$. If $v_i^{\wedge_i} < 0$, then one can prove the inequality

$$1 - \sum_{s=1}^{i} \lambda_s \tau_s \leq \frac{1}{\bar{\tau}^{\wedge_i}} \sum_{s=i}^{N} F_s \tau_s,$$

which in turn yields $1 - \hat{\rho}^{\wedge i} > 0$. By using (4.13), (4.12) and Theorem 4.2, it follows that the face $\wedge_{i+1}$ is ergodic. Conversely, if $v_i^{\wedge i} > 0$, then $\vec{v}^{\wedge i} > \vec{0}$ and the random walk $\mathcal{L}$ is transient (see [4]).

Thus, by induction, we have shown that either $\mathcal{L}$ is transient or the face $\wedge_N \equiv \{N\}$ is ergodic. In the latter case, the assumptions of the theorem, together with (4.13) and (4.12), yield $v_N^{\wedge_N} \equiv \lambda_N \bar{\tau}^{\wedge_N} - F_N > 0$, so that $\mathcal{L}$ is transient.   □

4.3. Remarks and problems.

1. In general, the second vector field requires to compute invariant measures of Markov chains in dimensions strictly smaller than N (the induced chains). This is rarely possible for $N > 3$, but there are some miracles, the most noticeable ones concerning Jackson networks and some polling systems [5].

2. No stochastic monotonicity argument seems to work, when the $\tau_k - \tilde{\tau}_k$'s are not all of the same sign, even for $P = \tilde{P}$. This monotonicity exists and is crucial in [1], [7] and [8].

3. The analysis of the vector field $\vec{v}^{\wedge}$ is interesting in itself and will be the subject of a future work. Let us just quote one negative property: when $\wedge$ is ergodic and $\vec{v}_k^{\wedge} < 0$, then $\wedge_1 = \wedge \setminus \{k\}$ can be non-ergodic, contrary to what happens in Jackson networks (see [5]). In the case $P = \tilde{P}$, we conjecture nonetheless that the dynamical system formed from the velocity vectors $\vec{v}^{\wedge}$ is strongly acyclic (in the sense that the same face is not visited twice).

5. The first moment of the queue length. This section shows that, when the system enjoys some symmetry properties, it becomes possible to derive the stationary mean queue length $\mathbb{E}[X_m \mid S = m]$ seen by the server at polling instants. To this end, ergodicity of $\mathcal{L}$ will be assumed, as well as the existence of second moments of all random variables of interest.

ASSUMPTION A_1. *The system has a* rotational *symmetry*

1. *for all $i, j \in \mathcal{S}$, $p_{i,j} = p_{j-i}$, $\tilde{p}_{i,j} = \tilde{p}_{j-i}$, $a_{i,j}(\vec{z}) = a_{j-i}(\vec{z})$ and $\tilde{a}_{i,j}(\vec{z}) = \tilde{a}_{j-i}(\vec{z})$;*
2. *for all $1 \leq s, d \leq N$,*

$$a_d(z_1, \ldots, z_N) = a_d(z_{1+s}, \ldots, z_{N+s}),$$
$$\tilde{a}_d(z_1, \ldots, z_N) = \tilde{a}_d(z_{1+s}, \ldots, z_{N+s}).$$

Note that this assumption is less restrictive than the ones appearing in the literature, even for state-independent situations. It allows in particular cyclic and random polling strategies. From A_1, we directly get, for all $1 \leq i, s \leq N$,

(5.1) $F_i(\vec{z}) = F_{i+s}(z_{1+s}, \ldots, z_{N+s}),$

(5.2) $\widetilde{F}_i(\vec{z}) = \widetilde{F}_{i+s}(z_{1+s}, \ldots, z_{N+s}).$

Consequently, $P(S = i) = F_i(\vec{e}) = 1/N$. In addition, $A(\vec{z})$ and $\tilde{A}(\vec{z})$ are *circulant* matrices—see Appendix A.2 for the results and notation which will be used from now on. It implies in particular that, for $k \in \mathcal{S}$, $\vec{v}_k$ is an eigenvector of $A(\vec{z})$ and $\tilde{A}(\vec{z})$ with respective eigenvalues

$$\mu_k(\vec{z}) \stackrel{\text{def}}{=} \sum_{d=1}^{N} p_d a_d(\vec{z}) \omega_k^d ,$$

$$\tilde{\mu}_k(\vec{z}) \stackrel{\text{def}}{=} \sum_{d=1}^{N} \tilde{p}_d \tilde{a}_d(\vec{z}) \omega_k^d .$$

Note that $\mu_N(\vec{z})$ (resp. $\tilde{\mu}_N(\vec{z})$) is the generating function of the number of arrivals during the transportation of one client (resp. a move without client). Moreover, $\mu_k \stackrel{\text{def}}{=} \mu_k(\vec{e})$ and $\tilde{\mu}_k \stackrel{\text{def}}{=} \tilde{\mu}_k(\vec{e})$ are eigenvalues of P and $\tilde{P}$. For *a priori* technical reasons we shall also need

ASSUMPTION A$_2$. *For any $k < N$, $\tilde{\mu}_k \neq 1$.*

These relations are not surprising in view of Proposition 3.2. Since the system is symmetrical, the mean number of customers arriving during a travel of the taxicab $\alpha_{i;q}$ and $\tilde{\alpha}_{i;q}$ as defined by (3.1)–(3.2) does not depend on i and q; we note them α and $\tilde{\alpha}$ and remark that

$$\alpha = \frac{\partial \mu_N}{\partial z_m}(\vec{e}),$$

$$\tilde{\alpha} = \frac{\partial \tilde{\mu}_N}{\partial z_m}(\vec{e}).$$

The second derivatives of $\mu_N(\vec{z})$ and $\tilde{\mu}_N(\vec{z})$ are defined as

(5.3) $$\alpha_{q,r}^{(2)} \stackrel{\text{def}}{=} \frac{\partial^2 \mu_N(\vec{z})}{\partial z_q \partial z_r}(\vec{e}),$$

(5.4) $$\tilde{\alpha}_{q,r}^{(2)} \stackrel{\text{def}}{=} \frac{\partial^2 \tilde{\mu}_N(\vec{z})}{\partial z_q \partial z_r}(\vec{e})$$

Note that these values depend only of the unsigned distance between q and r and that we can write $\alpha_{q,q+d}^{(2)} = \alpha_{q+d,q}^{(2)} \stackrel{\text{def}}{=} \alpha_d^{(2)}$, $\tilde{\alpha}_{q,q+d}^{(2)} = \tilde{\alpha}_{q+d,q}^{(2)} \stackrel{\text{def}}{=} \tilde{\alpha}_d^{(2)}$. It will be also convenient to introduce the first and second moments of the number of arrivals between two polling instants, that is

(5.5) $\bar{\alpha} \stackrel{\text{def}}{=} P(X_m > 0 \mid S = m)\alpha + P(X_m = 0 \mid S = m)\tilde{\alpha},$

(5.6) $\bar{\alpha}_d^{(2)} \stackrel{\text{def}}{=} P(X_m > 0 \mid S = m)\alpha_d^{(2)} + P(X_m = 0 \mid S = m)\tilde{\alpha}_d^{(2)}.$

Using the symmetry of the system and according to the properties of generating functions, we have for any $m \in \mathcal{S}$

(5.7) $$\mathbb{E}[X_m \mid S = m] = N \frac{\partial F_m}{\partial z_m}(\vec{e}).$$

By symmetry, this expression is independent from m. In order to derive $\mathbb{E}[X_m \mid S = m]$, we use the fact that the matrices $I - A\Delta(\vec{z})$ and $\widetilde{A}(\vec{z}) - A\Delta(\vec{z})$ are not invertible when $\vec{z} = \vec{e}$. This implies that the matrix $D(\vec{z})$ as defined in (2.2) is not continuous in a neighborhood of $\vec{z} = \vec{e}$ and provides a mean to compute the derivatives of $F(\vec{z})$ and $\widetilde{F}(\vec{z})$ for $\vec{z} = \vec{e}$. However in our setting, even the existence of $D(\vec{z})$ in a neighborhood of $\vec{e}$ is difficult to prove and the computations become really involved. The approach that we present here avoids the theoretical problems and simplifies the computations. In the next lemma, we show how it is possible to choose $\vec{z}$ such that the left member of Equation (2.1) can be rendered of "small" order w.r.t. the right one.

LEMMA 5.1. *Let $t \to \vec{z}_t$ be a function from $\mathbb{R}_- \to \mathcal{D}^N$ such that in a neighborhood of $t = 0$, we can write $\vec{z}_t = \vec{e} + t\dot{\vec{z}} + \frac{1}{2}t^2\ddot{\vec{z}} + \vec{o}(t^2)$, where $\dot{\vec{z}} = \dot{z}_1\vec{e}_1 + \cdots + \dot{z}_N\vec{e}_N = \dot{\zeta}_1\vec{v}_1 + \cdots + \dot{\zeta}_N\vec{v}_N \in \mathbb{C}^N$ and $\ddot{\vec{z}} = (\ddot{z}_1, \ldots, \ddot{z}_N) \in \mathbb{C}^N$. Assume that $\dot{z}_1 + \cdots + \dot{z}_N = 0$. Then there exists for $t > 0$ a vector $\vec{u}_t$ such that, for $1 \le k < N$,*

$$(5.8) \qquad {}^*\vec{u}_t\,[I - A\Delta(\vec{z}_t)]\vec{v}_k \;=\; o(t),$$

$$(5.9) \qquad {}^*\vec{u}_t\,[\widetilde{A}(\vec{z}_t) - A\Delta(\vec{z}_t)]\vec{v}_k \;=\; t\frac{1 - \tilde{\mu}_k}{1 - \mu_k}\dot{\zeta}_{N-k} + o(t),$$

and

$$
\begin{aligned}
{}^*\vec{u}_t\,[I - A\Delta(\vec{z}_t)]\vec{e} \;&=\; -t^2\sum_{l=1}^{N-1}\dot{\zeta}_l\dot{\zeta}_{N-l}\left[\frac{1}{1-\mu_l} + \frac{N}{2}\sum_{d=1}^{N}\alpha_d^{(2)}\omega_l^d\right]\\
(5.10) \qquad\qquad &\;+\; t^2\frac{1 - N\alpha}{2N}\sum_{q=1}^{N}\ddot{z}_q + o(t^2),
\end{aligned}
$$

$$
\begin{aligned}
{}^*\vec{u}_t\,[\widetilde{A}(\vec{z}_t) - A\Delta(\vec{z}_t)]\vec{e} \;&=\; -t^2\sum_{l=1}^{N-1}\dot{\zeta}_l\dot{\zeta}_{N-l}\left[\frac{1}{1-\mu_l} + \frac{N}{2}\sum_{d=1}^{N}(\alpha_d^{(2)} - \tilde{\alpha}_d^{(2)})\omega_l^d\right]\\
(5.11) \qquad\qquad &\;+\; t^2\frac{1 - N\alpha + N\tilde{\alpha}}{2N}\sum_{q=1}^{N}\ddot{z}_q + o(t^2).
\end{aligned}
$$

Proof. See Appendix A.3. $\square$

For the sake of simplicity, the computations have been carried out in the most natural way which apparently requires $\mu_k \ne 1$ for $k < N$. In fact, the reader can convince himself that all derivations could be achieved by rendering the right member of (2.1) small w.r.t. the left member. This would yield the same result without any restriction on μ_k, at the expense of a somewhat longer proof.

One problem in polling models with 1-limited service strategy is that the rank of the systems of equations that we can write is $\lfloor n/2 \rfloor$, which means that $\partial F/\partial z_m(\vec{e})$ cannot be computed. Fortunately, under the following assumption, we are able to compute the most important value, that is $\partial F_m / \partial z_m(\tilde{e})$.

ASSUMPTION A_3. *The routing matrices are such that*

$$[I - P][I - {}^*\widetilde{P}] = [I - {}^*P][I - \widetilde{P}].$$

It is not easy to describe all models satisfying A_3. The simplest one is the classical Markov polling obtained when $P = \widetilde{P}$, for which the results of Theorem 5.2 thereafter are greatly simplified. Another admissible model is when the routing probabilities depend only on the absolute distance between stations, that is when $P = {}^*P$ and $\widetilde{P} = {}^*\widetilde{P}$.

THEOREM 5.2. *Assume that A_1, A_2 and A_3 hold. Then, for any station m, the stationary probability that the server finds station m empty is given by*

$$(5.12) \qquad P(X_m = 0 \mid S = m) = \frac{1 - N\alpha}{1 - N\alpha + N\tilde{\alpha}}.$$

Moreover the mean number of customers found by the server when it arrives at station m is obtained by means of the following formula

$$
\begin{aligned}
\mathbb{E}[X_m \mid S = m] \;=\;& N\bar{\alpha} + \frac{N\tilde{\alpha}\bar{\alpha}}{1 - N\alpha} \sum_{l=1}^{N-1} \frac{1}{1 - \tilde{\mu}_l} \\
& + \frac{1 - N\alpha + N\tilde{\alpha}}{1 - N\alpha} \frac{N}{2} \bar{\alpha}_N^{(2)} + \frac{N\alpha - N\tilde{\alpha}}{1 - N\alpha} \frac{1}{2} \sum_{d=1}^{N} \bar{\alpha}_d^{(2)} \\
(5.13) \qquad & - \frac{N\tilde{\alpha}}{1 - N\alpha} \sum_{l=1}^{N-1} \frac{\mu_l - \tilde{\mu}_l}{1 - \tilde{\mu}_l} \frac{1}{2} \sum_{d=1}^{N} \bar{\alpha}_d^{(2)} \omega_l^d .
\end{aligned}
$$

Finally, the mean number of customers present at arbitrary station m at a polling instant can be expressed as

$$
\begin{aligned}
\mathbb{E}\,X_m \;=\;& \bar{\alpha} + \frac{\tilde{\alpha}}{1 - N\alpha} \sum_{l=1}^{N-1} \frac{1}{1 - \tilde{\mu}_l} \\
& + \frac{1 - N\alpha + N\tilde{\alpha}}{1 - N\alpha} \frac{N}{2} \bar{\alpha}_N^{(2)} + \frac{N\alpha - N\tilde{\alpha}}{1 - N\alpha} \frac{1}{2} \sum_{d=1}^{N} \bar{\alpha}_d^{(2)} \\
(5.14) \qquad & - \frac{1 - N\alpha + N\tilde{\alpha}}{1 - N\alpha} \sum_{l=1}^{N-1} \frac{\mu_l - \tilde{\mu}_l}{1 - \tilde{\mu}_l} \frac{1}{2} \sum_{d=1}^{N} \bar{\alpha}_d^{(2)} \omega_l^d .
\end{aligned}
$$

Proof. See Appendix A.3. $\qquad\qquad\qquad\qquad\qquad\qquad\qquad\qquad\qquad$ □

6. The mean waiting time of a customer. Theorem 5.2 is the key to compute of the waiting time of an arbitrary customer. In this section, we focus on the continuous-time case with compound Poisson arrivals and further assume that the arrivals at each station are independent. While Theorem 5.2 is valid in discrete time or with correlated arrivals, we do not give any formula in these cases, since the notation would become too involved.

The notation is as follows: customers arrive in i.i.d. batches of mean length b and second moment $b^{(2)}$ at the instants of a Poisson stream with intensity $\hat{\lambda}$ at each station. The intensity of the arrival process at each queue is $\lambda = \hat{\lambda} b$. The time between two polling instants when the first queue is not empty (resp. empty) has mean and second moment τ and $\tau^{(2)}$ (resp. $\tilde{\tau}$ and $\tilde{\tau}^{(2)}$). We must keep in mind that in general τ depends on P like $\alpha_{i;q}$ in Equation (3.1): if τ_d is the mean time between two consecutive polling instants at stations i and $i+d$, then we have $\tau = p_1\tau_1 + \cdots + p_N\tau_N$. However, in most symmetrical polling models that have been previously analyzed, the switchover times are equal and τ does not depend on P. We will say in this case that the stations are *equidistant*, which is obviously not the case in our original taxicab problem. The same holds for $\tau^{(2)}$, $\tilde{\tau}$ and $\tilde{\tau}^{(2)}$.

THEOREM 6.1.   *Assume that A_1, A_2 and A_3 hold. Then the mean stationary waiting time of a customer is given by the relation*

$$\mathbb{E}[W] = \frac{\tilde{\tau}}{1 - N\lambda\tau} \sum_{l=1}^{N-1} \left[\frac{1}{1-\tilde{\mu}_l}\right] + \frac{N\lambda\tau^{(2)}}{2(1-N\lambda\tau)} + \frac{\tilde{\tau}^{(2)}}{2\tilde{\tau}}$$

$$(6.1) \qquad + \frac{b^{(2)}-b}{2b}\left[\frac{\tau+(N-1)\tilde{\tau}}{1-N\lambda\tau} - \frac{\tilde{\tau}}{1-N\lambda\tau}\sum_{l=1}^{N-1}\frac{\mu_l - \tilde{\mu}_l}{1-\tilde{\mu}_l}\right].$$

Proof. See Appendix A.3.   □

One surprising consequence of this formula is that when the arrival process is Poisson and the stations are *equidistant*, the mean waiting time does not depend on the routing P. In other words, it does not depend on where the server goes after serving a customer.

Beside giving the mean waiting time for a customer in our taxicab model, Equation (6.1) contains in fact many known formulas for waiting times corresponding to various service and polling strategies. In the next subsections, we give some of these applications.

A notation closer to the classical queuing theory notation is necessary to make the link with known results. Let w (resp. $\tilde{w}$) be the mean walking time—or switchover time—from a non-empty (resp. empty) station and let σ be the mean service time required by the customers. We denote by $w^{(2)}$, $\tilde{w}^{(2)}$ and $\sigma^{(2)}$ the associated second moments.

6.1. A state-independent Markovian polling model. The first possible application of our model is the classical state independent polling model with 1-limited service strategy. In this model, we have $P = \widehat{P}$,

$\tau = w + \sigma$, $\tilde{\tau} = w$ and (6.1) becomes

$$\mathbb{E}[W] = \frac{w}{1 - N\lambda(w + \sigma)} \sum_{l=1}^{N-1} \frac{1}{1 - \mu_l} + \frac{N\lambda(w^{(2)} + 2w\sigma + \sigma^{(2)})}{2(1 - N\lambda(w + \sigma))} + \frac{w^{(2)}}{2w}$$
$$- \frac{b^{(2)} - b}{2b} \frac{Nw + \sigma}{1 - N\lambda(w + \sigma)}.$$

This formula is valid for all symmetric polling models with Markovian polling. It is interesting to note that, in the case of equidistant stations, there is only one term of the formula which depends on the routing. To compute this term, we remark that, if $\mathcal{P}(x)$ is the characteristic polynomial of P,

$$\sum_{l=1}^{N-1} \frac{1}{1 - \mu_l} = \sum_{l=1}^{N-1} \frac{d}{dx} \log(x - \mu_l) \bigg|_{x=1} = \frac{d}{dx} \log\left[\frac{\mathcal{P}(x)}{x - 1}\right]\bigg|_{x=1}.$$

The case of the cyclic polling is obtained by taking $p_1 = 1$ and $p_d = 0$ for $d \neq 1$. Then the eigenvalues of P are $\omega_1, \ldots, \omega_N$ and $\mathcal{P}(x) = x^N - 1$. So in this case, we have

$$\sum_{l=1}^{N-1} \frac{1}{1 - \mu_l} = \frac{N - 1}{2}.$$

Another classical polling strategy is the random polling with $p_d = 1/N$ for all d. In this case, we have $\mu_l = 0$ for $l < N$ and

$$\sum_{l=1}^{N-1} \frac{1}{1 - \mu_l} = N - 1.$$

The comparison between formulas for cyclic polling and random polling shows that the mean waiting time is smaller in the cyclic case. In fact this property can be generalized to all Markovian polling models.

LEMMA 6.2. *Among all Markovian polling strategies for symmetric 1-limited polling models with equidistant queues, the cyclic polling strategy minimizes the waiting time of the customers.*

Proof. This result is very easy to prove from Equation (6.1). We have

$$\sum_{l=1}^{N-1} \frac{1}{1 - \mu_l} = \sum_{l=1}^{N-1} \Re\left(\frac{1}{1 - \mu_l}\right)$$

$$= \sum_{l=1}^{N-1} \frac{1 - \Re(\mu_l)}{2(1 - \Re(\mu_l)) + |\mu_l|^2 - 1},$$

where $\Re(z)$ is the real part of z. Since $|\mu_l| \leq 1$ and $\Re(\mu_l) \leq 1$, we find that

$$\sum_{l=1}^{N-1} \frac{1}{1 - \mu_l} \geq \frac{N - 1}{2}.$$

The bound is attained if and only if for all l, $|\mu_l| = 1$. Since μ_l is the center of gravity of $\omega_l^1, \ldots, \omega_l^N$ with weights $p_1, \ldots, p_N$, this is only possible when for some $s \in \mathcal{S}$ we have $p_s = 1$ and $p_d = 0$ for $d \neq s$. When s is not a divider of N (to ensure that the routing matrix is irreducible), this is equivalent to cyclic polling.   □

This property also holds for discrete time systems and systems with correlated arrivals. The computation would be more difficult in the case where the stations are not equidistant.

6.2. The exhaustive and Bernoulli service strategies. An interesting application of our model is to show that some service strategies can be obtained by a proper choice of polling strategy. In this subsection, we show how we can apply our model to exhaustive and Bernoulli service strategies.

Consider the case where the stations are *equidistant*, except that the switchover time from a station to itself after a service is zero. Then, if we denote $p_N = 1 - \pi$, we have $\tau = \pi w + \sigma$, $\tau^{(2)} = \pi w^{(2)} + 2\pi w \sigma + \sigma^{(2)}$, $\tilde{\tau} = w$ and $\tilde{\tau}^{(2)} = w^{(2)}$. If we assume for the sake of simplicity that the arrival process is Poisson, (6.1) yields

$$\mathbb{E}[W] = \frac{w}{1 - N\lambda(\pi w + \sigma)} \sum_{l=1}^{N-1} \frac{1}{1 - \tilde{\mu}_l} + \frac{N\lambda(\pi w^{(2)} + 2\pi w \sigma + \sigma^{(2)})}{2(1 - N\lambda(\pi w + \sigma))} + \frac{w^{(2)}}{2w}.$$

The value of $\mathbb{E}[W]$ depends on P only through the value of p_N. One known model that is described by this equation is the Bernoulli strategy of Servi [14]: when the server has served a customer, it quits the queue with probability π and continues to serve it with probability $1 - \pi$. In the original model, the server polls the queues in cyclic order. Here, we have a Markovian Bernoulli polling model if we take $P = (1 - \pi)I + \pi \tilde{P}$. It is easy to check that this choice of P satisfies A_3. Moreover, $\mathbb{E}[W]$ is an increasing function of π that is minimal when $\pi = 0$. This case corresponds to an exhaustive service strategy with Markovian routing $\tilde{P}$: the server polls the same queue until it is empty and then moves to another queue using the routing matrix $\tilde{P}$. In this case, Equation (6.1) simply becomes

$$\mathbb{E}[W] = \frac{w}{1 - N\lambda\sigma} \sum_{l=1}^{N-1} \frac{1}{1 - \tilde{\mu}_l} + \frac{N\lambda\sigma^{(2)}}{2(1 - N\lambda\sigma)} + \frac{w^{(2)}}{2w}.$$

Using Lemma 6.2, we find that the above expression is minimal when $\tilde{P}$ describes a cyclic polling scheme. Further comparisons between different polling strategies can be found in Levy, Sidi and Boxma [11].

Appendix.

A.1. Second vector field. This appendix contains some basic definitions and results from the theory of dynamical systems. These definitions have been adapted for the Markov chain $\mathcal{L}$ defined in Section 2. We refer the reader to [12] and [4] for a more complete treatment of the subject.

Faces. For any $\wedge \subset \{1, \ldots, N\}$, define $B^\wedge \subset \mathbb{R}_+^N$ as

$$B^\wedge \stackrel{\text{def}}{=} \{(x_1, \ldots, x_n) : x_i > 0 \Leftrightarrow i \in \wedge\}.$$

$B^\wedge$ is the *face* of $\mathbb{R}_+^N$ associated to $\wedge$. Whenever not ambiguous, the face $B^\wedge$ will be identified with $\wedge$.

Induced chains. For any $\wedge \neq \mathcal{S}$ we choose an arbitrary point $\vec{a} \in B^\wedge \cap \mathbb{Z}_+^N$ and draw a plane $C^\wedge$ of dimension $N - |\wedge|$, perpendicular to $B^\wedge$ and containing $\vec{a}$. We define the *induced Markov chain* $\mathcal{L}^\wedge$, with state space $\mathcal{S} \times (C^\wedge \cap \mathbb{Z}_+^N)$ (by an obvious abuse in the notation, we shall write most of the time $\mathcal{S} \times C^\wedge$) and transition probabilities

$$\wedge P_{(s,\vec{x})(t,\vec{y})} = P_{(s,\vec{x})(t,\vec{y})} + \sum_{\vec{z} \neq \vec{y}} P_{(s,\vec{x})(t,\vec{z})}, \ \forall \vec{x}, \vec{y} \in C^\wedge, \ s, t \in \mathcal{S},$$

where the summation is performed over all $\vec{z} \in \mathbb{Z}_+^N$, such that the straight line connecting $\vec{z}$ and $\vec{y}$ is perpendicular to $C^\wedge$. It is important to note that this construction does not depend on $\vec{a}$.

ASSUMPTION A_4. *For any $\wedge$ the chain $\mathcal{L}^\wedge$ is irreducible and aperiodic. $\wedge$ is called* ergodic *(resp.* non-ergodic, transient*) according as $\mathcal{L}^\wedge$ is ergodic (resp. non-ergodic, transient).*

For an ergodic $\mathcal{L}^\wedge$, let $\pi^\wedge(s, \vec{x})$, $(s, \vec{x}) \in \mathcal{S} \times C^\wedge$ be its stationary transition probabilities. Moreover, let $\pi^\wedge(s)$ (resp. $\pi^\wedge(s, x_s > 0)$) be the stationary probability that the server is at station s (resp. and finds it nonempty).

First vector field. For any $(s, \vec{x}) \in \mathcal{S} \times \mathbb{Z}_+^N$, the *first vector field* is simply the mean drift of the random walk $\mathcal{L}$ at point $(s, \vec{x})$:

$$\vec{M}(s, \vec{x}) \stackrel{\text{def}}{=} \sum_{r \in \mathcal{S}, \ \vec{y} \in \mathbb{Z}_+^N} (\vec{y} - \vec{x}) P\Big[X(n+1) = \vec{y}, S(n+1) = r \ \Big| \ X(n) = \vec{x}, S(n) = s\Big].$$

Let $\vec{v}^\wedge \stackrel{\text{def}}{=} (v_1^\wedge, \ldots, v_N^\wedge)$ such that

$$v_i^\wedge = 0, \ i \notin \wedge,$$
$$v_i^\wedge = \sum_{(s,\vec{x}) \in \mathcal{S} \times C^\wedge} \pi^\wedge(s, \vec{x}) \vec{M}_i(s, \vec{x}), \ i \in \wedge.$$

Intuitively, one can imagine that the random walk starts from a point which is close to $\wedge$, but sufficiently far from all other faces $B^{\wedge'}$, with $\wedge \not\subset \wedge'$. After some time (sufficiently long, but less than the minimal distance from the above mentioned $B^{\wedge'}$), the stationary regime in the induced chain will be installed. In this regime, one can ask about the mean drift along $\wedge$: it is defined exactly by $\vec{v}^\wedge$. For $\wedge = \mathcal{S}$, we call $\wedge$ *ergodic*, by definition, and put

$$\vec{v}^{\{1,\dots,N\}} \equiv \sum_{s \in \mathcal{S}} \pi^{\{1,\dots,N\}}(s)\vec{M}(s,\vec{x}), \ \vec{x} \in B^{\{1,\dots,N\}}.$$

From now on, when speaking about the *components* of $\vec{v}^\wedge$, we mean the components $v_i^\wedge$ with $i \in \wedge$.

ASSUMPTION A$_5$. $v_i^\wedge \neq 0$, *for each* $i \in \wedge$.

Ingoing, outgoing and neutral faces. Let us fix $\wedge, \wedge_1$, so that $\wedge \supset \wedge_1, \wedge \neq \wedge_1$, that is to say $\overline{B}^\wedge \supset B^{\wedge_1}$ ($\overline{B}^\wedge$ is the closure of $B^\wedge$). Let $B^\wedge$ be ergodic. Thus $\vec{v}^\wedge$ is well defined. There are three possibilities for the direction of $\vec{v}^\wedge$ w.r.t. $B^{\wedge_1}$. We say that $B^\wedge$ is an *ingoing* (resp. *outgoing*) *face* for $B^{\wedge_1}$, if all the coordinates $v_i^\wedge$ for $i \in \wedge \setminus \wedge_1$ are negative (resp. positive). Otherwise we say that $B^\wedge$ is *neutral*. As an example we give simple sufficient criteria for a face to be ergodic.

The second vector field. To any point $\vec{x} \in \mathbb{R}_+^N$, we assign a vector $v(\vec{x})$ and call this function the *second vector field*. It can be multivalued on some non-ergodic faces. We put, for ergodic faces $B^\wedge$,

$$\vec{v}(\vec{x}) \equiv \vec{v}^\wedge, \ \vec{x} \in B^\wedge.$$

If $B^{\wedge_1}$ is non-ergodic, then at any point $\vec{x} \in B^{\wedge_1}$, $\vec{v}(\vec{x})$ takes all values $\vec{v}^\wedge$ for which $B^\wedge$ is an outgoing face with respect to $B^{\wedge_1}$. In other words, for $\vec{x}$ belonging to non-ergodic faces, with $\|\vec{x}\|$ sufficiently large,

$$\vec{x} + \vec{v}(\vec{x}) \ \in \mathbb{R}_+^N,$$

for any value $\vec{v}(\vec{x})$. If there is no such vector, we put $\vec{v}(\vec{x}) = 0$, for $\vec{x} \in B^{\wedge_1}$. Points $\vec{x} \in \mathbf{R}_+^N$, where $\vec{v}(\vec{x})$ is more than one-valued, are called *branch points*.

There are few interesting examples, for which only the first vector field suffices to obtain ergodicity conditions for the random walk of interest, but it is nevertheless the case for Jackson networks. In general, the second vector field must be introduced as shown in the following theorem, taken from [12] (and extended in [5] to the case of upward unbounded jumps).

THEOREM A.1. *Assume that there exists a nonnegative function f :* $\mathbb{R}_+^N \to \mathbb{R}_+$ *such that*

1. for some constant $c > 0$

$$f(\vec{x}) - f(\vec{y}) \leq c\|\vec{x} - \vec{y}\|;$$

2. there exists $\delta > 0$, $p > 0$ such that for all $x \in B^\wedge$

$$f(\vec{x} + v(\vec{x})) - f(\vec{x}) \leq -\delta.$$

Then the Markov chain $\mathcal{L}$ is ergodic.

A.2. Some simple results about circulant matrices. In this appendix, we recall some well-known properties of the circulant matrices which are used throughout this paper. These properties are given without proof, since they can easily be verified at hand. Throughout the paper, we take the convention that every subscript less than 1 or greater than N should be shifted into the correct range.

Let $(\vec{e}_1, \ldots, \vec{e}_N)$ denote the canonical basis of $\mathbb{C}^N$ and let $\vec{e} = (1, \ldots, 1)$. Moreover, we define $\omega_k \stackrel{\text{def}}{=} \exp(2\hat{\imath}\pi k/N)$, with $\hat{\imath}^2 = -1$.

DEFINITION A.2. *A circulant matrix M of size N is a matrix whose coefficients $m_{i,j}$ verify the relation:*

$$m_{i+k,j+k} = m_{i,j}, \text{ for all } 1 \le i, j, k \le N.$$

The following Lemma summarizes some key properties of these matrices:

LEMMA A.3. *Let M be a circulant matrix of the form:*

$$M = \begin{pmatrix} m_N & m_{N-1} & \cdots & m_1 \\ m_1 & m_N & \cdots & m_2 \\ \vdots & & \ddots & \vdots \\ m_{N-1} & m_{N-2} & \cdots & m_N \end{pmatrix}$$

Then for each $1 \le k \le N$, the vector $\vec{v}_k \stackrel{\text{def}}{=} \sum_{i=1}^{N} \omega_k^{-i}\vec{e}_i$ is an eigenvector of the matrix M with eigenvalue $\sum_{i=1}^{N} \omega_k^i m_i$. Moreover, $(\vec{v}_1, \ldots, \vec{v}_N)$ is an orthogonal basis of $\mathbb{C}^N$ in which $\vec{e}_1, \ldots, \vec{e}_N$ can be expressed as

$$\vec{e}_i = \frac{1}{N} \sum_{k=1}^{N} \omega_k^i \vec{v}_k.$$

One important feature of circulant matrices is that they share the same basis of eigenvectors. Note that, with the notation given before Lemma A.3, we have $\vec{v}_N = \vec{e}$. Circulant matrices enjoy other properties that are not used here: for example, the product of two circulant matrices is a circulant matrix and this product commutes.

A.3. Proofs of the results of Sections 5 and 6. The derivations in this section are essentially analytic.

Proof of Lemma 5.1. This proof uses specific properties of circulant matrices to perform a fine analysis of the behavior of $A(\vec{z})$, $\tilde{A}(\vec{z})$ and $\Delta(\vec{z})$ in the neighborhood of $\vec{z} = \vec{e}$. As pointed out later in the proof, we study these matrices only for $\vec{z} \in \mathcal{D}^N$, thus avoiding any analytical continuation. The basic relation used thereafter is a simple consequence of the definitions of $A(\vec{z})$ and $\Delta(\vec{z})$:

(A.1) $$A\Delta(\vec{z})\vec{v}_k = \mu_k(\vec{z})\vec{v}_k + \sum_{q=1}^{N}(\frac{1}{z_q} - 1)\omega_k^{-q} A_{\cdot q}(\vec{z}),$$

where $A_{\cdot q}(\vec{z})$ stands for the q-th column of $A(\vec{z})$ and can be written as

$$
\begin{aligned}
A_{\cdot q}(\vec{z}) &= \sum_{i=1}^{N} p_{i-q} a_{i-q}(\vec{z}) \vec{e}_i \\
&= \sum_{i=1}^{N} p_{i-q} a_{i-q}(\vec{z}) \frac{1}{N} \sum_{l=1}^{N} \omega_l^i \vec{v}_l \\
&= \sum_{l=1}^{N} \omega_l^q \mu_l(\vec{z}) \vec{v}_l.
\end{aligned}
$$

So, if we define

$$
\varepsilon_l(\vec{z}) \stackrel{\text{def}}{=} \frac{1}{N} \sum_{q=1}^{N} \left(\frac{1}{z_q} - 1 \right) \omega_l^q,
$$

(A.1) can be rewritten as

$$
\text{(A.2)} \quad [I - A\Delta(\vec{z})]\vec{v}_k = (1 - \mu_k(\vec{z}))\vec{v}_k - \sum_{l=1}^{N} \varepsilon_{l-k}(\vec{z}) \mu_l(\vec{z}) \vec{v}_l
$$

$$
\text{(A.3)} \ [\tilde{A}(\vec{z}) - A\Delta(\vec{z})]\vec{v}_k = (\tilde{\mu}_k(\vec{z}) - \mu_k(\vec{z}))\vec{v}_k - \sum_{l=1}^{N} \varepsilon_{l-k}(\vec{z}) \mu_l(\vec{z}) \vec{v}_l
$$

Since ${}^*\vec{e}[I - A\Delta(\vec{e})] = {}^*\vec{e}[\tilde{A}(\vec{e}) - A\Delta(\vec{e})] = \vec{0}$, we define, for any set of arbitrary complex numbers $(c_1, \ldots, c_{N-1})$, the vector $\vec{u}_t$ as follows:

$$
\vec{u}_t \stackrel{\text{def}}{=} \frac{1}{N} \vec{e} + \frac{t}{N} \sum_{l=1}^{N-1} c_l \vec{v}_l.
$$

With this definition, we have

$$
{}^*\vec{u}_t [I - A\Delta(\vec{z})]\vec{e} = \frac{{}^*\vec{e}}{N}[I - A\Delta(\vec{z})]\vec{e} + \frac{t}{N} \sum_{l=1}^{N-1} c_l {}^*\vec{v}_l [I - A\Delta(\vec{z})]\vec{e}
$$

$$
\text{(A.4)} \qquad = 1 - \mu_N(\vec{z}) - \varepsilon_0(\vec{z})\mu_N(\vec{z}) - t \sum_{l=1}^{N-1} c_l \varepsilon_l(\vec{z})\mu_l(\vec{z}).
$$

When t is small, we have the relations:

$$
\begin{aligned}
\mu_N(\vec{z}_t) &= 1 + t\alpha \sum_{q=1}^{N} \dot{z}_q + \frac{t^2}{2} \sum_{q,r=1}^{N} \alpha_{q,r}^{(2)} \dot{z}_q \dot{z}_r \\
\text{(A.5)} \qquad & + \frac{t^2 \alpha}{2} \sum_{q=1}^{N} \ddot{z}_q + o(t^2)
\end{aligned}
$$

$$\tilde{\mu}_N(\vec{z}_t) = 1 + t\tilde{\alpha} \sum_{q=1}^{N} \dot{z}_q + \frac{t^2}{2} \sum_{q,r=1}^{N} \tilde{\alpha}_{q,r}^{(2)} \dot{z}_q \dot{z}_r$$

(A.6)
$$+ \frac{t^2 \tilde{\alpha}}{2} \sum_{q=1}^{N} \ddot{z}_q + o(t^2)$$

$$\varepsilon_l(\vec{z}_t) = -\frac{t}{N} \sum_{q=1}^{N} \dot{z}_q \omega_l^q + \frac{t^2}{N} \sum_{q=1}^{N} \dot{z}_q^2 \omega_l^q$$

(A.7)
$$- \frac{t^2}{2N} \sum_{q=1}^{N} \ddot{z}_q \omega_l^q + o(t^2).$$

The formulas given in Appendix A.2 allow to relate $\dot{z}_1, \ldots, \dot{z}_N$ to $\dot{\zeta}_1, \ldots, \dot{\zeta}_N$:

$$\dot{\zeta}_k = \frac{1}{N} \sum_{q=1}^{N} \dot{z}_q \omega_k^q, \qquad \dot{z}_q = \sum_{l=1}^{N} \dot{\zeta}_q \omega_l^{-q}, \qquad \sum_{q=1}^{N} \dot{z}_q \dot{z}_{q+d} = N \sum_{l=1}^{N} \dot{\zeta}_l \dot{\zeta}_{N-l} \omega_l^d.$$

Using the remark after the definition of $\alpha_{q,r}^{(2)}$ and $\tilde{\alpha}_{q,r}^{(2)}$, we find that:

$$\sum_{q,r=1}^{N} \alpha_{q,r}^{(2)} \dot{z}_q \dot{z}_r = \sum_{d=1}^{N} \alpha_d^{(2)} \sum_{q=1}^{N} \dot{z}_q \dot{z}_{q+d} = N \sum_{l=1}^{N} \dot{\zeta}_l \dot{\zeta}_{N-l} \sum_{d=1}^{N} \alpha_d^{(2)} \omega_l^d.$$

Applying (A.5) and (A.7) to (A.4), we see that the first-order term in the expression (A.4) is $(\alpha - 1/N)t[\dot{z}_1 + \cdots + \dot{z}_N]$. So a necessary condition to have a formula like (5.10) is $\dot{z}_1 + \cdots + \dot{z}_N = 0$. Using this relation, we have

$$^*\vec{u}_t [I - A\Delta(\vec{z}_t)]\vec{e} = t^2 \sum_{l=1}^{N-1} c_l \dot{\zeta}_l \mu_l - t^2 \sum_{l=1}^{N} \dot{\zeta}_l \dot{\zeta}_{N-l}$$

$$- t^2 \frac{N}{2} \sum_{l=1}^{N} \dot{\zeta}_l \dot{\zeta}_{N-l} \sum_{d=1}^{N} \alpha_d^{(2)} \omega_l^d$$

(A.8)
$$+ t^2 \frac{1 - N\alpha}{2N} \sum_{q=1}^{N} \ddot{z}_q + o(t^2).$$

We have to check that it is possible to have $\vec{z}_t \in \mathcal{D}^N$ and $\dot{z}_1 + \cdots + \dot{z}_N = 0$. One easy way to satisfy these constraints is to ensure that all coordinates of $\vec{z}_t$ arrive to 1 tangentially to the unit circle as t goes to 0. This is the case when, for any q, $\dot{z}_q$ is an imaginary number and $\ddot{z}_q < 0$. Moreover, for $k < N$,

$$^*\vec{u}_t [I - A\Delta(\vec{z}_t)]\vec{v}_k = -\varepsilon_{N-k}(\vec{z}_t)\mu_N(\vec{z}_t) + tc_k(1 - \mu_k(\vec{z}_t))$$

$$-t \sum_{l=1}^{N-1} c_l \varepsilon_{l-k}(\vec{z}_t) \mu_l(\vec{z}_t)$$

$$(A.9) \qquad = t\dot{\zeta}_{N-k} + tc_k(1 - \mu_k) + o(t).$$

This shows that we get equations (5.8) and (5.10) from (A.8) and (A.9) if we take

$$c_k = -\frac{\dot{\zeta}_{N-k}}{1 - \mu_k}.$$

With this choice of $c_1, \ldots, c_{N-1}$, we obtain equations (5.11) and (5.9) in exactly the same way.   □

Proof of Theorem 5.2. The idea of this proof is to apply the results of Lemma 5.1 to the equation

$$(A.10) \qquad {}^*\vec{u}_t [I - A\Delta(\vec{z}_t)]F(\vec{z}_t) = {}^*\vec{u}_t [\tilde{A}(\vec{z}_t) - A\Delta(\vec{z}_t)]\tilde{F}(\vec{z}_t).$$

Define $\vec{z}_t$ as in Lemma 5.1. Then

$$F(\vec{z}_t) = F(\vec{e}) + t \sum_{r=1}^{N} \dot{z}_r \frac{\partial F}{\partial z_r}(\vec{e}) + \vec{o}(t).$$

Moreover, Equation (5.1) implies that $\partial F_i / \partial z_r(\vec{e}) = \partial F_{i-r+m}/\partial z_m(\vec{e})$ for any fixed $m \in \mathcal{S}$ and

$$\begin{aligned}
\frac{\partial F}{\partial z_r}(\vec{e}) &= \sum_{i=1}^{N} \frac{\partial F_i}{\partial z_r}(\vec{e})\vec{e}_i \\
&= \sum_{i=1}^{N} \left[\frac{\partial F_{i-r+m}}{\partial z_m}(\vec{e})\frac{1}{N}\sum_{k=1}^{N}\omega_k^i \vec{v}_k\right] \\
&= \sum_{k=1}^{N} \omega_k^{r-m}\left[\frac{1}{N}\sum_{i=1}^{N}\frac{\partial F_{i-r+m}}{\partial z_m}(\vec{e})\omega_k^{i-r+m}\right]\vec{v}_k \\
&= \sum_{k=1}^{N} \omega_k^{r-m}\frac{\partial \varphi_k}{\partial z_m}(\vec{e})\vec{v}_k,
\end{aligned}$$

where $(\varphi_1(\vec{z}), \ldots, \varphi_N(\vec{z}))$, defined as

$$\varphi_k(\vec{z}) \stackrel{\text{def}}{=} \frac{1}{N}\sum_{i=1}^{N} F_i(\vec{z})\omega_k^i,$$

are the coordinates of $F(\vec{z})$ in the basis $(\vec{v}_1, \ldots, \vec{v}_N)$. Finally,

$$(A.11) \qquad F(\vec{z}_t) = F(\vec{e}) + tN \sum_{k=1}^{N} \dot{\zeta}_k \omega_k^{-m}\frac{\partial \varphi_k}{\partial z_m}(\vec{e})\vec{v}_k + \vec{o}(t).$$

With a similar definition for $\tilde{\varphi}(\vec{z})$,

$$(A.12) \qquad \tilde{F}(\vec{z}_t) = \tilde{F}(\vec{e}) + tN \sum_{k=1}^{N} \dot{\zeta}_k \omega_k^{-m} \frac{\partial \tilde{\varphi}_k}{\partial z_m}(\vec{e})\vec{v}_k + \vec{o}(t).$$

We now apply Lemma 5.1 and Equations (A.11) and (A.12) to Equation (A.10) and use the fact that, by symmetry, $F(\vec{e}) = F_1(\vec{e})\vec{e}$ and $\tilde{F}(\vec{e}) = \tilde{F}_1(\vec{e})\vec{e}$

$$t^2 F_1(\vec{e}) \left\{ -\sum_{l=1}^{N-1} \dot{\zeta}_l \dot{\zeta}_{N-l} \left[\frac{1}{1-\mu_l} + \frac{N}{2} \sum_{d=1}^{N} \alpha_d^{(2)} \omega_l^d \right] + \frac{1-N\alpha}{2N} \sum_{q=1}^{N} \ddot{z}_q \right\}$$

$$= t^2 \tilde{F}_1(\vec{e}) \left\{ -\sum_{l=1}^{N-1} \dot{\zeta}_l \dot{\zeta}_{N-l} \left[\frac{1}{1-\mu_l} + \frac{N}{2} \sum_{d=1}^{N} (\alpha_d^{(2)} - \tilde{\alpha}_d^{(2)}) \omega_k^d \right] \right.$$

$$\left. + \frac{1-N\alpha+N\tilde{\alpha}}{2N} \sum_{q=1}^{N} \ddot{z}_q \right\}$$

$$+ t^2 N \sum_{k=1}^{N-1} \dot{\zeta}_k \dot{\zeta}_{N-k} \omega_k^{-m} \frac{1-\tilde{\mu}_k}{1-\mu_k} \frac{\partial \tilde{\varphi}_k}{\partial z_m}(\vec{e}) + o(t^2).$$

As $\vec{z}$ can be chosen freely, we can derive a first equality from this equation, namely

$$F_1(\vec{e}) \frac{1-N\alpha}{2N} = \tilde{F}_1(\vec{e}) \frac{1-N\alpha+N\tilde{\alpha}}{2N}.$$

Since $P(X_m = 0 \mid S = m) - \tilde{F}_1(\vec{e})/F_1(\vec{e})$, this yields Equation (5.12). Taking in account (5.6), we find

$$\sum_{k=1}^{N-1} \dot{\zeta}_k \dot{\zeta}_{N-k} \omega_k^{-m} \frac{1-\tilde{\mu}_k}{1-\mu_k} N \frac{\partial \tilde{\varphi}_k}{\partial z_m}(\vec{e})$$

$$(A.13) \qquad = -\sum_{l=1}^{N-1} \dot{\zeta}_l \dot{\zeta}_{N-l} \left[\frac{F_1(\vec{e}) - \tilde{F}_1(\vec{e})}{1-\mu_l} + \frac{1}{2} \sum_{d=1}^{N} \tilde{\alpha}_d^{(2)} \omega_l^d \right].$$

This equation contains in fact a system of linear equations that can be built by choosing $\vec{z}$. The problem is that in general the rank of this system is only $\lfloor n/2 \rfloor$. However, since μ_k (resp. μ_{N-k}, $\tilde{\mu}_k$, $\tilde{\mu}_{N-k}$) is the eigenvalue of $^*P$ (resp. P, $^*\tilde{P}$, $\tilde{P}$) associated to the eigenvector $\vec{v}_k$, A3 implies that, for $1 \le k < N$,

$$\frac{1-\mu_k}{1-\tilde{\mu}_k} = \frac{1-\mu_{N-k}}{1-\tilde{\mu}_{N-k}} \in \mathbb{R}.$$

As noted in the proof of Lemma 5.1, $\dot{z}_1, \ldots, \dot{z}_N$ are imaginary numbers and $\zeta_k \zeta_{N-k}$ is a negative real number for any $k < N$. Hence we can choose in (A.13)

$$\zeta_k \zeta_{N-k} = -\min\left[\frac{1-\mu_k}{1-\tilde{\mu}_k}, 0\right].$$

The combination of the resulting equation with a similar equation containing only the terms where $(1-\mu_k)/(1-\tilde{\mu}_k) < 0$ yields

(A.14)
$$\sum_{k=1}^{N-1} \omega_k^{-m} N \frac{\partial \tilde{\varphi}_k}{\partial z_m}(\vec{e})$$
$$= -\sum_{l=1}^{N-1} \frac{1-\mu_k}{1-\tilde{\mu}_k}\left[\frac{F_1(\vec{e}) - \tilde{F}_1(\vec{e})}{1-\mu_l} + \frac{1}{2}\sum_{d=1}^{N} \bar{\alpha}_d^{(2)} \omega_l^d\right].$$

We know that, by definition, $\tilde{F}_m(\vec{z})$ does not depend on z_m. Hence,

$$\frac{\partial \tilde{F}_m}{\partial z_m}(\vec{e}) = \sum_{k=1}^{N} \omega_k^{-m} \frac{\partial \tilde{\varphi}_k}{\partial z_m}(\vec{e}) = 0,$$

and, using Equation (A.14)

(A.15) $$N\frac{\partial \tilde{\varphi}_N}{\partial z_m}(\vec{e}) = \sum_{l=1}^{N-1}\left[\frac{F_1(\vec{e}) - \tilde{F}_1(\vec{e})}{1-\tilde{\mu}_l} + \frac{1-\mu_l}{1-\tilde{\mu}_l}\frac{1}{2}\sum_{d=1}^{N} \bar{\alpha}_d^{(2)} \omega_l^d\right].$$

The last step of the demonstration is to get $\partial F_m/\partial z_m(\vec{e})$ from these results. This can be done as in Lemma 5.1, but with $\vec{u}_t = \vec{e}/N$ and $\vec{z}_t$ chosen differently. Using Equations (A.2), (A.3) and (A.5)–(A.7) with the same notation as in Lemma 5.1, we have for $k < N$

$$\frac{^*\vec{e}}{N}[I - A\Delta(\vec{z}_t)]\vec{e} = 1 - \mu_N(\vec{z}_t) - \varepsilon_0(\vec{z}_t)\mu_N(\vec{z}_t)$$

$$= t(1-N\alpha)\dot{\zeta}_N - t^2\sum_{l=1}^{N}\dot{\zeta}_l\dot{\zeta}_{N-l} + t^2 N\alpha\dot{\zeta}_N^2$$

$$- t^2\frac{N}{2}\sum_{l=1}^{N}\dot{\zeta}_l\dot{\zeta}_{N-l}\sum_{d=1}^{N}\alpha_d^{(2)}\omega_l^d + o(t^2)$$

$$\frac{^*\vec{e}}{N}[I - A\Delta(\vec{z}_t)]\vec{v}_k = -\varepsilon_{N-k}(\vec{z}_t)\mu_N(\vec{z}_t) = t\dot{\zeta}_{N-k} + o(t)$$

$$\frac{^*\vec{e}}{N}[\tilde{A}(\vec{z}_t) - A\Delta(\vec{z}_t)]\vec{e} = \tilde{\mu}_N(\vec{z}_t) - \mu_N(\vec{z}_t) - \varepsilon_0(\vec{z}_t)\mu_N(\vec{z}_t)$$

$$= t(1-N\alpha+N\tilde{\alpha})\dot{\zeta}_N - t^2\sum_{l=1}^{N}\dot{\zeta}_l\dot{\zeta}_{N-l} + t^2 N\alpha\dot{\zeta}_N^2$$

$$+t^2\frac{N}{2}\sum_{l=1}^{N}\dot\zeta_l\dot\zeta_{N-l}\sum_{d=1}^{N}[\tilde\alpha_d^{(2)}-\alpha_d^{(2)}]\omega_l^d+o(t^2)$$

$$\frac{{}^*\vec e}{N}[\tilde A(\vec z_t)-A\Delta(\vec z_t)]\vec v_k \;=\; -\varepsilon_{N-k}(\vec z_t)\mu_N(\vec z_t)\;=\;t\dot\zeta_{N-k}+o(t)$$

Combining these equations with (A.11) and (A.12), we find

$$-t^2F_1(\vec e)\left\{\sum_{l=1}^{N}\dot\zeta_l\dot\zeta_{N-l}\left[1+\frac{N}{2}\sum_{d=1}^{N}\alpha_d^{(2)}\omega_l^d\right]-N\alpha\dot\zeta_N^2\right\}$$

$$+t^2N\sum_{k=1}^{N-1}\dot\zeta_k\dot\zeta_{N-k}\omega_k^{-m}\frac{\partial\varphi_k}{\partial z_m}(\vec e)+t^2N(1-N\alpha)\dot\zeta_N^2\frac{\partial\varphi_N}{\partial z_m}(\vec e)$$

$$=\;-t^2\tilde F_1(\vec e)\left\{\sum_{l=1}^{N}\dot\zeta_l\dot\zeta_{N-l}\left[1+\frac{N}{2}\sum_{d=1}^{N}[\alpha_d^{(2)}-\tilde\alpha_d^{(2)}]\omega_l^d\right]-N\alpha\dot\zeta_N^2\right\}$$

$$+t^2N\sum_{k=1}^{N-1}\dot\zeta_k\dot\zeta_{N-k}\omega_k^{-m}\frac{\partial\tilde\varphi_k}{\partial z_m}(\vec e)$$

(A.16) $$+t^2N(1-N\alpha+N\tilde\alpha)\dot\zeta_N^2\frac{\partial\tilde\varphi_N}{\partial z_m}(\vec e)+o(t^2).$$

This equation in turn gives for $(1-N\alpha)\dot\zeta_N^2=1$ and $\dot\zeta_l\dot\zeta_{N-l}=1$ if $1\le l<N$

$$N\sum_{k=1}^{N}\omega_k^{-m}\frac{\partial\varphi_k}{\partial z_m}(\vec e)\;=\;\frac{N\tilde\alpha}{1-N\alpha}N\frac{\partial\tilde\varphi_N}{\partial z_m}(\vec e)+N(F_1(\vec e)-\tilde F_1(\vec e))$$

$$+\frac{1}{2}\sum_{l=1}^{N-1}\sum_{d=1}^{N}\bar\alpha_d^{(2)}\omega_l^d+\frac{1}{1-N\alpha}\frac{1}{2}\sum_{d=1}^{N}\bar\alpha_d^{(2)}$$

This expression, together with (A.15) and (5.7), yields Equation (5.13). The derivation of (5.14) is done in a similar way: we use the fact that

$$\mathbb E X_m\;=\;\sum_{i=1}^{N}\frac{\partial F_i}{\partial z_m}(\vec e)\;=\;N\frac{\partial\varphi_N}{\partial z_m}(\vec e).$$

We use Equation (A.16) with $(1-N\alpha)\dot\zeta_N^2=1$ and $\dot\zeta_k\dot\zeta_{N-k}=0$ for $k<N$ and find

$$N\frac{\partial\varphi_N}{\partial z_m}(\vec e)=\frac{1-N\alpha+N\tilde\alpha}{1-N\alpha}N\frac{\partial\tilde\varphi_N}{\partial z_m}(\vec e)+F_1(\vec e)-\tilde F_1(\vec e)+\frac{1}{1-N\alpha}\frac{1}{2}\sum_{d=1}^{N}\bar\alpha_d^{(2)}.$$

This gives Equation (5.14) and concludes the proof of the theorem. $\square$

Proof of Theorem 6.1. We see easily that the mean numbers of clients arriving between polling times are respectively $\alpha = \lambda\tau$ and $\tilde\alpha = \lambda\tilde\tau$. Moreover, using Equations (5.3) and (5.4) and the properties of generating functions we find that, for $d < N$

$$
\begin{aligned}
\alpha_d^{(2)} &= \alpha^{(2)} = \lambda^2\tau^{(2)} \\
\alpha_N^{(2)} &= \alpha^{(2)} + \hat\lambda(b^{(2)} - b)\tau \\
\tilde\alpha_d^{(2)} &= \tilde\alpha^{(2)} = \lambda^2\tilde\tau^{(2)} \\
\tilde\alpha_N^{(2)} &= \tilde\alpha^{(2)} + \hat\lambda(b^{(2)} - b)\tilde\tau.
\end{aligned}
$$

The computation of waiting times uses the following classical argument: a non empty queue visited by the server can be decomposed into
- the head of line customer;
- the clients who arrived after him during his waiting time;
- clients who arrived in the same batch as the first client, but are not yet served; since, by renewal arguments, the mean size of this batch is $b^{(2)}/b$, the mean number of remaining clients is $(b^{(2)} - b)/2b$.

This can be written as

$$
\mathbb{E}[X_m \mid S = m, X_m > 0] = 1 + \lambda\,\mathbb{E}[W] + \frac{b^{(2)} - b}{2b}
$$

and, using the fact that $\sum_{d=1}^{N} \alpha_d^{(2)} w_l^d = \alpha_N^{(2)} - \alpha^{(2)}$,

$$
\begin{aligned}
\lambda b\,\mathbb{E}[W] &= \frac{\mathbb{E}[X_m \mid S = m]}{1 - P(X_m = 0 \mid S = m)} - 1 - \frac{b^{(2)} - b}{2b} \\
&= \frac{\tilde\alpha}{1 - N\alpha} \sum_{l=1}^{N-1} \frac{1}{1 - \mu_l} + \frac{1}{1 - N\alpha}\frac{N}{2}\alpha^{(2)} + \frac{1}{N\tilde\alpha}\frac{N}{2}\tilde\alpha^{(2)} \\
&\quad + \frac{b^{(2)} - b}{2b}\left[\frac{\alpha + (N-1)\tilde\alpha}{1 - N\alpha} - \frac{N\tilde\alpha}{1 - N\alpha}\frac{1}{N}\sum_{l=1}^{N-1}\frac{\mu_l - \tilde\mu_l}{1 - \tilde\mu_l}\right] \\
&= \frac{\lambda\tilde\tau}{1 - N\lambda\tau}\sum_{l=1}^{N-1}\frac{1}{1 - \mu_l} + \frac{N\lambda^2\tau^{(2)}}{2(1 - N\lambda\tau)} + \frac{\lambda\tilde\tau^{(2)}}{2\tilde\tau} \\
&\quad + \frac{b^{(2)} - b}{2b}\left[\frac{\lambda\tau + (N-1)\lambda\tilde\tau}{1 - N\lambda\tau} - \frac{\lambda\tilde\tau}{1 - N\lambda\tau}\sum_{l=1}^{N-1}\frac{\mu_l - \tilde\mu_l}{1 - \tilde\mu_l}\right].
\end{aligned}
$$

This gives Equation (6.1). $\qquad\qquad\Box$

REFERENCES

[1] A. A. BOROVKOV AND R. SCHASSBERGER, *Ergodicity of a polling network*, Stochastic Processes and their Applications, 50 (1994), pp. 253–262.

[2] O. J. BOXMA AND J. WESTSTRATE, *Waiting time in polling systems with Markovian server routing*, in Messung, Modellierung und Bewertung von Rechensysteme, Berlin, 1989, Proc. Conference Braunschweig, Springer-Verlag, pp. 89–104.

[3] H. S. BRADLOW AND H. F. BYRD, *Mean waiting time evaluation of packet switches for centrally controlled PBX's*, Performance Evaluation, 7 (1987), pp. 309–327.

[4] G. FAYOLLE, V. A. MALYSHEV, AND M. V. MENSHIKOV, *Topics in the Constructive Theory of Countable Markov Chains (Part I)*, Cambridge University Press, 1994.

[5] G. FAYOLLE, O. SHAPOVAL, AND A. A. ZAMYATIN, *Controlled random walks in Z_+^N and their applications to queueing networks*, Preprint, (1993). To appear.

[6] M. J. FERGUSON, *Mean waiting time for a token ring with station dependent overheads*, in Local Area & Multiple Access Networks, R. L. Pickholtz, ed., Computer Science Press, Rockville, Maryland, 1986, ch. 3, pp. 43–67.

[7] C. FRICKER AND M. R. JAÏBI, *Monotonicity and stability of periodic polling models*, Queueing Systems, Theory and Applications, 15 (1994), pp. 211–238.

[8] ———, *Stability of a polling model with a Markovian scheme*, Rapport de Recherche 2278, INRIA, Domaine de Voluceau, Rocquencourt B.P. 105 78153 Le Chesnay Cedex France, May 1994.

[9] L. KLEINROCK AND H. LEVY, *The analysis of random polling systems*, Operations Research, 36 (1988), pp. 716–732.

[10] H. LEVY AND M. SIDI, *Polling systems: Applications, modeling and optimization*, IEEE Transactions on Communications, 38 (1990), pp. 1750–1760.

[11] H. LEVY, M. SIDI, AND O. J. BOXMA, *Dominance relations in polling systems*, Queueing Systems, Theory and Applications, 6 (1990), pp. 155–172.

[12] V. A. MALYSHEV AND M. V. MENSHIKOV, *Ergodicity, continuity and analyticity of countable Markov chains*, Trans. Moscow. Math. Soc., 39 (1979), pp. 3–48.

[13] R. SCHASSBERGER, *Stability of polling networks with state-dependent server routing*, Preprint, (1993).

[14] L. D. SERVI, *Average delay approximation of M/G/1 cyclic service queues with Bernoulli schedules*, IEEE Journal on Selected Areas in Communications, SAC-4 (1986), pp. 813–822. Correction in Vol. Sac-5, No. 3, p. 547, 1987.

[15] M. M. SRINIVASAN, *Nondeterministic polling systems*, Management Science, 37 (1991), pp. 667–681.

[16] H. TAKAGI, *Queueing analysis of polling models: an update*, in Stochastic Analysis of Computer and Communication Systems, H. Takagi, ed., North Holland, Amsterdam, 1990, pp. 267–318.

STARLIKE NETWORKS WITH SYNCHRONIZATION
CONSTRAINTS

S.A. BEREZNER*, D.M. ROSE*, AND YU.M. SUHOV†

Abstract. We consider a circuit-switched starlike network with efficient work-conserving synchronization constraints. Nonpreemptive and preemptive forms of the resultant discipline will be analyzed. For such synchronization rules, waiting times might not be monotonic with respect to the initial workload so that an increase in initial conditions need not result in an increase in waiting times. In a network operating according to our discipline, waiting times will also depend on both past and future arrivals. Other work-conserving and non-work-conserving disciplines have been investigated in [2]–[4] and [6]–[8]. We compare and contrast our discipline with these on the basis of a number of criteria, including efficiency of operation. We also discuss the concept of a limiting picture considered in [2] and [6]. The discipline we introduce and the analytical methods we use could be extended to other networks. In particular, a number of results pertaining to a message-switched network are presented in this paper.

Key words. starlike network, circuit-switching, work-conserving synchronization constraints, nonpreemptive and preemptive service, stability, resequencing, message-switching

AMS(MOS) subject classifications. 60K25

1. Introduction. The concept of synchronization constraints ([3]) is a familiar one. When one encounters a large network system with complicated routing in which customers have intersecting paths ([4]), there is a need to generalize and synchronize the simpler class of disciplines, of which $FCFS$ is a well-known type.

In this paper, we introduce a new protocol (which incorporates synchronization constraints) to the list of disciplines which have already been introduced ([4], [2], [6] and [11]). Our new protocol combines desirable properties of the existing disciplines. Introduction of this protocol is motivated by the more efficient network operation which results.

We prove results about the uniform stability[1] of both the nonpreemptive and preemptive versions of our protocol which we compare and contrast with the various other disciplines according to a number of criteria, including their efficiency and the techniques used to analyze them mathematically. While [4] is concerned with a fairly general class of networks, the emphasis in our paper is on starlike networks which we now describe. (In the sequel $\mathbb{N}$ will denote the set of natural numbers $\{1, 2, 3, \ldots\}$ and $\mathbb{Z}$,

---

* Department of Mathematical Statistics, Faculty of Science, University of Natal, King George V Avenue, Durban 4001, Natal, South Africa.

† Statistical Laboratory, Department of Pure Mathematics and Mathematical Statistics, University of Cambridge, Cambridge CB2 1SB *and* St John's College, Cambridge CB2 1TP, England, UK.

[1] *Uniform stability* is explained in Remark 4.2 on page 320.

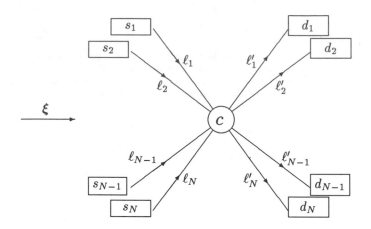

FIG. 1.1. *The Symmetric Starlike Network*

the set of integers. Also $x^+ = \max(0, x)$.)

Consider the symmetric circuit-switched starlike network with central node c (see Figure 1.1).

Denote the N peripheral sources by $s_i, i = 1, \ldots, N$ and the N peripheral destinations (addresses or receivers, depending on the context) by $d_j, j = 1, \ldots, N$. When there is no ambiguity, we will write s without a suffix for any one of the sources $s_1, \ldots, s_N$, and similarly with d. For the directed lines $s_i c$ and $c d_j$, write ℓ_i and ℓ'_j respectively.

A general arrival stream $\boldsymbol{\xi}$ decomposes into N IID (independent, identically distributed) stationary Poisson marked source flows $\{\xi_i, i = 1, \ldots, N\}$ on $[0, \infty)$, each of which is associated with a corresponding source s_i (and line ℓ_i). Batch arrivals are not allowed (point processes are simple). The total Poisson intensity of the stream $\boldsymbol{\xi}$ of arrivals to the network is given by λN, for some $\lambda > 0$. The (conditionally) IID marks of each RMPP (random marked point process) ξ_i are given by pairs (S, d) where S indicates the service time and d refers to the destination.

We assume that there are no calls in the system at time $t = 0$. More specifically, we will impose a zero initial condition on the network.

A process of interest, analogous to the process $\boldsymbol{\xi}$, is given by the family $\boldsymbol{\xi}' = \{\xi'_j, j = 1, \ldots, N\}$ of IID stationary Poisson RMPPs, where ξ'_j represents a destination flow associated with line ℓ'_j and destination d_j. Marks are given by pairs (S, s) where s, clearly, is the source of the call. It is evident that the RMPP families $\boldsymbol{\xi}$ and $\boldsymbol{\xi}'$ are equivalent ways of describing the network, i.e., the process of arrivals to the network is completely described by either one of $\boldsymbol{\xi}$ or $\boldsymbol{\xi}'$. Mark $s = s(y)$ has the discrete uniform (equiprobability) distribution given by $P[s = i] = \frac{1}{N}$ for $i \in 1, \ldots, N$ and similarly for d. Consequently any ξ'_j has the same intensity λ as does ξ_i. It is typically assumed that S (having distribution F), d and s are mutually independent.

A (marked) call or message from the total flow ξ is characterized by $y = [t, (S, s, d)]$ where $t = t(y)$ represents the time of arrival to the network of the call with vector of attributes y. In the sequel, *call y* should be understood to mean *the call with vector of attributes y*. The *total index* of y, $n(y)$, is defined to be a $\mathbb{N}$-valued function which gives the relative position of $t(y)$ in the sequence of network arrival times of all calls in ξ. The *source* and *destination indices* of y, $n_s(y)$ and $n_d(y)$, are analogously defined by substituting $\xi_{s(y)}$ and $\xi_{d(y)}$ respectively for ξ in the definition of $n(y)$.

Further notation and terminology will be introduced in the Preliminaries where we describe the disciplines mentioned earlier. Thereafter, in §3, these protocols are compared and contrasted. Some new theoretical results may be found in the fourth and fifth sections which are concerned with the nonpreemptive and preemptive forms of our discipline. Section 5 is followed by a brief analysis of message-switched starlike networks operating according to the three main disciplines considered. In the seventh and final section, some conclusions are drawn, and some open questions are posed.

2. Preliminaries. Our paper is devoted almost entirely to a circuit-switched model which is more challenging mathematically than its message-switched counterpart. Consider some call $y = [t, (S, s, d), n, n_s, n_d]$ arriving to source s of the circuit-switched network. Call y is required to wait for a time $w_1 = w_1(y)$ while all calls $\tilde{y}$ with $s(\tilde{y}) = s$ and $n_s(\tilde{y}) < n_s(y)$ leave the network. Once the system is free of these calls, y itself is allocated line ℓ_s at which point y has completed its first stage in the network. In accordance with the principles of circuit-switching, no other call may access channel ℓ_s until y has left the system.

In its second stage of processing, call y waits for a time of $w_2 = w_2(y)$ to be allocated its second channel, ℓ'_d, in order to complete the circuit $\ell_s\ell'_d$. Service for a period of S follows immediately thereafter, and then call y departs at time $\mathsf{dep}(y)$.

For most of the disciplines we consider, the fact that w_1 and w_2 are correctly defined (i.e., w_1 and w_2 are proper random variables) is obvious. In the nonpreemptive form of our discipline, this result is not immediate and a demonstration of the correctness will be provided in §4.

It is convenient to define the *source predecessor* of y, $p_s(y) = y' = [t', (S', s, d')]$, to be that call satisfying $n_s(y') = n_s(y) - 1$ $(n_s(y) > 1)$. Analogously to $p_s(y)$, define the *destination predecessor* of y to be that call $p_d(y) = y'' = [t'', (S'', s'', d)]$ satisfying $n_d(y'') = n_d(y) - 1$ $(n_d(y) > 1)$. More generally, $p_d^m(y), m \in \mathbb{Z}$ will represent the mth destination predecessor of y, i.e., that call with the destination index $n_d(y) - m$. It is appropriate to use the word *successor* if $m \in -\mathbb{N}$. The mth source predecessor $p_s^m(y)$ is defined in an analogous way. In particular $p_s^{m+1}(y) = p_s(p_s^m(y))$. Naturally, if the mth source (destination) predecessor of y is to exist, then we require that $m < n_s(y)$ $(m < n_d(y))$. Obviously

$$
\begin{aligned}
W(y) &= w_1(y) + w_2(y) \\
&= [W(y') + S(y') - (t(y) - t(y'))]^+ + w_2(y)
\end{aligned}
$$

(2.1)

(and $W(y) = w_2(y)$ when $n_s(y) = 1$).

REMARK 2.1 (A WORD ON NOTATION). Each of the symbols p_s, p_d, n_s and n_d should be regarded as a unit. For instance, if $d(\hat{y}) = \hat{d}$, we still write $p_d(\hat{y})$, and not $p_{\hat{d}}(\hat{y})$, for the destination predecessor of $\hat{y}$ (i.e., $p_d(y)$ may be interpreted as $p_{d(y)}(y)$). As a minor abuse of notation, s and d will also be used to represent a general (but fixed) source and destination respectively.

Conceptually, the least difficult discipline to apply in the second stage is $FCFS$ (abbreviation for **F**irst-**C**ome to the **S**econd Stage, **F**irst-**S**erved). If the channel ℓ_d' is free, then call $y = [t, (S, s, d)]$ may be processed immediately. Otherwise y must wait while all those calls $\{\bar{y}; d(\bar{y}) = d(y)\}$ already in the second stage are served. Although readily understood in practice, the $FCFS$ model creates mathematical difficulties. In [11], under the assumptions of a Poisson arrival stream *and* exponential service times, a necessary and sufficient condition was found for the existence and uniqueness of a stationary regime for any given N in a (finite) $FCFS$ network. This convergence could not be shown to be uniform in N, a fact which limited results for the infinite model.

The paper [6] was concerned with another discipline, which we shall denote by $FAFS$ (the abbreviation for **F**irst **A**rrived in the Network, **F**irst **S**erved). (In [2] the $FAFS$ message-switched network was analyzed and in [4], a very general circuit-switched $FAFS$ model was considered.) According to this protocol, $y = [t, (S, s, d)]$ is unable to have access to line ℓ_d' until the system is free of all calls $\tilde{y}$ with $d(\tilde{y}) = d$ and $n_d(\tilde{y}) < n_d(y)$.

In the current article we offer a third possible protocol—$FAAFS$, **F**irst **A**rrived in the Network and **A**llocated to **S**ervice, **F**irst Served—to govern the operation of a starlike (or more general) network. When call $y = [t, (S, s, d)]$ enters its second stage, it may find that ℓ_d' is free, in which case y may begin its service immediately. Otherwise, some call y_{n_0} must have access to channel ℓ_d'. If $t(y_{n_0}) < t(y)$, then y joins a (virtual) second-stage queue of calls awaiting access to ℓ_d'. In that queue, y is directly behind call y_k where $t(y_k) < t(y)$ and directly ahead of call y_{k+1} satisfying $t(y_{k+1}) > t(y)$.

On the other hand, if $t(y_{n_0}) > t$ two possibilities, determined by the priority service rule used, emerge:

(a) Under a *nonpreemptive* service rule, y proceeds to the head of the second-stage queue and waits while y_{n_0} and possibly other higher time-priority calls (arriving in the interim) complete their service but

(b) under a *preemptive* rule, y interrupts the service of y_{n_0}, which is forced to rejoin the queue. While y_{n_0} waits to be provided with service again, other calls may arrive to d. They then join the queue (possibly ahead of y_{n_0}) according to their respective times of arrival to the network. When y_{n_0} is processed once more, it may continue its service from the point of preemption (preemptive *resume* rule) or restart that service (preemptive *repeat* rule with or without resampling). Under an assumption of exponential service, the resume and (without-resampling) repeat rules produce (stochastically) equivalent waiting times.

We shall use the notation $FAAFS_N$ and $FAAFS_P$ to distinguish between the two priority rules. In particular, $FAAFS_P$ will refer to the $FAAFS$ discipline based on a preemptive resume rule.

REMARK 2.2. For a finite starlike network (of size N) operating under one of the three disciplines, it is easy to see that $\lambda N \mathbf{E}(S) < 1$ will ensure that the network will not "explode". It is, of course, of little practical value for any reasonably large network and is a bound which has a definite dependence on the size of the network.

We are now in a position to present a comparison of the various disciplines we described above.

3. The disciplines compared. In the Preliminaries we provided a review of the $FCFS$ and $FAFS$ disciplines and then introduced our new $FAAFS$ discipline which may further be divided into nonpreemptive and preemptive categories. The description and definitions of the various protocols already intimate that there are differences and similarities between them.

To formalize our comparison we define the *delayer* of y, written $\Delta(y)$, to exist if $W(y) > 0$ in which case

$$(3.1) \quad \Delta(y) = \begin{cases} p_s(y) & w_2(y) = 0 \\ \bar{y} : \mathsf{dep}(\bar{y}) = \max_{\tilde{y}:d(\tilde{y})=d(y)} \mathsf{dep}(\tilde{y}) < \mathsf{dep}(y) & w_2(y) > 0 \end{cases}$$

so that $\Delta(y)$ is the last call, from the same (source or destination) flow as call y, which delayed y before y itself was able to complete its service and leave the network. This means that if $w_2(y) > 0$ then

$$(3.2) \quad \Delta(y) = p_d^m(y), m \in \begin{cases} \{1\} \subset \mathbb{N} & FAFS \\ \mathbb{N} & FAAFS_P \\ \mathbb{Z} \setminus \{0\} & FAAFS_N \\ \mathbb{Z} \setminus \{0\} & FCFS \end{cases}$$

Hence $\{FAFS, FAAFS_P\}$ and $\{FAAFS_N, FCFS\}$ form two distinct groups of disciplines in that in the former one waiting times of a call depend on past arrivals alone, while in the latter group there is the potential

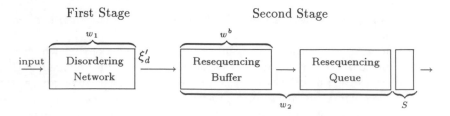

FIG. 3.1. *Resequencing Model for the Starlike Network*

for future dependence as well. We may phrase this fact differently as follows: when call y arrives to a starlike network operating under a $FAFS$ or $FAAFS_P$ discipline, $\Delta(y)$ has already entered the system. On the other hand, in a $FAAFS_N$ or $FCFS$ network, $\Delta(y)$ may not yet have arrived: this fact is mathematically problematic.

In order to present a more detailed comparison, it is convenient to summarize our network in terms of the resequencing model of [3]. (See Figure 3.1 in which we fix a flow ξ'_d.)

Figure 3.1 is a little deceptive in that it conveys the sense of network structures in series. A call in the second stage still has an impact on the operation of the disordering network. In effect, it leaves both the first and second stages only once it has departed from the network.

Consider the arrival of $y = [t, (S, s, d)]$ to the network. The disordering network is associated with the first stage of the queueing process, and is a feature common to the various disciplines of interest. We alluded to the concept of disordering towards the beginning of §2, where it was indicated that for calls in a specific destination flow, the relative times of arrival to the first stage are generally out of sequence with respect to those in the second stage.[2]

After its wait in the disordering network, y is allocated to the second stage but, as emphasized earlier, its presence is still felt in the first phase. In a network in which the $FCFS$ or our discipline operates, the waiting time in the resequencing buffer is determined by $w^b = 0$. In other words, calls pass instantaneously through the resequencing buffer. Practically-speaking, this means that whenever there are calls in the second stage, the server in the relevant destination must be busy attending to one of them. It is on these grounds that the term *work-conserving* is used.[3] The $FAFS$

---

[2] This establishes a connection with the disordering and *resequencing* (again, refer to [3]) encountered not only in computer-communication networks but also in other very practical situations such as in flexible manufacturing systems. In systems such as these, there is (simply put) a need to label units according to their "arrival times" in the initial stage of processing, and to re-establish that same order in subsequent stages. (For a more detailed discussion of this and related issues, refer to [12], [1] and the references therein.)

[3] Others might prefer to call this *full availability*.

network is not work-conserving, because $w^b > 0$ with positive probability: this immediately implies inefficiency. In the context of the resequencing model in Figure 3.1, a call is regarded as being part of the resequencing queue only after all higher time-priority calls with the same destination have passed through the resequencing buffer.

The $FAFS$ and $FAAFS$ disciplines have a property, which we shall call *order-conservation*, in common. The resequencing buffer and resequencing queue respectively have "nontrivial functions" in networks governed by these disciplines. In each of the cases, due regard is given to the times of arrival to the network. Only in the nonpreemptive $FAAFS$ situation (because of the inability for interruption) is there the possibility that a call is delayed directly by another call which arrived at the network after it.

Order-conservation may be associated with fairness so the $FAAFS$ class of disciplines may be seen to be fair *and* efficient whereas the remaining two disciplines have one or other property, but not both.

The $FAAFS$-type discipline, especially $FAAFS_P$, appears to be the most effective one because it is a suitable hybrid of $FCFS$ and $FAFS$. Unfortunately, though, the same cannot be said for mathematical tractability. This fact will become evident in the next two sections.

4. The nonpreemptive FAAFS discipline. We shall shortly state and prove Theorem 4.1 but first, some remarks are in order.

REMARK 4.1. The notion of a delayer was introduced in §3. The existence of that delayer for a call which must wait is guaranteed. We argue on the basis of the claim that only a finite number of customers is served before y_n $(n \in \mathbb{N})$ itself leaves the network. This is obviously true for y_1, who does not wait at all. So suppose the claim were true for all calls $y_1, \ldots, y_n$, having respective departure times $\mathsf{dep}_1, \ldots, \mathsf{dep}_n$, and let dep be the maximum of these times (clearly finite). It is possible that y_{n+1} leaves the network before dep. Otherwise, since all those calls which arrived to the network before it have already left by dep, it is also possible that y_{n+1} is in service at dep and by $\mathsf{dep} + S(y_{n+1})$, has already departed. This quantity is finite because service times are assumed to be finite with probability one. The remaining possibility is that y_{n+1} is still in the network at dep, but not yet entered service. By the nature of the $FAAFS$ discipline, this can only mean that some $p_d^{-m}(y_{n+1})$ $(m \in \mathbb{N})$ *is* in service at dep. Then y_{n+1} departs no later than the sum of dep and the service times of y_{n+1} itself and $p_d^{-m}(y_{n+1})$. Thus our claim is valid. It obviously follows that we may always associate a delayer with a call which is required to wait.

Theorem 4.1 will be proved using a zero initial condition. As we mentioned before, our result may, of course, be generalized to allow for a random finite number of customers present in the system at time 0, provided the state of the system at $t = 0$ is determined. In other words, the marks of all calls (hence the order of arrivals) as well as the position in the service facilities must be known.

For the theorem, it is convenient to define the virtual waiting time W_N^t of a (fictitious) call $y = [t, (S, s, d)]$ arriving to a network of size N at time t. Results given will be uniform in N, more especially it will be shown that there exists a $\lambda > 0$ such that waiting times W_N^t of fictitious customers, will be stable uniformly in N. The idea of stability stems from the era of [10]. In particular, we will write that our network is stable if there exists a proper distribution function $G(x)$ such that for all $t > 0$ $P(W_N^t \leq x) \geq G(x)$ uniformly in N. This means that the system does not explode below some positive value of the traffic intensity. Uniformity in N means that this critical value does not depend on the size of the network (i.e., the number N of source and destination channels). Stability will not, however, guarantee (weak) uniform convergence of $\{W_N^t\}$. For convenience, the waiting time of an arbitrary call $\tilde{y}$ will be denoted by $W_N(\tilde{y})$.

THEOREM 4.1. *Suppose that for some $\lambda > 0$*

$$(4.1) \qquad q = \inf_{a > 0} \mathbf{E} \, e^{aS} \frac{\lambda}{\lambda + a} < \frac{1}{2}$$

Then our starlike network is stable uniformly in the size of the network.

REMARK 4.2. The estimate (4.1) is the same as that in [6] for the $FAFS$ network which is not surprising when we take into account the greater efficiency of the $FAAFS$ model. The condition of uniform stability (uniform boundedness) may be expressed formally in the following way: $\forall \varepsilon > 0 \, \exists X = X(\varepsilon)$ (*not* depending on t and N) such that

$$(4.2) \qquad P\left(W_N^t > X\right) < \varepsilon$$

(*uniformly* in t and N). We will prove the theorem by considering arbitrary ε and finding an X satisfying this condition. It will be clear from the proof of the theorem that an inequality similar to (4.2) holds for joint waiting-time vectors. Also, it is possible to check (although we do not do it in this paper) that, if the service-time distribution F has a compact support, the whole sequence of waiting times associated with the calls which have arrived at a fixed peripheral source s_i is majorized by a stationary process, independently of N.

We will introduce some convenient concepts before proceeding to the proof of the theorem. Firstly, a *quasi-delayer* of a call y exists iff the delayer exists in which case

$$(4.3) \qquad \Delta_Q(y) = \begin{cases} p_s(y) & t[\Delta(y)] > t \\ \Delta(y) & \text{otherwise} \end{cases}$$

As usual $\Delta_Q^n(y) = \Delta_Q\left[\Delta_Q^{n-1}(y)\right]$, $n \in \mathbb{N}$ (with $\Delta_Q^0(y) = y$).

Suppose $\Delta_Q^n(y)$ exists for $n = 1, \ldots, m$ and $\Delta_Q^{m+1}(y)$ does not exist. Then we associate with call y a quasi-delayer path, $\gamma_Q(y)$, of length m — the sequence of calls $\Delta_Q^0(y) = y = \tilde{y}_0, \Delta_Q^1(y) = \tilde{y}_1, \ldots, \Delta_Q^m(y) = \tilde{y}_m$.

The important property of the calls in the sequence is that each call $\tilde{y}_i$ is either $p_s(\tilde{y}_{i-1})$ or $p_d^{k_i}(\tilde{y}_{i-1})$ for some $k_i \in \mathbb{N}$. An arbitrary sequence $y = \tilde{y}_0, \tilde{y}_1, \ldots, \tilde{y}_m$ of calls with such properties will be called a path $\gamma(y)$ of length m. Obviously not every path is a quasi-delayer path and every path (of the form mentioned) may be an actual quasi-delayer path. We have defined quasi-delayer and ordinary paths for the call y, but the definition extends to all other calls in exactly the same way. In correspondence with every path of length m is an m-dimensional vector $\boldsymbol{\theta}$ with components $\theta_i = s$ iff $\tilde{y}_i$ is a source predecessor of $\tilde{y}_{i-1}$, and $\theta_i = d$ iff $\tilde{y}_i$ has the same destination as $\tilde{y}_{i-1}$. Another m-dimensional vector, directly related to $\boldsymbol{\theta}$, is $\boldsymbol{k}$ with $k_i = 1$ if $\theta_i = s$ and some positive integer if $\theta_i = d$. In the latter case, k_i is the number of arrival intervals which exist between $\tilde{y}_i$ and $\tilde{y}_{i-1}$ for their particular destination flow, symbolically $k_i = n_d(\tilde{y}_i) - n_d(\tilde{y}_{i-1})$.

Further notation is in order. For message y, the family of paths $A_m = A_m(y)$ is given by

$$A_m = \begin{cases} \emptyset & m = 0 \\ \{\gamma(y) : \gamma(y) \text{ a path of length } m\} & m \in \mathbb{N} \end{cases}$$

and, for fixed $\boldsymbol{\theta}$ and $\boldsymbol{k}$,

$$(4.4) \qquad A_m(\boldsymbol{\theta}, \boldsymbol{k}) = A_m(\boldsymbol{\theta}, \boldsymbol{k}, y) = \{\gamma(y, \boldsymbol{\theta}^*, \boldsymbol{k}^*) : \boldsymbol{\theta}^* = \boldsymbol{\theta}, \boldsymbol{k}^* = \boldsymbol{k}\}$$

It is important to recognize that two paths in $A_m(\boldsymbol{\theta}, \boldsymbol{k})$ are distinguished by the characteristics t and S of the respective calls. We also note that

$$(4.5) \qquad\qquad A_m = \bigcup_{(\boldsymbol{\theta}, \boldsymbol{k}) \in \mathcal{M}_m} A_m(\boldsymbol{\theta}, \boldsymbol{k})$$

where $\mathcal{M}_m = \{(\boldsymbol{\theta}, \boldsymbol{k}) \in \{s, d\}^m \times \mathbb{N}^m\}$.

Note that paths in A_m have common length m, but may differ with respect to their characterizing vectors $(\boldsymbol{\theta}, \boldsymbol{k})$ as well as t and S.

It is also convenient to define

$$B_n^m = \left\{ \begin{array}{cc} \emptyset & n = 0 \\ \bigcup_{(\boldsymbol{\theta}, \boldsymbol{k}) \in \mathcal{M}_m : \sum k_i = n} A_m(\boldsymbol{\theta}, \boldsymbol{k}) \subseteq A_m & n \in \mathbb{N} \end{array} \right\} \forall m = 0, 1, 2, \ldots$$

and

$$(4.6) \qquad\qquad B_n = \bigcup_{m=1}^{\infty} B_n^m = \bigcup_{m=1}^{n} B_n^m$$

A path $\gamma(y) \in B_n$ is said to have range n. We note that for a path $\gamma(y) \in B_n^m$ the interval $t(y) - t(\tilde{y}_m)$ is a sum of n interarrival times and has the same distribution as the sum of n IID exponential random variables $\bar{\tau}_1, \ldots, \bar{\tau}_n$ with expectation λ^{-1}. By using the bar-symbol, we emphasize that we are considering the sequence of random variables $\bar{\tau}_1, \bar{\tau}_2, \ldots$ which are independent of the whole network and will be used to obtain the uniform estimates we need. For the same reason we will later also consider IID random variables $\bar{S}_1, \bar{S}_2, \ldots$ which are independent of the network and which have the same distribution F as the service times in the network.

The following important lemma gives the estimate for the probability that the waiting time was generated along the quasi-delayer path of range n. It will allow us to truncate the workload coming from the "distant past".

LEMMA 4.2. $P(W_N^t > 0, \gamma_Q(y) \in B_n) \leq c^n$ (where $c = 2q < 1$) uniformly in t and N.

Proof.

$$
\begin{aligned}
&P\left(W_N^t > 0, \gamma_Q(y) \in B_n\right) \\
= \ &P\left(W_N^t > 0, \gamma_Q(y) \in \bigcup_{m=1}^{n} B_n^m\right) \\
= \ &P\left(W_N^t > 0, \gamma_Q(y) \in \bigcup_{m=1}^{n} \bigcup_{(\boldsymbol{\theta},\boldsymbol{k}) \in \mathcal{M}_m : \sum k_i = n} A_m(\boldsymbol{\theta}, \boldsymbol{k})\right) \\
= \ &P\left(W_N^t > 0, \bigcup_{m=1}^{n} \bigcup_{(\boldsymbol{\theta},\boldsymbol{k}) \in \mathcal{M}_m : \sum k_i = n} [D_m(\boldsymbol{\theta}, \boldsymbol{k}), \gamma_Q(y) \in A_m(\boldsymbol{\theta}, \boldsymbol{k})]\right)
\end{aligned}
$$
(4.7)

where $D_m(\boldsymbol{\theta}, \boldsymbol{k})$ is an event for which

$$
\begin{aligned}
(4.8) \quad &S\left[p_{\theta_1}^{k_1}(y)\right] + S\left[p_{\theta_2}^{k_2}(p_{\theta_1}^{k_1}(y))\right] + \cdots + S\left[p_{\theta_m}^{k_m}(p_{\theta_{m-1}}^{k_{m-1}}(\ldots p_{\theta_1}^{k_1}(y)))\right] \\
&> t[y] - t\left[p_{\theta_m}^{k_m}(p_{\theta_{m-1}}^{k_{m-1}}(\ldots p_{\theta_1}^{k_1}(y)))\right]
\end{aligned}
$$

We need to explain the last equality in (4.7). The important property of the quasi-delayer path is that every call $\tilde{y}_i$ along it delays (in more general terms) the call $\tilde{y}_{i-1}$ which follows it. The validity of this is immediate if the quasi-delayer $\tilde{y}_i$ of $\tilde{y}_{i-1}$ is also its delayer (in this case $\tilde{y}_i$ delays $\tilde{y}_{i-1}$ in its second stage). If not, then we conclude from (4.3) that $t(\Delta(\tilde{y}_{i-1})) > t(\tilde{y}_{i-1})$, which, in accordance with the $FAAFS$ discipline, must mean that $\tilde{y}_{i-1}$ was delayed by $\tilde{y}_i$ in its first stage. In either case, the delays caused by the quasi-delayers along the path mean that we must at least have (4.8).

Now let us estimate the last term in (4.7). The probability that (4.8) holds for the real quasi-delayer path is smaller than the probability that it happens for at least one ordinary path (in other words, to have a real quasi-delayer path of a specific form, it should occur among ordinary paths of this form). It is here where we use the majorization of the real workload (arising from the real quasi- delayer path) by the workload that may originate from any possible path.

In doing so we may write

$$
P\left(W_N^t > 0, \bigcup_{m=1}^{n} \bigcup_{(\boldsymbol{\theta},\boldsymbol{k}) \in \mathcal{M}_m : \sum k_i = n} [D_m(\boldsymbol{\theta},\boldsymbol{k}), \gamma_Q(y) \in A_m(\boldsymbol{\theta},\boldsymbol{k})] \right)
$$

$$
\leq P\left(\bigcup_{m=1}^{n} \bigcup_{(\boldsymbol{\theta},\boldsymbol{k}) \in \mathcal{M}_m : \sum k_i = n} D_m(\boldsymbol{\theta},\boldsymbol{k}) \right)
$$

(4.9)

In view of (4.8), it is obvious that $P\left(\bigcup_{m=1}^{n} \bigcup_{(\boldsymbol{\theta},\boldsymbol{k}) \in \mathcal{M}_m : \sum k_i = n} D_m(\boldsymbol{\theta},\boldsymbol{k}) \right) =$

$P\left(\bigcup_{(\boldsymbol{\theta},\boldsymbol{k}) \in \mathcal{M}_n : \sum k_i = n} D_n(\boldsymbol{\theta},\boldsymbol{k}) \right)$ (because of the equality of the associated

events). Thus (4.9) yields

$$
P\left(W_N^t > 0, \bigcup_{m=1}^{n} \bigcup_{(\boldsymbol{\theta},\boldsymbol{k}) \in \mathcal{M}_m : \sum k_i = n} [D_m(\boldsymbol{\theta},\boldsymbol{k}), \gamma_Q(y) \in A_m(\boldsymbol{\theta},\boldsymbol{k})] \right)
$$

$$
\leq P\left(\bigcup_{(\boldsymbol{\theta},\boldsymbol{k}) \in \mathcal{M}_n : \sum k_i = n} D_n(\boldsymbol{\theta},\boldsymbol{k}) \right)
$$

$$
\leq \sum_{(\boldsymbol{\theta},\boldsymbol{k}) \in \mathcal{M}_n : \sum k_i = n} P\left(\sum_{j=1}^{n} \bar{S}_j > \sum_{j=1}^{n} \bar{\tau}_j \right)
$$

(4.10)

Here we use the fact that $t[y] - t\left[p_{\theta_m}^{k_m} (p_{\theta_{m-1}}^{k_{m-1}} (\ldots p_{\theta_1}^{k_1}(y))) \right]$ is distributed as the sum of $\sum k_i = n$ IID random variables $\bar{\tau}_j$ and service times of calls are distributed as IID random variables $\bar{S}_1, \bar{S}_2, \ldots$.

We are now in a position to apply a generalized version of Chebyshev's inequality, which was also used in [6], where it was considered as a derivation from Chernoff's inequality or the large deviation Cramer's Theorem. It states that every term in (4.10) is smaller than q^n where q is given by (4.1). This together with (4.7)–(4.10) results in the estimate

$$
P\left(W_N^t > 0, \gamma_Q(y) \in B_n \right) \leq \sum_{(\boldsymbol{\theta},\boldsymbol{k}) \in \mathcal{M}_n : \sum k_i = n} q^n
$$

(4.11)

$$
\leq (2q)^n
$$

which completes the proof of the lemma. □

Now we are in a position to prove the theorem. Bearing in mind that our aim is to establish the estimate (4.2) we proceed in stages, often applying the law of total probability.

Let $x > 0$ and n_0 be some positive integer. We have

$$
\begin{aligned}
P\left(W_N^t > x\right) &= P\left(W_N^t > x, \gamma_Q(y) \in \bigcup_{n=0}^{n_0} B_n\right) \\
&\quad + P\left(W_N^t > x, \gamma_Q(y) \in \bigcup_{n=n_0+1}^{\infty} B_n\right)
\end{aligned}
$$
(4.12)

The second term is, of course, less than

$$
\begin{aligned}
P\left(W_N^t > 0, \gamma_Q(y) \in \bigcup_{n=n_0+1}^{\infty} B_n\right) &\leq \sum_{n=n_0+1}^{\infty} P\left(W_N^t > 0, \gamma_Q(y) \in B_n\right) \\
&\leq \frac{(2q)^{n_0+1}}{1 - 2q} \\
&< \frac{\varepsilon}{2}
\end{aligned}
$$
(4.13)

for sufficiently large n_0; the second-last step clearly arises from (4.11).

The first term on the right-hand side of (4.12) requires more extensive analysis. In the lemma below we will prove that this term may be made less than ε_0 for arbitrary $\varepsilon_0 > 0$ (and so may be made less than $\frac{\varepsilon}{2}$).

LEMMA 4.3. *For any $\varepsilon_0 > 0$ and for any $n_0 \in \mathbb{N}$ uniformly in t and N there exists $X_0 = X_0(\varepsilon_0, n_0)$ such that $P\left(W_N^t > X_0, \gamma_Q(y) \in \bigcup_{n=0}^{n_0} B_n\right) < \varepsilon_0$.*

REMARK 4.3. We will prove the lemma by induction on n_0, but first we issue a note of caution. It is very tempting to think that one can majorize the service times of the n_0 delayers associated with the calls along the quasi-delayer path by the (IID) service times of the quasi-delayers together with those of the delayers, which should then be specified. The error in this approach is that the service time of our specific delayer is system-dependent; we lose the IID property we require for the majorants. In other words, when we define a specific call $\Delta(y)$ as the delayer of some call y, then the service time of the delayer has to satisfy a certain set of conditions. For instance, $\Delta(y)$ should not complete its service before the delayed call y is ready for service. This means that we are imposing special restrictions on the service time mark of the delayer.

Proof.

Base Step: Let $n_0=0$. In this case, the call does not wait at all and for any $\varepsilon_0 > 0$ we may take an arbitrary $X_0 > 0$ and we will have $P\left(W_N^t > X_0, \gamma_Q(y) \in B_0\right) = 0 < \varepsilon_0$.

Inductive Step: Assume the validity of our assumption up to some n_0. We now verify the result for $n_0 + 1$. We have

$$P\left(W_N^t > x, \gamma_Q(y) \in \bigcup_{n=0}^{n_0+1} B_n\right) = P\left(W_N^t > x, \gamma_Q(y) \in \bigcup_{n=0}^{n_0} B_n\right)$$

(4.14)
$$+ P\left(W_N^t > x, \gamma_Q(y) \in B_{n_0+1}\right)$$

where, by the inductive hypothesis, the first term may be made less than some $\varepsilon_0/3$, by choosing an appropriate x depending only on ε_0 and n_0.

Write the second term on the right of (4.14) as the following sum:

(4.15)
$$P\left(W_N^t > x, \gamma_Q(y) \in B_{n_0+1}, \Delta_Q(y) = \Delta(y)\right)$$
$$+ P\left(W_N^t > x, \gamma_Q(y) \in B_{n_0+1}, \Delta_Q(y) \neq \Delta(y)\right)$$

Let us consider the term for which $\Delta_Q(y) = \Delta(y)$. Two possibilities then exist. The quasi-delayer is $p_s(y)$ or $p_d^r(y)$ for some $r \in \mathbb{N}$ (in context, for $r \leq n_0 + 1$).

By the inductive hypothesis for every r there exists $X_1(\varepsilon_0, n_0) = X_1$ such that

(4.16)
$$P\left(W_N(p_d^r(y)) > X_1, \gamma_Q(\Delta_Q(y)) \in \bigcup_{n=0}^{n_0} B_n\right) < \frac{\varepsilon_0}{6(n_0 + 1)}$$

Also we can find $X_2 = X_2(\varepsilon_0, n_0)$ such that

(4.17)
$$P\left(\sum_{r=1}^{n_0+1} S\left(p_d^r(y)\right) + S(p_s(y)) > X_2\right) < \frac{\varepsilon_0}{6}$$

Thus for $x > X_1 + X_2$ we obtain the

$$P\left(W_N^t > x, \gamma_Q(y) \in B_{n_0+1}, \Delta_Q(y) = \Delta(y)\right)$$

(4.18)
$$\leq P\left(W_N(p_s(y)) > X_1, \gamma_Q(p_s(y)) \in \bigcup_{n=0}^{n_0} B_n\right) +$$
$$P\left(W_N(p_d^r(y)) > X_1, \gamma_Q(p_d^r(y)) \in \bigcup_{n=0}^{n_0} B_n, \text{ for some } r \leq n_o\right) +$$
$$P\left(\sum_{r=1}^{n_0+1} S\left(p_d^r(y)\right) + S(p_s(y)) > X_2\right)$$

Combining equations (4.16)–(4.18), we get that

(4.19)
$$P\left(W_N^t > x, \gamma_Q(y) \in B_{n_0+1}, \Delta_Q(y) = \Delta(y)\right) < \frac{\varepsilon_0}{3}$$

Next it is necessary to investigate the second term in the sum of (4.15). When $\Delta_Q(y) \neq \Delta(y)$, we gather from our definition of the quasi-delayer

((4.3)) that $\Delta_Q(y) = p_s(y)$. Our inductive argument enables us to conclude that

$$P\left(W_N(\Delta_Q(y)) + S(\Delta_Q(y)) > X_3, \gamma_Q(\Delta_Q(y)) \in \bigcup_{n=0}^{n_0} B_n,\right.$$

(4.20) $$\left.\Delta_Q(y) = p_s(y)\right) < \frac{\varepsilon_0}{9}$$

for some $X_3 > 0$. We can then find $K_0 = K_0(X_3)$ so that

(4.21) $$P\left(t\left(p_d^{-K_0}\right) - t(y) < X_3\right) < \frac{\varepsilon_0}{9}$$

and, in addition, some X_4 so that

(4.22) $$P\left(\sum_{j=1}^{K_0} S\left(p_d^{-j}(y)\right) > X_4\right) = P\left(\sum_{j=1}^{K_0} \bar{S}_j > X_4\right) < \frac{\varepsilon_0}{9}$$

So, letting $x > X_3 + X_4$, and combining (4.20)–(4.22), we obtain

(4.23) $$P\left(W_N^t > x, \gamma_Q(y) \in B_{n_0+1}, \Delta_Q(y) \neq \Delta(y)\right) < \frac{\varepsilon_0}{3}$$

Substitution of (4.19) and (4.23) into (4.15) gives us

(4.24) $$P\left(W_N^t > x, \gamma_Q(y) \in B_{n_0+1}\right) < \frac{2\varepsilon_0}{3}$$

We recall that all X_i are dependent only on n_0 and ε_0 (and not on t and N). Taking $X_0 = X_1 + X_2 + X_3 + X_4$, letting $x > X_0$ in (4.14) and using the estimate (4.24) altogether completes Lemma 4.3.

Finally, $\varepsilon_0 \leq \frac{\varepsilon}{2}$ produces the desired result in (4.12) in accordance with (4.13). $\square$

REMARK 4.4. The calculations provided in the proof look quite tedious, but the main idea is quite simple. What we do is to determine, for any small $\varepsilon > 0$, independently of the network operation (using only the indices of the messages in the associated flows), a finite number of calls which might contribute to the waiting time of a call y. This number is not dependent on the size of the network. With probability $1 - \varepsilon$ all the other calls will not contribute to the waiting time considered. As the service and interarrival times of the calls of interest are independent, we use them for majorizing the (actual) waiting times in the network.

5. The preemptive FAAFS discipline. Although the $FAAFS$-type discipline may be the most effective in practical terms, it is the most difficult of the three to analyze mathematically. In the first place, waiting-times are system- dependent: they are determined not only by the times of arrival $\{t\}$ to the network, but also by the shifted Poisson process $\{t + w_1\}$.

In the $FAAFS_N$ starlike network, the waiting time, $W(y)$, of some call y may be derived partially or completely from future arrivals. This dependence on the future required us to estimate the workload on the quasidelayer, rather than the true delayer, path (4.3). As a result, we were able to prove stability of the network under condition (4.1), but not convergence to a stationary regime.

In our analysis of the $FAAFS_P$ network we are faced with the problem of "indirect conditional future dependence". In its second-stage of processing, a call y will experience an initial wait (possibly 0), followed by a period of service which is likely to be interrupted before the service is complete. Consequently, the call y will again wait before completing the next segment of its service. This wait-service cycle for any call y is mathematically problematic because it results in there being a random number of calls which make a direct contribution to the workload of y.

Again, therefore, we are not in a position to prove convergence, and we must be satisfied with uniform stability. A proof of this fact could proceed similarly to that in §4, but (in order to avoid unnecessary formalities and notation) we choose an approach which relies on a result from [6].

THEOREM 5.1. *Under the condition (4.1), W_N^t is stable uniformly in N, i.e., there exists some random variable W_0 such that for all t $W_N^t \leq_d W_0$ uniformly in N.*

Proof.

Our proof will be based on the fact that for any realization of the arrival flow and any call y, $W_N^{FAAFS}(y) \leq W_N^{FAFS}(y)$. This result is important in its own right because it reinforces the argument of §3 where we indicated that the $FAAFS$ network operates more efficiently than its $FAFS$ counterpart. In view of the uniform convergence (in N) of W_N^{FAFS}, a fact which was proved in [6], we are then able to obtain that $W_N^{FAAFS}(y)$ is bounded uniformly in N by some W_0. Hence the result of our theorem, which is proved by induction.

Under zero initial conditions, it is obvious that the first call y_F arriving to the network need not wait at all. Hence $W_N^{FAAFS}(y_F) = 0 \leq W_N^{FAFS}(y_F)$.

In accordance with the inductive principle, let us assume that the result is valid for all calls $\tilde{y}$ of the form $p_{\theta_m}^{k_m}(p_{\theta_{m-1}}^{k_{m-1}}(\dots p_{\theta_1}^{k_1}(y)))$. Bearing this in mind, let us prove that the result is also valid for the call y itself.

The result is obvious if we have $W_N^{FAAFS}(y) = 0$, so suppose that $W_N^{FAAFS}(y) > 0$.

If $w_1^{FAAFS}(y) > 0$ and $w_2^{FAAFS}(y) = 0$, (where we have suppressed reference to the size N of the network) then, by definition, $\Delta^{FAAFS}(y) = p_s(y)$ and

$$
\begin{aligned}
W_N^{FAAFS}(y) &= w_1^{FAAFS}(y) \\
&= \{W_N^{FAAFS}[p_s(y)] + S[p_s(y)] - [t(y) - t[p_s(y)]]\}^+
\end{aligned}
$$

$$
\begin{aligned}
&\leq \quad \{W_N^{FAFS}[p_s(y)] + S[p_s(y)] - [t(y) - t[p_s(y)]]\}^+ \\
&= \quad w_1^{FAFS}(y) \\
(5.1) \qquad &\leq \quad W_N^{FAFS}(y)
\end{aligned}
$$

where we use the fact that $W_N^{FAAFS}[p_s(y)] \leq W_N^{FAFS}[p_s(y)]$.

Otherwise, if $w_2^{FAAFS}(y) > 0$, then $\Delta(y) = p_d^k(y)$ for some $k \in \mathbb{N}$, and we have

$$
\begin{aligned}
w_2^{FAAFS}(y) + S(y) \quad \leq \quad &\{t[p_d^k(y)] + W_N^{FAAFS}[p_d^k(y)] + S[p_d^k(y)]\} \\
&+ S(y) - [t(y) + w_1^{FAAFS}(y)]
\end{aligned}
$$

which implies that

$$
\begin{aligned}
W_N^{FAAFS}(y) \quad &\leq \quad W_N^{FAAFS}[p_d^k(y)] + S[p_d^k(y)] - [t(y) - t[p_d^k(y)]] \\
(5.2) \qquad &\leq \quad W_N^{FAFS}(y)
\end{aligned}
$$

In view of what was written earlier, the theorem is now complete. $\square$

It is important to emphasize that we are not in a position to prove uniform convergence (in N) for the preemptive $FAAFS$ model. A similar problem was encountered in [11] where a $FCFS$ starlike network was analyzed.

The analysis of the limiting picture ($N \to \infty$) for the $FAFS$ network is far more extensive than that in the case of the $FCFS$ and $FAAFS$ models. What *is* common to all three disciplines is the fact that if the limiting distribution $W^{\mathcal{D}}$ (where $\mathcal{D}$ represents any one of these disciplines) of the stationary distributions $\{W_N^{\mathcal{D}}, N \in \mathbb{N}\}$ exists, it should satisfy the following (limiting) stochastic equation (cf. (2.1))

$$
(5.3) \qquad W^{\mathcal{D}} =_d (W^{\mathcal{D}} + S - \tau)^+ + w_2^{\mathcal{D}}
$$

where, as usual, $w_2^{\mathcal{D}}$ is the waiting time in a queue where the times of arrival of all calls to the queue are shifted by the IID random variables $W_n^{\mathcal{D}} =_d W^{\mathcal{D}}$ and these calls are processed according to $\mathcal{D}$.

One of the main stumbling blocks in analyzing the equation (5.3) for our "infinite network" is $w_2^{\mathcal{D}}$. For the $FAFS$ model, the problem is solved by considering another, simpler equation, but which is equivalent to (5.3) and is of the form

$$
(5.4) \qquad W =_d \max(W' + S' - \tau', W'' + S'' - \tau'')^+
$$

where W, W' and W'' have the same distribution.

Greater mathematical tractability will characterize our next section on message-switched starlike networks.

6. Message-switched starlike networks. This section will be relatively short. There are several reasons for this. Firstly, it is not our intention to analyze the message-switched network as extensively as was the case with the circuit-switched case; more conclusions can be drawn than are presented here. Furthermore, the former model lends itself to more tractable analysis; the associated results are fairly immediate and based on the general theory of queues and networks (see, for example, [13]). We present this part mostly to re-emphasize how efficient our $FAAFS$ discipline is.

Although we will not be using the mathematical tools employed in the previous sections, we shall still briefly compare the mechanism of the $FAAFS$ network with that of the $FCFS$ and $FAFS$ networks as was done for the circuit-switching network. In the case of message-switching, the $FAFS$ network is the least trivial of the three models from the mathematical point of view. The mathematical analysis of the $FAFS$ network was presented in [2].

As in the circuit-switching case, one can always dispense with the requirements for symmetry or a *Poisson* arrival stream (at the risk of weakening the results obtained).

While the switching rule may have changed, the basic synchronization principles of the $FAAFS$ protocol still hold. A message y arriving to the network at source s is required to wait in a queue for a time of w_1 before being processed. Once all messages which arrived before y have left the first stage, the line ℓ_s becomes available to y, enabling it to be transmitted for S_1, after which y releases ℓ_s. The message begins its second stage (associated with some destination d) immediately thereafter.

The waiting time of message y in stage 2 is, as before, denoted by w_2, while its service time in that stage is written as S_2. The procedure which determines the position message y occupies in its destination queue and its access to the second-stage service facility is essentially the same as that in the second stage of a (nonpreemptive or preemptive) $FAAFS$ network.

The resequencing model described in §3 may also be used for the message-switching case. For instance, the buffer will provide a nontrivial resequencing function for the $FAFS$ discipline, but a trivial one in the case of $FCFS$ or our $FAAFS$. As before, therefore, the latter two protocols will be work-conserving ("subject to the switching rule") but the $FAFS$ one will not be. Other similarities and differences can, of course, be found.

For convenience, together with W, we will consider the queueing vector

$$q(t) = \left(q_1^{[1]}(t), \ldots, q_1^{[N]}(t), q_2^{[1]}(t), \ldots, q_2^{[N]}(t) \right)$$

where $q_i^{[j]}(t)$ represents the number of messages (including that in service) at time t in the jth queue ($j = 1, \ldots, N$) of the ith stage ($i = 1, 2$). Denote by $Q(t) = \sum_{j=1}^{N} [q_1^{[j]}(t) + q_2^{[j]}(t)]$ the number of messages in the network at time t. Finally, we shall use ρ_1 to represent the traffic intensity common to all channels in the first stage (similarly for ρ_2).

THEOREM 6.1.

(a) *For each of the three disciplines, FCFS, FAFS and (nonpre-emptive or preemptive) FAAFS, the first stage of the message-switching network operates precisely as the first stage of the net-work under the other two disciplines. In particular, each source queue in the first stage is simply of the form $GI/GI/1/\infty/FCFS$.*

(b) *Queueing vector processes $q^{FCFS}(t)$ and $q^{FAAFS}(t)$ are stochasti-cally equivalent.*

(c) *Suppose now that $\rho_1 < 1$ and $\rho_2 < 1$. Then*

(i) *$q^{FCFS}(t)$ and $q^{FAAFS}(t)$ converge (to the same) stationary queueing vector $q(t)$. If, in addition, the service times are ex-ponentially distributed and arrival streams are Poisson, then the limiting distribution is product-form $M/M/1/\infty/FCFS$, as if all the queues were independent.*

(ii) *for both the FCFS and FAAFS networks, waiting times con-verge in distribution, i.e., $(w_1^t, w_2^t)^{FCFS} \xrightarrow[\infty]{t} (w_1, w_2)^{FCFS}$ and $(w_1^t, w_2^t)^{FAAFS} \xrightarrow[\infty]{t} (w_1, w_2)^{FAAFS}$. Average station-ary waiting times are equal for these models: $\mathbf{E}w_i^{FCFS} = \mathbf{E}w_i^{FAAFS}$, $(i = 1, 2)$.*

All results are uniform in N, the number of channels (the result is thus still valid when $N \to \infty$).

We shall now briefly outline the details of the proof.

Proof.

(a) The first stage is not affected by the second, consequently one can-not distinguish between the first stages of the networks operating under the different protocols. It is important to realize, though, that the two stages are not independent. Nonetheless, if $\rho_1 < 1$, each queue will be of form $GI/GI/1/\infty/FCFS$.

(b) The stochastic equivalence follows simply by recognizing two facts. First (all) marks are (conditionally) independent. Secondly, after leaving the first stage in the $FCFS$ or $FAAFS$ network, the mes-sage will occupy some position in the second-stage queue (without being delayed in a buffer). Thus the $q^{FCFS}(t)$ and $q^{FAAFS}(t)$ systems are stochastically indistinguishable.

(c) The convergence to $q(t)$ follows from (b), as well as (a). We con-clude that there exists $Q = \lim_{t \to \infty} Q(t)$ for each of the $FCFS$ and $FAAFS$ protocols. Using Little's formula (for example, see [13]), we obtain the equalities $\mathbf{E}Q^{FAAFS} = \lambda \mathbf{E}T^{FAAFS}$ and similarly for the $FCFS$ expectations. Also, from (b) $\mathbf{E}Q^{FCFS} = \mathbf{E}Q^{FAAFS}$ so that $\mathbf{E}T^{FCFS} = \mathbf{E}T^{FAAFS}$. With the obvious equality of mean service times and $\mathbf{E}w_1^{FCFS} = \mathbf{E}w_1^{FAAFS}$ from (a), we obtain the equality $\mathbf{E}w_2^{FCFS} = \mathbf{E}w_2^{FAAFS}$. The result on limiting distributions for the exponential case is immediate. For the more general case, stability is obvious; the convergence of waiting times relies on the relationship between W and Q, general queueing network theory (see [13]) and the innovations event (renewal) method of Borovkov ([5]). $\square$

One can see that it is possible to provide considerably more results in this particular case. However that was not our intention; we just used this simple model to make a precise comparison with the other disciplines. Such a comparison is not usually possible in a more general setting.

7. Concluding remarks. We will now summarize our results and ideas about the starlike networks operating under the disciplines considered in our article.

There are obviously several ways in which these network models could be modified. Several modifications have already been mentioned. To these we may add the idea of other disciplines — based on different synchronization constraints — as described in [9]. For instance, it is possible to consider service in random order for the second stage of the network.

Another possibility is to extend our methods to two-directional traffic models, in which the nodes (and channels) could have a dual role, both as sources and destinations. Additional results for the message-switched network of §6 could have been given. It is also possible to consider non-symmetric networks or non-Poisson arrival flows. We may also weaken some other conditions imposed on our model. One of our main intentions, however, was to compare various disciplines on the standard starlike network, which is sufficiently well understood.

It is worth re-emphasizing the significance of the uniformity of our results. We prove that our bound is valid, regardless of the number of channels in the network. As indicated previously, because of the complexity of the $FAAFS$ discipline, we are not in a position to prove convergence. For our model, as well as for the $FCFS$ model, the question of uniform convergence to a "limiting picture" still remains open. Another set of open questions arises from the consideration of limiting equations where the issue of uniqueness and the nature of the distribution which solves the equation is still to be understood fully. The uniform estimates we have been able to obtain, however, are sufficiently delicate to extend our results to various types of networks.

REFERENCES

[1] F. BACCELLI, E. GELENBE, AND B. PLATEAU. An end-to-end approach to the re-sequencing problem. *Journal of the Association for Computing Machinery,* **31**(3):474–485, 1984.

[2] F. BACCELLI, F.I. KARPELEVICH, M.YA. KELBERT, A.A. PUHALSKII, A.N. RYBKO, AND YU.M. SUHOV. *A mean-field limit for a class of queueing networks.* Journal of Statistical Physics, **66**(3–4):803–825, 1992.

[3] F. BACCELLI AND A.M. MAKOWSKI. *Queueing models for systems with synchronization constraints.* Proceedings of the Institute of Electrical and Electronics Engineers, **77**(1):138–161, 1989.

[4] S.A. BEREZNER AND V.A. MALYSHEV. *The stability of infinite-server networks with random routing.* Journal of Applied Probability, **26**(2):363–371, 1989.

[5] A.A. BOROVKOV. *Asymptotic Methods in Queueing Theory.* Wiley, 1984.

[6] F.I. KARPELEVICH, M.YA. KELBERT, AND YU.M. SUHOV. *Higher-order Lindley equations.* Stochastic Processes and their Applications, **53**:65–96, 1994.

[7] M.YA. KELBERT, R.P. KOPEIKA, R.N. SHAMSIEV, AND YU.M. SUKHOV. *Perturbation theory approach for a class of hybrid switching networks with small transit flows.* Advances in Applied Probability, **22**(1):211–229, 1990.

[8] M.YA. KELBERT AND YU.M. SUHOV. *One class of star-shaped communication networks with packet switching.* Problems of Information Transmission, **15**(4):286–300, 1980.

[9] M.YA. KELBERT AND YU.M. SUKHOV. *Mathematical theory of queuing networks.* Journal of Soviet Mathematics, **50**(3):1527–1600, 1990.

[10] R.M. LOYNES. *The stability of a queue with non-independent inter-arrival and service times.* Proceedings of the Cambridge Philosophical Society, **58**(3):497–520, 1962.

[11] A.N. RYBKO AND V.A. MIKHAILOV. *Range of carrying capacity for communication networks with channel switching.* Problemy Peredachi Informatsii, **22**(1):66–84, 1986.

[12] N. VENKATASUBRAMANYAN AND J.L. SANDERS. *Order restoration in queues.* Queueing Systems, **12**:409–429, 1992.

[13] J. WALRAND. *An Introduction to Queueing Networks.* Prentice-Hall, 1988.

MATRIX PRODUCT-FORM SOLUTIONS FOR MARKOV CHAINS AND A LCFS QUEUE

RAYMOND W. YEUNG* AND BHASKAR SENGUPTA†

Abstract. We consider a bivariate Markov chain in which one of the variables can take values in a countable set, which is arranged in the form of a tree. The other variable takes values from a finite set. Each node of the tree can branch out into d other nodes. The steady state solution of this Markov chain has a matrix product-form, which can be expressed as a function of d matrices $R_1, \ldots, R_d$. We then use this to solve a multi-class LCFS queue, in which the service times have PH-distributions and arrivals are according to the Markov modulated Poisson process. In this discipline, when a customer's service is preempted in phase j (due to a new arrival), the resumption of service at a later time could take place in a phase which depends on j.

1. Introduction . In this paper, we generalize the theory of matrix-geometric solutions developed by Neuts [5]. We consider a discrete time bivariate Markov chain $\{(X_\xi, N_\xi), \xi \geq 0\}$ in which the values of X_ξ are represented by the nodes of a d-ary tree, and N_ξ takes integer values between 1 and m. We refer to X_ξ as the *node* and N_ξ as the *auxiliary variable* of the Markov chain at time ξ. This Markov chain is used to model a multi-class last-come-first-served queueing system with a generalized preemptive resume (LCFS-GPR) discipline.

A d-ary tree is a tree for which each node has d children. The root of the tree is represented by the symbol 0. In representing the remaining nodes of the tree (values taken by X_ξ), we use strings of integers, where each integer is between 1 and d. Thus the node representing the kth child of the root has a representation of k, which is a string of length 1. Similarly, the node representing the ζth child of k has a representation $k\zeta$ with a length of 2, and so on. Using this convention, the root can also be represented by an empty string whose length is 0. The length of the string representing node K is denoted by $|K|$, and we use $K > (=) 0$ to denote all strings K which satisfy $|K| > (=) 0$. In the rest of the paper, for a string $K > 0$, we let $K = k_1 k_2 \ldots k_{|K|}$. Throughout this paper, we use lower case letters to represent integers and upper case letters to represent strings of integers when referring to nodes of the tree.

We now define an algebra using the strings defined above. Even though this algebra may seem a bit unnatural at first, the use of this algebra makes it easy to write down the transition probabilities later. Let $f(J, i)$ $(0 \leq i \leq |J|)$ be a sub-string of J consisting of the last i numbers of J. For $i = 0$, $f(J, i)$ refers to the empty string. We use "+" to denote concatenation on the right and "−" to denote deletion from the right; the

---

* Department of Information Engineering, The Chinese University of Hong Kong, Shatin, N.T., Hong Kong; e-mail:whyeung@ie.cuhk.hk.

† C&C Research Labs., NEC USA, 4 Independence Way, Princeton, NJ 08540; e-mail:bhaskar @ccrl.nj.nec.com.

evaluations of "+" and "−" are from left to right. To illustrate the use of these notations, suppose $J = 1231$ and $K = 23$. Then, $f(J, 2) = 31$, $J + 1 = 12311$, $J + K = 123123$, $J - f(J, 2) = 12$, $J - f(J, 2) + 3 = 123$, etc.

From any node J, transitions with non-negative probability are allowed to the root, the children of J and the children of all ancestors of J. Let us define a set $S(J) = \{(i, \zeta) : 0 \leq i \leq |J| \text{ and } 1 \leq \zeta \leq d\}$. For $(i, \zeta) \in S(J)$, $J - f(J, i) + \zeta$ refers to a node which is either a child of an ancestor of J (if $i > 0$) or a child of J (if $i = 0$). Note that J is a child of the parent of J, and for $J \neq 0$, $J - f(J, 1) + f(J, 1)$ refers to J itself. In a similar fashion, any ancestor K of J (other than the root) is a child of the parent of K itself. Thus, any ancestor of J (other than the root) as well as J can be represented as a child of an ancestor of J. An interesting fact which can be verified easily is that for any node which is a child of J or a child of an ancestor of J, there exists a unique value of (i, ζ) representing this node by $J - f(J, i) + \zeta$.

We now specify the transition probabilities of the Markov chain $\{(X_\xi, N_\xi), \xi \geq 0\}$. Suppose $(X_\xi, N_\xi) = (J, j)$. Then, only the following two types of transitions are allowed with non-zero probability:

1. $(X_{\xi+1}, N_{\xi+1}) = (J - f(J, i) + \zeta, \nu)$ with probability $a_{j,\nu}^{f(J,i),\zeta}$ where $(i, \zeta) \in S(J)$.

2. $(X_{\xi+1}, N_{\xi+1}) = (0, \nu)$ with probability $b_{j,\nu}^{(J)}$.

Since all other transitions are of zero probability, these transition probabilities satisfy the condition

$$(1.1) \qquad \sum_{\nu=1}^{m} \left(\sum_{(i,\zeta) \in S(J)} a_{j,\nu}^{f(J,i),\zeta} + b_{j,\nu}^{(J)} \right) = 1.$$

We define the $m \times m$ matrices $A^{f(J,i),\zeta}$ and $B^{(J)}$, whose (j, ν)th elements are $a_{j,\nu}^{f(J,i),\zeta}$ and $b_{j,\nu}^{(J)}$, respectively. Then, (1.1) can be written as

$$(1.2) \qquad \left(\sum_{(i,\zeta) \in S(J)} A^{f(J,i),\zeta} + B^{(J)} \right) e = e,$$

where e is an $m \times 1$ column vector with all the components equal to 1.

At this point, it is easy to understand why this Markov chain exhibits a *skip-free to the right* property. Imagine that we start the chain at node 0 and look at the sample path that takes the chain to some node J. The skip-free to the right property ensures that it is impossible to move from node 0 to node J without visiting *all ancestors* of J at least once. More generally, assume that we start the Markov chain in node K and look at the sample path to some node $J + \zeta$, where K is not a descendent of J. This sample path must visit node J at least once. Clearly, this is a generalization of the corresponding notion for the set of non-negative integers. Further, this property is crucial to the development of the theory in the next section.

This completes the description of the Markov chain $\{(X_\xi, N_\xi), \xi \geq 0\}$. The remainder of this paper contains 2 more sections. In section 2, we prove the matrix product-form solution. In section 3, we discuss an application to a LCFS-GPR queue. Note that a stability condition for this Markov chain and other extensions have been derived by Yeung and Sengupta [9]. Further, a generalization of the matrix M/G/1 paradigm to tree-like structures has been done by Takine et al [7].

2. The matrix product-form solution . Define the taboo probability (Chung [2], Neuts [5]) $_J P^{(n)}_{(J,j);(J+K,\nu)}$, $K > 0$ as the probability that starting from state (J, j), the Markov chain is in state $(J + K, \nu)$ after n transitions without visiting node J in between. Now $_J P^{(n)}_{(J,j);(J+K,\nu)}$ is the sum of the probabilities of all the possible sample paths starting from (J, j) and ending in $(J + K, \nu)$ after n steps without visiting node J in between, and each of these sample paths does not involve any transition from any node $J + K'$, $K' > 0$ to a child of an ancestor of J by the skip free to the right property. We then see from the structure of $\{(X_\xi, N_\xi), \xi \geq 0\}$ that $_J P^{(n)}_{(J,j);(J+K,\nu)}$ does not depend on J. Then $r^{(K)}_{j,\nu}$, the expected number of visits to $(J + K, \nu)$ (given that $(X_0, N_0) = (J, j)$) before visiting node J again, is given by

$$r^{(K)}_{j,\nu} = \sum_{n=0}^{\infty} {}_J P^{(n)}_{(J,j);(J+K,\nu)},$$

because $_J P^{(n)}_{(J,j);(J+K,\nu)}$ is also recognized as the expected number of visits to $(J + K, \nu)$ at the nth transition without visiting node J in between. Let $R^{(K)}$ denote an $m \times m$ matrix whose (j, ν)th element is $r^{(K)}_{j,\nu}$. For the special case $K = \zeta$ for $1 \leq \zeta \leq d$, we define $R_\zeta = R^{(\zeta)}$ to simplify the notations.

LEMMA 2.1. *For $K > 0$,*

(2.1)
$$R^{(K)} = R_{k_1} R_{k_2} \cdots R_{k_{|K|}}.$$

Proof. This is proved by induction on the length of K. For $|K| = 1$, the lemma is true by definition. Suppose $(X_0, N_0) = (J, j)$ and assume that (2.1) is true for some K with $|K| \geq 1$, and consider $K + \zeta$ where $1 \leq \zeta \leq d$. By the skip free to the right property, from node J to node $J + K + \zeta$, the Markov chain must visit node $J + K$ at least once in between. Conditioning on the time ξ and the auxiliary variable h of the last visit to node $J + K$ before visiting $(J + K + \zeta, \nu)$ at time n, we have

$$_J P^{(n)}_{(J,j);(J+K+\zeta,\nu)} = \sum_{h=1}^{m} \sum_{\xi=0}^{n} {}_J P^{(\xi)}_{(J,j);(J+K,h)} \, _J P^{(n-\xi)}_{(J+K,h);(J+K+\zeta,\nu)}$$

$$(2.2) \qquad = \sum_{h=1}^{m} \sum_{\xi=0}^{n} {}_J P_{(J,j);(J+K,h)}^{(\xi)} {}_J P_{(J,h);(J+\zeta,\nu)}^{(n-\xi)}.$$

Thus,

$$\begin{aligned} r_{j,\nu}^{(K+\zeta)} &= \sum_{n=0}^{\infty} \sum_{h=1}^{m} \sum_{\xi=0}^{n} {}_J P_{(J,j);(J+K,h)}^{(\xi)} {}_J P_{(J,h);(J+\zeta,\nu)}^{(n-\xi)} \\ &= \sum_{h=1}^{m} \sum_{\xi=0}^{\infty} {}_J P_{(J,j);(J+K,h)}^{(\xi)} \sum_{n'=0}^{\infty} {}_J P_{(J,h);(J+\zeta,\nu)}^{(n')} \\ &= \sum_{h=1}^{m} r_{j,h}^{(K)} r_{h,\nu}^{(\zeta)}. \end{aligned}$$

Therefore,

$$R^{(K+\zeta)} = R^{(K)} R^{(\zeta)},$$

and hence,

$$R^{(K+\zeta)} = R_{k_1} R_{k_2} \cdots R_{k_{|K|}} R_\zeta$$

by the induction hypothesis, proving the lemma. $\square$

LEMMA 2.2. *For* $1 \leq \zeta \leq d$, R_ζ *satisfies*

$$(2.3) \qquad R_\zeta = A^{0,\zeta} + \sum_{\gamma=1}^{d} \sum_{K \geq 0} \left(R_\gamma \prod_{\eta=1}^{|K|} R_{k_\eta} \right) A^{f(\gamma+K,|K|+1),\zeta}.$$

Proof. Consider ${}_J P_{(J,j);(J+\zeta,\nu)}^{(n)}$. This probability is equal to 0 for $n = 0$. For $n = 1$,

$$(2.4) \qquad {}_J P_{(J,j);(J+\zeta,\nu)}^{(1)} = a_{j,\nu}^{0,\zeta}.$$

For $n \geq 2$, if the Markov chain is in node $J + \zeta$ after n transitions without visiting node J in between, the Markov chain must be in node $J + \gamma$ after one transition for some $1 \leq \gamma \leq d$, and hence after $n - 1$ transitions, the Markov chain must be in node $J + \gamma + K$ for some $K \geq 0$. Conditioning on γ, K and h, the auxiliary variable of the Markov chain after $n - 1$ transitions, we have

$$(2.5) \quad {}_J P_{(J,j);(J+\zeta,\nu)}^{(n)} = \sum_{h=1}^{m} \sum_{\gamma=1}^{d} \sum_{K \geq 0} {}_J P_{(J,j);(J+\gamma+K,h)}^{(n-1)} a_{h,\nu}^{f(\gamma+K,|K|+1),\zeta}.$$

Therefore,

$$r_{j,\nu}^{(\zeta)} = \sum_{n=1}^{\infty} {}_J P_{(J,j);(J+\zeta,\nu)}^{(n)}$$

$$= a_{j,\nu}^{0,\varsigma} + \sum_{n=2}^{\infty} \sum_{h=1}^{m} \sum_{\gamma=1}^{d} \sum_{K \geq 0} {}_J P_{(J,j);(J+\gamma+K,h)}^{(n-1)} a_{h,\nu}^{f(\gamma+K,|K|+1),\varsigma}$$

$$= a_{j,\nu}^{0,\varsigma} + \sum_{\gamma=1}^{d} \sum_{K \geq 0} \sum_{h=1}^{m} \left(\sum_{n=1}^{\infty} {}_J P_{(J,j);(J+\gamma+K,h)}^{(n)} \right) a_{h,\nu}^{f(\gamma,|K|+1),\varsigma}$$

$$= a_{j,\nu}^{0,\varsigma} + \sum_{\gamma=1}^{d} \sum_{K \geq 0} \sum_{h=1}^{m} r_{j,h}^{(\gamma+K)} a_{h,\nu}^{f(\gamma+K,|K|+1),\varsigma}$$

and (2.3) follows by virtue of lemma 2.1. $\square$

LEMMA 2.3. *If* $\{(X_\xi, N_\xi), \xi \geq 0\}$ *is positive recurrent, for* $1 \leq \varsigma \leq d$, R_ς *can be obtained as* $\lim_{N \to \infty} R_\varsigma[N]$ *from the recursion*

$$(2.6) \quad R_\varsigma[N+1] = A^{0,\varsigma} + \sum_{\gamma=1}^{d} \sum_{K \geq 0} \left(R_\gamma[N] \prod_{\eta=1}^{|K|} R_{k_\eta}[N] \right) A^{f(\gamma+K,|K|+1),\varsigma}$$

with $R_\varsigma[0]$ *being the zero matrix, and* $\Psi_\varsigma = R_\varsigma$ *is the minimal non-negative solution of*

$$(2.7) \qquad \Psi_\varsigma = A^{0,\varsigma} + \sum_{\gamma=1}^{d} \sum_{K \geq 0} \left(\Psi_\gamma \prod_{\eta=1}^{|K|} \Psi_{k_\eta} \right) A^{f(\gamma+K,|K|+1),\varsigma},$$

where Ψ_ς *is an* $m \times m$ *matrix.*

Proof. We first show that $R_\varsigma[N]$ is entrywise non-decreasing for all $1 \leq \varsigma \leq d$ by induction on N. For $N = 1$,

$$R_\varsigma[1] = A^{0,\varsigma} \geq R_\varsigma[0].$$

Assume $R_\varsigma[N] \geq R_\varsigma[N-1]$ for some $N \geq 1$. Then

$$R_\varsigma[N+1] - R_\varsigma[N] = \sum_{\gamma=1}^{d} \sum_{K \geq 0} \left(R_\gamma[N] \prod_{\eta=1}^{|K|} R_{k_\eta}[N] - R_\gamma[N-1] \right.$$
$$\left. \times \prod_{\eta=1}^{|K|} R_{k_\eta}[N-1] \right) A^{f(\gamma+K,|K|+1),\varsigma}.$$

We omit the straightforward proof that $R_\varsigma[N] \geq R_\varsigma[N-1]$ for all $1 \leq \varsigma \leq d$ implies the matrix expression in the parenthesis above is entrywise non-negative. This proves that $R_\varsigma[N]$ is entrywise non-decreasing in N.

We now show that $\lim_{N \to \infty} R_\varsigma[N]$ exists. Let $\hat{R}_\varsigma$, $1 \leq \varsigma \leq d$ be any non-negative solution of (2.7). We prove by induction that $\hat{R}_\varsigma - R_\varsigma[N]$ is entrywise non-negative for $N \geq 0$. This is obviously true for $N = 0$.

Assume it is true for some $N \geq 0$. Then

$$\hat{R}_\zeta - R_\zeta[N+1] = \sum_{\gamma=1}^{d} \sum_{K \geq 0} \left(\hat{R}_\gamma \prod_{\eta=1}^{|K|} \hat{R}_{k_\eta} - R_\gamma[N] \prod_{\eta=1}^{|K|} R_{k_\eta}[N] \right)$$
$$\times A^{f(\gamma+K,|K|+1),\zeta}.$$

Again, we omit the proof of the fact that the induction hypothesis implies the matrix expression in the parenthesis above is entrywise non-negative. Therefore, $\hat{R}_\zeta \geq R_\zeta[N]$ for all $N \geq 0$. In particular, $\hat{R}_\zeta \geq R_\zeta[N]$ for all $N \geq 0$. If $\{(X_\xi, N_\xi), \xi \geq 0\}$ is positive recurrent, the entries of R_ζ for all $1 \leq \zeta \leq d$ are finite (Neuts [5]). Since $R_\zeta[N]$ is entrywise non-decreasing in N, this implies $R_\zeta[\infty] = \lim_{N \to \infty} R_\zeta[N]$ exists.

Taking the limit as $N \to \infty$ in (2.6) and applying the monotone convergence theorem, we see that $R_\zeta[\infty]$, $1 \leq \zeta \leq d$ is a solution of (2.7). Since $\hat{R}_\zeta \geq R_\zeta[N]$ for all $N \geq 0$, $\hat{R}_\zeta \geq R_\zeta[\infty]$. Thus we conclude that $R_\zeta[\infty]$, $1 \leq \zeta \leq d$ is the minimal non-negative solution of (2.7).

It remains to show that $R_\zeta \leq R_\zeta[\infty]$. Toward this end, we consider

$$\tilde{r}_{j,\nu}^{(K)}[N] = \sum_{n=0}^{N} J P_{(J,j);(J+K,\nu)}^{(n)}$$

and define the matrix $\tilde{R}^{(K)}[N] = \left[\tilde{r}_{j,\nu}^{(K)}[N] \right]$. Note that $\lim_{N \to \infty} \tilde{R}^{(K)}[N] = R^{(K)}$. We first show by induction on $|K|$ that

$$(2.8) \qquad \tilde{R}^{(K)}[N] \leq \tilde{R}_{k_1}[N] \tilde{R}_{k_2}[N] \cdots \tilde{R}_{k_{|K|}}[N],$$

where $\tilde{R}_\zeta[N] = \tilde{R}^{(\zeta)}[N]$. For $|K| = 1$, (2.8) is clearly true. Assume (2.8) is true for some K with $|K| \geq 1$. Then from (2.2),

$$\tilde{r}_{j,\nu}^{(K+\zeta)}[N] = \sum_{n=0}^{N} \sum_{h=1}^{m} \sum_{\xi=0}^{n} J P_{(J,j);(J+K,h)}^{(\xi)} J P_{(J,h);(J+\zeta,\nu)}^{(n-\xi)}$$

$$= \sum_{h=1}^{m} \sum_{\xi=0}^{N} J P_{(J,j);(J+K,h)}^{(\xi)} \sum_{n'=0}^{N-\xi} J P_{(J,h);(J+\zeta,\nu)}^{(n')}$$

$$\leq \sum_{h=1}^{m} \sum_{\xi=0}^{N} J P_{(J,j);(J+K,h)}^{(\xi)} \sum_{n'=0}^{N} J P_{(J,h);(J+\zeta,\nu)}^{(n')}$$

$$= \sum_{h=1}^{m} \tilde{r}_{j,h}^{(K)}[N] \tilde{r}_{h,\nu}^{(\zeta)}[N].$$

Therefore,

$$\tilde{R}^{(K+\zeta)}[N] \leq \tilde{R}^{(K)}[N] \tilde{R}_\zeta[N]$$
$$\leq \tilde{R}_{k_1}[N] \tilde{R}_{k_2}[N] \cdots \tilde{R}_{k_{|K|}}[N] \tilde{R}_\zeta[N],$$

and (2.8) follows. Now from (2.4) and (2.5),

$$\tilde{r}_{j,\nu}^{(\zeta)}[N] = \sum_{n=1}^{N} {}_J P_{(J,j);(J+\zeta,\nu)}^{(n)}$$

$$= a_{j,\nu}^{0,\zeta} + \sum_{n=2}^{N}\sum_{h=1}^{m}\sum_{\gamma=1}^{d}\sum_{K\geq 0} {}_J P_{(J,j);(J+\gamma+K,h)}^{(n-1)} a_{h,\nu}^{f(\gamma+K,|K|+1),\zeta}$$

$$= a_{j,\nu}^{0,\zeta} + \sum_{\gamma=1}^{d}\sum_{K\geq 0}\sum_{h=1}^{m} \left(\sum_{n=1}^{N-1} {}_J P_{(J,j);(J+\gamma+K,h)}^{(n)} \right) a_{h,\nu}^{f(\gamma+K,|K|+1),\zeta}$$

$$= a_{j,\nu}^{0,\zeta} + \sum_{\gamma=1}^{d}\sum_{K\geq 0}\sum_{h=1}^{m} \tilde{r}_{j,h}^{(\gamma+K)}[N-1] a_{h,\nu}^{f(\gamma+K,|K|+1),\zeta}$$

Rewriting this in matrix form, we have

$$\tilde{R}^{(\zeta)}[N] = A^{0,\zeta} + \sum_{\gamma=1}^{d}\sum_{K\geq 0} \tilde{R}^{(\gamma+K)}[N-1] A^{f(\gamma+K,|K|+1),\zeta}$$

$$\leq A^{0,\zeta} + \sum_{\gamma=1}^{d}\sum_{K\geq 0} \tilde{R}_\gamma[N-1]\tilde{R}_{k_1}[N-1]\cdots\tilde{R}_{k_{|K|}}[N-1]$$

(2.9)
$$\times A^{f(\gamma+K,|K|+1),\zeta}$$

by (2.8). We now prove by induction on N that

(2.10) $$\tilde{R}_\zeta[N] \leq R_\zeta[N]$$

for $N \geq 1$. For $N = 1$,

$$\tilde{R}_\zeta[1] = A^{0,\zeta} = R_\zeta[1].$$

Assume (2.10) is true for some $N \geq 1$. Then by (2.9),

$$\tilde{R}_\zeta[N+1] \leq A^{0,\zeta} + \sum_{\gamma=1}^{d}\sum_{K\geq 0} \tilde{R}_\gamma[N]\tilde{R}_{k_1}[N]\cdots\tilde{R}_{k_{|K|}}[N]A^{f(\gamma+K,|K|+1),\zeta}$$

$$\leq A^{0,\zeta} + \sum_{\gamma=1}^{d}\sum_{K\geq 0} R_\gamma[N]R_{k_1}[N]\cdots R_{k_{|K|}}[N]A^{f(\gamma+K,|K|+1),\zeta}$$

$$= A^{0,\zeta} + \sum_{\gamma=1}^{d}\sum_{K\geq 0} \left(R_\gamma[N]\prod_{\eta=1}^{|K|} R_{k_\eta}[N] \right) A^{f(\gamma+K,|K|+1),\zeta}$$

$$= R_\zeta[N+1],$$

proving (2.10). Letting $N \to \infty$, we have

$$R_\zeta \leq R_\zeta[\infty],$$

completing the proof. ☐

We now state a lemma which is instrumental to the proof of theorem 2.5 (to follow). The proof of this lemma is exactly the same as the first part of the proof of theorem 1.2.1 in Neuts [5], but our lemma holds in a more general setting. This lemma gives the relation between the stationary distribution of any state and the stationary distributions of any set of states not containing this state in a positive recurrent Markov chain.

LEMMA 2.4. *For a positive recurrent Markov chain, let A be a set of states and j be any state not in A. For any state i in A, define*

$$r_{ij} = \sum_{n=0}^{\infty} {}_A P_{i;j}^{(n)}.$$

Thus r_{ij} is the expected number of visits to j before visiting A again given that the Markov chain starts at i. Let π_i be the stationary distribution of state i. Then,

$$\pi_j = \sum_{i \in A} \pi_i r_{ij}.$$

In part (c) of the theorem to follow, we use the term "the transition matrix of $\{(X_\xi, N_\xi), \xi \geq 0\}$ watched in the set of states $(0, j)$". This means that we imbed a process $\{N_{\xi_n}, n \geq 0\}$ in the Markov chain $\{(X_\xi, N_\xi), \xi \geq 0\}$ such that $X_{\xi_n} = 0$. Thus, ξ_n is the time of the nth visit of the node to the root. The resulting process $\{N_{\xi_n}, n \geq 0\}$ is itself a Markov chain. For the special case of $d = 1$, this has been observed by Tweedie [6] and also by Asmussen [1].

THEOREM 2.5. *Let $\pi^{(K)} = [\pi_1^{(K)} \cdots \pi_m^{(K)}]$ be the stationary distribution of the auxiliary variables in node K, $K \geq 0$. If $\{(X_\xi, N_\xi), \xi \geq 0\}$ is positive recurrent, then,*

(a) $\pi^{(K+\zeta)} = \pi^{(K)} R_\zeta$, $1 \leq \zeta \leq d$,

(b) the eigenvalues of $\sum_{\zeta=1}^{d} R_\zeta$ lie inside the unit disk,

(c) the transition matrix of $\{(X_\xi, N_\xi), \xi \geq 0\}$ watched in the set of states $(0, j)$, $1 \leq j \leq m$ is given by

$$B = \sum_{K \geq 0} R^{(K)} B^{(K)},$$

where $B = [b_{j,\nu}]$ is an $m \times m$ matrix and

(d) the vector $\pi^{(0)}$ is a left eigenvector of B normalized by

$$\pi^{(0)} \left(I - \sum_{\zeta=1}^{d} R_\zeta \right)^{-1} e = 1.$$

Proof. An application of lemma 2.4 gives

$$\pi_j^{(K+\zeta)} = \sum_{h=1}^{m} \pi_h^{(K)} r_{h,j}^{(\zeta)},$$

or in matrix form

$$\pi^{(K+\zeta)} = \pi^{(K)} R_\zeta,$$

proving (a). For (b), we consider the expected number of transitions before the first return to the root given that $(X_0, N_0) = (0, j)$. This quantity is finite because $\{(X_\xi, N_\xi), \xi \geq 0\}$ is positive recurrent. It is given by the jth component of the vector $\sum_{K>0} R^{(K)} e$, which is finite if and only if

$$\sum_{K \geq 0} R^{(K)} = I + \sum_{\zeta=1}^{d} R_\zeta + \left(\sum_{\zeta=1}^{d} R_\zeta \right)^2 + \cdots$$

is finite, or if all the eigenvalues of $\sum_{\zeta=1}^{d} R_\zeta$ lie inside the unit disk.

For (c), suppose $(X_0, N_0) = (0, j)$ and consider the probability that the Markov chain $\{(X_\xi, N_\xi), \xi \geq 0\}$ is in $(0, \nu)$ when it first returns to the root. This happens after one transition with probability $b_{j,\nu}^{(0)}$. If this happens after n transitions, where $n \geq 2$, then after $n - 1$ transitions, the Markov chain $\{(X_\xi, N_\xi), \xi \geq 0\}$ is in state (K, h) for some $K > 0$ and $1 \leq h \leq m$ without visiting the root in between. Therefore,

$$
\begin{aligned}
b_{j,\nu} &= b_{j,\nu}^{(0)} + \sum_{n=2}^{\infty} \sum_{K>0} \sum_{h=1}^{m} {}_0 P_{(0,j);(K,h)}^{(n-1)} b_{h,\nu}^{(K)} \\
&= b_{j,\nu}^{(0)} + \sum_{K>0} \sum_{h=1}^{m} \left(\sum_{n=1}^{\infty} {}_0 P_{(0,j);(K,h)}^{(n)} \right) b_{h,\nu}^{(K)} \\
&= b_{j,\nu}^{(0)} + \sum_{K>0} \sum_{h=1}^{m} r_{j,h}^{(K)} b_{h,\nu}^{(K)}.
\end{aligned}
$$

In matrix form, this becomes

$$B = B^{(0)} + \sum_{K>0} R^{(K)} B^{(K)} = \sum_{K \geq 0} R^{(K)} B^{(K)},$$

proving (c).

It is now clear that $\pi^{(0)}$ is a left eigenvector of B. Thus we have

$$\sum_{K \geq 0} \pi^{(0)} R^{(K)} e = \pi^{(0)} \left(\sum_{K \geq 0} R^{(K)} \right) e = 1$$

by (a). Let $U = \sum_{K \geq 0} R^{(K)}$. Then,

$$
U = R^{(0)} + \sum_{\zeta=1}^{d} R_\zeta \left(\sum_{K \geq 0} R^{(K)} \right)
$$

$$
= I + \left(\sum_{\zeta=1}^{d} R_\zeta \right) U,
$$

or

$$
U = \left(I - \sum_{\zeta=1}^{d} R_\zeta \right)^{-1},
$$

proving (d). □

3. Application to a LCFS-GPR queue . In this section, we consider a multi-class LCFS queue in which arrivals occur according to a Markov modulated Poisson process (MMPP) and service times have a PH-distribution. The discipline allows resumption of service in a phase different from the one in which a customer is preempted. We assume that there are c classes of customers and for class l customers, there are p_l phases in the PH-distribution. For class l customers $(1 \leq l \leq c)$, the service time distribution has a representation (τ_l, S_l), where τ_l is a $1 \times p_l$ vector and S_l is a $p_l \times p_l$ matrix. Let the uth element of τ_l be denoted by $\tau_l(u)$ and the (u, v)th element of S_l be denoted by $S_l(u, v)$. Further, let S_l^0 denote a $p_l \times 1$ vector such that $S_l e_l + S_l^0 = 0$ where e_l is a $p_l \times 1$ vector of ones. Let the uth element of S_l^0 be denoted by $S_l^0(u)$ for $1 \leq u \leq p_l$. The arrival process can be in one of m states. The state of the MMPP is governed by an $m \times m$ generator matrix denoted by D, whose (j, ν)th element is $D_{j,\nu}$. Further, the arrival rate (of all customers) in state j of the MMPP is given by λ_j $(j = 1, \ldots, m)$. Given that an arrival has occured in state j of the MMPP, let q_{lj} denote the probability that the arriving customer belongs to class l $(\sum_{l=1}^{c} q_{lj} = 1$ for $j = 1, \ldots, m)$. Finally, if a customer C of class l is in service in phase u and an arrival occurs, then customer C is preempted and resumes service at some later time in phase v with probability $P_l(u, v)$. We assume the actual change in phase of service occurs only when service is resumed and that customer C continues to be in phase u until resumption of service. Let P_l denote a $p_l \times p_l$ matrix whose elements are $P_l(u, v)$. Note that this formulation includes the preemptive resume discipline by choosing the matrix P_l to be the identity matrix for all l. It also includes the preemptive restart discipline by choosing every row of P_l to be equal to τ_l. Obviously, the use of the matrices P_l for $1 \leq l \leq c$ allows the use of more general forms of preemption than the two special cases given above, and each class of customers can have a different preemptive resume discipline.

On account of the richness of the discipline, the symmetric queue (Kelly [3]) is not a special case of the LCFS-GPR queue. However, our model enables one to solve more general LCFS queues than those solvable by symmetric queues.

We now cast this problem in the framework of the Markov chain studied in the earlier section. In order to do so, we first let $d = \sum_{l=1}^{c} p_l$. The node carries the information about the order, class, and phase of service for all customers in the system, and the auxiliary variable carries the information about the state of the MMPP. If node K in the tree has a representation given by $k_1 k_2 \ldots k_{|K|}$, then each of the k_n contains information about the class and phase of service of the nth oldest customer present in the system. Thus, k_1 represents l_1 (the class of the oldest customer) and v_1 (the phase in which this customer last received service); similarly for the other customers. Since we have chosen $d = \sum_{l=1}^{c} p_l$, the representation of the node completely describes the class and phase in which each customer last received service. By this method, $k_{|K|}$ contains information about the class and phase of service of the newest customer (who also happens to be the one in service).

Before we continue, we introduce some more notations to enable us to write the transitions of this Markov chain in an unambiguous manner. First, we represent k_n by $\{l_n, v_n\}$ whenever we want to make the class (l_n) and phase of service (v_n) of the nth customer explicit. An example of this is to write $a_{j,\nu}^{f(J,i),k}$ as $a_{j,\nu}^{f(J,i),\{l,v\}}$ whenever k can be represented by $\{l, v\}$. Further, for any node K, we define $\psi[f(K,1)]$ to mean $l_{|K|}$ and $\varphi[f(K,1)]$ to mean $v_{|K|}$. Similarly, we define $\psi'[f(K,2)]$ to mean $l_{|K-1|}$ and $\varphi'[f(K,2)]$ to mean $v_{|K-1|}$. This notation expresses the class and phase of service of the last two customers in a manner such that their dependence on $f(K,1)$ and $f(K,2)$ is made explicit.

We attempt to obtain the steady state results (time average) of this queue in a direct fashion. So, we assume that there is a Poisson process at a high rate θ ($\theta \geq \max_{l,v,j}\{-S_l(v,v) - D_{j,j} + \lambda_j\}$) and we imbed a Markov chain at the epochs of this Poisson process. That way, the results obtained from the Markov chain are indeed the steady state (time average) quantities from the PASTA theorem (Wolff [8]). The following types of transitions are now possible in this system from state (K,j), when $|K| \geq 2$.

1. We consider the transition in which the customer in service changes phase from $\varphi[f(K,1)]$ to v. We have

$$a_{j,j}^{f(K,1),\{\psi[f(K,1)],v\}} = \theta^{-1} S_{\psi[f(K,1)]}(\varphi[f(K,1)], v)$$

for $v = 1, \ldots, p_{\psi[f(K,1)]}$.

2. We now consider the probability of an arrival of class l who starts service in phase v. For this, we have

$$a_{j,j}^{f(K,0),\{l,v\}} = \theta^{-1} \lambda_j q_{lj} \tau_l(v)$$

where $1 \leq v \leq p_l$ and $1 \leq l \leq c$.

3. The transition probability of a state change from j to ν of the MMPP for $\nu \neq j$ is given by

$$a_{j,\nu}^{f(K,1),f(K,1)} = \theta^{-1} D_{j,\nu}.$$

4. For a departure of one customer followed by a resumption of service of the next customer in phase v, the transition probability is given by

$$a_{j,j}^{f(K,2),\{\psi'[f(K,2)],v\}} = \theta^{-1} S_{\psi[f(K,1)]}^{0}(\varphi[f(K,1)]) P_{\psi'[f(K,2)]}(\varphi'[f(K,2)], v)$$

for $1 \leq v \leq p_{\psi'[f(K,2)]}$. Note that the right hand side contains $f(K,1)$, which is itself a function of $f(K,2)$.

5. Since the probability of a transition to node 0 is zero, we have $b_{j,\nu}^{(K)} = 0$ for $1 \leq \nu \leq m$.

6. The null transition that leaves the state unchanged is now given by

$$a_{j,j}^{f(K,1),f(K,1)} = 1 - \sum_{v=1}^{p_{\psi[f(K,1)]}} a_{j,j}^{f(K,1),\{\psi[f(K,1)],v\}} - \sum_{l=1}^{c}\sum_{v=1}^{p_l} a_{j,j}^{f(K,0),\{l,v\}}$$
$$- \sum_{\nu \neq j} a_{j,\nu}^{f(K,1),f(K,1)} - \sum_{v=1}^{p_{\psi'[f(K,2)]}} a_{j,j}^{f(K,2),\{\psi'[f(K,2)],v\}}.$$

We now consider the transitions from the state (K, j), when $|K| = 1$. Transition types 1, 2 and 3 remain the same as above. Transition type 4 is not applicable, and transition type 5 is given by

$$b_{j,j}^{(K)} = \theta^{-1} S_{\psi[f(K,1)]}^{0}(\varphi[f(K,1)]) \quad \text{and} \quad b_{j,\nu}^{(K)} = 0 \quad \text{if} \quad \nu \neq j.$$

Finally, transition type 6 remains the same as before.

Now, the transitions from state $(0, j)$ are possible only when an arrival occurs or when a state change of the MMPP occurs. This means that transition type 1 is not applicable. Transition type 2 remains the same as before. Transitions type 3 and 4 are not applicable, whereas transition type 5 is given by

$$b_{j,\nu}^{(0)} = \theta^{-1} D_{j,\nu} \quad \text{for} \quad \nu \neq j.$$

Finally, transition type 6 is given by

$$b_{j,j}^{(0)} = 1 - \sum_{l=1}^{c}\sum_{v=1}^{p_l} a_{j,j}^{f(0,0),\{l,v\}} - \sum_{\nu \neq j} b_{j,\nu}^{(0)}.$$

We now discuss the stability of the LCFS-GPR queue in which the arrivals are according to a Poisson process at a rate of λ. For this special case, let q_l denote the probability that an arriving customer belongs to class l ($l = 1, \ldots, c$). Before we do so, let us point out that the problem described above can be studied as a single class problem (by increasing the number of phases suitably) in which the service time has a PH-distribution with a representation (τ, S) where

$$
(3.1) \qquad S = \begin{bmatrix} S_1 & & \\ & \ddots & \\ & & S_c \end{bmatrix}
$$

$$
(3.2) \qquad \tau = [q_1 \tau_1 \ \cdots \ q_c \tau_c].
$$

Here, S is a $d \times d$ matrix and τ is a $1 \times d$ vector. We now define a $d \times d$ matrix P as

$$
(3.3) \qquad P = \begin{bmatrix} P_1 & & \\ & \ddots & \\ & & P_c \end{bmatrix}.
$$

With this representation, the phase of service automatically tracks the class of the customer. The notations for this problem are the same as the general problem except that the subscripts denoting the class of a customer and the state of the MMPP are now dropped. Since a customer resumes service in phase v if it is preempted in phase u with probability $P(u, v)$, the Poisson (memoryless) arrivals could be viewed as an additional way in which a change of phase of service could take place. Thus, the total rate at which a change of phase (from u to v) of service occurs is given by $S(u, v) + \lambda P(u, v)$ for $v \neq u$. So, we now define a $d \times d$ matrix $\hat{S}$ whose elements are $\hat{S}(u, v) = S(u, v) + \lambda P(u, v)$ for $v \neq u$ and $\hat{S}(u, u) = S(u, u) - \lambda \sum_{v \neq u} P(u, v)$ for $1 \leq u \leq d$. Alternatively, we can write

$$
\hat{S} = S + \lambda(P - I).
$$

Note that $\hat{S} = S$ when $P = I$. Then, the *actual amount of service rendered* to a customer has a PH-distribution with representation $(\tau, \hat{S})$ with mean $\hat{\mu}^{-1} = -\tau \hat{S}^{-1} e$. Once we have characterized the distribution of the actual amount of service rendered to a customer, the stability condition does not depend on the order of service. It now follows from Loynes [4] that the queue is stable if and only if $\lambda \hat{\mu}^{-1} < 1$. Note that this argument does not carry over to the MMPP arrivals because the phase in which service is resumed depends on the state of the MMPP in a very complex manner.

REFERENCES

[1] S. ASMUSSEN, *Applied probability and queues*, Wiley, New York (1987).

[2] K. L. CHUNG, *Markov chains with stationary transition probabilities*, 2nd ed., Springer-Verlag, New York (1967).

[3] F. P. KELLY, *Reversibility and stochastic networks*, Wiley, New York (1979).

[4] R. M. LOYNES, The stability of a queue with non-independent interarrival and service times, *Proc. Cambridge Philos. Soc.*, **58** (1962), 497–520.

[5] M. F. NEUTS, *Matrix-geometric solutions in stochastic models: An algorithmic approach*, Johns Hopkins, Baltimore (1981).

[6] R. L. TWEEDIE, Operator-geometric stationary distributions for Markov chains, with applications to queueing models, *AAP*, **14** (1982), 368–391.

[7] T. TAKINE, B. SENGUPTA AND R. W. YEUNG, A generalization of the matrix M/G/1 paradigm for Markov chains with a tree structure, *submitted for publication* (1994).

[8] R. A. WOLFF, Poisson arrivals see time averages, *Operations Research*, **30** (1982), 223–231.

[9] R. W. YEUNG AND B. SENGUPTA, Matrix product-form solutions for Markov chains with a tree structure, to appear in *AAP*, **26** (4) (Dec., 1994).

LARGE DEVIATION ANALYSIS OF QUEUEING SYSTEMS

PAUL DUPUIS* AND RICHARD S. ELLIS[†]

Abstract. In recent work we have succeeded in proving the large deviation principle for a general class of jump Markov processes that model queueing systems. The present paper highlights this result by giving the main ideas and omitting technicalities. Two key steps in the proof are to represent the large deviation probabilities as the minimal cost functions of associated stochastic optimal control problems and to study the limits of these probabilities using a subadditivity–type argument. This procedure leads to a characterization of the rate function that can be used either to evaluate it explicitly in the cases where this is possible or to compute it numerically in the cases where an explicit evaluation is not possible.

1. Introduction. The purpose of this paper is to highlight a new approach to the large deviation analysis of queueing systems which is developed in the paper [6]. Our aim here is to present the main ideas without giving all the technicalities needed for a rigorous treatment. Using this approach, not only do we prove the large deviation principle for a very wide class of such systems, but also derive an explicit formula for the rate function in a number of cases of interest, including tandem queues, open and closed Jackson networks, and certain two–dimensional stable queues. Such explicit formulas will be given in the paper [7]. Because of the generality of the model that we introduce, it should come as no surprise that an explicit formula for the rate function is in many other cases not available. Even when the rate function cannot be identified explicitly, our methods can also be used to calculate the rate function numerically. This aspect of our approach is currently under investigation.

To date the large deviation analysis of queueing systems has been restricted to those having special structures. In the case of tandem queues, contraction mapping techniques and the Skorohod problem have been used. These techniques originated in [1], were extended to domains with corners in [3,10], and have been applied to queueing systems in [13,16]. Using partial differential equation methods and in particular viscosity solutions, the paper [11] proves the large deviation principle for a class of open Jackson networks. A closed queueing model is treated in [12]. In addition, the results of Chapter 6 of [5] allow one to treat a general class of stable queueing systems in two dimensions that are modeled by discrete–time Markov chains [8]. Other one and two dimensional models and numerous applications are treated in the book [14]. In terms of generality, the results derived

* Lefschetz Center for Dynamical Systems, Division of Applied Mathematics, Brown University, Providence, RI 02912. This research was supported in part by the National Science Foundation (NSF-DMS-9115762) and Air Force Office of Scientific Research (F49620-93-1-0264).

† Department of Mathematics and Statistics, University of Massachusetts, Amherst, MA 01003. This research was supported in part by the National Science Foundation (NSF-DMS-9123575).

in [6] and discussed below represent a vast improvement. In [6] we proved the large deviation principle for an extensive class of lattice–based queueing systems in arbitrary dimensions under the assumption that the queueing systems are modeled by jump Markov processes. As in our forthcoming book [5], we made important use of ideas from stochastic optimal control theory, representing the large deviation probabilities as the minimal cost functions of associated stochastic optimal control problems. The methods used in [6] are not restricted to this framework and in fact can be adapted in a straightforward way to a number of important related models, including Markov–modulated processes, related non–Markovian processes, and discrete–time models.

The large deviation literature on Markov processes is extensive. However, the vast majority of papers study processes having the property that the components of their generators depend smoothly upon the spatial parameter. The jump Markov processes that we treat are examples of Markov processes violating this smoothness condition. We refer to them as having "discontinuous statistics." In [9], in [4], and in Chapter 6 of [5] we have carried out the large deviation analysis of several classes of Markov processes with discontinuous statistics. In the last two works we considered two halfspaces of smooth statistical behavior separated by a hyperplane across which the statistical behavior changes discontinuously, and we obtained an explicit formula for the rate function.

The results summarized in the present paper concern lattice–based jump Markov processes with discontinuous statistics that generalize the type of discontinuity previously considered in [4] and in Chapter 6 of [5]. These processes are defined in terms of much more complicated geometries involving three or more regions of constant statistical behavior separated by an arbitrary finite number of intersecting hyperplanes of codimension one across which the statistical behavior can change discontinuously. In the special case of two halfspaces one can show that the jump Markov processes are superexponentially close to a sequence of processes of the form considered in Chapter 6 of [5] and that because of this the rate function for the corresponding queueing system takes the same form as in that chapter. As we discuss in detail in the introduction to [6], in the case of the general queueing processes that we consider, the difficulty in determining an explicit formula for the rate function along intersections of two or more hyperplanes arises from the very hard problem of determining necessary and sufficient conditions for stability that are valid for this more complicated geometry.

The queueing systems that we analyze have as state space a subset S of the d–dimensional integer lattice $\mathbb{Z}^d$, where $d \geq 2$. They may incorporate a variety of input disciplines and service disciplines together with other features such as feedback. The precise assumptions needed to prove the large deviation principle are given below in Conditions 2.1 and 2.4. These assumptions are reasonably weak. Conditions 2.1 has three components: that the state space $\mathcal{S}$ is partitioned into finitely many subsets called *facets*, that the jump intensities of the jump Markov processes modeling the queueing systems are constant in each facet, and that only finitely many of the jump intensities are positive. Condition 2.4 is a communication condition. The large deviation principle for jump Markov processes satisfying Conditions 2.1 and 2.4 is stated in Theorem 3.1. In our paper [6] we prove an extension of this large deviation principle, allowing jump Markov processes for which the jump intensities have some additional state dependence. Namely, we allow discontinuous statistics that are defined by a state dependent jump intensity function that is continuous, and not necessarily constant, within each facet. We refer the reader there for details. Examples of facets that arise in applications are "wedges," halflines, and the faces of the nonnegative orthant in $\mathbb{R}^d$. See, for example, Section 2 of [6], in which we discuss a modified join–the–shorter–queue model.

Here are the main steps in the proof of the large deviation principle stated in Theorem 3.1.

- Step 1 is to obtain local large deviation estimates for "tubes" centered at paths mapping $[0, 1]$ into $\mathbb{R}^d$ and having constant velocities. These paths have the form $\{x + \beta t, t \in [0, 1]\}$. The local large deviation estimates are stated in Proposition 2.5.
- Step 2 is to combine Step 1 and the Markov property to deduce a local large deviation principle for tubes centered at a class of piecewise linear, continuous paths mapping $[0, 1]$ into $\mathbb{R}^d$. For such paths φ with starting points x we obtain a local rate function $I_x(\varphi)$.
- Step 3 is to combine Step 2 and an approximation argument to prove the full large deviation principle (Theorem 3.1). The rate function is the lower semicontinuous regularization of the function I_x obtained in Step 2.

In the proof of Step 1 suitable large deviation probabilities are represented in terms of the minimal cost functions of associated stochastic optimal control problems, and a subadditivity–type argument is used to show that the minimal cost functions satisfy appropriate limits from which properties of the rate function can be deduced. This is one of our main contributions. The representation of large deviation probabilities as minimal cost functions is also the starting point for the development of efficient computational methods, which will be treated elsewhere.

Our paper [6], which is summarized here, shares some features with [5]. In both works, a key step in the proof of the large deviation principle is to represent the large deviation probabilities as the minimal cost functions of a sequence of stochastic optimal control problems. However, rather than use weak convergence methods to calculate the limits of the minimal cost functions as we do in [5], the existence of the limits is shown using a subadditivity–type argument. Our experience in [5] leads us to think that the weak convergence methods are convenient for treating problems in which an explicit formula for the rate function is available. Because of the generality of the setup considered in [6] and here, in many cases an explicit formula for the rate function cannot be found. Hence the weak convergence methods of [5] are not used to prove the large deviation principle for the general queueing system, although they are used when evaluating the rate function in particular cases (see [7]).

In Section 2 we introduce a class of jump Markov processes that model queueing systems and state certain local large deviation estimates for scaled versions of these processes. In order to prove the existence of a large deviation principle, we first prove local large deviation estimates via a subadditivity–type argument. Key ingredients are representation formulas for the large deviation probabilities as minimal cost functions of stochastic optimal control problems. These formulas are also stated in Section 2. In Section 3 we state the full large deviation principle. In Section 4 we describe other related models for which the proof of a large deviation principle is very close to the proof for the model described in Section 2.

2. Definition of the local rate function.
This section consists of two subsections. In Subsection 2.1 we introduce the class of queueing systems for which the large deviation principle will be formulated in Section 3. The definition of the rate function in this large deviation principle depends on a local rate function that arises when we consider certain local large deviation estimates. The latter estimates are the subject of Subsection 2.2. In order to prove them, we represent the large deviation probabilities as the minimal cost functions of associated stochastic optimal control problems (Theorem 2.3).

2.1. A description of the queueing system.
We consider a queueing system modeled by a stationary jump Markov process. In order to describe the process, we take as given the following data:

- *The state space S of the process, which is a subset of $\mathbb{Z}^d$ and contains the origin 0.*
- *The jump intensities at each point $x \in S$. These are of the form $\{r(x, v), v \in \mathbb{Z}^d\}$, where $r(x, v) \geq 0$ for all x and v and where $r(x, v) > 0$ only if $x + v \in S$. The quantity $r(x, v)$ gives the jump intensity from x to $x + v$.*

We call $r(\cdot, \cdot)$ the jump intensity function. Additional assumptions on this function are given in Conditions 2.1 and 2.4.

Let $\{X(t), t \in [0, \infty)\}$ be a stationary jump Markov process with state space S and jump intensities $\{r(x, v), x \in S, v \in \mathbb{Z}^d\}$. For $x \in S$ and f any bounded real–valued function on $\mathbb{Z}^d$ these processes can be characterized by their generators, which are of the form

$$\mathcal{L}f(x) \doteq \sum_{v \in \mathbb{Z}^d} r(x, v) \left[f(x + v) - f(x) \right].$$

For $0 \leq t_1 < t_2 < \infty$ and A a subset of $\mathbb{R}^d$ we define $\mathcal{D}([t_1, t_2] : A)$ to be the space of functions that map $[t_1, t_2]$ into A and that are continuous from the right and have limits from the left. The space $\mathcal{D}([t_1, \infty) : A)$ is defined similarly. The jump Markov process $\{X(t), t \in [0, \infty)\}$ takes values in $\mathcal{D}([0, \infty) : \mathcal{S})$.

The structure of the queueing systems to be considered in this section depends on a finite collection of nonzero vectors $\{w^{(i)}, i = 1, 2, \ldots, \kappa\}$, where each $w^{(i)} \in \mathbb{R}^d$ and each $w^{(i)}$ has rational coordinates. In terms of these vectors we define facets of $\mathbb{R}^d$ associated with the given model. As we explain in Condition 2.1, the facets determine the regions of constant statistical behavior of the queueing system. More precisely, the jump intensities are stipulated to be constant in the intersection of each facet with the state space $\mathcal{S}$.

For each $i \in \{1, 2, \ldots, \kappa\}$ let $\circ_i$ represent one of the three symbols $>, =, <$. We label each facet by the notation $\mathcal{F}^{(\circ_1, \ldots, \circ_\kappa)}$ to indicate that the facet is the intersection of the sets $\{y \in \mathbb{R}^d : \langle y, w^{(i)} \rangle \circ_i 0\}$, where $\langle \cdot, \cdot \rangle$ denotes the inner product on $\mathbb{R}^d$. In other words,

$$(2.1) \quad \mathcal{F}^{(\circ_1, \ldots, \circ_\kappa)} \doteq \{y \in \mathbb{R}^d : \langle y, w^{(i)} \rangle \circ_i 0 \text{ for all } i = 1, 2, \ldots, \kappa\}.$$

Note that if $y \in \mathcal{F}^{(\circ_1, \ldots, \circ_\kappa)}$, then $\alpha y \in \mathcal{F}^{(\circ_1, \ldots, \circ_\kappa)}$ for every $\alpha > 0$. Thus each facet $\mathcal{F}^{(\circ_1, \ldots, \circ_\kappa)}$ is a cone. Assumptions concerning these sets and the jump intensity function $r(x, v)$ are given in the next condition. Because of the structure of the state space $\mathcal{S}$ as expressed in part (a) of Condition 2.1, the assumption that the jump intensities are constant in each of the facets is a radial homogeneity condition. For each x the support of $r(x, \cdot)$ is defined to be the set

$$\Sigma(x) \doteq \{v \in \mathbb{Z}^d : r(x, v) > 0\}.$$

CONDITION 2.1. *We consider a finite collection of nonzero vectors* $\{w^{(i)}, i = 1, 2, \ldots, \kappa\}$ *in* $\mathbb{R}^d$ *such that each* $w^{(i)}$ *has rational components. We define the sets* $\mathcal{F}^{(\circ_1, \ldots, \circ_\kappa)}$ *in terms of these vectors by equation (2.1). The sets* $\{\mathcal{F}^{(\circ_1, \ldots, \circ_\kappa)} \cap \mathcal{S}\}$ *are a partition of* $\mathcal{S}$. *We refer to these sets as the facets of* $\mathcal{S}$. *For each sequence of symbols* $(\circ_1, \ldots, \circ_\kappa)$ *we assume the following:*

(a) *If* $\mathcal{F}^{(\circ_1, \ldots, \circ_\kappa)} \cap \mathcal{S} \neq \emptyset$, *then* $\mathcal{F}^{(\circ_1, \ldots, \circ_\kappa)} \cap \mathbb{Z}^d \subset \mathcal{S}$.

(b) $r(x, v)$ *is independent of* $x \in \mathcal{F}^{(\circ_1, \ldots, \circ_\kappa)} \cap \mathcal{S}$.

(c) *For* $x \in \mathcal{F}^{(\circ_1, \ldots, \circ_\kappa)} \cap \mathcal{S}$ *the support* $\Sigma(x)$ *consists of a finite number of vectors.*

Part (a) of Condition 2.1 states that if any point in a facet is contained in $\mathcal{S}$, then the entire intersection of that facet with $\mathbb{Z}^d$ is also contained in $\mathcal{S}$. In particular, all positive integral multiples of any point in $\mathcal{F}^{(\circ_1, \ldots, \circ_\kappa)} \cap \mathcal{S}$ are also in $\mathcal{S}$. We will need the following direct consequence of Condition 2.1:

$$(2.2) \quad r(\alpha x, v) = r(x, v) \text{ if } \alpha > 0 \text{ and } \alpha x \in \mathcal{S}.$$

For any $\alpha > 0$ we extend the definition of $r(\cdot, \cdot)$ from $\mathcal{S} \times \mathbb{Z}^d$ to $(\alpha\mathcal{S}) \times \mathbb{Z}^d$ by setting $r(\alpha x, v) \doteq r(x, v)$ whenever $x \in \mathcal{S}$ and $v \in \mathbb{Z}^d$. Because of the radial homogeneity of the jump intensity function, this extension is well defined. The extension will allow us to define $r(\cdot, \cdot)$ for certain scaled lattices that we introduce next.

For $n \in \mathbb{N}$ we introduce the scaled state spaces

$$\mathcal{S}^n \doteq \left\{ \frac{1}{n} y : y \in \mathcal{S} \right\}$$

and let Γ be the linear subspace

$$\Gamma \doteq \{\beta \in \mathbb{R}^d : \langle \beta, w^{(i)} \rangle = 0 \text{ for all } i = 1, 2, \dots, \kappa\} = \mathcal{F}^{(=,\dots,=)}.$$

Condition 2.1 and the fact that $0 \in \mathcal{S}$ imply that the scaled state spaces $\{\mathcal{S}^n, n \in \mathbb{N}\}$ are "asymptotically dense" in Γ in the sense that

$$(2.3) \quad \Gamma = \left\{ \beta \in \mathbb{R}^d : \beta = \lim_{n \to \infty} \beta^n \text{ for some sequence } \{\beta^n, n \in \mathbb{N}\} \right.$$
$$\left. \text{with each } \beta^n \in \Gamma \cap \mathcal{S}^n \right\}.$$

In Proposition 2.5 in the next subsection we will state local large deviation estimates that involve vectors β lying in this set Γ.

2.2. A local rate function for the model. In terms of the model of the previous subsection, we now introduce a sequence of scaled jump Markov processes for which the large deviation principle will be stated. We then represent the corresponding large deviation probabilities in terms of the minimal cost functions of a sequence of associated stochastic optimal control problems. In order to obtain the large deviation estimates, we will need a communication condition for the jump Markov processes that is used in [6] as a controllability condition for the associated controlled jump Markov processes. This communication/controllability condition is given in Condition 2.4.

Let $\{X(t), t \in [0, \infty)\}$ be the jump Markov process with state space $\mathcal{S} \subset \mathbb{Z}^d$, starting point $X(0) = y \in \mathcal{S}$, and jump intensities $r(x, v)$ satisfying Condition 2.1. For $n \in \mathbb{N}$ we define the processes $\{X^n(t), t \in [0, 1]\}$ by the formula

$$X^n(t) = \frac{1}{n} X(nt).$$

For each fixed n this process is a stationary jump Markov process having state space $\mathcal{S}^n$ and taking values in $\mathcal{D}([0, 1] : \mathcal{S}^n)$. Using equation (2.2), one easily verifies that for $y \in \mathcal{S}^n$ the generator of the process is given by the formula

$$\mathcal{L}^n f(y) = n \sum_{v \in \mathbb{Z}^d} r(y, v) \left[f(y + v/n) - f(y) \right],$$

where f is any bounded real–valued function on $\mathcal{S}^n$.

Let β be any vector in $\mathbb{R}^d$. For $n \in \mathbb{N}$, $y \in \mathcal{S}^n$, and $\varepsilon > 0$, we define the conditional probability

$$p^n(y; \beta, \varepsilon) \doteq P_y^n \left\{ \sup_{s \in [0,1]} \|X^n(s) - s\beta\| < \varepsilon \right\},$$

where P_y^n denotes probability conditioned on $X^n(0) = y$. Whenever β lies in the subspace $\Gamma = \mathcal{F}^{(=,\ldots,=)}$, Proposition 2.5 guarantees the existence of a convex function L mapping Γ into $[0, \infty)$ such that

$$
\begin{aligned}
(2.4) \qquad & \lim_{\varepsilon \to 0} \lim_{\delta \to 0} \liminf_{n \to \infty} \inf_{\{y \in \mathcal{S}^n : \|y\| \leq \delta\}} \left(-\frac{1}{n} \log p^n(y; \beta, \varepsilon) \right) \\
& = \lim_{\varepsilon \to 0} \lim_{\delta \to 0} \limsup_{n \to \infty} \sup_{\{y \in \mathcal{S}^n : \|y\| \leq \delta\}} \left(-\frac{1}{n} \log p^n(y; \beta, \varepsilon) \right) \\
& = L(\beta).
\end{aligned}
$$

These limits will lead to the statement of the large deviation principle in Section 3. The quantity $L(\beta)$ is uniquely determined by these limits.

In order to prove the limits in formula (2.4), it is convenient to introduce the functions

$$(2.5) \qquad q^n(y; \beta, \varepsilon) \doteq -\frac{1}{n} \log p^n(y; \beta, \varepsilon),$$

where we set $-\log 0 \doteq \infty$. Theorem 2.3 will represent $q^n(y; \beta, \varepsilon)$ as the minimal cost function of a control problem. This representation of the large deviation probabilities $q^n(y; \beta, \varepsilon)$ is a key step in the proof of the limits (2.4).

A Stochastic Control Representation. In Theorem 2.3 we will introduce a stochastic control representation for the functions $q^n(y; \beta, \varepsilon)$. This representation is used in [6] to establish the existence of the limits in formula (2.4). The stochastic control representation consists of a controlled process and a cost structure.

Controlled Process. Fix $n \in \mathbb{N}$. A measurable function $u^n(y, v, t)$ mapping $\mathcal{S}^n \times \mathbb{Z}^d \times [0, 1]$ into $[0, \infty)$ is said to be a control. It will be sufficient to consider only controls with the property that for $(y, v, t) \in \mathcal{S}^n \times \mathbb{Z}^d \times [0, 1]$

$$(2.6) \qquad u^n(y, v, t) > 0 \text{ implies } r(y, v) > 0.$$

Let $u^n(y, v, t)$ be a control satisfying property (2.6). The controlled process is a nonstationary jump Markov process $\{\Xi^n(t), t \in [0, 1]\}$ with state space $\mathcal{S}^n$ and, for each $y \in \mathcal{S}^n$, $v \in \mathbb{Z}^d$, and $t \in [0, 1]$, the jump intensity

$u^n(y, v, t)$ from y to $y + v/n$ at time t. The generators of the controlled processes are given by the formula

$$(2.7) \qquad (\bar{\mathcal{L}}^n f)(y, t) \doteq n \sum_{v \in \mathbb{Z}^d} u^n(y, v, t) \left[f(y + v/n) - f(y) \right],$$

where f is any bounded real–valued function on $\mathcal{S}^n$. Although the process Ξ^n and the generators $\bar{\mathcal{L}}^n$ depend on the control u^n, we do not indicate this dependence in the notation.

It turns out that the optimal control in the representation for $q^n(y; \beta, \varepsilon)$ is not uniformly bounded for all y, v, and t. In fact, the lack of a uniform upper bound is a feature of any control with finite cost under the cost structure to be introduced below. Since we do not require that the controls be bounded for all (y, v, t), we must take care to ensure the existence of the controlled processes. We will say that the control u^n has an associated controlled process starting at y at time t if the following hold: there exists a Markov process $\{\Xi^n(t), t \in [0, 1]\}$ on some probability space such that $\Xi^n(0) = y$ with probability 1 and

$$\varphi(\Xi^n(t), t) - \varphi(y, 0) - \int_0^t \left(\frac{\partial \varphi(\Xi^n(s), s)}{\partial s} + \bar{\mathcal{L}}^n \varphi(\Xi^n(s), s) \right) ds$$

is a martingale in t for $t \in [0, 1]$ for any bounded function $\varphi : \mathcal{S}^n \times [0, 1] \mapsto \mathbb{R}$ for which $\varphi(y, t)$ is continuously differentiable in t for each value of $y \in \mathcal{S}^n$. We also require that in the sense of distributions there be only one process $\{\Xi^n(t), t \in [0, 1]\}$ satisfying these two conditions. For example, if the control u^n is bounded, then such a process $\{\Xi^n(t), t \in [0, 1]\}$ exists and is unique. This leads to the following definition.

DEFINITION 2.2. *If a control u^n satisfies property (2.6) and if it has an associated controlled process $\{\Xi^n(t), t \in [0, 1]\}$, then the control will be called an admissible control.*

Since in the sequel we will consider only admissible controls, any control will automatically be understood to be an admissible control.

Cost structure. For a path $\varphi \in D([0, 1] : \mathcal{S}^n)$, a vector $\beta \in \mathbb{R}^d$, and a number $\varepsilon > 0$, we define the exit time

$$(2.8) \qquad \pi(\varphi; \beta, \varepsilon) \doteq \inf\{t \in [0, 1] : \|\varphi(t) - t\beta\| \geq \varepsilon\}$$

and follow the convention that the infimum over the empty set is ∞. Thus $\pi(\varphi; \beta, \varepsilon) = \infty$ if and only if $\|\varphi(t) - t\beta\| < \varepsilon$ for all $t \in [0, 1]$. For $a \in \mathbb{R}$ we also define the nonnegative function

$$(2.9) \qquad \ell(a) \doteq \begin{cases} a \log a - a + 1 & \text{if } a \geq 0 \\ \infty & \text{if } a < 0 \end{cases}$$

and for $w \in [0,1] \cup \{\infty\}$ the function

$$(2.10) \qquad g(w) \doteq \begin{cases} 0 & \text{if } w = \infty \\ \infty & \text{if } w \in [0,1]. \end{cases}$$

The function ℓ has superlinear growth in the sense that

$$\lim_{a \to \infty} \ell(a)/a = \infty.$$

The definition of the stochastic control problem is completed by defining the minimal cost function

$$(2.11) \qquad \begin{aligned} V^n(y; \beta, \varepsilon) \doteq \inf \bar{E}_y^n \bigg\{ \int_0^{\pi(\Xi^n; \beta, \varepsilon) \wedge 1} \sum_{v \in \mathbb{Z}^d} \\ r(\Xi^n(t), v) \, \ell(u^n(\Xi^n(t), v, t)/r(\Xi^n(t), v)) dt + g(\pi(\Xi^n; \beta, \varepsilon)) \bigg\}. \end{aligned}$$

The infimum in this formula is taken over all controls and associated controlled processes, $\bar{E}_y^n$ denotes expectation conditioned on $\Xi^n(0) = y$, and when $r = u = 0$, we set $r\,\ell(u/r) = 0$. The integrand depending on t which appears in the definition of $V^n(y; \beta, \varepsilon)$ is called the *running cost* while the term $g(\pi(\Xi^n; \beta, \varepsilon))$ is called the *exit cost*. For a given control u^n the expected value appearing on the right hand side of equation (2.11) is called the *total cost* associated with u^n.

The next theorem shows that the function $q^n(y; \beta, \varepsilon)$ equals the minimal cost function $V^n(y; \beta, \varepsilon)$. The theorem is proved in the appendix of [6]. In our manuscript [5] we derive analogous representation formulas and apply them to the study of a variety of large deviation phenomena for several classes of Markov processes.

THEOREM 2.3. *We assume Condition 2.1. Then for each $n \in \mathbb{N}$, $y \in \mathcal{S}^n$, $\beta \in \mathbb{R}^d$, and $\varepsilon > 0$, we have*

$$q^n(y; \beta, \varepsilon) = V^n(y; \beta, \varepsilon).$$

We next state an important assumption on the processes $\{X^n(t), t \in [0,1]\}$. The condition is stated for the original unscaled system so that it is easier to interpret.

CONDITION 2.4. **(Communication/Controllability Condition).** *Given $x \in \mathcal{S}$, let $N(x)$ be the set of points of the form $\{x + v : r(x, v) > 0\}$; thus $N(x) = x + \Sigma(x)$, where $\Sigma(x)$ denotes the support of $r(x, \cdot)$. The set $N(x)$ can be thought of as a "neighborhood" of points that are reachable from x in a single jump. We assume that there exists a positive integer K such that for each pair of points x and y in $\mathcal{S}$ there exists a positive integer J satisfying $J \leq K\|x - y\|$ and a sequence $\{x_0, x_1, \ldots, x_J\}$ such that $x_0 = x$, $x_J = y$, and $x_{j+1} \in N(x_j)$ for each $j = 0, 1, \ldots, J - 1$.*

This condition can be thought of as a communication condition for the original jump Markov process, and it is applied in [6] as a controllability condition for the controlled jump Markov processes $\{\Xi^n(t), t \in [0,1], n \in N\}$. The condition implies that with positive probability each point $y \in S$ can be reached from any other point $x \in S$ by a sequence $\{x_0, x_1, \ldots, x_J\}$ that progresses through the sets $N(x_j)$ and also that the length of this sequence is bounded above by a fixed constant K times the distance from x to y.

We introduce the set

$$S^\infty \doteq \left\{ \xi \in \mathbb{R}^d : \xi = \lim_{n \to \infty} \xi^n \text{ for some sequence } \{\xi^n, n \in N\} \right.$$

$$\left. \text{with each } \xi^n \in S^n \right\}.$$

Together, Conditions 2.1 and 2.4 imply that S^∞ is the effective "limit state space" of the process. By this we mean the following: given any point $\xi \in S^\infty$ and any $\delta > 0$, then for all sufficiently large n there exists at least one point in S^n that is within δ of y and that communicates with all other points in S^n. We can now state the main result of this section, part (a) of which gives local large deviation estimates for "tubes" centered at paths mapping $[0,1]$ into $\mathbb{R}^d$ and having constant velocities. The rate function L appearing in these local estimates will be used later on in order to define the rate function appearing in the large deviation principle stated in Theorem 3.1.

PROPOSITION 2.5. *Assume Conditions 2.1 and 2.4 and consider the linear subspace*

$$\Gamma \doteq \{\beta \in \mathbb{R}^d : \langle \beta, w_i \rangle = 0 \text{ for all } i = 1, 2, \ldots, \kappa\} = \mathcal{F}^{(=,\ldots,=)}.$$

The following conclusions hold.

(a) *Let a vector $\beta \in \Gamma$ be given. Then there exists a quantity $L(\beta) \in [0, \infty)$ such that the following limits hold:*

$$\lim_{\varepsilon \to 0} \lim_{\delta \to 0} \liminf_{n \to \infty} \inf_{\{y \in S^n : \|y\| \leq \delta\}} \left(-\frac{1}{n} \log p^n(y; \beta, \varepsilon) \right)$$

(2.12) $$= \lim_{\varepsilon \to 0} \lim_{\delta \to 0} \limsup_{n \to \infty} \sup_{\{y \in S^n : \|y\| \leq \delta\}} \left(-\frac{1}{n} \log p^n(y; \beta, \varepsilon) \right)$$

$$= L(\beta).$$

(b) *The function L is a finite convex function on Γ.*

Idea of the proof. The large deviation estimates in part (a) of this proposition are proved in Section 3 of [6] via a subadditivity–type argument using the representation of the large deviation probabilities given in Theorem 2.3. Here is an outline of the main ideas that are involved.

Fix $\beta \in \Gamma$. For $n \in \mathbb{N}$ and integers $m \geq n$, we would like to prove the following relationship between $q^n(y; \beta, \varepsilon)$ and $q^m(y; \beta, \varepsilon)$: given $\eta > 0$ there exists $\varepsilon_0 > 0$ such that for any $\varepsilon \in (0, \varepsilon_0)$ there is an $N < \infty$ with the property that if $n \geq N$ and if $m \geq n$ is sufficiently large, then

$$(2.13) \qquad 0 \leq q^m(y; \beta, \varepsilon) \leq q^n(y; \beta, \varepsilon) + \eta.$$

This is the type of bound that arises in proving the existence of limits involving subadditive functions (see [15, Lem. 3.11]). It turns out that the communication/controllability condition given in Condition 2.4 implies that the quantity

$$\limsup_{\delta \to 0} \; \limsup_{n \to \infty} \; \sup_{\|y\| \leq \delta} \; q^n(y; \beta, \varepsilon)$$

is bounded uniformly in $\varepsilon > 0$. Suppose that formula (2.13) is valid. It is then elementary to show that the limits in part (a) of Proposition 2.5 exist as finite numbers and are equal. Hence we are reduced to proving formula (2.13).

The proof of (2.13) relies on the representation formula stated in Theorem 2.3. Roughly speaking, the argument is as follows. Let $\varepsilon > 0$ and $n \in \mathbb{N}$ be fixed and suppose for simplicity that $y = 0$ and $\beta = 0$. Let m be much larger than n and define $\Delta \doteq n/m$. To simplify the notation, let us assume that there exists an optimal control in the stochastic control representation for $q^n(0; 0, \varepsilon)$ and let us call it u^n; $u^n = u^n(x, v, t)$, $x \in \mathcal{S}^n$, $v \in \mathbb{Z}^d$, and $t \in [0, 1]$. We would like to apply this control in the representation for $q^m(0; 0, \varepsilon)$. We do this via the definition

$$u^m(x, v, t) \doteq u^n(xm/n, v, tm/n) \text{ for } x \in \mathcal{S}^m, v \in \mathbb{Z}^d, t \in [0, n/m].$$

Recall that the exit cost appearing in the definition of $V^n(0; 0, \varepsilon)$ equals ∞ unless $\|\Xi^n(t)\| < \varepsilon$ for all $t \in [0, 1]$ with probability one. Hence the finiteness of the total cost associated with u^n implies that under this control the same bound $\|\Xi^n(t)\| < \varepsilon$ holds for all $t \in [0, 1]$ with probability one. Because of this bound, the control u^m (which is defined only for $t \in [0, n/m)$) is associated with a controlled process Ξ^m that satisfies $\|\Xi^m(t)\| < \varepsilon n/m$ for all $t \in [0, n/m]$. One can easily verify that the expected integrated running cost for this control on the interval $[0, n/m]$ is precisely $V^n(0; 0, \varepsilon) \, n/m$. One is therefore tempted to apply repeatedly a translated version of the control u^m to each interval of the form $[j\Delta, j\Delta + \Delta)$ and thereby obtain a total cost $V^n(0; 0, \varepsilon)$. If this procedure were possible, then the definition of V^m as an infimum over all controls would imply that $V^m(0; 0, \varepsilon) \leq V^n(0; 0, \varepsilon)$, which is even stronger than (2.13).

The problem with this argument is that the translated versions of the control u^m are not necessarily appropriate to apply on intervals that are different from the first interval $[0, \Delta)$. The reason is that the controlled process Ξ^m need not be at the origin at the times $j\Delta, j \in \mathbb{N}$. Consider, for example, the position of Ξ^m at time Δ. Concerning this, all that we are guaranteed is that $\|\Xi^n(\Delta)\| < \varepsilon n/m$. However, it turns out that Condition 2.4 can be interpreted as a controllability condition, which can be stated as follows (see Lemma 3.5 in [6] for details). There exist numbers $c_1 \in (0, \infty)$, $c_2 \in (0, \infty)$, and $c_3 \in (0, \infty)$, and for any $\eta > 0$ there exists a control $\bar{u}^m$ and an associated controlled process Ξ^m such that if $\Xi^m(0) \in \mathcal{S}^m$ satisfies $\|\Xi^m(0)\| < c_1\eta$, then

- $\Xi^m(c_2\eta) = 0$,
- $\sup_{t \in [0, c_2\eta]} \|\Xi^m(t)\| \le c_3\|y\|$,
- the expected integrated running cost on the interval $[0, c_2\eta]$ under this control is less than $\eta n/m$.

We can now define a composite control on the interval $[0, \Delta + c_2\eta)$ via the formula

$$\tilde{u}^m(y, v, t) = \left\{ \begin{array}{ll} u^m(y, v, t) & \text{for } t \in [0, \Delta) \\ \bar{u}^m(y, v, t) & \text{for } t \in [\Delta, \Delta + c_2\eta). \end{array} \right.$$

If Ξ^m is the controlled process that is associated with $\tilde{u}^m$ and that starts at 0 at time 0, then it has the properties

- $\Xi^m(\Delta + c_2\eta) = 0$,
- $\sup_{t \in [0, \Delta + c_2\eta]} \|\Xi^m(t)\| \le (1 \wedge c_3)\varepsilon n/m$,
- the expected integrated running cost on the interval $[0, \Delta c_2\eta]$ under this control is less than $(V^n(0; 0, \varepsilon) + \eta)n/m$.

Because of the first property, a translated version of the control $\tilde{u}^m$ can be applied repeatedly, thereby producing a control that can be used in the definition of $V^m(0; 0, \varepsilon)$. If we choose m such that $c_3n/m < 1$, then under this control the exit cost in the definition of $V^m(0; 0, \varepsilon)$ will be zero and the properties of the control just listed will imply the desired formula (2.13).

This completes our discussion of local large deviation estimates. In the next section we formulate the large deviation principle for jump Markov processes satisfying Conditions 2.1 and 2.4.

3. Statement of the large deviation principle. In Theorem 3.1 we state the large deviation principle for the class of processes introduced in the previous section. In our paper [6] the large deviation principle is stated and proved for a broader class of processes that allow some additional state dependencies in the jump intensities. We refer the reader there for details. By using localization methods we can extend the range of applicability of Theorem 3.1 to include a number of other queueing systems, such as systems with finite buffers and scaled closed queueing networks. An example of how this is done is given in Example 4.5 in [6].

In order to state the theorem we must introduce, for each point $\tilde{x} \in \mathcal{S}^\infty$, a new set of facets determined by $\tilde{x}$. The identification of the facets determined by $\tilde{x}$ is needed so that we can define a "localized" version of the queueing process which is appropriate for the point $\tilde{x}$. In order to explain why this is necessary, we temporarily consider the special case of a three dimensional queueing system in which the facets are given as follows: the interior of the nonnegative orthant; the three open faces of the nonnegative orthant; the three edges $\{ae_i : a > 0\}$, $i = 1, 2, 3$, where e_1, e_2 and e_3 are the standard basis vectors in $\mathbb{R}^3$; and the origin. From the established theory of large deviations for sample paths for Markov processes, we expect the rate function for X^n to take the form $\int_0^1 L(\phi(t), \dot{\phi}(t)) \, dt$ for an appropriate function $L(\tilde{x}, \beta)$. We would like to use the results of the previous section to define this function. As we will see below, we need consider only those values of β that lie in the smallest linear subspace that contains the facet in which $\tilde{x}$ lies. Suppose, for example, that the point $\tilde{x}$ lies in an open face F of the nonnegative orthant. Let β lie in the smallest linear subspace that contains this face and call this subspace $\Gamma(\tilde{x})$. Since F is an open face, if we consider the process X^n over a sufficiently small time interval $[0, \delta]$, then the probability that the process visits any facet except those whose closure intersects F is superexponentially small; in the present setting the only such facet is the interior of the nonnegative orthant. Because of this superexponentially small probability, for the purposes of estimating the probability that $X^n(t)$ remains in a small neighborhood of $\tilde{x} + t\beta$ for $t \in [0, \delta]$, we can essentially ignore the definition of the process on any facet whose closure does not intersect F. Our goal is to exploit this fact so that in estimating this probability we can reduce to the setting of the previous section. The introduction of the facets determined by $\tilde{x}$ and the definition of a localized jump intensity function associated with $\tilde{x}$ are the means by which we accomplish this goal. Conveniently, the subspace $\Gamma(\tilde{x})$, which identifies those values of β for which we need to define $L(\tilde{x}, \beta)$, corresponds to the subspace Γ of the previous section, which consists exactly of those vectors for which the limits in part (a) of Proposition 2.5 are proved.

We now describe the localization procedure in precise terms. Consider a queueing system that satisfies Condition 2.1. Thus the system is defined in terms of a fixed set of nonzero vectors $\{w^{(i)}, i = 1, 2, \ldots, \kappa\}$. For each point $\tilde{x} \in \mathcal{S}^\infty$ we define the natural subset of $\{1, 2, , \ldots, \kappa\}$ that is associated with this point by the formula

$$\Phi(\tilde{x}) \doteq \{i \in \{1, 2, \ldots, \kappa\} : \langle w^{(i)}, \tilde{x} \rangle = 0\}.$$

If $\Phi(\tilde{x}) \neq \emptyset$, then we define the linear subspace

$$\Gamma(\tilde{x}) \doteq \{\beta \in \mathbb{R}^d : \langle \beta, w^{(i)} \rangle = 0 \text{ for all } i \in \Phi(\tilde{x})\}.$$

If $\Phi(\tilde{x}) = \emptyset$, then $\Gamma(\tilde{x})$ is defined to be $\mathbb{R}^d$. In general, $\Gamma(\tilde{x})$ is the smallest linear subspace of $\mathbb{R}^d$ containing the facet in which $\tilde{x}$ lies.

For each $\tilde{x} \in \mathcal{S}^\infty$ let $\kappa(\tilde{x})$ be the number of elements in $\Phi(\tilde{x})$. If $\kappa(\tilde{x}) = 0$, then $\Phi(\tilde{x}) = \emptyset$. If $\kappa(\tilde{x}) \geq 1$, then by relabeling, if necessary, the vectors $\{w^{(i)}, i = 1, 2, \ldots, \kappa\}$, we can assume without loss of generality that $\Phi(\tilde{x}) = \{1, 2, ..., \kappa(\tilde{x})\}$. When $\kappa(\tilde{x}) \geq 1$, we define the new facets

$$(3.1) \quad \tilde{\mathcal{F}}^{(\circ_1, \ldots, \circ_{\kappa(\tilde{x})})} \doteq \{y \in \mathbb{R}^d : \langle y - \tilde{x}, w^{(i)} \rangle \circ_i 0 \text{ for all } i = 1, \ldots, \kappa(\tilde{x})\},$$

where each o_i represents one of the symbols $>, =, <$. If $\kappa(\tilde{x}) = 0$, then we
define $\tilde{\mathcal{F}}^{(o_1,\ldots,o_{\kappa(\tilde{x})})} \doteq I\!\!R^d$. In this notation $\Gamma(\tilde{x})$ is the facet $\tilde{\mathcal{F}}^{(=,\ldots,=)}$, which
is $\tilde{\mathcal{F}}^{(o_1,\ldots,o_{\kappa(\tilde{x})})}$ with each of the symbols o_i taking the value $=$. We call each
new facet $\tilde{\mathcal{F}}^{(o_1,\ldots,o_{\kappa(\tilde{x})})}$ a facet determined by $\tilde{x}$. The facets $\{\tilde{\mathcal{F}}^{(o_1,\ldots,o_{\kappa(\tilde{x})})}\}$
determined by $\tilde{x}$ give a partition of $I\!\!R^d$. Each of the facets determined by
$\tilde{x}$ either equals an original facet or is obtained from an original facet by
taking away one or more "distant" boundaries.

For $\tilde{x} \in \mathcal{S}^\infty$, y lying in a facet $\tilde{\mathcal{F}}^{(o_1,\ldots,o_{\kappa(\tilde{x})})}$, and $v \in Z\!\!\!Z^d$ a new jump
intensity function $\tilde{r}(\tilde{x}; y, v)$ is defined by the formula

$$\tilde{r}(\tilde{x}; y, v) = r(z, v)$$

where z is any point such that $z - x$ lies in the same facet $\tilde{\mathcal{F}}^{(o_1,\ldots,o_{\kappa(\tilde{x})})}$ as
y. This localized jump intensity function is constant on each of the facets
$\tilde{\mathcal{F}}^{(o_1,\ldots,o_{\kappa(\tilde{x})})}$ determined by $\tilde{x}$ and so satisfies Condition 2.1 with state space
depending on $\tilde{x}$. We denote this $\tilde{x}$–dependent state space by $\mathcal{S}_{\tilde{x}}$. $\mathcal{S}_{\tilde{x}}$ equals
the union of the sets $\tilde{\mathcal{F}}^{(o_1,\ldots,o_{\kappa(\tilde{x})})} \cap Z\!\!\!Z^d$, where the union is taken over those
facets determined by $\tilde{x}$ which have nonempty intersection with the original
state space $\mathcal{S}$.

Fix $\tilde{x} \in \mathcal{S}^\infty$ and for $n \in I\!\!N$ consider the scaled state space

$$\mathcal{S}_{\tilde{x}}^n \doteq \left\{ \frac{1}{n} y : y \in \mathcal{S}_{\tilde{x}} \right\}.$$

We define $\{\tilde{X}_{\tilde{x}}(t), t \in [0, 1]\}$ to be the stationary jump Markov process
having the state space $\mathcal{S}_{\tilde{x}}$ and the jump intensity function $\tilde{r}(\tilde{x}; y, v)$. In
terms of this process we define the scaled jump Markov process $\{\tilde{X}_{\tilde{x}}^n(t), t \in [0, 1]\}$ by the formula

$$\tilde{X}_{\tilde{x}}^n(t) \doteq \frac{1}{n} \tilde{X}_{\tilde{x}}(nt).$$

The state space of this scaled process is $\mathcal{S}_{\tilde{x}}^n$. For f any bounded real–valued
function on $\mathcal{S}_{\tilde{x}}^n$, the generator of the scaled process is given by the formula

$$\tilde{\mathcal{L}}_{\tilde{x}}^n f(\xi) = n \sum_{v \in Z\!\!\!Z^d} \tilde{r}(\tilde{x}; \xi, v) \left[f(\xi + v/n) - f(\xi) \right].$$

Both Conditions 2.1 and 2.4 of Section 3 are met for the processes
$\{\tilde{X}_{\tilde{x}}^n(t), t \in [0, 1], n \in I\!\!N\}$. For $y \in \mathcal{S}_{\tilde{x}}^n$, numbers t_1 and t_2 satisfying $0 \le t_1 \le t_2 \le 1$, $\beta \in Z\!\!\!Z^d$, and $\varepsilon > 0$, let $\tilde{p}_{\tilde{x}}^n(y, t_1, t_2; \beta, \varepsilon)$ denote the conditional
probability analogous to the conditional probability $p^n(y; \beta, \varepsilon)$ of Section 3,
except that the process $\{\tilde{X}_{\tilde{x}}^n(t), t \in [t_1, t_2]\}$ replaces the process $\{X^n(t), t \in [0, 1]\}$ there. In other words

$$\tilde{p}_{\tilde{x}}^n(y, t_1, t_2; \beta, \varepsilon) \doteq P_{\tilde{x}; y, t_1}^n \left\{ \sup_{s \in [t_1, t_2]} \left\| \tilde{X}_{\tilde{x}}^n(s) - (s - t_1)\beta \right\| < \varepsilon \right\},$$

where $P^n_{\tilde{x};y,t_1}$ denotes probability conditioned on $\tilde{X}^n_{\tilde{x}}(t_1) = y$. We now apply Proposition 2.5 and a simple time–rescaling argument to conclude, for each $\tilde{x} \in \mathcal{S}^\infty$, the existence of a convex function $L(\tilde{x}, \cdot)$ mapping $\Gamma(\tilde{x})$ into $[0, \infty)$ such that for each $\beta \in \Gamma(\tilde{x})$

$$
\begin{aligned}
(3.2) \quad &\lim_{\varepsilon \to 0} \lim_{\delta \to 0} \liminf_{n \to \infty} \inf_{\{y \in \mathcal{S}^n_{\tilde{x}} : \|y\| \le \delta\}} \left(-\frac{1}{n} \log \tilde{p}^n_{\tilde{x}}(y, t_1, t_2; \beta, \varepsilon) \right) \\
&= \lim_{\varepsilon \to 0} \lim_{\delta \to 0} \limsup_{n \to \infty} \sup_{\{y \in \mathcal{S}^n_{\tilde{x}} : \|y\| \le \delta\}} \left(-\frac{1}{n} \log \tilde{p}^n_{\tilde{x}}(y, t_1, t_2; \beta, \varepsilon) \right) \\
&= (t_2 - t_1) \cdot L(\tilde{x}, \beta).
\end{aligned}
$$

The function $L(\tilde{x}, \beta)$ is uniquely determined from these limits. A variational representation for $L(\tilde{x}, \beta)$ follows from Theorem 2.3.

Before we can state the large deviation principle, we need a few more definitions. $\mathcal{D}([0, 1] : \mathbb{R}^d)$ denotes the space of functions that map $[0, 1]$ into $\mathbb{R}^d$ and that are continuous from the right and have limits from the left. We topologize $\mathcal{D}([0, 1] : \mathbb{R}^d)$ by the Skorohod metric, with respect to which it is a complete separable metric space [2, Sect. 14]. Let $\mathcal{T}([0, 1] : \mathbb{R}^d)$ denote the subset of $\mathcal{D}([0, 1] : \mathbb{R}^d)$ consisting of piecewise linear functions φ mapping $[0, 1]$ into $\mathbb{R}^d$ and having only finitely many discontinuities. For $t \in (0, 1)$ not a point of discontinuity of φ, $\dot{\varphi}(t)$ denotes the derivative of φ with respect to t. Let $\mathcal{T}^\infty([0, 1] : \mathbb{R}^d)$ denote the subset of $\mathcal{T}([0, 1] : \mathbb{R}^d)$ consisting of functions φ for which $\varphi(s) \in \mathcal{S}^\infty$ for every $s \in [0, 1]$. For $x \in \mathcal{S}^\infty$ and $\varphi \in \mathcal{T}^\infty([0, 1] : \mathbb{R}^d)$ satisfying $\varphi(0) = x$, we define the quantity

$$
(3.3) \qquad I_x(\varphi) \doteq \int_0^1 L(\varphi(t), \dot{\varphi}(t)) \, dt.
$$

Since φ is piecewise linear, for all but finitely many $t \in [0, 1]$ we have $\dot{\varphi}(t) \in \Gamma(\varphi(t))$. Hence $I_x(\varphi)$ is well defined. In addition, since $L(\xi, \beta)$ is finite for $\xi \in \mathcal{S}^\infty$ when $\beta \in \Gamma(\xi)$, $I_x(\varphi)$ is finite for $x \in \mathcal{S}^\infty$ and $\varphi \in \mathcal{T}^\infty([0, 1] : \mathbb{R}^d)$ satisfying $\varphi(0) = x$. For all other $x \in \mathbb{R}^d$ and $\varphi \in \mathcal{D}([0, 1] : \mathbb{R}^d)$ we set $I_x(\varphi) \doteq \infty$. In particular

$$
I_x(\varphi) = \infty \text{ for all } \varphi \in \mathcal{T}([0, 1] : \mathbb{R}^d) \setminus \mathcal{T}^\infty([0, 1] : \mathbb{R}^d).
$$

Given any set A in $\mathcal{D}([0, 1] : \mathbb{R}^d)$ we write $I_x(A)$ to denote $\inf_{\psi \in A} I_x(\psi)$.

Let $\sigma(\cdot, \cdot)$ denote the Skorohod metric on $\mathcal{D}([0, 1] : \mathbb{R}^d)$, and for $\psi \in \mathcal{D}([0, 1] : \mathbb{R}^d)$ and $\alpha > 0$ we define the open ball

$$
B_\sigma(\psi, \alpha) \doteq \{\zeta \in \mathcal{D}([0, 1] : \mathbb{R}^d) : \sigma(\zeta, \psi) < \alpha\}.
$$

The rate function in the large deviation principle to be stated just below is the lower semicontinuous regularization of the function I_x. Specifically, for $x \in \mathbb{R}^d$ and any function $\psi \in \mathcal{D}([0,1] : \mathbb{R}^d)$ we define the quantity

$$(3.4) \qquad J_x(\psi) \doteq \lim_{\alpha \to 0} \left(\inf_{\{y \in \mathbb{R}^d : \|y - x\| < \alpha\}} I_y(B_\sigma(\psi, \alpha)) \right).$$

Finally, for any set A in $\mathcal{D}([0,1] : \mathbb{R}^d)$ we write $J_x(A)$ to denote $\inf_{\psi \in A} J_x(\psi)$.

Here is the main theorem of the paper, the large deviation principle for the jump Markov processes $\{X^n(t), t \in [0,1], n \in \mathbb{N}\}$ having state space $\mathcal{S}$ and jump intensity function $r(x, v)$. While the theorem is formulated for the space $\mathcal{D}([0,1] : \mathbb{R}^d)$, there is an analogous statement for the space $\mathcal{D}([0,T] : \mathbb{R}^d)$ for any $T \in (0, \infty)$.

THEOREM 3.1. *We assume Conditions 2.1 and 2.4. For each $x \in \mathcal{S}^\infty$ let $L(x, \cdot)$ be the function mapping $\Gamma(x)$ into $[0, \infty)$ that satisfies the limits in equation (3.2). For $\psi \in \mathcal{D}([0,1] : \mathbb{R}^d)$ we define the function $J_x(\psi)$ by formulas (3.3) and (3.4). Then for each $x \in \mathcal{S}^\infty$ the following conclusions hold.*

(a) The function J_x defined in equation (3.4) is nonnegative and lower semicontinuous on $\mathcal{D}([0,1] : \mathbb{R}^d)$. Furthermore for any compact set V in $\mathbb{R}^d$ and any $M \in [0, \infty)$ the set

$$\bigcup_{x \in V} \{\psi \in \mathcal{D}([0,1] : \mathbb{R}^d) : J_x(\psi) \le M\}$$

is a compact set in $\mathcal{D}([0,1] : \mathbb{R}^d)$.

(b) For $\varphi \in \mathcal{T}^\infty([0,1] : \mathbb{R}^d)$ and $x \doteq \varphi(0) \in \mathbb{R}^d$, we have $J_x(\varphi) = I_x(\varphi) < \infty$. For $\varphi \in \mathcal{T}([0,1] : \mathbb{R}^d) \setminus \mathcal{T}^\infty([0,1] : \mathbb{R}^d)$ and $x \doteq \varphi(0) \in \mathbb{R}^d$, we have $J_x(\varphi) = I_x(\varphi) = \infty$.

(c) For any open set G in $\mathcal{D}([0,1] : \mathbb{R}^d)$ and each $x \in \mathbb{R}^d$ we have the large deviation lower bound

$$\lim_{\delta \to 0} \liminf_{n \to \infty} \inf_{\{y \in \mathcal{S}^n : \|y - x\| \le \delta\}} \frac{1}{n} \log P^n_{y,0}\{X^n \in G\} \ge -J_x(G).$$

(d) For any closed set F in $\mathcal{D}([0,1] : \mathbb{R}^d)$ and each $x \in \mathbb{R}^d$ we have the large deviation upper bound

$$\lim_{\delta \to 0} \limsup_{n \to \infty} \sup_{\{y \in \mathcal{S}^n : \|y - x\| \le \delta\}} \frac{1}{n} \log P^n_{y,0}\{X^n \in F\} \le -J_x(F).$$

Theorem 3.1 is proved in Sections 5 and 6 of [6]. We conjecture that if ψ is absolutely continuous, ψ satisfies $\psi(0) = x$, and $\psi(s)$ lies in $\mathcal{S}^\infty$ for each $s \in [0,1]$, then $J_x(\psi)$ is given by a formula as in equation (3.3) and for all other $\psi \in \mathcal{D}([0,1] : \mathbb{R}^d)$ $J_x(\psi) = \infty$. While this is known to be true in a number of special cases, its validity in general is an interesting open question.

4. Concluding remarks. In this final section we describe a few related models for which the proof of a large deviation principle is very close to the proof for the model given in Section 2 under Conditions 2.1 and 2.4.

Markov-modulated processes. Let Θ be a finite set of points $\{\theta_j, j = 1, 2, ..., J\}$. Suppose that instead of a jump Markov process on $\mathcal{S}$ we consider a jump Markov process $(X(t), \xi(t))$ on $\mathcal{S} \times \Theta$. Let us assume that for f a bounded real–valued function on $\mathcal{S} \times \Theta$ the generator of this process takes the form

$$\mathcal{L}f(x, \theta_j) \doteq \sum_{v \in \mathbb{Z}^d} r(x, v; \theta_j) \left[f(x + v, \theta_j) - f(x, \theta_j) \right]$$

$$+ \sum_{k=1}^J s(j, k) \left[f(x, \theta_k) - f(x, \theta_j) \right].$$

Thus the jump intensity $s(j, k)$ of the Θ–valued component $\xi(t)$ does not depend on the value of $X(t)$ while the jump intensity $r(x, v; \theta_j)$ is "modulated" by the process $\xi(t)$. Let us also assume that the Markov process $\{\xi(t), t \in [0, \infty)\}$ is ergodic. The process $\{X(t), t \in [0, \infty)\}$ can be used to model non-Markovian types of behavior. As before, we would like to understand the large deviation behavior of the scaled processes $X^n(t) \doteq X(nt)/n$. If the assumptions of Condition 2.1 are satisfied by $r(x, v; \theta_j)$ for each $j \in \{1, ..., J\}$, and if the analogue of Condition 2.4 holds with the neighborhoods now defined by the formula

$$N(x, \theta_j) \doteq \{(y, \theta_k) : k = j \text{ and } y = x + v \text{ with } r(x, v; \theta_j) > 0,$$

$$\text{or } y = x \text{ and } s(j, k) > 0\},$$

then Theorem 3.1 will continue to hold. With regard to the stochastic control representation given by equation (2.11), the new $\xi(t)$–component modifies the existing running cost and also adds a new term. The form of the representation for a discrete–time analogue of this model is given in Section 3.8 of [5]. These changes will of course affect the particular form of the rate function. However, as far as the proof of Theorem 3.1 is concerned the only differences are additional notational nuisances.

Although it does not commonly appear in queuing models at the present time, one could also consider (under a uniform ergodicity assumption) the case where Θ is a compact subset of some Polish space. Finally, we note that in parallel with the type of additional state dependency allowed in Section 3.8 of [5], it is also possible to allow the jump rates $s(j, k)$ to depend on the value of $X^n(t)$.

Discrete–time processes. Apart from a slightly more exotic looking stochastic control representation (see Section 3.6 of [5]), the treatment of discrete–time models that satisfy the obvious analogues of Conditions 2.1 and 2.4 is very much the same as in the continuous–time setting. In formulating Conditions 2.1 and 2.4 we have tried to balance a succinct list of required conditions with relatively broad applicability. This was done with the continuous–time case especially in mind.

Unfortunately, Condition 2.1 rules out models that occur naturally in both discrete time and continuous time but that can still be dealt with by our approach. To illustrate, consider the case of a discrete–time Markov process that takes values in $S \doteq \{x \in \mathbb{Z}^2 : x = (x_1, x_2) \text{ with } x_1 \geq 0\}$ and has the transition probability function $p(x, y), x \in S, y \in S$. Let us suppose (as would seem reasonable) that $\kappa = 1$ and $w^{(1)} = (1, 0)$. Thus, the only two facets of interest are $\mathcal{F}^{(=)} \doteq \{x : x_1 = 0\}$ and $\mathcal{F}^{(>)} \doteq \{x : x_1 > 0\}$. Let $N(x) \doteq \{y \in S : p(x, y) > 0\}$. For discrete–time models it is often the case that $N(x)$ contains points other than those that differ by at most 1 from x in any coordinate. In particular, it may be the case for points $x \in \mathbb{Z}^2$ with $x_1 > 1$ that $N(x)$ contains points y with $y_1 < x_1 - 1$. If this holds, then such a property automatically violates the discrete–time analogue of Condition 2.1 since the transition probabilities at points $x \in \mathcal{F}^{(>)}$ with $x_1 = 1$ cannot possibly equal the transition probabilities at points $x \in \mathcal{F}^{(>)}$ with $x_1 > 1$.

However, if the neighborhoods $\{N(x) - x, x \in S\}$ are uniformly bounded, then it is still possible to prove a version of Proposition 2.5 and hence establish Theorem 3.1. This is because the crucial geometric property that is needed to establish Proposition 2.5 is actually slightly weaker that Condition 2.1. Specifically, what is used is that the transition probability function must be translation invariant along the linear space Γ. In the example under consideration Γ equals $\{x \in \mathbb{Z}^2 : x_1 = 0\}$, and this translation invariance would require that

$$p(x, y) = p(z, y) \text{ for all } x, z \text{ in } S \text{ with } x_1 = z_1 \text{ and for all } y \in S.$$

This allows some variation of $p(x, \cdot)$ with respect to the x_2 coordinate. The difficulty with formulating a condition that exploits this weakening is due to the fact that conditions cannot easily be given in terms of the unscaled system. However, a formulation that is sufficient to cover many interesting discrete–time models is as follows.

We let $S, S^n, S^\infty, \kappa, \{w^{(i)}, i = 1, .., \kappa\}$ be as in Section 2. For $x \in S$ define $N(x) \doteq \{y \in S : p(x, y) > 0\}$. The transition probability that is appropriate for the scaled lattice S^n is $p^n(x, y) \doteq p(nx, ny)$. Let $\{X_i^n, i \in \mathbb{N}\}$ denote the discrete–time Markov chain with state space S^n and transition probability function $p^n(\cdot, \cdot)$, and define a continuous–time interpolation by the formula

$$X^n(t) \doteq X_i^n \text{ for } t \in [i/n, (i+1)/n).$$

The next condition replaces Condition 2.1.

CONDITION 4.1. *For each sequence of symbols* $(\circ_1, ..., \circ_\kappa)$ *we assume the following:*

(a) *If* $\mathcal{F}^{(\circ_1, ..., \circ_\kappa)} \cap S \neq \emptyset$, *then* $\mathcal{F}^{(\circ_1, ..., \circ_\kappa)} \cap \mathbb{Z}^d \subset S$.

(b) There exist $c \in (0, \infty)$ and $C \in (0, \infty)$ such that for all $x \in S$ the set $N(x)$ has cardinality less than C, for all $x \in S$ and $y \in N(x)$ $\|x - y\| \leq C$, and for all $y \in N(x)$ $p(x, y) > c$.

(c) Fix any point $\tilde{x} \in S^\infty$ and let $(\circ_1, ..., \circ_\kappa)$ denote the symbols such that $\tilde{x} \in \mathcal{F}^{(\circ_1, ..., \circ_\kappa)}$. There exist $\varepsilon > 0$ and $N < \infty$ such that if $n \geq N$, if x and z in S^n satisfy $\|x - \tilde{x}\| \leq \varepsilon$ and $\|z - \tilde{x}\| \leq \varepsilon$, and if $x - z \in \mathcal{F}^{(\circ_1, ..., \circ_\kappa)}$, then $p^n(x, \cdot) = p^n(z, \cdot)$.

Then under Condition 4.1 and the natural analogue of Condition 2.4, Theorem 3.1 continues to hold. If desired, one can also accommodate state dependent models as in Section 3.8 of [5] and non-Markovian models such as the Markov-modulated processes introduced earlier in this section.

REFERENCES

[1] R. Anderson and S. Orey. Small random perturbations of dynamical systems with reflecting boundary. *Nagoya Math. J.*, 60:189–216, 1976.

[2] P. Billingsley. *Convergence of Probability Measures*. John Wiley & Sons, New York, 1968.

[3] P. Dupuis. Large deviations analysis of reflected diffusions and constrained stochastic approximation algorithms in convex sets. *Stochastics*, 21:63–96, 1987.

[4] P. Dupuis and R. S. Ellis. Large deviations for Markov processes with discontinuous statistics, II: Random walks. *Probab. Th. Relat. Fields*, 91:153–194, 1992.

[5] P. Dupuis and R. S. Ellis. A weak convergence approach to the theory of large deviations. Technical report, Lefschetz Center for Dynamical Systems, Brown University, 1993. Preprint # 93-6. This manuscript will be published as a book by John Wiley & Sons.

[6] P. Dupuis and R. S. Ellis. The large deviation principle for a general class of queueing systems, I. *Trans. Amer. Math. Soc.*, 1995. To appear.

[7] P. Dupuis and R. S. Ellis. The large deviation principle for a general class of queueing systems, II: explicit formulas for the rate function. Technical report, Brown University and University of Massachusets., 1995. In preparation.

[8] P. Dupuis, R. S. Ellis, and G. Kieffer. The large deviation principle for two–dimensional stable systems. Technical report, Brown University and University of Massachusetts., 1995. In preparation.

[9] P. Dupuis, R. S. Ellis, and A. Weiss. Large deviations for Markov processes with discontinuous statistics, I: General upper bounds. *Ann. Probab.*, 19:1280–1297, 1991.

[10] P. Dupuis and H. Ishii. On Lipschitz continuity of the solution mapping to the Skorokhod problem, with applications. *Stochastics and Stochastic Reports*, 35:31–62, 1991.

[11] P. Dupuis, H. Ishii, and H. M. Soner. A viscosity solution approach to the asymptotic analysis of queueing systems. *Ann. Probab.*, 18:226–255, 1990.

[12] R. Sh. Liptser. Large deviations for a simple closed queueing model. *Queueing Systems*, 14:1–31, 1993.

[13] Y. W. Park. *Large Deviation Theory for Queueing Systems*. PhD thesis, Virginia Polytechnic Institute and State University, 1991.

[14] A. Shwartz and A. Weiss. *Large Deviations for Performance Analysis: Queues, Communication, and Computing*. Chapman and Hall, New York., 1995. To appear.

[15] D. W. Stroock. *An Introduction to the Theory of Large Deviations*. Springer–Verlag, New York, 1984.

[16] P. Tsoucas. Rare events in series of queues. *J. Appl. Probab.*, 29:168–175, 1992.

STATIONARY TAIL PROBABILITIES IN EXPONENTIAL SERVER TANDEMS WITH RENEWAL ARRIVALS*

A. GANESH[†] AND V. ANANTHARAM[†]

Abstract. The problem considered is that of estimating the tail stationary probability for two exponential server queues in series fed by renewal arrivals. We compute the tail of the marginal queue length distribution at the second queue. The marginal at the first queue is known by the classical result for the $GI/M/1$ queue. The approach involves deriving necessary and sufficient conditions on the paths of the arrival and virtual service processes in order to get a large queue size at the second queue. We then use large deviations estimates of the probabilities of these paths, and solve a constrained convex optimization problem to find the most likely path leading to a large queue size. We find that the stationary queue length distribution at the second queue has an exponentially decaying tail, and obtain the exact rate of decay.

Key words. Tandem queues, large deviations.

AMS(MOS) subject classifications. 60K

1. Introduction. Networks of queues are widely used to model communications systems. Bit streams representing either voice, video or data, arrive at the nodes of the network, and need to be transmitted to other nodes, either directly or via intermediate nodes. The bit streams are usually broken up into cells or packets for transmission. The arrival process is typically modeled as a stochastic process; the packet sizes may also be random. Packets need to queue at the nodes to await transmission, because link capacities are finite. Questions of interest in the design of communication networks relate to the probability of packet loss due to inadequate storage capacity at a node of the network, or the probability of large queueing delays within the network, which may be unacceptable in applications involving voice or video transmission. Estimating these probabilities requires knowledge of the tail stationary distribution in the associated queueing network model. Such distributions are known at present only for a few special types of networks.

There has recently been considerable interest in the use of large deviations techniques to estimate tail stationary probabilities. The problem of fast simulation of tail probabilities in $M/M/1$ tandems was studied in [14]. The model is of exponential servers in series with Poisson arrivals. The system is started empty, and the problem considered is that of estimating the probability that the total number in system exceeds a large number, N, before the system is again empty after having had customers. Large deviations ideas were used to relate the problem heuristically to a variational problem involving the most likely path to the build-up of large

---

* Research supported by the NSF under PYI award NCR 8857731, and IRI 9005849, by an AT&T Foundation award, and by Bellcore Inc.

† School of Electrical Engineering, Cornell University, Ithaca, NY 14853.

queue sizes. This variational problem was solved for a system of two queues in tandem. The solution was extended to an arbitrary number of tandem queues in [9]. An alternative approach using the time reversal of Markov chains was followed in [3] to find the most likely path to large queue size. The heuristic derived in [14] was made rigorous in [15] by relating the path of the queue length process to that of the 'free' process, which is obtained as the difference of the arrival and service processes at each queue. The connection was made explicit for two queues in series, and the general idea sketched for an arbitrary number of queues in series, though the detailed analysis becomes cumbersome as the number of queues increases. More recently, the problem of estimating the tail distribution in a single queue with several multiplexed traffic streams at the input was considered in [7]. Solutions were obtained for fairly general arrival and service distributions. In [6], the tail stationary probability is estimated in intree networks of deterministic service nodes, for quite general arrival processes.

In this paper, we consider the problem of estimating the tail of the stationary distribution in a tandem of exponential server queues with renewal arrivals. We suspect that some of the techniques developed here will be relevant in extending the result to more general network models. We use ideas from large deviations theory to estimate the most likely path along which the system builds up to a large queue size. More precisely, we use large deviations results to estimate the probability of a path leading to large queue sizes, and solve a constrained optimization problem to find the most likely such path. We find that the stationary queue length distributions decay exponentially, and obtain the exact exponential rate of decay.

2. The basic idea. In this section, we illustrate the solution techniques by using them to analyze the standard $GI/M/1$ queue. We omit proofs for the most part. The approach in this section is heuristic, the rigorous analysis is shown in the following sections when we study the system of two such queues in series.

We begin by describing the model. Customers arrive into a queue according to a renewal process, i.e., with independent, identically distributed ($i.i.d.$) inter-arrival times. The customers are served by a single server in their order of arrival, i.e., the service discipline is first-come-first-served ($FCFS$). It is assumed that the service requirements of the various customers are $i.i.d.$ exponential random variables with mean $1/\mu$, and also that these are independent of the customer arrival times. On completion of service, the customers leave the system. This is the classical $GI/M/1$ queueing model. We are interested in the queue length process, i.e., in the number of customers in the queue at any given time.

Let $\tau_i, i = \ldots - 1, 0, 1, \ldots$ denote a sequence of arrival epochs with $\tau_0 = 0$, and let $T_i = \tau_i - \tau_{i-1}$ define the corresponding sequence of inter-arrival times. Then the T_i are $i.i.d.$ random variables. We use T to denote a random variable with the common distribution of the T_i, and F to denote the corresponding probability distribution function. That is, $F(t) = P(T \leq t)$. Let $X(t)$ denote the queue length at time t, and $X_n = X(\tau_n)$ denote the queue length seen by the customer arriving at time τ_n, excluding the arriving customer. We define the queue length process to be continuous from the left.

We assume that $\mu ET > 1$. This is known as the stability criterion. The quantity μ defined above as the reciprocal of the mean service time is called the service rate. It is known that the stability assumption is necessary and sufficient for the queue length process to approach a unique steady state (or stationary) distribution in this model. We shall assume that the evolution of the system starts at time $-\infty$, so that the system has reached stationarity by any finite time.

Let $V(t)$ denote the virtual service process, a Poisson process of rate μ. The evolution of the queue is completely described by the arrival process and the virtual service process as follows : the virtual service process may be thought of as generating tokens according to a Poisson process of rate μ. If a token is generated at a time that the queue is non-empty, then the customer at the head of the queue picks up the token and leaves. Thus, this results in an actual service. If the queue is empty when a token is generated, then the token is thrown away. That is, the virtual service is wasted. Such a construction of the actual service process is possible because of the memoryless nature of the exponential service time distribution. We also let $V_i = V(\tau_{i+1}) - V(\tau_i)$ denote the number of virtual services in the inter-arrival period $[\tau_i, \tau_{i+1})$.

We wish to compute the stationary distribution of the queue length process seen by arrivals. That is, with the system started at time $-\infty$, we want to compute $P(X_n = k)$ for an arbitrary n, say $n = 0$, and for all possible values of k. This is a well known result for the model we have described, which is the $GI/M/1$ queue. We shall first describe the solution and briefly outline how it is traditionally obtained. Then, we shall illustrate our technique by using it to derive the same result for this model. In later sections, we shall then apply this technique to extend the result to a system of two queues in series, which cannot be tackled by the traditional methods.

Define the Laplace transform of the inter-arrival distribution as

$$F^*(s) = E[e^{-sT}] = \int_0^\infty e^{-st} dF(t)$$

Since T is a positive random variable, there is $-\infty \leq \sigma \leq 0$ such that $F^*(s)$ is finite on (σ, ∞), and infinite on $(-\infty, \sigma)$ (if $-\infty < \sigma$). $F^*(\sigma)$ could be either finite or infinite. $F^*(s)$ is a convex, decreasing function of s, and in fact strictly convex and strictly decreasing if T is not identically zero, as we assume. Finally, $\log F^*(s)$ is also a convex function of s.

Consider the fixed point equation

(2.1) $$x = F^*(\mu - \mu x)$$

Clearly, $x = 1$ is a solution. Furthermore,

$$\frac{d}{dx}F^*(\mu - \mu x)\bigg|_{x=1} = -\mu\frac{d}{ds}F^*(s)\bigg|_{s=0} = \mu ET > 1$$

where the inequality follows from the stability criterion. That is, the slope of the curve $y = F^*(\mu - \mu x)$ exceeds that of the line $y = x$, at $x = 1$. At $x = 0$, we have $F^*(\mu - \mu x) > 0 = x$. Observe that $F^*(\mu - \mu x)$ is a continuous function of $x \in (0, 1)$. Therefore, (2.1) has a solution $\nu \in (0, 1)$. Further, by the strict convexity of $F^*(s)$, (2.1) has no solutions other than ν and 1. It is known that the stationary distribution for a $GI/M/1$ queue is given by

(2.2) $$P(X_0 \geq N) = \nu^N$$

We now show how the above result is obtained by the traditional method, which is to set up and solve the flow balance equations. Let a_k denote the probability of there being k virtual services in one inter-arrival time. This can be evaluated explicitly using the assumption of exponential service times. Thus

$$a_k = P(V_1 = k) = \int_0^\infty \frac{(\mu t)^k}{k!} e^{-\mu t} dF(t)$$

Let $\pi_k = P(X_0 = k)$ denote the stationary probability of there being exactly k customers in queue. The flow balance equations can be written as

$$\pi_j = \sum_{i=j-1}^{\infty} \pi_i a_{i+1-j}, \quad j > 0$$

$$\pi_0 = \sum_{i=0}^{\infty} \pi_i \left(\sum_{j \geq i+1} a_j\right)$$

It is now easy to check, using the expression for a_k above, that $\pi_k = (1 - \nu)\nu^k$, $k \geq 0$ solves the flow balance equations above. It also satisfies $\sum_{k=0}^{\infty} \pi_k = 1$. Thus, π_k as defined above is the stationary queue length distribution. But this is exactly the same as (2.2).

Next, we show how the same result can be obtained using large deviations techniques. We shall employ similar methods later to extend the result to two queues in series.

Consider a path of the process leading to the event $X_0 \geq N$. There must be a last time before τ_0 on this path, call it τ_{-n}, at which the queue is empty. That is, the customer arriving at time τ_{-n} sees an empty queue, and all subsequent arrivals up to the one at τ_0 see a non-empty queue on arrival. Since the queue is non-empty during $(\tau_{-n}, \tau_0]$, every virtual service during this period results in an actual service. Hence, the queue length at the end of this period is given by the total number of arrivals during this period, which is n, less the total number of departures, which is the same as the number of virtual services. In other words,

$$X_0 = \sum_{i=-n}^{-1} (1 - V_i)$$

We have thus shown that

$$X_0 \geq N \;\Rightarrow\; \exists n > 0 \;:\; \sum_{i=-n}^{-1} (1 - V_i) \geq N$$

The reverse implication is easy to see because, for any $n > 0$, the number of arrivals during $(\tau_{-n}, \tau_0]$ is n, and the number of departures is at most $\sum_{i=-n}^{-1} V_i$. Thus,

$$(2.3) \qquad P(X_0 \geq N) \;=\; P\left(\bigcup_{n>0} \left[\sum_{i=-n}^{-1} (1 - V_i) \geq N \right] \right)$$

$$\leq \; \sum_{n=1}^{\infty} P\left(\sum_{i=-n}^{-1} (1 - V_i) \geq N \right)$$

We shall next find the value of n which maximizes the summand above, or rather, an approximate expression for it which we give below. We shall claim that this maximum value is roughly the same as the value of the sum. We claim that this maximum value is also a rough lower bound on $P(X_0 \geq N)$. Since we are taking the liberty of waving our hands in this section, these claims will not be proved.

We define the type, or empirical distribution, of the sequence $\{V_i, i = -1, \ldots, -n\}$ to be the probability distribution $\mathbf{p}$ with p_k denoting the relative frequency of k in the sequence. That is,

$$p_k = \frac{1}{n} \sum_{i=-n}^{-1} 1_{\{V_i = k\}}, \quad k \geq 0$$

Then, it is a well known result in large deviations theory that the probability of $\{V_i, i = -1, \ldots, -n\}$ having the empirical distribution $\mathbf{p}$ is approximately equal to $\exp[-n \sum_{k=0}^{\infty} p_k \log(p_k/a_k)]$. Here a_k was defined earlier, and is the true distribution of V_i.

Now, in order for the V_i to satisfy $\sum_{i=-n}^{-1} (1 - V_i) \geq N$, their empirical distribution $\mathbf{p}$ should satisfy $n \sum_{k=0}^{\infty} (1 - k) p_k \geq N$. Hence, the problem of finding n to maximize the summand in (2.3) may be approximated by the following constrained optimization problem.

$$\inf \quad f(\mathbf{p}, n) \ = \ n \sum_{k=0}^{\infty} p_k \log \frac{p_k}{a_k}$$

subject to

$$g_0(\mathbf{p}, n) \ = \ \sum_{k=0}^{\infty} p_k - 1 = 0$$

$$g_1(\mathbf{p}, n) \ = \ N - n \sum_{k=0}^{\infty} (1 - k) p_k \leq 0$$

Observe that we also need non-negativity constraints on n and the p_k, but we have not included these explicitly. It will turn out that these constraints are not tight at the optimizing point. Since the purpose here is simply to illustrate the ideas involved, we have chosen to drop these constraints to keep things simpler. Also note that the optimum is required over positive integer values of n. However, we shall solve the optimization problem over positive real values of n. We introduce Lagrange multipliers y_0 and y_1 corresponding to the two constraints, and use standard techniques for finding necessary conditions for a local optimum.

Differentiating with respect to p_k gives

$$(2.4) \qquad n \left[1 + \log \frac{p_k}{a_k} \right] + y_0 - y_1 n (1 - k) = 0, \quad k = 0, 1, \ldots$$

Differentiating with respect to n gives

$$(2.5) \qquad \sum_{k=0}^{\infty} p_k \left[\log \frac{p_k}{a_k} - y_1 (1 - k) \right] = 0$$

In addition, we have the conditions

$$(2.6) \qquad g_0(\mathbf{p}, n) = 0, \quad g_1(\mathbf{p}, n) \leq 0, \quad y_1 \geq 0, \quad y_1 g_1(\mathbf{p}, n) = 0$$

Observe from (2.4) and (2.5) that

$$y_0 = -n \sum_{k=0}^{\infty} p_k = -n$$

where the second equality follows from $g_0(\mathbf{p}, n) = 0$. Hence, (2.4) implies that

$$p_k = a_k z^{k-1} \quad \text{where } z = \exp(-y_1)$$

Therefore, by the expression for a_k given earlier, the requirement $\sum_{k=0}^{\infty} p_k = 1$ translates to

$$F^*(\mu - \mu z) = z$$

But, as pointed out earlier, the above equation, which is the same as (2.1), has exactly two solutions, namely $z = 1$ and $z = \nu$. If $z = 1$, then $p_k = a_k$, and it is easy to check using the stability criterion that it is not possible to satisfy $g_1(\mathbf{p}, n) \leq 0$ with any positive value of n. It is also possible to check that, if $z = \nu$, then we can find $n > 0$ such that $g_1(\mathbf{p}, n) = 0$. Finally, taking $y_1 = -\log \nu$, it is easy to check that all the conditions (2.4)-(2.6) are satisfied. Thus, these values of n and p_k satisfy the necessary conditions for achieving a local optimum. We claim without proof that, in fact, the global optimum is achieved for these values of n and p_k.

We now compute the optimum value of the function $f(\mathbf{p}, n)$. By (2.4), $f(\mathbf{p}, n) = n y_1 \sum_{k=0}^{\infty}(1 - k) p_k$. Since n was chosen to achieve $g_1(\mathbf{p}, n) = 0$, and $y_1 = -\log \nu$, it follows that $f(\mathbf{p}, n) = -N \log \nu$ is the global optimum value.

Finally, note that $f(\mathbf{p}, n)$, as defined above, is the negative logarithm of the large deviations approximation for $P\left(\sum_{i=1}^{n}(1 - V_i) \geq N\right)$. Therefore, by our claim that the maximum over n of the above probability is a good approximation for $P(X_0 \geq N)$, we get

$$P(X_0 \geq N) \approx \nu^N$$

But this is the same as the exact result in (2.2), obtained by solving the flow balance equations.

Our arguments in this section involved approximations which require justification, and claims which were not proved. The purpose was to illustrate the basic idea behind our approach in as simple a manner as possible. In the next section, we shall formulate the problem for two queues in series, and provide a rigorous solution in later sections.

3. The problem formulation. The model we consider consists of two queues in series. Customers arrive into the first queue according to a renewal process, move to the second queue after receiving service, and leave the system after being served at the second queue. There is a single server at each queue, and the service discipline at each queue is first-come-first-served (FCFS). As in the previous section, we let $\tau_i, i = \ldots, -1, 0, 1, \ldots$ denote a sequence of arrival epochs with $\tau_0 = 0$, and define $T_i = \tau_i - \tau_{i-1}$ to be the corresponding sequence of inter-arrival times. Then, the T_i are i.i.d. random variables, and we let T denote a random variable with their common distribution, denoted F. We let F^* denote the Laplace transform of F. The service times of the customers at the two queues are i.i.d. exponential random variables, with mean $1/\mu_1$ at the first queue and $1/\mu_2$ at the second queue. The arrival process and the service processes at the two queues are mutually independent.

Let $X(t) = (X^1(t), X^2(t)), t \in \mathbb{R}$ denote the queue length process. Here $X^1(t)$ and $X^2(t)$ denote the queue lengths in the first and second queues respectively at time t. We are interested in the queue length process seen by arrivals. Denote $X_n = X(\tau_n)$ to be the queue lengths seen by the

n^{th} arrival, $X_n = (X_n^1, X_n^2)$. Then, X_n is a Markov chain because of the memoryless property of the service time distributions. We construct the process $X(t)$ as described below.

We let $V(t) = (V^1(t), V^2(t)), t \in \mathbb{R}$ denote a pair of Poisson processes with rates μ_1, μ_2, defined on the same sample space as the arrival process. These are taken to represent the virtual service processes at the first and second queue respectively. It is assumed that the arrival process, the virtual service process at the first queue, and the virtual service process at the second queue are mutually independent. A virtual service at either queue results in an actual service if the corresponding queue is non-empty, and is wasted otherwise. Such a construction of the service process is possible because of the memoryless nature of the exponential distribution. We also define $V_n = V(\tau_{n+1}) - V(\tau_n)$ to be the number of virtual services at the two queues between the nth and $(n+1)^{\text{th}}$ arrival epochs.

We assume that $\mu_1 ET > 1$ and $\mu_2 ET > 1$, that is, the service rate at each queue is faster than the arrival rate. This is the stability criterion, and is needed to ensure that the queue sizes do not blow up. Under this assumption, it is known (see [12]) that the waiting time process for each arriving customer, and hence the queue length process, can be constructed from the above description; the queue lengths and waiting times are finite a.s., and the queue length and waiting time processes are stationary and ergodic. We make the additional assumption that the paths of the queue length process are continuous from the left. That suffices to completely and uniquely specify the queue length processes (up to sets of measure zero).

We wish to compute the stationary distribution of the queue length process seen by arrivals. We shall confine ourselves to computing the marginals of this distribution. That is, with the system started at $-\infty$, we want to compute $P(X_n^1 = j)$ and $P(X_n^2 = k)$, for all j, k, and an arbitrary n, say $n = 0$.

Consider the pair of equations

$$(3.1) \qquad\qquad x = F^*(\mu_i - \mu_i x), \quad i = 1, 2$$

Recall that equation (2.1) was seen to have ν and 1 as its only solutions, with $\nu < 1$, under the stability assumption $\mu ET > 1$. Since we have assumed that $\mu_i ET > 1$ for $i = 1, 2$, we likewise obtain that each equation in (3.1) above has two solutions, 1 and ν_i, where $\nu_i < 1$ for $i = 1, 2$.

The stationary distribution of an embedded $GI/M/1$ queue is known explicitly (see [4]); thus

$$P(X_0^1 \geq N) = \nu_1^N$$

In this paper, we compute the asymptotics of $P(X_0^2 \geq N)$. Our result is summarized in the theorem below.

Theorem 1 : The stationary distribution of the queue length of the second queue, as seen by arrivals, satisfies

$$\lim_{N \to \infty} \frac{1}{N} \log P(X_0^2 \geq N) = \beta$$

where β is defined as follows :

(1) : If $\mu_1 \geq \mu_2$ and $\dfrac{\mu_1}{\mu_2 \nu_2} \dfrac{d}{dx} F^*(\mu_2 - \mu_2 x)|_{x=\nu_2} > 1$, then $\beta = \log \nu_2$.

(2) : If $\mu_1 \leq \mu_2$ and $\dfrac{\mu_2}{\mu_1 \nu_1} \dfrac{d}{dx} F^*(\mu_1 - \mu_1 x)|_{x=\nu_1} > 1$, then $\beta = \log \dfrac{\mu_1 \nu_1}{\mu_2}$.

(3) : If the conditions in neither (1) nor (2) are satisfied, then $\beta = \log v_2$, where (v_1, v_2) is the unique solution of the pair of equations

(3.2) $$F^* \ (\mu_1 - \mu_1 x_1 + \mu_2 - \mu_2 x_2) = x_1 x_2$$

and

(3.3) $$\frac{d}{dx} \ F^* \ (\mu_1 - \mu_1 x + \mu_2 - \mu_2 x_2) \Big|_{x = x_1} = x_2$$

which satisfies the requirements

(3.4) $v_1 \geq 1$, $v_1 \leq v_1 v_2 < 1$, $\dfrac{d}{dx} F^*(\mu_1 - \mu_1 v_1 + \mu_2 - \mu_2 x) \Big|_{x = v_2} < v_1$

The existence and uniqueness of the solution of equations (3.2) and (3.3) satisfying (3.4) can be proved, but will be omitted from this report.

In the next two sections, we sketch the main steps in the proof of the above theorem. For details, see [10] and the forthcoming paper, [11]. The proof consists of deriving upper and lower bounds on the logarithm of the desired probability as $N \to \infty$.

4. The upper bound. Consider a tandem of two queues with exponential servers and renewal arrivals as described in the last section. We are interested in estimating $P(X_0^2 \geq N)$, where we use the notation $X_n = X(\tau_n)$. The above probability is the same as the stationary probability at the arrival instants. We describe below certain necessary conditions for the event $\{X_0^2 \geq N\}$ and use an estimate of their probability to obtain an upper bound on $P(X_0^2 \geq N)$.

It was shown in [12] that, under the stability assumption, each queue empties infinitely often, with probability one. Thus, on the path leading to $X^2(\tau_0) \geq N$, there is a last time $t \leq \tau_0$ at which the second queue is empty, and $t > -\infty$ almost surely. More precisely, define

$$t = \inf\{s : X^2(u) > 0 \ \forall \ u \in (s, \tau_0]\}$$

It follows that every virtual service at the second queue in $[t, \tau_0)$ results in an actual service. We define

$$n = \sup\{k : t \leq \tau_{-k}\}$$

Now, it is clear from the definition of t and left continuity of the paths of $X(t)$ that $X^2(t) = 0$. Hence, we have the following upper bounds for the queue length of the second queue at time τ_0.

$$X^2(\tau_0) \leq V^1([t, \tau_{-n})) - V^2([t, \tau_{-n})) + \sum_{i=-n}^{-1} (V_i^1 - V_i^2)$$

$$X^2(\tau_0) \leq X^1(\tau_{-n-1}) + 1 - V^2([t, \tau_{-n})) + \sum_{i=-n}^{-1} (V_i^1 - V_i^2)$$

$$X^2(\tau_0) \leq X^1(\tau_{-n-1}) + 1 - V^2([t, \tau_{-n})) + \sum_{i=-n}^{-1} (1 - V_i^2)$$

The first equation uses the fact that the total number of arrivals into the second queue during $[t, \tau_0)$ is bounded above by the number of virtual services at the first queue during this period. The second equation uses the total number in the first queue at time τ_{-n-1}, plus the arrival at τ_{-n-1}, as an upper bound on the number of arrivals into the second queue during $[t, \tau_{-n})$; the corresponding upper bound for the interval $[\tau_{-n}, \tau_0)$ is provided by the number of virtual services at the first queue during this period. In the third equation, the number of arrivals into the second queue in $[t, \tau_0)$ is bounded above by the number at the first queue at time τ_{-n-1} plus the number of arrivals into the first queue in $[\tau_{-n-1}, \tau_0)$. All three equations use the fact that $X^2(t)$ is zero, and that the number of actual services at the second queue during $[t, \tau_0)$ is the same as $V^2([t, \tau_0))$, the number of virtual services.

Hence, for a given number N, a necessary condition for $X_0^2 \geq N$ is that there exist an $n \geq 0$, a time $t \in [\tau_{-n-1}, \tau_{-n})$, and $K, L, M \geq 0$ such that

(4.1) $$V^1([t, \tau_{-n})) = K, \quad V^2([t, \tau_{-n})) = L$$

(4.2) $$X^1(\tau_{-n-1}) + 1 = M$$

and

(4.3) $$K - L + \sum_{i=-n}^{-1} (V_i^1 - V_i^2) \geq N$$

$$M - L + \sum_{i=-n}^{-1} (V_i^1 - V_i^2) \geq N$$

$$M - L + \sum_{i=-n}^{-1} (1 - V_i^2) \geq N$$

Observe that the events in (4.1), (4.2), (4.3) are independent. The reason is that the event in (4.1) depends only on the virtual service processes in the interval $[\tau_{-n-1}, \tau_{-n})$, the event in (4.2) depends only on the virtual service processes before time τ_{-n-1}, and the event in (4.3) depends only on the virtual service processes after time τ_{-n}; their independence then follows from the Poisson nature of the virtual service processes.

An upper bound on $P(X_0^2 \geq N)$ is therefore given by the probability of the union over all $(n, K, L, M) \geq 0$ of the intersection of the events in (4.1),(4.2) and (4.3). By the independence of said events, the probability of the intersection is the product of the individual probabilities. The probability of the union may be bounded above by the sum of the corresponding probabilities. We thus get

(4.4) $P \ (X_0^2 \geq N) \leq$

$$\sum_{n,K,L,M=0}^{\infty} P \left(\exists t \in [\tau_{-n-1}, \tau_{-n}) : V^1([t, \tau_{-n})) = K, V^2([t, \tau_{-n})) = L \right)$$

$$\cdot \ P \ (X^1(\tau_{-n-1} + 1 = M)$$

$$\cdot \ P \left(\begin{array}{ccc} K - L + \sum_{i=-n}^{-1}(V_i^1 - V_i^2) & \geq & N \\ M - L + \sum_{i=-n}^{-1}(V_i^1 - V_i^2) & \geq & N \\ M - L + \sum_{i=-n}^{-1}(1 - V_i^2) & \geq & N \end{array} \right)$$

We now estimate the probability of each term in the product above. Observe that the probability of the first term is bounded above by the probability that there exists a time $t < \tau_{-n}$ such that $V^1([t, \tau_{-n})) = K$ and $V^2([t, \tau_{-n})) = L$. Since V^1, V^2 are Poisson processes, this is the same as the probability of getting K heads and L tails in $K + L$ tosses of a biased coin with $P(\text{head}) = \mu_1/(\mu_1 + \mu_2)$. Thus,

(4.5) $P\left(\exists t \in [\tau_{-n-1}, \tau_{-n}) \ : \ V^1([t, \tau_{-n})) = K, \quad V^2([t, \tau_{-n})) = L \right)$

$$\leq \binom{K + L}{K} \left(\frac{\mu_1}{\mu_1 + \mu_2} \right)^K \left(\frac{\mu_2}{\mu_1 + \mu_2} \right)^L$$

The probability of the second term is given by the known stationary distribution for a single $GI/M/1$ queue, since τ_{-n-1} is just an arbitrary arrival time and the system is in stationarity. Therefore,

(4.6) $P\left(X^1(\tau_{-n-1}) + 1 = M \right) = (1 - \nu_1)\nu_1^{M-1} \, 1\{M \geq 1\}$

The probability of the third term in the product can be bounded above using the following theorem from large deviations.

Definition : The Kullback-Leibler distance $D(\mathbf{q}; \mathbf{p})$ between two probability distributions $\mathbf{q}$ and $\mathbf{p}$ on a countable set I is defined as

$$D(\mathbf{q}; \mathbf{p}) = \sum_{i \in I} q_i \log \frac{q_i}{p_i}$$

where $0 \log 0$ and $0 \log(0/0)$ are defined to be 0. If I has cardinality two, then the distributions are completely specified by giving the probability of one of the points, and we shall make use of the notation

$$D_2(\beta; \alpha) = \beta \log \frac{\beta}{\alpha} + (1 - \beta) \log \frac{1 - \beta}{1 - \alpha}$$

Sanov's theorem : Let $X_1, \ldots, X_n$ be i.i.d. random variables with common distribution $\mathbf{p}$, taking values in a countable set $\mathcal{X}$. Denote the type (or empirical distribution) of $X_1, \ldots, X_n$ to be the probability distribution on $\mathcal{X}$ defined as

$$\mathbf{q}(x) = \frac{1}{n} \sum_{i=1}^{n} 1\{X_i = x\}$$

Let $P_n(\mathcal{A})$ denote the probability that $X_1, \ldots, X_n$ has its type lying in a convex set $\mathcal{A}$ of probability distributions on $\mathcal{X}$. Then

$$P_n(\mathcal{A}) \leq \exp\left(-n \inf_{q \in \mathcal{A}} D(\mathbf{q}; \mathbf{p})\right)$$

Proof : See [5].

Observe that the third term in the product in (4.4) is the probability that the type of the random vector $(V_i, -n \leq i \leq -1)$ lies in a certain convex set. Hence, an upper bound on this probability can be obtained from Sanov's theorem. In order to do so, we first compute the distribution of the random variable $V_i = (V_i^1, V_i^2)$, which takes values in $\mathbb{Z}_+^2$. We have

$$(4.7) \quad a_{jk} \triangleq P(V_0^1 = j, V_0^2 = k) = \int_0^\infty \frac{(\mu_1 t)^j}{j!} \frac{(\mu_2 t)^k}{k!} e^{-(\mu_1 + \mu_2)t} dF(t)$$

Let $\mathcal{A}$ denote a set of probability distributions on $\mathbb{Z}_+^2$ described as follows.

$$(4.8) \qquad \mathcal{A} = \{\mathbf{p} : \quad \sum_{j,k=0}^{\infty} j p_{jk} < \infty, \quad \sum_{j,k=0}^{\infty} k p_{jk} < \infty$$

$$K - L + n \sum_{j,k} (j - k) p_{jk} \geq N$$

$$M - L + n \sum_{j,k} (j - k) p_{jk} \geq N$$

$$M - L + n \sum_{j,k} (1 - k) p_{jk} \geq N\}$$

It is clear that $(V_i, -n \leq i \leq -1)$ satisfies the constraints in (4.4) if and only if its type lies in the set $\mathcal{A}$ defined above (the finiteness conditions on the sums are automatically satisfied by the empirical distributions since these involve only a finite number of the V_i). It is also easy to see that $\mathcal{A}$ is a convex set. Hence, by Sanov's theorem, the probability of the last term in the product in (4.4) is bounded above by $\exp(-n \inf_{\mathbf{p} \in \mathcal{A}} D(\mathbf{p}; \mathbf{a}))$.

We have shown that $P(X_0^2 \geq N)$ is bounded above by an infinite sum over $(n, K, L, M) \in \mathbb{Z}_+^4$, and derived an upper bound on the summand for each value of n, K, L, M. Next, we shall maximize this upper bound, or rather, its logarithm, over $(n, K, L, M) \in \mathbb{R}_+^4$. This is clearly no smaller than the maximum over $\mathbb{Z}_+^4$. Finally, we shall derive an upper bound on the infinite sum in terms of the maximum value mentioned above. The rest of this section leads up to the formulation of the optimization problem. The details regarding its solution are omitted, see [10], [11].

We observe that the probability of the first term in the product, given in (4.5), can be bounded above as follows :

$$(4.9) \quad P\left(\exists t \in [\tau_{-n-1}, \tau_{-n}) \quad : \quad V^1([t, \tau_{-n})) = K, \ V^2([t, \tau_{-n})) = L\right)$$
$$\leq \ \exp[-(K + L)D_2\left(\frac{K}{K+L}; \frac{\mu_1}{\mu_1 + \mu_2}\right)]$$

Indeed, this is the same as the expression in (4.5) if K or L is zero, and if neither is zero, the inequality follows from Stirling's formula (or by using Sanov's theorem).

Now, using the estimates computed above, we can rewrite (4.4) as

$$(4.10) \qquad P(X_0^2 \geq N) \leq \sum_{n, K, L, M = 0}^{\infty} \exp[-\hat{f}(n, K, L, M)]$$

where

$$(4.11) \quad \hat{f}(n, K, L, M) \ = (K + L)D_2\left(\frac{K}{K+L}; \frac{\mu_1}{\mu_1 + \mu_2}\right) + \chi(M)$$
$$- \log(\frac{1 - \nu_1}{\nu_1}) - M \log \nu_1 + n \inf_{\mathbf{p} \in \mathcal{A}} D(\mathbf{p}; \mathbf{a})$$

Here $\chi(M)$ denotes the function which takes value $+\infty$ at $M = 0$ and zero at all other M, and the set $\mathcal{A}$ is defined in (4.8) (note that the definition of $\mathcal{A}$ involves the values of n, K, L, and M, though this dependence has not been made explicit in the notation). In order to find the maximum value of the summand in (4.10), we need to compute the infimum of $\hat{f}(n, K, L, M)$ over all $(n, K, L, M) \in \mathbb{Z}_+^4$. For notational convenience, we shall drop the term $\log(\frac{1 - \nu_1}{\nu_1})$, remembering that it gives rise to a multiplicative constant in the final estimate. We shall also compute the infimum over $(n, K, L, M) \in \mathbb{R}_+^4$. Clearly, this is no larger than the infimum over $\mathbb{Z}_+^4$, and hence provides an upper bound for the maximum value of the summand in (4.10). Likewise, we drop the term $\chi(M)$, noting that this too does not increase the estimate of the infimum of f.

We are thus left with the following problem.

$$(4.12) \quad \inf f(\mathbf{p}, n, K, L, M) \quad \triangleq \quad nD(\mathbf{p}; \mathbf{a}) - M \log \nu_1$$
$$+ \quad (K + L)D_2 \left(\frac{K}{K + L}; \frac{\mu_1}{\mu_1 + \mu_2} \right)$$

subject to the constraints

$$(4.13) \qquad (\mathbf{p}, n, K, L, M) \in \ell^1_+(\mathbb{Z}^2_+) \times (\mathbb{R}^4_+)$$

$$\sum_{j,k=0}^{\infty} p_{jk} = 1, \quad \sum_{j,k=0}^{\infty} j p_{jk} < \infty, \quad \sum_{j,k=0}^{\infty} k p_{jk} < \infty$$

$$g_1(\mathbf{p}, n, K, L, M) = K - L + n \sum_{j,k}(j - k)p_{jk} \geq N$$

$$g_2(\mathbf{p}, n, K, L, M) = M - L + n \sum_{j,k}(j - k)p_{jk} \geq N$$

$$g_3(\mathbf{p}, n, K, L, M) = M - L + n \sum_{j,k}(1 - k)p_{jk} \geq N$$

Here $\ell^1(\mathbb{Z}^2_+)$ denotes the space of absolutely summable functions on $\mathbb{Z}^2_+$ and $\ell^1_+(\mathbb{Z}^2_+)$ denotes the subset consisting of the non-negative valued functions.

The solution of the above optimization problem is rather involved, and is omitted; for details, see [10] and [11]. A key step is to replace $\mathbf{p}$ by $n\mathbf{p}$ which makes the problem a convex constrained optimization problem. This can be analyzed by defining a suitable Lagrangian. Here, we just present the result.

Theorem 2 : The global minimum value of the function f defined in (4.12) above, subject to the constraints in (4.13), is given by $\inf f = -\beta N$, for β as defined in Theorem 1. Note that β is a constant that depends only on the system parameters F, μ_1, μ_2, and not on N.

Note that, by definition of f, the maximum value of the summand in (4.4) is bounded above by $(\nu_1/1 - \nu_1) \exp[-\inf f]$, where $\inf f$ refers to the constrained infimum of f, and is given by Theorem 2. Next, we need to obtain an upper bound on the sum in (4.4). The argument for doing so involves the details of the solution of the optimization problem, and is therefore omitted. Suffice it to say that we can obtain the upper bound

$$(4.14) \qquad P(X_0^2 \geq N) \leq cN^4 \exp[\beta N]$$

where $\inf f = -\beta N$, and the quantity β was defined in Theorem 1. Finally, we need to obtain a lower bound on $P(X_0^2 \geq N)$, which we sketch in the next section.

5. The lower bound. The solution of the optimization problem that was sketched in the last section yields values, n^0, $\mathbf{p}^0$ and M^0 at which the optimum occurs. It turns out that $K^0 = L^0 = 0$. These values have the following interpretation. The probability distribution $\mathbf{p}^0$ is the twisted distribution which describes the most likely evolution of the system to large queue size. That is, in order for the queue size to build up, it is necessary that the virtual service processes during each inter-arrival period have the distribution $\mathbf{p}^0$, rather than their original distribution $\mathbf{a}$, that this behavior be sustained over n^0 inter-arrival periods, and that we start with M^0 customers in the first queue. It is observed from the solution to the optimization problem that $\mathbf{p}^0$ does not depend on N, whereas n^0 and M^0 are linear in N. We shall make use of this intuition in proving the lower bound.

We shall derive sufficient conditions for the event $\{X_0^2 \geq N\}$, and estimate their probability. Define

$$(5.1) \qquad m_1 = \sum_{j,k=0}^{\infty} j p_{jk}^0 \qquad m_2 = \sum_{j,k=0}^{\infty} k p_{jk}^0$$

We shall make use of the fact that $m_1 \geq 1$, and $m_2 \leq 1$, which is a consequence of the solution to the optimization problem.

Let $\epsilon > 0$ be given. Consider the system at time $\tau_{-(1+\epsilon)n^0}$ having at least $(1+\epsilon)M^0$ customers in the first queue, i.e., $X_{-(1+\epsilon)n^0}^2 \geq (1+\epsilon)M^0$. Suppose that the virtual service process during $[\tau_{-(1+\epsilon)n^0}, \tau_0)$ evolves in a tube around a straight line with gradient (m_1, m_2). More precisely, let $\delta > 0$ be a given constant, and suppose that the evolution of the system during the period $[\tau_{-(1+\epsilon)n^0}, \tau_0)$ satisfies the following constraints.

$$(5.2) \frac{m_1 k}{(1+\epsilon)n^0} - \delta < \frac{1}{(1+\epsilon)n^0} \sum_{i=-(1+\epsilon)n^0}^{-(1+\epsilon)n^0+k-1} V_i^1 < \frac{m_1 k}{(1+\epsilon)n^0} + \delta$$

$$\forall \qquad k \in \{1, \ldots, (1+\epsilon)n^0\}$$

$$(5.3) \frac{m_2 k}{(1+\epsilon)n^0} - \delta < \frac{1}{(1+\epsilon)n^0} \sum_{i=-(1+\epsilon)n^0}^{-(1+\epsilon)n^0+k-1} V_i^2 < \frac{m_2 k}{(1+\epsilon)n^0} + \delta$$

$$\forall \qquad k \in \{1, \ldots, (1+\epsilon)n^0\}$$

We now show that, if these constraints are satisfied for $\delta = c\epsilon$ with c sufficiently small, then $X_0^2 \geq N$.

Let $S_k = (S_k^1, S_k^2)$ denote the actual number of services at the first and second queue respectively during the interval $[\tau_k, \tau_{k+1})$. Clearly, then

$$(5.4) \qquad X_0^2 > \sum_{i=1}^{(1+\epsilon)n^0} (S_{-i}^1 - S_{-i}^2)$$

We shall obtain a lower bound on the sum of the S^1_{-i} using (5.2), an upper bound on the sum of the S^2_{-i} using (5.3), and thereby get a lower bound on X^2_0.

The number of actual services at the second queue during $[\tau_{-(1+\epsilon)n^0}, \tau_0)$ is bounded above by the number of virtual services during this same period. Therefore, if (5.3) holds, we have

$$(5.5) \qquad \sum_{i=1}^{(1+\epsilon)n^0} S^2_{-i} \leq (1+\epsilon)(m_2 + \delta)n^0$$

Suppose the first queue is never empty during the period $(\tau_{-(1+\epsilon)n^0}, \tau_0]$. Then, it is clear that every virtual service results in an actual service. Hence, it follows from (5.2) that

$$(5.6) \qquad \sum_{i=1}^{(1+\epsilon)n^0} S^1_{-i} \geq (1+\epsilon)(m_1 - \delta)n^0$$

Next, suppose the first queue does empty during $(\tau_{-(1+\epsilon)n^0}, \tau_0]$. Let $\tau_{-(1+\epsilon)n^0+\kappa}$ denote the last time that the first queue is empty during this period. Note that $1 \leq \kappa \leq (1+\epsilon)n^0$. The first queue is never empty during $(\tau_{-(1+\epsilon)n^0+\kappa}, \tau_0]$, and so the number of services at the first queue during this period is the same as the number of virtual services. The latter, by (5.2), is at least $(1+\epsilon)(m_1 - 2\delta)n^0 - m_1\kappa$. The number of services at the first queue during $[\tau_{-(1+\epsilon)n^0}, \tau_{-(1+\epsilon)n^0+\kappa})$ is equal to the number of external arrivals during this time plus the number originally in the queue at time $\tau_{-(1+\epsilon)n^0}$, since the first queue is empty at the end of this period. But this is equal to $\kappa + (1+\epsilon)M^0$. Thus, we get

$$\sum_{i=1}^{(1+\epsilon)n^0} S^1_{-i} \geq (1+\epsilon)M^0 + (1+\epsilon)(m_1 - 2\delta)n^0 + (1 - m_1)\kappa$$

It is known from the solution to the optimization problem that $m_1 \geq 1$. We also noted above that $\kappa < (1+\epsilon)n^0$. Combining these observations with the above expression, we get

$$(5.7) \qquad \sum_{i=1}^{(1+\epsilon)n^0} S^1_{-i} \geq (1+\epsilon)M^0 + (1+\epsilon)(1 - 2\delta)n^0$$

Observe from (5.4)-(5.6) that

$$(5.8) \qquad X^2_0 \geq (1+\epsilon)(m_1 - m_2 - 2\delta)n^0$$

or

$$(5.9) \qquad X^2_0 \geq (1+\epsilon)[M^0 + (1 - m_2 - 3\delta)n^0]$$

Observe that we must have $g_i(\mathbf{p}^0, n^0, K^0, L^0, M^0) \leq 0$ for $i = 1, 2, 3$, where the g_i were defined in (4.13). As noted above, $K^0 = L^0 = 0$ by the solution to the optimization problem. Hence, by the definition of m_1 and m_2 in (5.1), we obtain the conditions below corresponding to $g_1(\mathbf{p}^0, n^0, K^0, L^0, M^0) \leq 0$ and $g_3(\mathbf{p}^0, n^0, K^0, L^0, M^0) \leq 0$ respectively.

$$(m_1 - m_2)n^0 \geq N$$
$$M^0 + (1 - m_2)n^0 \geq N$$

Therefore, it is clear from (5.8) and (5.9) that either

$$(5.10) \qquad X_0^2 \geq (1 + \epsilon)(N - 2\delta cN)$$

or

$$(5.11) \qquad X_0^2 \geq (1 + \epsilon)(N - 3\delta cN)$$

where $c = n^0/N$ is a constant, by the solution to the optimization problem. It is now clear that, if $\delta \leq \epsilon/[3c(1+\epsilon)]$, then $X_0^2 \geq N$. This is exactly what we set out to prove.

Let $P(\epsilon, \delta, N)$ denote the probability that the virtual service process satisfies the conditions in (5.2) and (5.3). We write N rather than n^0 in the notation since the relation between N and n^0 is explicitly given by the solution to the optimization problem. We have $n^0 = cN$, for a constant c that depends only on the system parameters, namely F, μ_1 and μ_2. A lower bound on the probability that the virtual service process satisfies these constraints is given by well-known results in large deviations theory. Given any $\epsilon > 0$, and a constant $c > 0$, we have

$$(5.12) \qquad \liminf_{N \to \infty} \frac{1}{N} \log P(\epsilon, c\epsilon, N) \geq -\frac{(1+\epsilon)n^0}{N} D(\mathbf{p}^0; \mathbf{a})$$

Here $\mathbf{p}^0$ and n^0 are given by the solution to the optimization problem, and it is known that n^0/N is a constant that does not depend on N. Likewise $\mathbf{p}^0$ does not depend on N. Finally, $\mathbf{a}$, defined in (4.7), is the original distribution of the virtual service process. A proof of the above inequality can be obtained by a change of measure argument as in Theorem 2.2.3 of [8], for details see [10], [11]. For a proof of the above inequality under the additional assumption that F^* is finite in a neighborhood of 0, see Lemma 5.1.6 of [8], or [13].

We have shown above that, if $X^1_{-(1+\epsilon)n^0} \geq (1 + \epsilon)M^0$, and (5.2) and (5.3) hold for δ sufficiently small, then $X_0^2 \geq N$. By the known stationary distribution for the first queue,

$$(5.13) \qquad P\big(X^1_{-(1+\epsilon)n^0} \geq (1 + \epsilon)M^0\big) = \nu_1^{(1+\epsilon)M^0}$$

Furthermore, this last event is independent of the event that (5.2) and (5.3) hold, since (5.2) and (5.3) involve only the virtual service process after time

$\tau_{-(1+\epsilon)n^0}$. Hence, by (5.12) and (5.13), we get

$$(5.14) \qquad \liminf_{N \to \infty} P(X_0^2 \geq N) \geq \frac{1+\epsilon}{N} \left[M^0 \log \nu_1 - n^0 D(\mathbf{p}^0; \mathbf{a}) \right]$$

Now, observe from the definition of f in (4.12), and the fact that $K^0 = L^0 = 0$, that the term in brackets above is precisely $-\inf f$, the optimum value in the constrained optimization problem. Also recall from Theorem 2 that $\inf f = -\beta N$, for β as defined in Theorem 1. Finally, since $\epsilon > 0$ is arbitrary, we let $\epsilon \to 0$ in (5.14) to obtain

$$(5.15) \qquad \liminf_{N \to \infty} \frac{1}{N} \log P(X_0^2 \geq N) \geq \beta$$

Combining the lower bound in (5.15) with the upper bound obtained in the previous section in (4.14), and with the value of β given by Theorem 2, the proof of Theorem 1 is complete.

6. Conclusion. We considered the problem of estimating the tail of the stationary queue length distribution in a tandem of exponential server queues with renewal arrivals. We showed that the marginal distribution at each queue decays exponentially, and obtained the exact exponential rate of decay. That is, we showed that, if X_n^i, $i = 1, 2$, denotes the queue length at the ith queue seen by the nth arrival, then, in stationarity

$$\lim_{N \to \infty} \frac{1}{N} \log P(X_n^2 \geq N) = \beta$$

where β is defined in Theorem 1. By classical results for the $GI/M/1$ queue, we also have

$$\lim_{N \to \infty} \frac{1}{N} \log P(X_n^1 \geq N) = \log \nu_1$$

where ν_1 is defined below (3.1).

A result of this sort impacts several related problems. For instance, in the tandem above, consider how to optimally allocate a total of N buffers so as to minimize some cost function associated with the time to buffer overflow. We suggest the following rule of thumb : allocate $p_1 N$ buffers to the first queue and $p_2 N$ to the second, where $p_1 + p_2 = 1$, such that $p_1 \log \nu_1 = p_2 \beta$. It can be shown that, for any reasonable cost function associated with the time to buffer overflow, the above allocation is approximately optimal in the sense that, if $N_1(N), N_2(N)$ is the exact optimal allocation for that cost function, then

$$\lim_{N \to \infty} \frac{N_1(N)}{N} = p_1 \qquad \lim_{N \to \infty} \frac{N_2(N)}{N} = p_2$$

The technique for proving this is parallel to that in [1], [2].

The problem of extending our result to an arbitrary number of queues in tandem remains open. It would also be of interest to extend the result to more general service distributions, and to consider arrival and service processes that are autocorrelated.

REFERENCES

[1] V. ANANTHARAM, "The Optimal Buffer Allocation Problem", *IEEE Transactions on Information Theory*, Vol. 35, No. 4, pp. 721-725, 1989.

[2] V. ANANTHARAM AND A. GANESH, "Correctness within a constant of an Optimal Buffer Allocation Rule of Thumb", *IEEE Transactions on Information Theory*, Vol. 40, No. 3, pp. 871-882, 1994.

[3] V. ANANTHARAM, P. HEIDELBERGER AND P. TSOUCAS, "Analysis of Rare Events in Continuous Time Markov Chains via Time Reversal and Fluid Approximations", *IBM Research Report RC 16280*, 1990.

[4] S. ASMUSSEN, *Applied Probability and Queues*, John Wiley and Sons, 1987.

[5] R. BLAHUT, *Principles and Practice of Information Theory*, Addison Wesley, 1987.

[6] C.S. CHANG, "Sample Path Large Deviations and Intree Networks", *IBM RC 19118*, 1993.

[7] G. DE VECIANA AND J. WALRAND, "Effective Bandwidths : Call admission, Traffic policing and Filtering for ATM networks", *Preprint*, 1993.

[8] A. DEMBO AND O. ZEITOUNI, *Large Deviations Techniques and Applications*, Jones and Bartlett, 1993.

[9] M.R. FRATER, T.M. LENNON AND B.D.O. ANDERSON, "Optimally efficient estimation of the statistics of rare events in queueing networks", *IEEE Transactions on Automatic Control*, Vol. 36, pp. 1395-1405, 1991.

[10] A. GANESH, *PhD thesis*, Cornell University, 1994.

[11] A. GANESH AND V. ANANTHARAM, "Tail Stationary Probabilities in Exponential Server Tandems with Renewal Arrivals", *Preprint*.

[12] R.M. LOYNES, "The stability of queues with non-independent inter-arrival and service times", *Proceedings of the Cambridge Philosophical Society* , Vol. 58, pp. 497-520, 1962.

[13] A. MOGULSKII, "Large deviations for trajectories of multi-dimensional random walks", *Theory of Probability and its Applications*, Vol. 21, pp. 300-315, 1976.

[14] S. PAREKH AND J. WALRAND, "A Quick Simulation Method for Excessive Backlogs in Networks of Queues", *IEEE Transactions on Automatic Control* , Vol. 34, No. 1, pp. 54-66, 1989.

[15] P. TSOUCAS, "Rare Events in Series of Queues", *Journal of Applied Probability*, Vol. 29, pp. 168-175, 1992.

LARGE DEVIATIONS FOR THE INFINITE SERVER QUEUE IN HEAVY TRAFFIC

PETER W. GLYNN*

Abstract. In this paper, we establish large deviations approximations to tail probabilities of the queue-length r.v. in an infinite-server queue in heavy traffic. These large deviations approximations complement the existing Gaussian approximations developed for such systems using weak convergence theory on function spaces. We also describe a simulation-based algorithm for numerically computing such tail probabilities that takes advantage of the large deviations theory developed here.

Key words. infinite server queue, large deviations, simulation

1. Introduction. This paper is concerned with developing large deviations approximations for tail probabilities of infinite server queues in heavy traffic. By heavy traffic, we refer to systems in which the offered load is high, so that the system typically contains many customers.

The theory developed here is basically a many-server analog to the Cramer-Lundberg type approximations that are commonly used to study the single-server queue. This paper has two parts. In Section 2, we develop a large deviations approximation for the tail probabilities associated with the number of customers in the system at time t when the system is in heavy traffic. We also provide a large deviations argument to offer asymptotic justification for the approximation. Section 3 is concerned with simulation-based algorithms for numerically computing the tail probabilities that take advantage of the large deviations ideas presented in Section 2.

An expanded version of this paper will appear elsewhere.

2. Large deviations for $Q(t)$. We start by giving a precise description of the $GI/G/\infty$ queue. Suppose that $(A_k : k \geq 0)$ is a non-decreasing sequence in which A_k corresponds to the arrival epoch of the k'th customer. If $Q(0) = 0$ and V_j is the "time-in-system" of the j'th customer, then the number of customers $Q(t)$ in the system at time t is given by

$$Q(t) = \sum_{k=1}^{\infty} I(A_k \leq t < A_k + V_k).$$

Let $N(t) = \max\{n \geq 0 : A_n \leq t\}$ be the number of customers to arrive to the system in $[0, t]$. Then, $Q(t)$ can be re-expressed in terms of $N(\cdot)$ as

$$Q(t) = \sum_{k=1}^{N(t)} I(A_k + V_k > t).$$

---

* Dept. of Operations Research, Stanford University, Stanford, CA 94305-4022. This research was supported by the National Science Foundation under grant DDM-9101580 and the Army Research Office under Contract No. DAAL03-91-G-0319.

We will be using the Gärtner-Ellis theorem to study the large deviations behavior of the r.v. $Q(t)$. In order to apply this result, the moment generating function of $Q(t)$ must be computed. This calculation is particularly simple when the sequence $V = (V_n : n \geq 1)$ is i.i.d. and independent of $N = (N(t) : t \geq 0)$; we therefore impose this condition throughout the paper. Let $F(x) = P(V \leq x)$ and $\bar{F}(x) = 1 - F(x)$.

PROPOSITION 2.1. *Suppose that* $\varphi(\theta, t) = E \exp(\theta Q(t))$. *Then,*

$$(2.1) \quad \varphi(\theta, t) = E \exp\left(\int_{[0,t]} \log(e^\theta \, \bar{F}(t - x) + F(t - x)) \; N(dx) \right).$$

Proof. Note that

$$
\begin{aligned}
\varphi(\theta, t) \;=\;& E\{ E[\exp(\theta Q(t)) \mid N] \} \\[4pt]
=\;& E\left\{ E\left[\exp\left(\theta \sum_{k=1}^{N(t)} I(A_k + V_k > t) \right) \,\middle|\, N \right] \right\} \\[4pt]
=\;& E \prod_{k=1}^{N(t)} E[\exp(\theta \, I(V_k > t - A_k)) \mid N] \\[4pt]
=\;& E \prod_{k=1}^{N(t)} (e^\theta \bar{F}(t - A_k) + F(t - A_k)) \\[4pt]
=\;& E \exp\left(\sum_{k=1}^{N(t)} \log(e^\theta \bar{F}(t - A_k) + F(t - A_k)) \right) \\[4pt]
=\;& E \exp\left(\int_{[0,t]} \log(e^\theta \bar{F}(t - x) + F(t - x)) \, N(dx) \right).
\end{aligned}
$$

$\square$

Large deviations theory is intended here to provide refined approximations, relative to the central limit theory already developed, for the tail probabilities of the r.v. $Q(t)$. By "central limit theory", we refer here to the idea that if the arrival rate is high, then various limit theorems make rigorous the approximation

$$(2.2) \qquad\qquad Q(t) \overset{\mathcal{D}}{\approx} E\,Q(t) + \sqrt{\mathrm{var}Q(t)} \; N(0,1),$$

where $\overset{\mathcal{D}}{\approx}$ denotes approximate equality in distribution, and $N(0,1)$ is a standard normal r.v.; see Iglehart (1965), Borovkov (1967), (1984), Newell (1973), Whitt (1982), and Glynn and Whitt (1991) for details. In particular, the approximation to $P(Q(t) > x)$ suggested by (2.2) is typically good whenever the arrival rate is high, and $|x - EQ(t)|/\sqrt{\mathrm{var}Q(t)}$ is of moderate magnitude. (Note that there is no requirement here that t be large, in order that (2.2) yield a good approximation.) A reasonable heuristic for judging when the arrival rate is high is to look for situations in which the standard deviation $\sqrt{\operatorname{var}Q(t)}$ is small relative to $EQ(t)$.

Large deviations is intended to provide improved approximations to $P(Q(t) > x)$ when the arrival rate is high, and $|x - EQ(t)|/\sqrt{\operatorname{var}Q(t)}$ is large. Let $\psi(\theta, t) = \log \varphi(\theta, t)$ and suppose that θ^* is the root (assumed to exist uniquely) of $\psi'(\theta^*, t) = x$. Then, large deviations suggests the rough approximation

$$(2.3) \qquad P(Q(t) > x) \approx \exp(-\theta^* x + \psi(\theta^*, t)).$$

Our goal is to now describe several asymptotic regimes in which (2.3) is valid, and describe the precise sense in which the approximation holds.

To accomplish this, we consider a sequence of systems in which the arrival rate is sent to infinity. Let $N_n = (N_n(t) : t \geq 0)$ be the arrival process to the n'th system, so that $N_n(t)$ represents the number of customers to arrive to the n'th system in $[0, t]$. We leave the service time sequence $V = (V_n : n \geq 1)$ unchanged as a function of n, and let $Q_n(t)$ be the corresponding number of customers in the n'th system at time t, so that

$$Q_n(t) = \sum_{k=1}^{N_n(t)} I(A_{k,n} + V_k > t),$$

where $A_{k,n} = \min\{l \geq 0 : N_n(t) \geq k\}$.

There are two fundamentally different ways of modeling the notion that the arrival rate to the system is high. Most of the "heavy traffic" literature assumes the presence of a single "fast" source. The easiest mathematical means of studying this situation is to fix a given arrival process N, and to speed up the arrival rate via the sequence of processes

$$N_n(t) = N(nt).$$

Here, $A_{k,n} = A_k/n$ and the arrival rate associated with $N_n(\cdot)$ is roughly n times that of $N(\cdot)$. We will make the following assumption about the joint cumulant generating function of the increments of N.

A1. *There exists a finite-valued function ψ_N such that for $0 = t_0 < t_1 < \ldots < t_m = t$ and $(\theta_1, \ldots, \theta_m) \in \Re^m$,*

$$\frac{1}{n} \log E \exp\left(\sum_{i=1}^{m} \theta_i [N(nt_i) - N(nt_{i-1})] \right) \to \sum_{i=1}^{m} \psi_N(\theta_i)(t_i - t_{i-1})$$

as $n \to \infty$.

This assumption is satisfied by many different arrival processes; see Dembo and Zajic (1993). To offer some insight into this assumption, we describe the function ψ_N in a couple of important applications settings:

Example 2.2. Suppose that the arrival process is renewal, so that A_k can be represented as $A_k = U_1 + \ldots + U_k$, where $(U_k : k \geq 1)$ is i.i.d. Then, under suitable regularity conditions on the U_k's (see Glynn and Whitt (1994)),

$$\psi_N(\theta) = -\kappa^{-1}(-\theta),$$

where $\kappa(\theta) = \log(E \exp(\theta U_1))$.

Example 2.3. Suppose now that the arrival process is a Markov-modulated Poisson process, so that there exists an S-valued continuous-time Markov chain $X = (X(t) : t \geq 0)$ and a function $f : S \rightarrow (0, \infty)$ such that the arrival rate of the Poisson process at time t is $f(X(t))$. Assume that S is finite and X is irreducible. Then, if B is the generator of X, set

$$B(\theta) = B + D(\theta)$$

where $D(\theta) = \text{diag}((e^\theta - 1)f(x) : x \in S)$. Here, $\psi_N(\theta)$ is the eigenvalue of $B(\theta)$ having maximum real part (which turns out necessarily to be real).

In any case, to study the large deviations behavior of $Q_n(t)$ via the Gärtner-Ellis theorem, we must obtain the limit behavior of the cumulant generating function of $Q_n(t)$ defined by $\psi_n(\theta, t) = \log E \exp(\theta Q_n(t))$.

THEOREM 2.4. *If A1 holds, then*

(2.4) $$\frac{1}{n}\psi_n(\theta, t) \rightarrow \int_0^t \psi_N(\log(e^\theta \bar{F}(x) + F(x)))\, dx$$

as $n \rightarrow \infty$.

Proof. Suppose $\theta \geq 0$. Then, $h(x) = \log(e^\theta \bar{F}(t - x) + F(t - x))$ is non-decreasing in x. We therefore obtain the bounds

$$\frac{1}{n}\log E \exp(\sum_{k=0}^{m-1} h(\frac{kt}{m})\,[\,N(\frac{n(k+1)t}{m}) - N(\frac{nkt}{m})\,])$$

$$\leq \frac{1}{n}\log E \exp(\int_{[0,t]} \log(e^\theta \bar{F}(t - x) + F(t - x))\, N_n(dx))$$

$$\leq \frac{1}{n}\log E \exp(\sum_{k=0}^{m-1} h(\frac{(k+1)t}{m})\,[\,N(\frac{n(k+1)t}{m}) - N(\frac{nkt}{m})\,]).$$

Letting $n \rightarrow \infty$ and using Proposition 2.1 as well as A1, we get

$$\frac{t}{m}\sum_{k=0}^{m-1} \psi_N(h(\frac{kt}{m})) \leq \liminf_{n\to\infty} \frac{1}{n}\psi_n(\theta, t)$$

(2.5) $$\leq \limsup_{n\to\infty} \frac{1}{n}\psi_n(\theta, t) \leq \frac{t}{m}\sum_{k=0}^{m-1} \psi_N(h(\frac{(k+1)t}{m})).$$

Now, note that by setting $n = 1$ in A1, it is clear that ψ_N, being the limit of convex functions, must necessarily be convex and hence continuous. It follows that the extreme members of the inequality (2.5) are Riemann approximations to the integral of the continuous function $\psi_N \circ h$. By letting $m \to \infty$, we therefore conclude that

$$\frac{1}{n}\psi_n(\theta, t) \to \int_0^t \psi_N(h(x))dx,$$

as desired. For $\theta < 0$, a similar argument works.   □

Let

$$\psi_{Q(t)}(\theta) = \int_0^t \psi_N(\log(e^\theta \bar{F}(x) + F(x)))\, dx.$$

Suppose that $\psi_{Q(t)}(\cdot)$ is differentiable and that there exists a unique θ^* such that $\psi'_{Q(t)}(\theta^*) = x$. Then, the Gärtner-Ellis theorem implies that

$$\frac{1}{n}\log P(Q_n(t) > xn) \to -\theta^* x + \psi_{Q(t)}(\theta^*)$$

as $n \to \infty$; see, for example, p.14–15 of Bucklew (1990). On the other hand, if $x_n = xn$, (2.4) establishes that

$$\frac{1}{n}\log(\exp(-\theta^* x_n + \psi_n(\theta^*, t))) \to -\theta^* x + \psi_{Q(t)}(\theta^*).$$

Furthermore, since the ψ_n's are convex, we may differentiate through (2.4) (see Rockafellar (1970)), yielding

$$\frac{1}{n}\psi'_n(\theta, t) \to \psi'_{Q(t)}(\theta)$$

as $n \to \infty$. Since $\psi'_n(\cdot, t)$ is non-decreasing, we can conclude that $\theta^*_n \to \theta^*$ as $n \to \infty$, where θ^*_n is the root of $\psi'_n(\theta^*_n, t) = x_n$. It follows that

$$(2.6) \qquad \frac{1}{x_n}\log[\, P(Q_n(t) > x_n)/\exp(-\theta^*_n x_n + \psi_n(\theta^*_n, t))\,] \to 0$$

as $n \to \infty$. We view (2.6) as a rigorous statement of the mathematical sense in which the right-hand side of (2.3) is a valid approximation to $P(Q(t) > x)$ (when $x/EQ(t)$ is of moderate magnitude, and not close to unity).

A second setting in which (2.3) is valid is when the high arrival rate is due to the presence of many sources, each of moderate size. Mathematically, this situation can be captured by fixing a given arrival process N, and letting $N(1, \cdot), N(2, \cdot), \ldots$ be i.i.d. copies of N. Here, we set

$$N_n(t) = \sum_{i=1}^n N(i, t)$$

for $t \geq 0$. Since

$$Q_n(t) \overset{\mathcal{D}}{=} \sum_{i=1}^{n} \int_{[0,t]} I(V_{ij} > t - x) \, N(i, dx)$$

($\overset{\mathcal{D}}{=}$ denotes "equality in distribution"), where $(V_{ij} : i \geq 1, j \geq 1)$ is a family of i.i.d. copies of V_1, it is evident that $Q_n(t)$ is a sum of n i.i.d. r.v.'s. To obtain a large deviations result in this setting is purely a matter of applying the existing theory for i.i.d. r.v.'s. This large deviations result implies that the approximation (2.3) can be reasonable when many sources are present (so that $EQ(t)/\sqrt{\mathrm{var}Q(t)}$ is large) and the tail probability $P(Q(t) > x)$ is such that $x/EQ(t)$ is of moderate magnitude and not close to unity; the argument is similar to that leading to (2.6).

As one might expect, these results can easily be extended to large deviations approximations for tail probabilities of the steady-state r.v. $Q(\infty)$. To apply (2.3), note that

$$\varphi(\theta, +\infty) = E \exp\left(\int_{[0,\infty]} \log(e^\theta \bar{F}(x) + F(x)) \, \tilde{N}(dx) \right),$$

where $\tilde{N} = (\tilde{N}(t) : t \geq 0)$ is a time-stationary version of the time-reversal of the arrival process N; details will be supplied elsewhere.

3. Simulation implications. A glance at (2.6) suggests that the right-hand side of (2.3) is typically, at best, only a crude approximation to $P(Q(t) > x)$. In particular, we note that the approximation can be multiplied by any constant factor, without any impact on the logarithmic asymptotic (2.6). Hence, the right-hand side of (2.3) should perhaps only be viewed as an order-of-magnitude guess as to the size of $P(Q(t) > x)$.

Given this state of affairs, it seems reasonable to consider efficient numerical algorithms for computing the tail probability $P(Q(t) > x)$. We shall describe here some "importance sampling" ideas, based on the large deviations theory already developed, for efficiently simulating this tail probability.

The theory of rare event simulation suggests that the tail probability $P(Q(t) > x)$ can be efficiently simulated under the change-of-measure

(3.1) $$P^*(d\omega) = \exp(\theta^* x - \psi(\theta^*, t)) \, P(d\omega),$$

where θ^* is as defined in (2.3). Furthermore, the level of variance reduction incurred by using (3.1) increases under the same asymptotic regimes as those validating the approximation (2.3). Of course, the computational efficiency of such a simulation algorithm depends not only on the variance but also on the computational ease associated with producing simulation runs under the modified probability measure P^*. Generating variates under P^* is, in this setting, quite challenging as it typically fails to induce any Markovian structure on the dynamical behavior of the simulated process.

However, there exists a modification to P^*, call it $\tilde{P}$, which enjoys the same asymptotic variance reducing properties as does P^*, and which is easily simulatable. The probability measure $\tilde{P}$ is easiest to describe in the setting in which the process N is a non-delayed renewal process, and we shall therefore specialize to this important subclass of problems. Let $U_n = A_n - A_{n-1}$, $\kappa(\theta) = \log E \exp(\theta U_i)$ (assumed to converge for all θ), and put $\psi(\theta) = -\kappa^{-1}(-\theta)$. The assumption that $\kappa(\theta)$ is finite for all θ is a technical restriction that we impose merely to easily describe the algorithm and result, and should in no way be viewed as necessary.

The following simulation algorithm implicitly describes $\tilde{P}$. The quantities Q and L defined in the algorithm correspond to the r.v.'s $Q(t)$ and $dP/d\tilde{P}$, as generated under $\tilde{P}$.

Algorithm 3.1.

1. $A \leftarrow 0$, $Q \leftarrow 0$, $L \leftarrow 1$.

2. Generate U from the distribution
$\exp(-\psi(\log(e^{\theta^*}\bar{F}(t-A)+F(t-A)))x+\log(e^{\theta^*}\bar{F}(t-A)+F(t-A)))$
$P(U_i \in dx)$.

3. $L \leftarrow L \exp(\psi(\log(e^{\theta^*}\bar{F}(t-A)+F(t-A)))U)(e^{\theta^*}\bar{F}(t-A)+F(t-A))^{-1}$.

4. $A \leftarrow A+U$.

5. If $A > t$, return Q, L.

6. Generate the $0-1$ r.v. I from Bernoulli $(e^{\theta^*}\bar{F}(t-A)(e^{\theta^*}\bar{F}(t-A)+F(t-A))^{-1})$.

7. $L \leftarrow L e^{-\theta^* I}(e^{\theta^*}\bar{F}(t-A)+F(t-A))$.

8. $Q \leftarrow Q+I$.

9. Go to 2.

This algorithm is intended to work efficiently when the system is in "heavy traffic", as a result of a single "fast" renewal arrival source. Full asymptotic justification for this algorithm will appear elsewhere; the algorithm turns out to be asymptotically "optimal" in a certain sense.

We note that when "heavy traffic" is achieved as a result of the presence of many i.i.d. arrival sources, rare event simulation algorithms for tail probabilities of sums of i.i.d. r.v.'s comes into play; for details, see, for example, Bucklew (1990). This theory suggests that each of the sources should be simulated independently. The change-of-measure for the individual sources takes the same form as that specified above in the "fast source" setting.

REFERENCES

[1] Borovkov, A. A. (1967) On limit laws for service processes in multi-channel systems. *Siberian Math. J.* **8**, 746–763.

[2] Borovkov, A. A. (1984) *Asymptotic Methods in Queueing Theory.* Wiley, New York.

[3] Bucklew, J. A. (1990) *Large Deviation Techniques in Decision, Simulation, and Estimation.* J. Wiley and Sons, New York.

[4] Dembo, A. and Zajic, T. (1993) Large deviations: from empirical mean and measure to partial sums process. Preprint.

[5] Glynn, P. W. and Whitt, W. (1991) A new view of the heavy-traffic limit theorem for infinite-server queues. *Adv. Appl. Prob.* **23**, 188–209.

[6] Glynn, P. W. and Whitt, W. (1994) Large deviation behavior of counting processes and their inverses. *Queueing Systems* **17**, 107–128.

[7] Iglehart, D. L. (1965) Limit diffusion approximations for the many server queue and the repairman problem. *J. Appl. Prob.* **2**, 429–441.

[8] Newell, G. F. (1973) *Approximate Stochastic Behavior of n-Server Service Systems with Large n.* Lecture Notes in Economics and Mathematical Systems **87**, Springer-Verlag, Berlin.

[9] Rockafellar, R.T. (1970) *Convex Analysis.* Princeton University Press, Princeton, N.J.

[10] Whitt, W. (1982) On the heavy-traffic limit theorem for $GI/G/\infty$ queues. *Adv. Appl. Prob.* **14**, 171–190.

TRAFFIC MODELING FOR HIGH-SPEED NETWORKS: THEORY VERSUS PRACTICE

WALTER WILLINGER*

Abstract. Statistical analyses of large sets of traffic measurements from working packet networks show that, from a statistical view point, traditional modeling assumptions such as Poisson packet arrivals, exponential service and Markovian structures have little to do with reality. Instead, the analyses provide convincing evidence for the presence of features in actual network traffic such as long-range dependence ("Joseph Effect"), the infinite variance syndrome ("Noah Effect") and self-similarity ("fractals"). We present some results of our traffic data analysis and discuss their implications for traffic modeling of high-speed communications systems. In particular, we point out directions in which traditional traffic models should be extended (i) to describe basic characteristics observed in measured traffic more accurately and (ii) to be more relevant for practical applications in modern telecommunications.

Key words. Infinite variance, long-range dependence, self-similarity, fractal traffic.

1. Introduction. Traffic is the driving force of communications systems, and traffic models are of crucial importance for assessing their performance. In practice, stochastic models of traffic streams are relevant to network traffic engineering and performance analysis, to the extent that they are able to predict system performance measures to a reasonable degree of accuracy. The fundamental systems, of which traffic is a major ingredient, are queueing systems. Traditional traffic models have often been devised and selected for the analytical tractability they induce in the corresponding queueing system. However, a practitioner's confidence in a given traffic model is greatly diminished if the model is only able to crudely approximate basic statistics but cannot capture visually dominant features of empirical traffic collected from a variety of working communications systems. While originally the validity and efficacy of models for modern high-speed network traffic was difficult to assess due to the unavailability of empirical data, recently very large sets of traffic measurements from working packet networks have become available (e.g., from Common Channel Signaling Networks (CCSNs) at 56 kbps, from Integrated Services Digital Networks (ISDN) at 1.5 Mbps, and from Ethernet local area network (LANs) at 10 Mbps). More importantly, statistical analyses of these enormous traffic data sets (see for example, [6], [21], [17]) have revealed features in measured network traffic that (i) have gone unnoticed by the teletraffic literature, (ii) show that from a statistical view point traditional traffic models have little in common with empirical data from modern high-speed networks, and (iii) seem to have serious implications for the design, management and control of modern telecommunications systems.

---

* Bellcore, 445 South Street, Room 2P-372, Morristown, NJ 07960-6438, email: walter@bellcore.com

The most striking finding from these traffic data studies is that, in a statistical sense, one can sharply distinguish between empirical network traffic data and traffic generated from traditional models. To illustrate, traditional traffic modeling typically assumes that call holding times are *light-tailed*, i.e., that they have exponentially decaying tail probabilities (e.g., exponential distribution), implying the existence of all moments. In contrast, we show in Section 2 that there exists strong empirical evidence that call holding times are *heavy-tailed* in the sense that the corresponding tail probabilities decay hyperbolically, that is, like a power, possibly with infinite mean and/or variance (e.g., Pareto distribution). Another example concerns the fundamental structure of the traffic processes that represent the number of packets or cells per time unit. Traditional traffic processes have in common that they are Markovian or, more generally, *short-range dependent* in nature, that is, their autocorrelations decay exponentially fast. On the other hand, we illustrate in Section 3 that measurements from modern networks give rise to empirical traffic processes that are generally non-Markovian in nature and exhibit *long-range dependence*. In other words, empirical traffic processes are characterized by slowly decaying autocorrelations (hyperbolic or power decay) which, in turn, result in *self-similar* or, to use a more popular term, "fractal" traffic.

Although fractal traffic exhibits properties that are drastically different from those of traffic generated from traditional models, it is nevertheless possible to clearly identify the point-of-departure from traditional traffic modeling that results in fractal characteristics such as long-range dependence. To this end, we report in Section 4 on some preliminary results that show how infinite variance properties (or, using Mandelbrot's terminology, the *Noah Effect*) can provide a "phenomenological" explanation for the observed long-range dependence (or, using again Mandelbrot's language, the *Joseph Effect*) in measured network traffic. Using the popular "on-off" models for representing the traffic generated by individual sources, we illustrate that by simply replacing the traditional exponential distributions for the lengths of the "on" and "off" periods by distributions which have infinite variance, the superposition of such "on-off" processes will exhibit fractal behavior rather than the traditional Markovian or short-range dependent characteristic. As a result, we demonstrate that by including the Noah Effect into traditional traffic modeling, the resulting traffic models promise to be more relevant for practical application in modern telecommunications systems. At the same time, the practitioners' confidence in traffic modeling is regained because the resulting traffic processes are capable of resembling empirical traffic.

2. Finite vs. infinite variance: The Noah effect. A prime example of the discrepancy between traditionally made modeling assumptions and empirically observed properties in measured data concerns the lengths of telephone calls or *call holding times* (CHTs, for short). Standard text books in probability continue to use CHT as a standard example of a random variable that is *exponentially* distributed. Empirical studies in support of the exponential distribution for CHTs typically date back to about 1960 or earlier and are still considered relevant or adequate, despite the technological advances (e.g., introduction of fax and modems) and the economic changes (e.g., higher penetration of basic phone service, different pricing structures) in telephony during the past decades. The availability of large sets of recent traffic measurements from CCSNs that use the Signaling System Number 7 (SS7) protocol for communication and operates at a speed of 56 Kbps, has enabled us to re-examine some of the traditionally made modeling assumptions. The data consist of individual time-stamped SS7 messages (to the millisecond accuracy) and allows, for example, for the retrospective extraction of packets corresponding to call arrivals and call disconnects; as a result, the arrival process of telephone calls (not shown here) as well as the corresponding CHTs can be studied. The data were collected from a number of different working CCS subnetworks, include traffic from a large variety of links, and cover monitoring periods ranging from hours to days and weeks. This variety within the available traffic data results in findings that are largely independent of time, network load and network configuration. For details about the CCS/SS7 network, the data collection, and an in-depth analysis of CCSN traffic, see [6].

To illustrate the results on CHTs, we use data collected from a working CCS subnetwork during the 4-day period from 1/20/93-1/24/43. Figure 2.1 (a) shows the Q-Q-plot of CHTs (in seconds) for calls that started between 9:30 am and 5:50 pm on 1/22/93, against a standard exponential distribution. It is critical that the selected sample of calls not be truncated, i.e., that we consider all calls that started in a given time period, and not those that started and ended in a given time period. In the case at hand, there were 302,225 such calls, and their durations ranged from 0.001 seconds to 29.5 hours. This figure illustrates very convincingly that the exponential distribution is a very poor fit for the observed CHTs; in particular, the Q-Q-plot shows that measured CHTs exhibit a drastically different (right and left) tail behavior than that predicted by the exponential distribution. In order to investigate the nature of the (right) tail of the empirical CHT distribution in more detail, we recall that a probability distribution F is said to be *heavy tailed* if

$$(2.1) \qquad 1 - F(x) \sim x^{-\alpha} L(x), \text{ as } x \to \infty \quad (\alpha > 0),$$

where L is a slowly varying function (at infinity). Note that if $\alpha < 2$, then F has infinite variance or exhibits the *Noah Effect* [18] (e.g., Pareto distribution with $\alpha < 2$); if $\alpha < 1$, then F has infinite mean. Thus, if an empirical distribution function $\hat{F}(x)$ is heavy tailed in this sense, then a plot, on log-log scale, of $1 - \hat{F}(x)$ vs. x will yield an approximately straight line for large x-values, with a slope of about $-\alpha$. Indeed, Figure 2.1 (b) exhibits such structure, with a slope estimate (obtained by a simple least

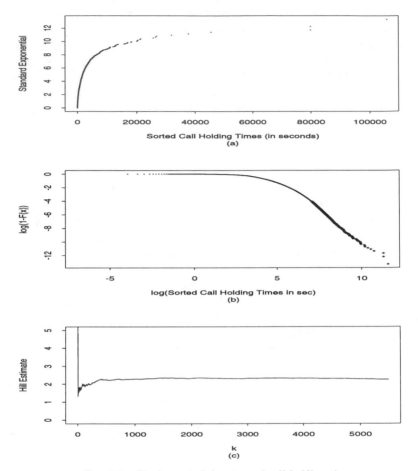

FIG. 2.1. *The heavy-tailed nature of call holding times.*

squares fit using the "brushed" points) of about -2.0. Typically, between 1% - 5% of the CHTs contribute to this observed tail heaviness of $\hat{F}$. A more rigorous and widely studied method for estimating α is given in [9]. If $X_1, X_2, \ldots, X_n$ are the observed CHTs and $X_{1,n}, X_{2,n}, \ldots, X_{n,n}$ denote the corresponding order statistics, then *Hill's estimate* of α is given by

$$(2.2) \quad \hat{\alpha}_n = (1/k \sum_{i=0}^{i=k-1} (\log X_{n-i,n} - \log X_{n-k,n}))^{-1} \quad (0 < k \leq n).$$

Figure 2.1 (c) shows the Hill estimator as a function of k, where the k largest order statistics are used to calculate $\hat{\alpha}_n$. Note in particular that the estimate is quite stable, with an $\hat{\alpha}_n$-value slightly above 2.0, for k-values ranging from 500 to about 5000 (i.e., representing up to about 2% of the

CHTs). Moreover, the Hill estimator agrees with our conclusion drawn from Figure 2.1 (b): the CHT distribution is heavy tailed and, in fact, at the borderline of having infinite variance. A similar analysis (not shown here) of night-time traffic data indicates that the CHT distribution during low traffic periods is also heavy tailed (with typically between 10% - 50% of all CHTs contributing to the tail heaviness); in fact, the corresponding Hill estimator hovers around 1.0, raising the question of an infinite mean CHT distribution. We will return to infinite variance random variables and the Noah Effect in Section 4.

3. Short-range vs. long-range dependence: The Joseph effect. Stochastic models of packet network traffic currently considered in the teletraffic literature are almost exclusively Markovian in nature or, more generally, result in *short-range dependent* traffic processes $X = (X_k : k \geq 0)$, where X_k denotes the number of packets in the k-th time unit. That is, traditional packet traffic models give rise to covariance stationary traffic processes X (mean μ and variance σ^2) with autocorrelation functions $r(k), k \geq 0$, that decay exponentially fast, i.e.,

$$(3.1) \qquad r(k) \sim L_1(k)\rho^k, \text{ as } k \to \infty, \ 0 < \rho < 1,$$

where L_1 is slowly varying (in the following, we assume for simplicity that L_1 is asymptotically constant). In other words, if for each $m = 1, 2, 3, \ldots$, $X^{(m)} = (X_k^{(m)} : k = 1, 2, 3, \ldots)$ denotes the new covariance stationary process (with autocorrelation function $r^{(m)}$) obtained by averaging the original process X over non-overlapping blocks of size m, i.e., $X_k^{(m)} = m^{-1}(X_{km-m+1} + \cdots + X_{km}), k \geq 0$, then the *aggregated processes* $X^{(m)}$ derived from a traditional packet process X tend to covariance stationary white noise; that is, they satisfy

$$(3.2) \qquad \text{var}(X^{(m)}) \sim am^{-1}, \text{ as } m \to \infty,$$

(here and below, a denotes some finite positive constant) and

$$(3.3) \qquad r^{(m)}(k) \to 0, \text{ as } m \to \infty \ (k \geq 1).$$

These traditional traffic models have served admirably in advancing traffic engineering and understanding performance issues, primarily in traditional telephony. However, the emergence of modern high-speed packet networks (LANs, Gigabit networks, ATM networks) combines drastically new and different transmission and switching technologies with dramatically heterogeneous mixtures of services and applications. As a result, the use of traditional traffic processes for modeling traffic in modern high-speed telecommunications systems has come under intense scrutiny, especially in the absence of validations of these models against actual data. This observation and the recent availability of large sets of traffic measurements from working packet networks have brought about renewed interest in traffic modeling and have driven the development of new traffic models utilizing non-traditional paradigms (see, for example, [14]).

To illustrate, we recall the analysis in [16] of large sets of high-resolution Ethernet LAN traffic measurements at the packet level and the visually convincing evidence that Ethernet LAN traffic behaves drastically different from traffic generated from traditional models (see Figure 3.1, ibid). For 27 consecutive hours of monitored Ethernet traffic, Figure 3.1 (a)-(e) depicts a plot sequence of time series of packet counts (i.e., number of Ethernet packets per time unit) for 5 different choices of time units. Starting with a time unit of 100 seconds in Plot (a), each subsequent plot is obtained from the previous one by increasing the time resolution (decreasing the time unit) by a factor of 10 and then focusing on a randomly chosen subinterval (indicated by a darker shade in each of the plots). The time unit corresponding to the finest time scale is 10 milliseconds in Plot (e); this plot is "jittered" in order to avoid the visually irritating quantization effect associated with such high resolution, that is, a small amount of noise has been added to the actual observed arrival rate. Observe that all plots are visually "similar" to each other, i.e., the arrival rates measured over larger time scales (hours, minutes) are indistinguishable (in a distributional sense) from those measured over smaller time scales (seconds, milliseconds). In particular, no natural length of a "burst" is discernible: at every time scale ranging from milliseconds to minutes and hours, bursts consist of bursty subperiods separated by less bursty subperiods.

This scale-invariant or "self-similar" feature of Ethernet traffic is drastically different from both conventional telephone traffic and from traffic generated from traditional stochastic models of packet traffic. The latter models typically give rise to plots of packet counts which are indistinguishable from white noise after aggregating the original time series over a few hundred milliseconds, as illustrated by the plot sequence (a')-(e') in Figure 3.1; this sequence was obtained by successive aggregation as in the empirical plot sequence (a)-(e), except that it arose from synthetic traffic generated from a comparable compound Poisson process with the same average packet size and arrival rate as the empirical data. While the choice of a compound Poisson process is admittedly not very sophisticated, even more complex Markovian arrival streams would produce plot sequences comparable to the sequence (a')-(e') in Figure 3.1. Thus, the procedure resulting in Figure 3.1 provides a surprisingly simple method for sharply distinguishing between empirical packet traffic and traditional model-generated traffic, thereby motivating the use of self-similar stochastic processes for traffic modeling purposes.

Using the same notation introduced earlier in this section, we follow Cox [4] and call the discrete time traffic process X (*exactly second-order*) *self-similar* with self-similarity parameter H if for all $m = 1, 2, 3, \ldots$,

$$(3.4) \qquad\qquad \text{var}(X^{(m)}) = \sigma^2 m^{-\beta}$$

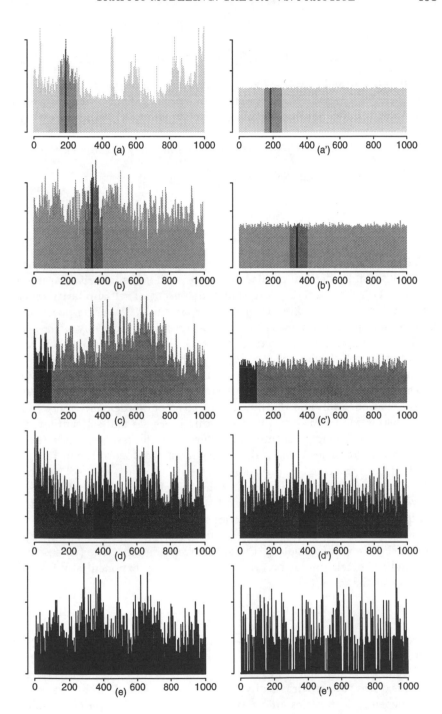

FIG. 3.1. *Self-similar Ethernet traffic (a)-(e), and synthetic traffic generated from a traditional model (a')-(e').*

and

(3.5) $r^{(m)}(k) = 1/2\delta^2(|k|^{2-\beta}), \quad k \geq 1,$

with $0 < \beta < 1$ and $\beta = 2 - 2H$, and where $\delta^2(\cdot)$ denotes the second central difference operator. An example of an exactly second-order self-similar process with self-similarity parameter H is *fractional Gaussian noise* with parameter $1/2 < H < 1$ (see [19]). The traffic process X is called *(asymptotically second-order) self-similar* with self-similarity parameter H if

(3.6) $\text{var}(X^{(m)}) \sim am^{-\beta}, \text{ as } m \to \infty$

and

(3.7) $r^{(m)}(k) \to 1/2\delta^2(|k|^{2-\beta}), \text{ as } m \to \infty \quad (k \geq 1).$

Fractional ARIMA(p, d, q) models (see [8], [10]) are examples of asymptotically second-order self-similar processes with self-similarity parameter $H = d + 1/2, 0 < d < 1/2$. Note that the concept of self-similarity relates statistical properties of X to their counterparts in $X^{(m)}$, using power laws or invariance properties, either for all $m \geq 1$, or asymptotically as $m \to \infty$.

Intuitively, the most striking feature of (exactly or asymptotically) second-order self-similar processes is that their aggregated processes possess a non-degenerate autocorrelation structure, as $m \to \infty$. This behavior is in stark contrast to traditional traffic models, whose aggregated processes satisfy relations (3.2) and (3.3), i.e., tend to second-order pure noise. This sharp distinction between traditional traffic processes and self-similar traffic data is convincingly captured in Figure 3.1: the plot sequence (a)-(e) suggests a more or less identical non-degenerate autocorrelation function for all of the aggregated empirical processes; in contrast, the plot sequence (a')-(e') shows the pure white noise behavior of the aggregated processes derived from a traffic traces generated from traditional model, i.e., identical but degenerate autocorrelation functions of the $X^{(m)}$'s for $m > 100$.

Finally, notice that due to the asymptotic equivalence (for large k) of differencing and differentiating, the autocorrelation function $r(k), k \geq 0$, of a second-order self-similar process X cannot decay exponentially fast but has instead the form

(3.8) $r(k) \sim k^{-\beta}L(k), \text{ as } k \to \infty,$

where $0 < \beta < 1$ and L is slowly varying. Stochastic processes satisfying relation (3.8) are said to exhibit *long-range dependence* or *the Joseph Effect* [20]. Long-range dependent processes are non-Markovian in nature and give rise to features that are drastically different from those of traditional short-range dependent processes. In particular, note that the latter give rise to a summable autocorrelation function $0 < \sum_k r(k) < \infty$, while long-range dependence implies non-summability of the correlations, i.e., $\sum_k r(k) = \infty$.

Long-range dependence can be seen as one of a number of different manifestations (see, for example [4]) of the fact that the underlying covariance stationary traffic process is (asymptotically or exactly) second-order self-similar. Consequently, the problem of inference for self-similar processes can be approached from a number of different angles utilizing both time domain and frequency domain techniques. For an in-depth statistical analysis of the self-similar nature of the Ethernet LAN traffic measurements, see [15], [16] and [17]. For other examples of the omnipresence of long-range dependence in traffic measurements from modern packet networks, see [6], [2], [7].

4. Explaining the Joseph effect via the Noah effect. The fact that features of network traffic such as self-similarity and long-range dependence have gone unnoticed in the teletraffic literature and yet, seem to be ubiquitous in traffic measurements from modern packet networks, provokes the ever-present question about a physical "explanation" for these empirically observed properties in measured traffic data. To this end, we revisit the Noah Effect considered in Section 2 and recall a construction by Mandelbrot [18] as expounded in [25], originally cast in an economic framework involving commodity prices.

Briefly, let $\{I_k\}_{k\geq0}$ be a sequence of i.i.d. integer-valued random variables ("inter-renewal times") satisfying relation (2.1) with $1 < \alpha < 2$. For example, the stable (Pareto) distribution with parameter $1 < \alpha < 2$ satisfies the "heavy-tail" condition (2.1). Furthermore, let $\{G_k\}_{k\geq0}$ be an i.i.d. sequence ("rewards"), independent of $\{I_k\}$, with $E[G_k] = 0$ and $E[G_k^2] < \infty$. Consider the stationary (delayed) renewal sequence $\{S_k\}_{k\geq0}$ defined by $S_k = S_0 + \sum_{j=1}^{k} I_j$, $k \geq 1$, with an appropriately chosen S_0. A discrete-time (renewal) reward process, $W = \{W_k\}_{k\geq0}$, can then be defined by $W_k = \sum_{n=0}^{k} G_n 1_{(S_{n-1},S_n]}(k)$. Notice that W is stationary in the sense that its finite-dimensional distributions are invariant under time shifts. By aggregating M i.i.d. copies, $W^{(1)}, W^{(2)}, \cdots, W^{(M)}$ of W, one obtains the process of interest, $W^* = \{W_k^*(M)\}_{k\geq0}$, given by $W_0^*(M) = 0$ and $W_k^*(M) = \sum_{n=1}^{k} \sum_{m=1}^{M} W_n^{(m)}$, $k > 0$. It can be shown [18,25] that for k and M both large and $k \ll M$, the process W^* behaves like a fractional Brownian motion. More precisely, the process W^*, properly normalized, converges to the integrated version of fractional Gaussian noise, the notion of convergence being that of finite-dimensional distributions. Thus, the increment process of W^* behaves asymptotically like fractional Gaussian noise.

In the context of aggregate packet traffic, a simple variant of Mandelbrot's construction provides an intuitively appealing argument for the visually obvious (see Figure 3.1) and statistically significant (see [17]) self-similarity property of the aggregate Ethernet LAN traffic in terms of the behavior of individual Ethernet users. In its simplest form, this variation of Mandelbrot's technique states that if an individual traffic source (e.g.,

Ethernet host) is of the "on-off" type, that is, it goes through "active" periods (during which it generates packets (or bytes) at regular intervals) and "inactive" periods (when no packets are generated), and if the lengths of the active/inactive periods are i.i.d. (and independent from one another) and have infinite variance (i.e., satisfy (2.1) with $1 < \alpha < 2$), then aggregating many such sources produces traffic that is self-similarity in the limit (as the number of sources increases). Clearly, the convergence result for superpositions of these "on-off" models relies heavily on the Noah Effect which assumes that with non-negligible probability, the active/inactive periods last very long. On the one hand, this property is in stark contrast to traditional "on-off" models whose on- and off-periods typically follow an exponential distribution (or more generally, have "light tails") and whose superposition process is short-range dependent. On the other hand, the Noah Effect seems rather plausible in light of the way a typical Ethernet host (workstation user, file server, router) contributes to the overall traffic on an Ethernet, and it has been observed in a similar setting but for different set of traffic measurements by Meier-Hellstern et al. [21].

To illustrate the Noah Effect in the context of the Ethernet traffic measurements, we extracted individual user traffic (using the origination address in the header of each packet) from the Ethernet traffic data collected between 4:25 pm and 5:25 pm on 8/29/89. Approximately 100 hosts spoke up during that time period, and Figure 4.1 shows the results of our analysis for source#10 which contributed about .5% to the overall traffic seen on the Ethernet during this hour. Other sources resulted in very similar looking pictures. Although the plots in Figure 4.1 were obtained using a subjective criterion for the definition of an "inactive" period (namely, any period longer than $t = 0.1$ seconds during which no packets arrive is called an off-period), we also experimented with other values of t (ranging from seconds to milliseconds) and observed a very similar behavior. The first 2 plots in Figure 4.1 depict, on log-log scale, the complementary cumulative empirical distribution functions $1 - \hat{F}(x)$ for the lengths of "active" and "inactive" periods, respectively, vs. x. There were 11,312 observations for the "active" and "inactive" periods. The plots exhibit a convincing straight line behavior for large x-values (typically, more than 20% of the observations follow the tail behavior assumed by relation (2.1)), with slopes typically between -1.0 and -2.0. In fact, the 2 plots at the bottom of Figure 4.1 show that the corresponding Hill estimators are stable for a wide range of k-values, with values consistently below 2.0. Thus, in the context of the Ethernet traffic measurements, there is strong empirical evidence in favor of Mandelbrot's construction that aims at describing the Joseph Effect observed in aggregate traffic via the Noah Effect exhibited by each individual component process. For a more detailed analysis of the Ethernet traffic measurements at the source level, a rigorous treatment of the above-mentioned variant of Mandelbrot's construction, and implications of these findings for generating self-similar network traffic, and for designing

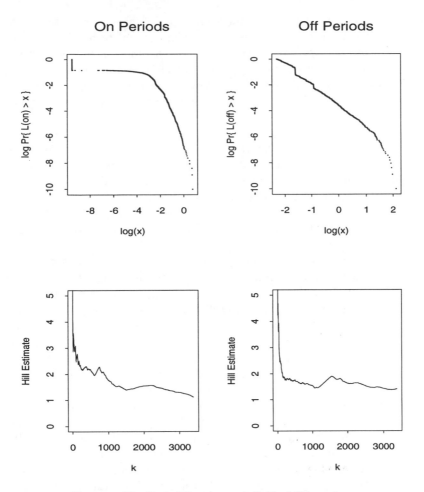

FIG. 4.1. *The Noah Effect for an individual Ethernet source.*

and controlling modern high-speed packet networks, see [26]. For related traffic studies that aim at "explaining" self-similar features of aggregate wide area network traffic at the application level, see [23] and [12].

5. Discussion. The ability to sharply distinguish, in a statistical sense, between measured traffic data from modern packet networks and traffic generated from traditional models (i) clearly challenges traditional traffic modeling approaches, (ii) gives rise to new and interesting traffic engineering problems for modern high-speed telecommunications systems, and (iii) opens up novel areas of mathematical research in queueing theory and performance analysis involving long-range dependent traffic streams. In the following, we list a number of items that range from statistical issues related to the analysis of data with long-range dependence to the study of queues with "fractal" input streams and that are of practical importance for traffic engineering and performance analysis for current and future communications networks. The list is by no means complete but is meant to illustrate possible ways in which traditional traffic modeling approaches and non-traditional fractal modeling techniques can benefit from one another.

Statistical issues: Statistical analyses of data that suggest the presence of the Joseph and/or Noah Effects often seem to provoke subsequent studies that try to "explain" the allegedly observed long-range dependence or heavy tails in the data by some other means (in the area of hydrology, such a discussion can, for example, be found in [11]). Such studies are useful for gaining more insight into the questions of long-range vs. short-range dependence and light-tailed vs. heavy-tailed modeling. However, given the size and quality of the data available in the telecommunications setting, and given the more sophisticated statistical techniques available today for checking for the Joseph/Noah Effect in large data sets, it is unlikely to expect that long-range dependence or heavy-tailed distributions can be explained by simple artifacts of the data (e.g., certain types of non-stationarity), even more so in light of the results mentioned in Section 4.

Stochastic modeling issues: The practical importance for traffic modeling of the findings reported in the earlier mentioned traffic studies is that they show the two directions in which traditional traffic models have to be extended in order to account for empirically observed traffic characteristics: in one direction, "light-tailed" modeling needs to be extended to allow for "heavy-tailed" distributions; in the other direction, short-range dependent models need to be generalized so as to account for the Joseph Effect. Given such an extension of the "current world of traffic models", *parsimonious* modeling becomes again possible (using, for example, fractional Gaussian or stable noise-based models or fractional ARIMA processes). Indeed, in the case at hand, parsimonious modeling becomes a necessity, due to the large number of parameters needed to fit a traditional model to a large time series that is, in fact, fractal in nature. For example, modeling long-range dependence via short-range dependent processes is equivalent to approximating a hyperbolically decaying autocorrelation function by a sum of exponentials. Although this is always possible mathematically, the number of requisite parameters will tend to infinity in the sample size; furthermore, physically meaningful interpretations of the ensuing parameters become increasingly difficult. In contrast, long-range dependence can be parsimoniously modeled via self-similar processes using only a single parameter. A similar statement holds for heavy-tailed modeling. Time series modeling that combines the Joseph and Noah Effects is currently an active area of research (e.g., see [24]).

Parameter estimation and model selection: In practice, the successful application of traffic models relies, among other factors, on having readily available statistical methods for "real-time" parameter estimation for these processes. Although parameter estimation techniques for self-similar processes are known, statistically rigorous methods often turn out to be computationally too intensive to be useful in practice when faced with large data sets. The problem of real-time and efficient parameter estimation for very large sets of self-similar data opens up new areas of research in the interface between statistics and statistical computing. For example, recent work by Beran and Terrin [3] shows how - assuming a high-performance distributed computing environment - an existing parameter estimation technique can be adapted and results in a fast estimation method with good statistical properties, even for very large time series. For more details, see [1,3,17].

Synthetic traffic generation: Another important issue for the success of traffic modeling in teletraffic practice is the ability to quickly generate long traces of synthetic traffic from a chosen traffic model. While exact methods for generating traces from some basic self-similar processes exist (e.g., see [10]), they are in general only appropriate for short traces and become impractical when the sample size becomes exceedingly large (a few thousand observations). The latter is typical for telecommunications applications and requires the development of new methodologies, often tailor-made for certain high-performance computers. For a discussion of fast new algorithms for generating long traces of self-similar traffic on a parallel machine with many processors, see [17]. Once implemented, it will be instructive to see how the traffic generated by these new algorithms will compare with Figure 3.1.

Queueing analysis with fractal inputs: Only direct arguments, detailing the impact on network performance, will convince practitioners of the value of fractal traffic models for traffic engineering purposes and network performance analysis. An essential issue is the sensitivity of network design and control actions to traffic characteristics such as the Joseph and Noah Effects. However, queueing analysis with fractal input streams represents a new area of research and practically no analytic results are available at this time (see, however, [22,5] where it is shown that long-range dependent traffic violates the basic assumption for call admission policies that are based on the popular concept of "effective bandwidth"). A promising alternative approach to direct queueing analysis for fractal inputs is based on the idea of approximating fractal traffic processes by a sequence of traditional processes so as to capture more and more of the fractal characteristics exhibited by empirically observed network traffic (see for example, [13]). Since the approximating (highly parameterized) models are typically amenable to numerical solutions, it should be possible to gain this way a better understanding of the implications of the fractal nature of traffic on queueing and network performance.

6. Acknowledgements. Traffic data analysis from today's networks is a team effort that has benefited from numerous colleagues at Bellcore and elsewhere. Special thanks go to Diane Duffy, Ashok Erramilli, Will Leland, Allen McIntosh, Bob Sherman and Dan V. Wilson form Bellcore, Jan Beran (University of Zürich) and Murad Taqqu (Boston University).

REFERENCES

[1] BERAN, J., *Statistics for Long-Memory Processes*, Chapman and Hall, New York, 1994 (to appear).

[2] BERAN, J., SHERMAN, R., TAQQU, M.S. AND WILLINGER, W., *Long-Range Dependence in Variable-Bit-Rate Video Traffic*, IEEE Trans. Communications (accepted for publication subject to revision), 1994.

[3] BERAN, J. AND TERRIN, N., *Estimation of the Long-Memory Parameter, based on a Multivariate Central Limit Theorem*, J. Time Series Anal. 1994 (to appear).

[4] COX, D.R., *Long-Range Dependence: A Review*, in: Statistics: An Appraisal (H.A. David and H.T. David, Eds.), The Iowa State University, Ames, Iowa, pp. 55–74, 1984.

[5] DUFFIELD, N.G. AND O'CONNELL, N., *Large Deviations and Overflow Probabilities for the General Single-Server Queue, with Applications*, Dublin Institute for Advanced Studies, preprint, 1993.

[6] DUFFY, D.E., MCINTOSH, A.A., ROSENSTEIN, M. AND WILLINGER, W., *Statistical Analysis of CCSN/SS7 Traffic Data from Working Subnetworks*, IEEE J. Select. Areas Communications 12, pp. 544–551, 1994.

[7] GARRETT, M.W. AND WILLINGER, W., *Analysis, Modeling and Generation of Self-Similar VBR Video Traffic*, Proc. of the ACM Sigcomm'94, London, UK, pp. 269–280, 1994.

[8] GRANGER, C.W.J. AND JOYEUX, R., *An Introduction to Long-Memory Time Series Models and Fractional Differencing*, J. Time Series Anal. 1, pp. 15–29, 1980.

[9] HILL, B.M., *A Simple General Approach to Inference about the Tail of a Distribution*, Ann. Statist. 3, pp. 1163–1174, 1974.

[10] HOSKING, J.R.M., *Fractional Differencing*, Biometrika 68, pp. 165–176, 1981.

[11] KLEMES, V., *The Hurst Phenomenon: A Puzzle?*, Water Resources Research 10, pp. 675–688, 1974.

[12] KLIVANSKY, S., MUKHERJEE, A. AND SONG, C., *Factors Contributing to Self-Similarity over NSFNet*, preprint, 1994.

[13] LI, S.-Q. AND HWANG, C.-L., *Queue Response to Input Correlation Functions: Continuous Spectral Analysis*, IEEE/ACM Trans. on Networking 1, pp. 678–692, 1993.

[14] JAGERMAN, D.L., MELAMED, B. AND WILLINGER, W., *Stochastic Modeling of Traffic Processes*, in: Frontiers in Queueing: Models, Methods and Problems (J. Dshalalow, Ed.), CRC Press, 1994 (to appear).

[15] LELAND, W.E., TAQQU, M.S., WILLINGER, W. AND WILSON, D.V., *On the Self-Similar Nature of Ethernet Traffic*, Proc. of the ACM Sigcomm'93, San Francisco, California, pp. 183–193, 1993.

[16] LELAND, W.E., TAQQU, M.S., WILLINGER, W. AND WILSON, D.V., *On the Self-Similar Nature of Ethernet Traffic (Extended Version)*, IEEE/ACM Trans. on Networking 2, pp. 1–15, 1994.

[17] LELAND, W.E., TAQQU, M.S., WILLINGER, W. AND WILSON, D.V., *Self-Similarity in High-Speed Packet Traffic: Analysis and Modeling of Ethernet Traffic Measurements*, Statistical Science, 1994 (to appear).

[18] MANDELBROT, B.B., *Long-Run Linearity, Locally Gaussian Processes, H-Spectra and Infinite Variances*, Intern. Econom. Review 10, pp. 82–113, 1969.

[19] MANDELBROT, B.B. AND VAN NESS, J.W., *Fractional Brownian Motions, Fractional Noises and Applications*, SIAM Review 10, pp. 422–437, 1968.

[20] MANDELBROT, B.B. AND TAQQU, M.S., *Robust R/S Analysis of Long Run Serial Correlations*, Proc. 42nd Session ISI, vol. XLVIII, Book 2, 1979, pp. 69–99.

[21] MEIER-HELLSTERN, K., WIRTH, P.E., YAN, Y.-L. AND HOEFLIN, D.A., *Traffic Models for ISDN Data Users*, Proc. ITC-13, Copenhagen, Denmark, pp. 167–172, 1991.

[22] NORROS, I., *A Storage Model with Self-Similar Input*, Queueing Systems, 1994 (to appear).

[23] PAXSON, V. AND FLOYD, S., *Wide-Area Traffic: The Failure of Poisson Modeling* Proc. of the ACM Sigcomm'94, London, UK, pp. 257-268, 1994.

[24] SAMORODNITSKY, G. AND TAQQU, M.S., *Stable Non-Gaussian Random Processes: Stochastic Models with Infinite Variance*, Chapman and Hall, New York, 1994.

[25] TAQQU, M.S. AND LEVY, J.B., *Using Renewal Processes to Generate Long-Range Dependence and High Variability*, in: Dependence in Probability and Statistics (E. Eberlein and M.S. Taqqu, Eds.), Progress in Prob. and Stat. 11, Birkhauser, Boston, 1986, pp. 73-89.

[26] WILLINGER, W., TAQQU, M.S., SHERMAN, R. AND WILSON, D.V., *Analysis of Ethernet Traffic Measurements at the Source Level*, 1994 (in preparation).

ADMISSION CONTROL FOR ATM NETWORKS [*]

IVY HSU[†] AND JEAN WALRAND[†]

Abstract. We start with a brief review of ATM (Asynchronous Transfer Mode) technology and the control actions that networks can take. We then present methods to estimate the small loss probabilities in ATM networks. Small losses are achieved by averaging traffic over time in a large buffer, or by multiplexing a large number of sources when the buffer is small or negligible. In the first case (large buffer), we discuss the notions of effective and decoupling bandwidths. In the second case (many sources), we show that a good estimate of the loss probability can be obtained using a theorem of Bahadur and Rao.

We propose a call admission strategy in which connections with tight delay constraints, such as interactive audio/video, are given service priority. The delay constraints limit the acceptable buffer size. Thus the many sources asymptotic is appropriate in estimating the loss rate of such service. Effective and decoupling bandwidths are then used for the control of connections that can tolerate longer delays due to large buffers.

1. Introduction. One objective of ATM (Asynchronous Transfer Mode) networks is to transport a wide range of information, such as voice, video, and data, with diverse characteristics and quality of service requirements.

Internet transports messages without guarantee nor control on the delay and the throughput of transmissions. In fact, Internet transports messages without being aware that these messages are part of a connection. Consequently, Internet does not use information about the temporal characteristics of the connections to control the quality of service that it offers to users. Modifications of the Internet protocols are being developed to try to fix those limitations. Although some networking experts believe that Internet can be upgraded to gradually offer the benefits expected from future networks, many researchers see the advantages of the ATM technology and understand that these advantages can be gained without making the Internet applications or most of its infrastructure obsolete. This paper is concerned with ATM networks.

The basic components of ATM networks include transmission links, ATM switches, ATM interface boards installed in user equipment, and specialized servers. We will review some of these basic components in section 2. Today, you can buy local ATM switches and ATM interface boards from a dozen vendors and use them to set up local ATM networks. The cost per connected workstation is rapidly dropping to a level comparable to that of FDDI even though the ATM network can have a much larger throughput and smaller delays. A few vendors sell large ATM switches that are used by public carriers to build metropolitan and wide area networks.

---

[*] This research is supported in part by a grant of the Networking and Communications Research Program of the National Science Foundation.

[†] Department of EECS, University of California, Berkeley, CA 94720. Email addresses: ihsu@eecs.berkeley.edu, WLR@eecs.berkeley.edu

The services of ATM networks, in addition to the usual Internet applications, will include interactive TV, video conferencing, high-speed bridging of LANs, video and high-fidelity broadcasts. Many of these services benefit from higher rates (e.g., Mosaic) while others necessitate them and require low delays (e.g., video conferences).

After reviewing the basic features of ATM networks in section 2, we turn our attention to control methods in section 3. Our next topic is the analysis of small overflow probabilities. In section 4, we focus on the case of large buffers. In section 5, we analyze the case of many sources. In section 6, we propose a admission control strategy based on the results of sections 4 and 5. We conclude the paper in section 6 with a few remarks.

See [13] for a detailed presentation of the technology of high-speed networks and of their analysis, design, and control. That text is complemented by an interactive CD-ROM that contains illustrations and animations of the main algorithms, protocols, and devices.

2. ATM networks. ATM networks transport information in fixed-sized packets of 53 bytes, called *cells*, along *virtual circuits*. Virtual circuit transport means that all the cells of the same connection are identified as such and follow the same path in the network. Examples of connections include transmissions of video or audio signals and file transfers. The fixed size of cells simplifies the hardware. The small size was selected to limit the time needed to assemble cells that transport voice.

Physically, the ATM network consists of switches and user equipment connected by optical fibers and by high-speed links. An ATM switch has a number of input and output lines. When a cell arrives, the switch reads its virtual circuit number and sends the cell to the corresponding output link. The switch is equipped with buffers that can store cells when they occasionally arrive faster than they can be transmitted. ATM switches differ in how their buffers and routing fabric are arranged.

A local ATM network may consist of a single ATM switch located in some wiring closet to which a number of workstations are attached via spare telephone twisted pairs. Pairs may be used in groups of 3 or 4 to achieve a transmission rate of 100 Mbps. Alternatively, optical fibers may be installed to connect the devices. Each workstation is equipped with an ATM interface board. The interface board packages the messages or video or audio signals to be transmitted into ATM cells and reassembles the ATM cells that it receives into messages or signals. To start a connection, the workstation must exchange some signaling information with the switch. That information, also transported in cells, specifies the characteristics of the connections (e.g., average rate and burstiness) and the desired quality of service (e.g., maximum delay and acceptable loss rate). The switch, by monitoring the ongoing connections, decides whether it can accept the new connection and offer the desired quality of service while maintaining the quality of service it promised to the ongoing connections.

We try to understand how the switch can make such a determination in sections 4-5. In a typical installation, 64 workstations can be connected to the central switch by links at 100 Mbps (or higher rates) each. Potentially, hundreds of simultaneous connections are in progress. Some of the connections are video conferences at 1.5 Mbps each, others are real-time transfers of animations produced by simulations on high-performance workstations, others are transfers of high-resolution medical images, and other still are connections between bridges attached to FDDIs or other LANs.

A metropolitan area ATM network can consist of a single ATM switch that belongs to some public carrier (e.g., a local phone company). Users are attached to the switch by 1.5 Mpbs, 45 Mbps, 155 Mbps, or possibly higher-rate lines, either twisted pairs, coaxial cable, or optical fiber. The local access between the users and the public carrier can be provided by a cable TV operator, by the phone company, or by a third party. The metropolitan network connects local ATM networks, other LANs, specialized machines (e.g., video-servers to set-top boxes).

A wide area ATM network consists of a mesh-like network of ATM switches. The backbone consists of optical fibers at 155 Mbps, 622 Mbps, or multiples. This backbone connects the metropolitan networks we discussed earlier.

A few scenarios are plausible for the implementation of ATM networks. Some installations upgrade their LANs from Ethernets to local ATM networks to provide video conferencing or high-speed access to specialized file- and compute-servers or for implementing parallel processing across their networks. Other users subscribe to high-speed services such as the switched multi-megabit data service (SMDS) of a public carrier for connecting LANs. Residential customers subscribe to interactive TV and to videophone services from the cable TV or phone company. Eventually, these services are transported by an ATM wide area network that piggybacks improved Internet applications. We can barely imagine the impact of a truly multimedia version of Internet.

There is not yet a universal agreement on quality of service parameters. Values that are often cited are a maximum delay of about 300 ms for interactive voice/video and of a few seconds for interactive database services. The acceptable loss rate ranges from 10^{-8} for data to a few percent for video and audio. To be specific, let us choose a delay of about 200 ms for interactive audio/video with a loss rate of 10^{-5}. Let us call this application "video." For remote access to databases, let us choose a delay of 1 second and a loss rate of 10^{-7}. We call this application "database." Keep in mind that these sets of requirements are our working hypotheses and that they are not standardized values.

To place the delay values in some perspective, note that a 53-byte cell takes about 3 microseconds to be transmitted by a 155 Mbps link so that a delay of 30 ms corresponds to 10 ^{4} cell transmission times. If the connection goes through 10 links at 155 Mbps each, then the cell can be queued behind

1,000 other cells at each node and still face a queueing delay of less than 30 ms through the network. This should guarantee a maximum total end-to-end delay less than our target 200 ms. In addition, the cell faces a propagation delay (about 5 μs per km) and processing delays (a few 100 μs per node). The delay jitter (i.e., fluctuations) is caused by variations in the queueing delays. Another important cause of delay is the packetization delay: the time it takes to assemble a cell. If the source produces bits at the rate of R bps, then it takes $48 \times 8/R$ seconds to assemble the 48 bytes that are in a 53-byte cell. (The remaining 5 bytes of the cell are occupied by the header.) For a standard (64-kbps) voice stream, this packetization delay is 6 ms. For higher bit rate streams, the packetization delay is essentially negligible.

These elementary considerations show that the delay requirement for our *video* application is satisfied if the queueing does not exceed 1,000 cells per node. For our *database* application, the queueing should not exceed 20,000 cells per node.

Cells are lost because of transmission errors and when buffers overflow. The cell loss rate due to transmission errors is roughly equal to the bit error rate multiplied by 424 (= 53 bytes). Thus, if the BER is 10^{-12}, then the cell error rate is about 4×10^{-10}. The probability that a buffer overflows can be made negligible by using a very large buffer. However, this defeats the objective of keeping the delays small.

Remembering our previous analysis, we see that a reasonable objective for *video* is to control the network so that buffers that can hold about 1,000 cells overflow with a probability less than 10^{-6} (assuming 10 nodes). For *database* the objective is for buffers that can hold 20,000 cells to overflow with a probability less than 10^{-8}. We will refer to these values to check the conclusions of our analysis. In this preliminary discussion, we ignore the interactions of video and database services. Presumably, we give priority to video when both services share a buffer. Another possibility is to segregate buffers and serve them in parallel with different fractions of the bandwidth. We discuss that point later.

3. Control actions. What can the network do to control its operations? The basic control actions are admission control, routing, flow control, traffic shaping, bandwidth and buffer allocation.

Admission control: When a connection is requested with its traffic descriptors and quality of service requirements, the network decides whether to accept or reject the connection. The network determines if it has the necessary resources available to meet the requirements of the new connection while maintaining those of the ongoing connections.

Routing: The connections are transported as virtual circuits. The network selects a path for the new connection. Note that the admission control and routing are coupled problems since the network accepts a call only if it can find a suitable path.

Flow control: While transporting a connection, a network node may decide to postpone the transmission of cells. The objective of the node is to regulate its output to avoid swamping a neighbor.

Traffic shaping: The source may smooth out its cell stream before sending them to the network. This traffic shaping makes the stream easier to transmit because it reduces the amount of storage the stream requires in the network.

Bandwidth allocation: A switch can decide which cell it transmits next on a given fiber. For instance, the switch may give priority to *video* traffic over *database* traffic.

Buffer allocation: When a buffer shared by *video* and *database* cells is full and a new database cell arrives, the switch may discard a video cell to make room for a database cell.

A key question when designing control mechanisms is the information available to the agent choosing the control actions. A related question is the potential obsolescence of the available information: dated information may not be useful. To appreciate this question, consider two possible strategies for traffic shaping: open-loop and link-feedback. One open-loop traffic shaping technique for video signal uses a multi-resolution coding that separates coarse resolution information and finer resolution information in different cells. The mechanism marks the fine resolution cells so that the network switch can discard them when it becomes congested. The possibility of discarding a significant fraction of cells when the switch becomes congested substantially reduces the likelihood of loosing coarse resolution cells because of overflows. A link-feedback strategy uses two different types of video compression: one that produces a coarse output with a small bit rate and a high-quality compression with a large bit rate. The network switch indicates to the user when it becomes congested and the algorithm switches to the low resolution mode. This feedback method is potentially more responsive to the network congestion and avoids sending cells that have to be flushed anyway. However, if the propagation time between the switch and the user is very long, then the feedback mechanism may be too sluggish to be effective.

Similar considerations apply to flow control. Say that the source throttles its output to avoid congesting the network. Two extreme methods can be used: window flow control and rate control. When using window flow control, the destination signals back to the source when it receives cells. This information enables the source to limit the number of its cells inside the network to a specified value. The rate control is open-loop and limits the burstiness of the stream. A simple mechanism for doing this is the so-called *leaky bucket*. (A variation is the generic cell rate algorithm GCRA recommended by the ATM forum. See [2].) Essentially, the algorithm grants permits at a fixed rate but only a specified number of permits can be accumulated. To transmit a cell, the stream must use up one of its permits. The algorithm limits the average rate of cells to the rate at which it provides permits and it also limits the size of a burst of cells to the number of permits that can be accumulated. Intuition suggests that window flow control becomes ineffective when the buffer-delay product of the stream is large. For instance, consider a source that produces cells at 1 Gbps. Assume that the connection goes coast-to-coast with a total delay of about 25 ms. The bandwidth-delay product is then about 25 Mbits. If it limits the number of cells to fewer than the equivalent of 25 Mbits, the window flow control slows down the connection. If it limits that number to 25 Mbits, then there is no guarantee that the cells will not pile up in a few nodes, so that the control is not effective. For such sources, a leaky bucket control may prove more effective. In practice, a 45-Mbps source (HDTV) is already quite fast and it is not sure that window flow control cannot be made effective for such rates. However, common wisdom seems to currently favor open-loop rate control.

There is a compromise between quality of service and number of connections that can be accommodated. One method for reaching a satisfactory compromise is through pricing mechanisms. Various pricing mechanisms are being explored. We do not discuss this important issue here for lack of space.

With this general overview behind us, we can begin our exploration of analytical methods. Before turning to this exploration, let us make a few general observations. Our objective will be to develop methods that network engineers can use to improve networks. Specifically, we focus on methods that estimate cell loss rates. In the paper, we explain how to estimate the loss rate if one has a good model of the traffic. In practice, such good models may be obtained by running an estimation algorithm as the network operates. One can also design control algorithms that learn desirable control actions without attempting to estimate parameters of traffic models. We believe that the analysis that we present can guide the development of such adaptive control algorithms. Finally, keep in mind that the performance evaluation of ATM networks is a relatively new subject and although some useful results are available much remains to be done.

Most sources produce *bursty* cell streams. That is, the peak instantaneous rate of such streams is much larger than their long-term average rate. Allocating bandwidth according to the peak rate results in low resource utilization. For connections that can tolerate some amount of loss, a significant saving in bandwidth can be achieved via *time-averaging* of fluctuations of such streams and by *statistical multiplexing* of many traffic streams. Such averaging smoothes out the fluctuations of the streams. Time averaging occurs in large buffers and is the subject of section 4. Statistical multiplexing is discussed in section 5.

4. Large buffers. In this section, we consider a situation where the loss rate is kept small by using a large buffer. The buffer stores bursts of cells that arrive faster than they can be transmitted. It is unlikely that the bursts are frequent enough to make the buffer overflow.

Except for very simple source models (e.g., Poisson or a Markov modulate process with few states), it is difficult to analyze exactly the small loss rate at a large buffer. The cause of the difficulty is that the state space of a Markov model of the source and buffer system is large and this fact makes the numerical analysis complex. Because of that complexity, and with the objective of deriving tractable results, we turn to an asymptotic analysis of the loss rate as the buffer becomes large. Not surprisingly, the loss rate becomes smaller as the buffer increases. When the buffer is large, the loss rate is well approximated by an exponential function of the buffer size. Roughly, the loss rate is approximately $\exp\{-BI(C)\}$ where B is the buffer size (in cells, say) and $I(C)$ is some increasing function of the transmitter rate C and, obviously, of the statistics of the traffic. We argue that we should choose C large enough so that $\exp\{-BI(C)\}$ is small enough. For our video source, we want $\exp\{-BI(C)\} \approx 10^{-6}$ when $B = 1,000$. (See section 2.) Thus, we want $I(C) \approx 1\%$. For our database source, we want $\exp\{-BI(C)\} \approx 10^{-8}$ for $B = 20,000$, so that $I(C) \approx 0.1\%$. We designate this target value of $I(C)$ by δ. Thus, $\delta = 1\%$ for video and $\delta = 0.1\%$ for database. (Once again, recall that these are working hypotheses and not standards.)

Now suppose that there are J types of traffic, and n_j sources of type j are multiplexed onto an output link. We want

$$\lim_{B \to \infty} \frac{1}{B} \log P(W \geq B) \leq -\delta,$$

where W is the stationary buffer occupancy. It is shown [12,7,9,11] that under appropriate assumptions, this constraint can be satisfied when

$$(4.1) \qquad \sum_{j \in J} n_j \alpha_j(\delta) \leq C,$$

where C is the total output link rate, and $\alpha_j(\delta)$ is the *effective bandwidth* for the type j source corresponding to δ. The most general result is in [12,7] where the above conclusion is shown to hold for a very large class of traffic sources.

Equation (4.1) allows a simple policy for call acceptance that is analogous to that of the traditional circuit-switch networks, since the effective bandwidth for each call can be determined independently of the other types of calls. Furthermore, since $\alpha_j(\delta)$ lies between the mean and peak rates of the source, the difference between the peak rate and $\alpha_j(\delta)$ is the bandwidth saving through multiplexing.

The effective bandwidth $\alpha(\delta)$ of a source that produces a random number $A(t)$ of cells in t seconds can be calculated as

$$(4.2) \qquad \alpha(\delta) = \frac{\Lambda(\delta)}{\delta}$$

where

(4.3) $$\Lambda(\delta) = \lim_{t \to \infty} \frac{1}{t} \log E\{e^{\delta A(t)}\}.$$

When δ is small enough, one may be able to justify the following approximation:

$$\log E\{e^{\delta A(t)}\} \approx \log[1 + \delta E\{A(t)\} + \frac{\delta^2}{2} E\{A(t)^2\}] \approx \delta E\{A(t)\} + \frac{\delta^2}{2} E\{A(t)^2\},$$

which results in

(4.4) $$\alpha(\delta) \approx \lambda + \frac{1}{2}\delta D^2$$

where

$$\lambda := \lim_{t \to \infty} \frac{1}{t} E\{A(t)\} \text{ and } D^2 := \lim_{t \to \infty} \frac{1}{t} E\{A(t)^2\}.$$

In the above definitions and in (4.4), λ is the average rate of the stream and D^2 is called its *dispersion*. This simple approximation for $\delta \ll 1$ shows that the effective bandwidth increases with the burstiness of the stream and indicates that an approximate measure of burstiness (for large buffers and small δ) is the dispersion. Note that $\delta \ll 1$ means that we are willing to lose quite a few cells so that the second moments of the stream are good predictors of the loss rate, as one might guess from a functional central limit theorem. When δ is larger, then the losses are determined by the tail behavior and the higher moments cannot be neglected in the calculation of the effective bandwidth.

Formulas or algorithms are available for calculating the effective bandwidth of a large class of models. See e.g. [12]. Methods for on-line estimation of the effective bandwidth are the subject of current research and so are adaptive techniques for selecting a suitable value of C.

The above result deals with a single buffer and may be usable for a local ATM network. When traffic goes through multiple buffers, the situation becomes more complex. When streams share a buffer, they interact and modify each other's statistics and effective bandwidth. At first the problem appears intractable: the statistics of a stream depend on those of all the streams it interacted with and the same is true for the latter streams. Fortunately, a simplification occurs. One can show that if the transmitter rate C of a buffer is large enough, then a stream preserves its effective bandwidth as it goes through the buffer. Specifically, stream j preserves its effective bandwidth if C is larger than the sum of $\alpha_j^*(\delta)$ and the average rate of all the other streams that share the buffer. Here, $\alpha_j^*(\delta)$ is the *decoupling bandwidth* of stream j. Formulas for calculating that decoupling bandwidth are given in [7] where the applications of that result to call admissions are discussed.

The above approach works well only if the buffer is large. Numerical and simulation experiments show that admission control based on the notions of effective and decoupling bandwidth may be too conservative. What is happening is that the method is based on the estimate of the exponential rate of decay of the loss probability and it ignores the pre-exponential factor which may be very small.

In the next section, we analyze the case of many sources.

5. Many sources. Many real-time applications, such as voice and video, are subject to tight delay constraints and cannot allow large buffers. Over the small buffer region, the effective bandwidths are significantly over-estimated, leading to inefficient bandwidth utilization. Thus it is necessary to consider the small buffer case separately.

The precise boundary that separates the large buffer case from the small buffer case is not easy to determine. Our experiments suggest the following rule of thumb for ON-OFF Markov sources. The buffer is large if it can store at least a few bursts from a source. Here, a burst is understood as the average number of cells that a source produces when it is ON. For instance, if we model a variable bit rate video source (say MPEG 2) as being ON-OFF, then the average duration of the ON period corresponds to the average duration of a scene of the movie where the data compression is not very effective. The average duration of a typical movie scene is a few seconds, say 6 seconds. During that time, the source will produce thousands of cells. Indeed, the active period produces bits at a rate that is a few times the average rate, say 4 Mbps. During these 6 active seconds one source produces about 24 Mbits, i.e., about 6×10^4 cells. In section 2, we argued that a reasonable buffer for video was about 1,000 cells, shared by a number of video sources. Thus, we are led to conclude that the video buffer is a small buffer. Intuitively, the fluctuations of the video signal are slow and we cannot smooth them out with a buffer of an acceptable size. To avoid having to design the network for the peak rate of traffic, we can multiplex many sources.

In this section we analyze the case of small buffers. If the number of virtual circuits routed through a link is large, the combined input rate from all the virtual circuits rarely exceeds a value larger than the mean. Indeed, as we will see, the buffer overflow probability is roughly inversely exponential in the number of virtual circuits N. We refer to this multiplexing approach as the *many sources asymptotic*, as opposed to the large buffer asymptotic.

Consider then a large number N of sources that share a buffer served by a transmitter with rate Nc. Thus, c is the service rate per source. The buffer can store Nb cells, i.e., b cells per source. We want to analyze the buffer overflow probability for large N.

We decompose our analysis into two parts. In the first part, we consider the case $b = 0$ (zero buffer). In the second part, we examine the effect of

$b > 0$ (small buffer).

5.1. Zero buffer. The idea of not using a buffer may seem absurd, specially in view of the relatively low cost of memories. However, we will see that this case is most interesting and leads to somewhat counter-intuitive results.

We denote by Y_j the stationary rate of source j, for $j = 1, \ldots, N$. We assume that the sources are independent and identically distributed, so that the random variables $\{Y_j, j = 1, \ldots, N\}$ are i.i.d. Note that we are ignoring the dynamics of the rates and are focusing instead on their instantaneous distribution. The transmitter is unable to keep up with the traffic produced by the N sources as soon as $Y_1 + \cdots + Y_N > Nc$. We can estimate the fraction of time that this situation occurs by using Cramer's theorem [4],

$$(5.1) \qquad P(Y_1 + \cdots + Y_N > Nc) \approx e^{-NI(c)},$$

where $I(c) \equiv \sup_\theta [\theta c - \varphi(\theta)]$, and $\varphi(\theta) \equiv \log E[\exp(\theta Y_1)]$.

For ON-OFF Markov sources with birth rate λ, death rate μ, and ON-rate a, we note that Y_j takes values in $\{0, a\}$ with $P(Y_j = a) = 1 - P(Y_j = 0) = \lambda/(\lambda + \mu) \equiv \rho$. Consequently,

$$\varphi(\theta) = \log\left(\frac{\mu}{\lambda + \mu} + \frac{\lambda}{\lambda + \mu} e^{a\theta}\right).$$

The value of θ that maximizes $I(c)$ is $\theta_c = \frac{1}{a}\log(\frac{c\mu}{\lambda(a-c)})$, at which $I(c) = \frac{c}{a}\log(\frac{c/a}{\rho}) + (1 - \frac{c}{a})\log(\frac{1-c/a}{1-\rho})$.

We will use a refinement of Cramer's theorem due to Bahadur and Rao [3] (see also [6], section 3.7). Numerical experiments show that the increase in complexity of the algebra results in much improved estimates. Using the Bahadur and Rao theorem, we find (see Appendix)

$$(5.2) \qquad P(Y_1 + \cdots + Y_N > Nc) \approx \frac{1}{\sqrt{2\pi}\sigma\theta_c\sqrt{N}}e^{-NI(c)}.$$

In the above expression,

$$\sigma^2 = \frac{M''(\theta_c)}{M(\theta_c)} - c^2$$

where

$$M(\theta) := E[\exp(\theta Y_1)]$$

and θ_c achieves the maximum in

$$I(c) = \sup_\theta [\theta c - \varphi(\theta)].$$

As a numerical example, we use the following values for two-state sources, each with states 1 and 0: $1/\lambda = 45$ sec, $1/\mu = 5$ sec, $a_1 = 3$ Mbits/sec, $a_0 = 1$ Mbits/sec, and $c = 1.55$ Mbits/sec. This two-state source is a constant source with rate a_0 plus the ON-OFF source define earlier with $a = a_1 - a_0$.

Note that the number of sources that are in state 1 has a binomial distribution. Thus the probability that the aggregate input rate exceeds the output rate can be represented exactly as

$$\sum_{k \geq N(c-a_0)/(a_1-a_0)}^{N} \binom{N}{k} \rho^k (1-\rho)^{N-k}.$$

For large N, we approximate $N!$ by Stirling's formula: $N! \sim N^N e^{-N} \sqrt{2\pi N}$, where $A(N) \sim B(N)$ means $\lim_{N \to \infty} \frac{A(N)}{B(N)} = 1$.

This is compared with the approximation using Cramer's theorem given in Eqn. 5.1 and the refined approximation in Eqn. 5.2. The result is plotted over a range of values for N in Figure 5.1. Note that Eqn. 5.2 not only is much closer to the exact value than Eqn. 5.1, but also approximates the exact value well even for very small N.

The Bahadur-Rao theorem enables us to analyze the case of multi-rate sources. That theorem can also be used to analyze the overflows of mixtures of different types of sources. To do this, say that we have 50% of sources of type Y and 50% of sources of type Z. We can then construct an hybrid source W that is of type $Y + Z$. The analysis of such a situation reveals that the value of c required to achieve a small loss probability cannot be written as the sum of the necessary rates for the Y-sources and the Z-sources. Thus, unfortunately, for small buffers there is no additive result similar to the effective bandwidth. Our preliminary work indicates that the boundary of the acceptable region is convex. It appears that the "closer" Y and Z are in terms of the peak and average rates, the closer this boundary is to a hyperplane.

5.2. Small buffer. In this section we explore the effect of the small buffer b per source on the loss probability. We combine the analysis of Weiss [14] with our application of the Bahadur-Rao theorem.

Recall the model. There are N i.i.d. sources that share a buffer with capacity Nb and served at rate Nc.

Following Weiss, we argue that the buffer overflows because of the conjunction of two events. First, the sources become active at the same time, so that their total rate reaches Nc. Second, their total rate continues to exceed Nc long enough so that the buffer accumulates Nb cells and starts overflowing. We calculated the probability of the first event in the previous section, using the Bahadur-Rao theorem. In [14], Weiss calculates that, for ON-OFF sources, the likelihood of the second event is approximately given

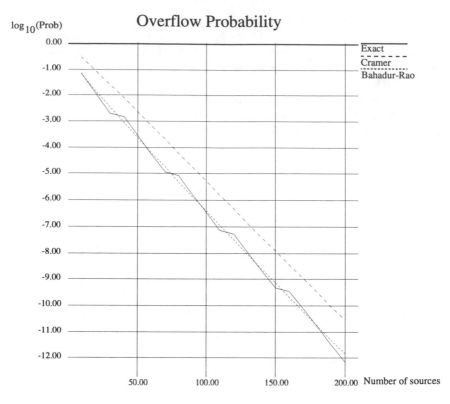

FIG. 5.1. *Exact and approximate values of the probability that the aggregate input rate exceeds the output link rate.*

by

$$\exp(-NC_2\sqrt{b})$$

when N is large. The value of C_2 depends on the parameters of the sources. For the ON-OFF sources defined earlier,

$$C_2 = \sqrt{\frac{1}{a}(\lambda(1 - \frac{c}{a}) + \mu\frac{c}{a}\log(\frac{\mu c}{\lambda(a-c)}) - 2(\mu\frac{c}{a} - \lambda(1 - \frac{c}{a})))}.$$

Combining this calculation with the result of the previous section, we conclude that the loss probability is approximately given by

$$\frac{1}{\sqrt{2\pi}\sigma\theta_c\sqrt{N}} \exp\{-N[I(c) + C_2\sqrt{b}]\}.$$

We compare the approximation with Anick's exact derivation [1]. For the same two-state sources as above, we fix $N = 100$, and plot the results over a range of values for b in Figure 5.2. Note that in this case the overflow probability does not improve very much by the presence of a small buffer.

Overflow Probability

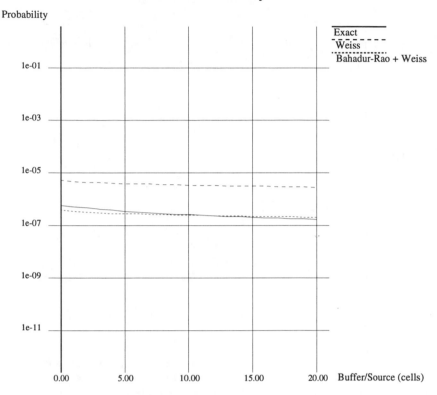

FIG. 5.2. *Exact and approximate values of the buffer overflow probability.*

For comparison, Figure 5.3 shows the buffer overflow probability when the transition rates of the Markov sources are ten times faster ($1/\lambda = 4.5$ sec, $1/\mu = 0.5$ sec). Note two observations:

(1) The benefit of placing a buffer at the output line is much greater in this case. This is because when the source alternates between the two states at a faster rate, the expected duration over which the aggregate input rate exceeds the output rate is much shorter. Indeed, define $K \equiv \lceil N(c - a_0)/(a_1 - a_0) \rceil$ and $A \equiv \{K, K + 1, \ldots, N\}$. Then by solving the first-step equations, the expected sojourn time in A can be expressed as

$$E[\text{sojourn time in } A] = \sum_{i=0}^{N-K} \left[\frac{1}{(N - i)\mu} \prod_{j=i+1}^{N-K} \frac{j\lambda}{(N - j)\mu} \right],$$

which is inversely proportional to μ.

(2) The approximation $\exp(-N\sqrt{b}C_2)$ diverges more quickly from the exact values in this case.

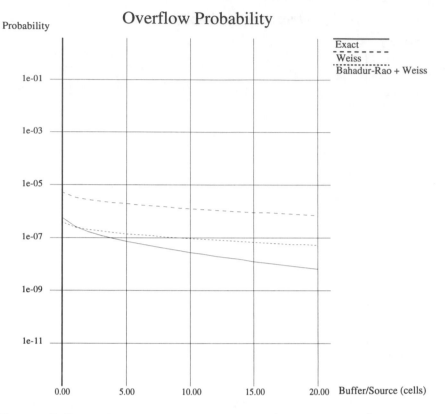

FIG. 5.3. *Buffer overflow probability for sources with* $1/\lambda = 4.5$ *sec and* $1/\mu = 0.5$ *sec.*

These two observations point out that one must normalize the time scale in order to determine if the buffer can be considered "small." Take the time unit to be the average time that a source spends in the ON state ($= 1/\mu$), and the data unit the number of bits generated in one time unit in the ON state ($=a/\mu$). Then the normalized buffer size is $b\mu/a$. Therefore the system in Figure 5.3 has ten times the normalized buffer than that in Figure 5.2.

6. Admission control. In this section, we propose guidelines for admission control in an ATM network. We are aware that these guidelines need to be refined and we propose them mainly as a way of summarizing the results of the paper.

Consider the problem of designing and controlling a wide area ATM network that supports video and database services in addition to best-effort traffic. Best-effort traffic (e.g., email) is carried by the network with no service quality guarantee. Recall our definitions of video and database services in section 2.

We simplify the design by assigning priority to video traffic. We design the video traffic assuming zero buffers. The motivation for this assumption is that buffers would only reduce the loss rate by a small amount, as we saw in section 5.2. Say that we use links at 155 Mbps. We calculate values of N and c so that $Nc = 155Mbps$ and so that

$$\frac{1}{\sqrt{2\pi}\sigma\theta_c\sqrt{N}}e^{-NI(c)} \approx 10^{-6}.$$

Of course, we need a satisfactory model of the video sources to calculate these values of N and c. Once this calculation is performed, we know how many video sources can be transmitted through each one of our switches.

The admission and routing of video calls is then reduced to the corresponding problems for circuit-switched networks. Indeed, we know the capacity N of each link. A call is blocked if a path cannot be found that has spare capacity. In fact, we can benefit from the lessons learned for the telephone network and implement an alternate routing strategy with trunk reservations.

We then propose to analyze the database traffic as follows. If N video sources can go through the switch, the average available rate for the non-video traffic is $C' := N(c - m)$, where m is the average rate per video connection $(m < c)$. In a first approximation, we consider this rate as fixed. We propose to solve the admission and routing problem for the database traffic by using the effective bandwidth and decoupling bandwidth heuristic. The justification for this approach is that the acceptable delays are much larger and therefore, so are the queue lengths.

Finally, the best effort traffic is carried with the left-over capacity, at the lowest priority.

7. Conclusion. In this paper we reviewed methods that estimate the small loss probabilities in ATM networks.

We explained that these small losses are achieved by averaging traffic over time in a large buffer or by multiplexing a large number of sources.

In the first case (large buffer), we discussed the notions of effective and decoupling bandwidths.

In the second case (many sources), we analyzed the zero buffer and the small buffer cases separately. We explained that the loss rate can be estimated by assuming zero buffers and that a good estimate can be derived using a theorem of Bahadur and Rao.

Finally, we proposed a strategy for call admission control based on the previous results. The network is first designed for video traffic by using the zero-buffer estimates. That traffic is admitted and routed using alternate routing with trunk reservations. The next priority is for database traffic and the network is designed and controlled using the effective and decoupling bandwidth ideas. This control amounts to a second level of alternate routing and trunk reservation. Finally, the best effort traffic is allocated the left over capacity.

8. Appendix: Bahadur-Rao theorem.

Let $\nu(\cdot)$ be the distribution of the i.i.d. random variables $Y_1, Y_2, \ldots$. Define $M(\theta) = E[\exp(\theta Y_1)]$, and the "tilted" probability distribution

$$\nu_\gamma(dx) \equiv \frac{\exp(\gamma x)\nu(dx)}{\int \exp(\gamma x)\nu(dx)} = \frac{\exp(\gamma x)\nu(dx)}{M(\gamma)}.$$

Thus the moment generating function of this distribution is

$$\int \exp(\theta x)\nu_\gamma(dx) = \frac{M(\theta + \gamma)}{M(\gamma)}.$$

Taking the inverse transform of $[\frac{M(\theta+\gamma)}{M(\gamma)}]^N$,

$$\nu_\gamma^{*N}(dx) \equiv \text{convolution of } N \text{ copies of } \nu_\gamma(dx) = \frac{\exp(\gamma x)\nu^{*N}(dx)}{M(\gamma)^N}.$$

Thus

$$P(\frac{1}{N}(Y_1 + \cdots + Y_N) \ge c) = \int_{Nc}^\infty \nu^{*N}(dx)$$

$$= M(\gamma)^N \int_0^\infty e^{-\gamma(x+Nc)}\nu_\gamma^{*N}(dx + Nc).$$

Substitute $\gamma = \theta_c$,

$$P(\frac{1}{N}(Y_1 + \cdots + Y_N) \ge c) = e^{-NI(c)} \int_0^\infty e^{-\theta_c x}\nu_{\theta_c}^{*N}(dx + Nc).$$

Note that the mean of $\nu_{\theta_c}(\cdot)$ is $\frac{M'(\theta_c)}{M(\theta_c)} = c$, and its variance is $\sigma^2 = \frac{M''(\theta_c)}{M(\theta_c)} - c^2$. Therefore $\nu_{\theta_c}^{*N}(\cdot + Nc)$ has zero mean and variance $N\sigma^2$. Approximate it by $\mathcal{N}(0, N\sigma^2)$ and let $y = x/\sqrt{N}\sigma$,

$$\int_0^\infty e^{-\theta_c x}\nu_{\theta_c}^{*N}(dx + Nc) \approx \frac{1}{\sqrt{2\pi}\sqrt{N}\sigma} \int_0^\infty e^{-\theta_c x}e^{-\frac{x^2}{2N\sigma^2}}dx$$

$$= \frac{1}{\sqrt{2\pi}}e^{\frac{1}{2}(\theta_c\sqrt{N}\sigma)^2} \int_0^\infty e^{-\frac{1}{2}(y+\theta_c\sqrt{N}\sigma)^2}dy$$

$$= \frac{1}{\sqrt{2\pi}}e^{\frac{1}{2}(\theta_c\sqrt{N}\sigma)^2} \int_{\theta_c\sqrt{N}\sigma}^\infty e^{-y^2/2}dy.$$

Since $\int_u^\infty \exp(-y^2/2)dy \sim \frac{1}{u}\exp(-u^2/2)$ as $u \to \infty$ ([8], Theorem 1.3), we further obtain the following approximation:

$$(8.1) \qquad P(\frac{1}{N}(Y_1 + \cdots + Y_N) \ge c) \approx \frac{1}{\sqrt{2\pi}\sigma\theta_c\sqrt{N}}e^{-NI(c)}.$$

Acknowledgments

This paper summarizes the work of many researchers. The authors are particularly grateful to Professors Courcoubetis, de Veciana, Kesidis, Varaiya, and Weber for their contributions and suggestions.

REFERENCES

[1] D. ANICK, D. MITRA, AND M. M. SONDHI, "Stochastic Theory of a Data-Handling System with Multiple Source," *Bell System Technical Journal*, vol. 61, no. 8 (1982), pp. 1871–1894.

[2] THE ATM FORUM, *ATM User-Network Interface Specification: Version 3.0*, PTR Prentice-Hall (1993).

[3] R. R. BAHADUR AND R. R. RAO, "On Deviations of the Sample Mean," *Ann. Math. Statis.*, vol. 31, (1960), pp. 1015–1027.

[4] J. A. BUCKLEW, *Large Deviation Techniques in Decision, Simulation, and Estimation*, Wiley (1990).

[5] C. COURCOUBETIS AND R. WEBER, "Buffer Overflow Asymptotics for a Switch Handling Many Traffic Sources," *preprint* (1993).

[6] A. DEMBO AND O. ZEITOUNI, *Large Deviations Techniques and Applications*, Jones and Bartlett (1993).

[7] G. DE VECIANA, *Design Issues in ATM Networks: Traffic Shaping and Congestion Control*, PhD thesis, Dept. of EECS, Univ. of California, Berkeley (1993).

[8] R. DURRETT, *Probability: Theory and Examples*, Wadsworth and Brooks (1991).

[9] R. J. GIBBENS AND P. J. HUNT, "Effective Bandwidths for the Multi-Type UAS Channel," *Queueing Systems*, no. 9 (1991), pp. 17-28.

[10] J. Y. HUI, "Resource Allocation for Broadband Networks," *IEEE Journal on Selected Areas in Communications*, vol. 6, no. 9 (1988), pp. 1598–1608.

[11] F. P. KELLY, "Effective Bandwidths at Multi-Class Queues," *Queueing Systems*, no. 9 (1991), pp. 5–16.

[12] G. KESIDIS, J. WALRAND, AND C.S. CHANG, "Effective Bandwidths for Multiclass Markov Fluids and other ATM Sources," *IEEE/ACM Trans. Networking* (1993).

[13] P. VARAIYA AND J. WALRAND, *High-Speed Networks - Building the Information Superhighway*, to be published by Morgan Kaufman (1995).

[14] A. WEISS, "A New Technique for Analyzing Large Traffic Systems," *Adv. Appl. Prob.*, vol. 18 (1986), pp. 506–532.